The Physical Earth

Scientiam non dedit natura semina scientiae nobis dedit
"Nature has given us not knowledge itself, but the seeds thereof."
Seneca

The Joy of Knowledge Encyclopaedia is affectionately dedicated to the memory of John Beazley 1932–1977, Book Designer, Publisher,and Co-Founder of the publishing house of Mitchell Beazley Limited, by all his many friends and colleagues in the company.

The Joy of Knowledge Library

General Editor: James Mitchell
With an overall preface by Lord Butler, Master of Trinity College, University of Cambridge

The Mitchell Beazley Joy of Knowledge Library

The Physical Earth

Introduced by William A. Nierenberg,
PhD, MA, BS (Physics)

Director, Scripps Institution of Oceanography, University of California

MITCHELL BEAZLEY

The Joy of Knowledge Library

Editorial Director	**Frank Wallis**
Creative Director	**Ed Day**
Project Director	**Harold Bull**

Volume editors
Science and The Universe John Clark
  Lawrence Clarke
The Natural World Ruth Binney
The Physical Earth Erik Abranson
  Dougal Dixon
Man and Society Max Monsarrat
History and Culture 1 & 2 John Tusa
  Roger Hearn
Time Chart Jane Kenrick
Man and Machines John Clark
Fact Index Stephen Elliott
  Stanley Schindler
  John Clark

Art Director Rod Stribley
Production Editor Helen Yeomans
Assistant to the Project
Director Graham Darlow
Associate Art Director Anthony Cobb
Art Buyer Ted McCausland
Co-editions Manager Averil Macintyre
Printing Manager Bob Towell
Information Consultant Jeremy Weston

Sub-Editors Don Binney
  Arthur Butterfield
  Charyn Jones
  Jenny Mulherin
  Shiva Naipaul
  David Sharpe
  Jack Tresidder
Proof-Readers Jeff Groman
  Anthony Livesey
Researchers Peter Furtado
  Malcolm Hart
  Peter Kilkenny
  Ann Kramer
  Lloyd Lindo
  Heather Maisner
  Valerie Nicholson
  Elizabeth Peadon
  John Smallwood
  Jim Somerville

Senior Designer Sally Smallwood
Designers Rosamund Briggs
  Mike Brown
  Lynn Cawley
  Nigel Chapman
  Pauline Faulks
  Nicole Fothergill
  Juanita Grout
  Ingrid Jacob
  Carole Johnson
  Chrissie Lloyd
  Aean Pinheiro
  Andrew Sutterby
Senior Picture Researchers Jenny Golden
  Kate Parish
Picture Researchers Phyllida Holbeach
  Philippa Lewis
  Caroline Lucas
  Ann Usborne

Assistant to
the Editorial Director Judy Garlick
Assistant to
the Section Editors Sandra Creese
Editorial Assistants Joyce Evison
  Miranda Grinling
Production Controllers Jeremy Albutt
  John Olive
  Anthony Bonsels
Production Assistants Nick Rochez
  John Swan

The Joy of Knowledge Encyclopaedia
© Mitchell Beazley Encyclopaedias Limited 1976

The Joy of Knowledge The Physical Earth
© Mitchell Beazley Encyclopaedias Limited 1977

Artwork © Mitchell Beazley Publishers Limited
1970, 1971, 1972, 1973, 1974, 1975 and 1976
© Mitchell Beazley Encyclopaedias Limited 1976
© International Visual Resource 1972

ISBN 0 85533 107 0

Typesetting by Filmtype Services Limited, England
Photoprint Plates Ltd, Rayleigh, Essex, England

Printed in England by Balding + Mansell

Major contributors and advisers to The Joy of Knowledge Library

Fabian Acker CEng, MIEE, MIMarE; Professor H.C. Allen MC; Leonard Amey OBE; Neil Ardley BSc; Professor H.R.V. Arnstein DSc, PhD, FIBiol; Russell Ash BA(Dunelm), FRAI; Norman Ashford PhD, CEng, MICE, MASCE, MCIT; Professor Robert Ashton; B.W. Atkinson BSc, PhD; Anthony Atmore BA; Professor Philip S. Bagwell BSc(Econ), PhD; Peter Ball MA; Edwin Banks MIOP; Professor Michael Banton; Dulan Barber; Harry Barrett; Professor J.P. Barron MA, DPhil, FSA; Professor W.G. Beasley FBA; Alan Bender PhD, MSc, DIC, ARCS; Lionel Bender BSc; Israel Berkovitch PhD, FRIC, MIChemE; David Berry MA; M.L. Bierbrier PhD; A.T.E. Binsted FBBI (Dipl); David Black; Maurice E.F. Block BA, PhD(Cantab); Richard H. Bomback BSc (London), FRPS; Basil Booth BSc(Hons), PhD, FGS, FRGS; J. Harry Bowen MA(Cantab), PhD(London); Mary Briggs MPS, FLS; John Brodrick BSc (Econ); J.M. Bruce ISO, MA, FRHistS, MRAeS; Professor D.A. Bullough MA, FSA, FRHistS; Tony Buzan BA(Hons) UBC; Dr Alan R. Cane; Dr J.G. de Casparis; Dr Jeremy Catto MA; Denis Chamberlain; E.W. Chanter MA; Professor Colin Cherry DSc(Eng), MIEE; A.H. Christie MA, FRAI, FRAS; Dr Anthony W. Clare MPhil(London), MB, BCh, MRCPI, MRCPsych; Sonia Cole; John R. Collis MA, PhD; Professor Gordon Connell-Smith BA, PhD, FRHistS; Dr A.H. Cook FRS; Professor A.H. Cook FRS; J.A.L. Cooke MA, DPhil; R.W. Cooke BSc, CEng, MICE; B.K. Cooper; Penelope J. Corfield MA; Robin Cormack MA, PhD, FSA; Nona Coxhead; Patricia Crone BA, PhD; Geoffrey P. Crow BSc(Eng), MICE, MIMunE, MInstHE, DIPTE; J.G. Crowther; Professor R.B. Cundall FRIC; Noel Currer-Briggs MA, FSG; Christopher Cviic BA(Zagreb), BSc(Econ, London); Gordon Daniels BSc(Econ, London), DPhil(Oxon); George Darby BA; G.J. Darwin; Dr David Delvin; Robin Denselow BA; Professor Bernard L. Diamond; John Dickson; Paul Dinnage MA; M.L. Dockrill BSc(Econ), MA, PhD; Patricia Dodd BA; James Dowdall; Anne Dowson MA(Cantab); Peter M. Driver BSc, PhD, MIBiol; Rev Professor C.W. Dugmore DD; Herbert L. Edlin BSc, Dip in Forestry; Pamela Egan MA(Oxon); Major S.R. Elliot CD, BComm; Professor H.J. Eysenck PhD, DSc; Dr Peter Fenwick BA, MB, BChir, DPM, MRCPsych; Jim Flegg BSc, PhD, ARCS, MBOU; Andrew M. Fleming MA; Professor Antony Flew MA(Oxon), DLitt(Keele); Wyn K. Ford FRHistS; Paul Freeman DSc(London); G.E. Fussell DLitt, FRHistS; Kenneth W. Gatland FRAS, FBIS; Norman Gelb BA; John Gilbert BA(Hons, London); Professor A.C. Gimson; John Glaves-Smith BA; David Glen; Professor S.J. Goldsack BSc, PhD, FINSTP, FBCS; Richard Gombrich MA, DPhil; A.F. Gomm; Professor A. Goodwin MA; William Gould BA(Wales); Professor J.R. Gray; Christopher Green PhD; Bill Gunston; Professor A. Rupert Hall LittD; Richard Halsey BA(Hons, UEA); Lynette K. Hamblin BSc; Norman Hammond; Professor Thomas G. Harding PhD; Richard Harris; Dr Randall P. Harrison; Cyril Hart MA, PhD, FRICS, FIFor; Anthony P. Harvey; Nigel Hawkes BA(Oxon); F.P. Heath; Peter Hebblethwaite MA(Oxon), LicTheol; Frances Mary Heidensohn BA; Dr Alan Hill MC, FRCP; Robert Hillenbrand MA, DPhil; Professor F.H. Hinsley; Dr Richard Hitchcock; Dorothy Hollingsworth OBE, BSc, FRIC, FIBiol, FIFST, SRD; H.P. Hope BSc (Hons, Agric); Antony Hopkins CBE, FRCM, LRAM, FRSA; Brian Hook; Peter

Howell BPhil, MA(Oxon); Brigadier K. Hunt; Peter Hurst BDS, FDS, LDS, RSCEd, MSc(London); Anthony Hyman MA, PhD; Professor R.S. Illingworth MD, FRCP, DPH, DCH; Oliver Impey MA, DPhil; D.E.G. Irvine PhD; L.M. Irvine BSc; Anne Jamieson cand mag(Copenhagen), MSc(London); Michael A. Janson BSc; Professor P.A. Jewell BSc(Agric), MA, PhD, FIBiol; Hugh Johnson; Commander I.E. Johnston RN; I.P. Jolliffe BSc, MSc, PhD, CompICE, FGS; Dr D.E.H. Jones ARCS, FCS; R.H. Jones PhD, BSc, CEng, MICE, FGS, MASCE; Hugh Kay; Dr Janet Kear; Sam Keen; D.R.C. Kempe BSc, DPhil, FGS; Alan Kendall MA(Cantab); Michael Kenward; John R. King BSc(Eng), DIC, CEng, MIProdE; D.G. King-Hele FRS; Professor J.F. Kirkaldy DSc; Malcolm Kitch; Michael Kitson MA; B.C. Lamb BSc, PhD; Nick Landon; Major J.C. Larminie QDG, Retd; Diana Leat BSc(Econ), PhD; Roger Lewin BSc, PhD; Harold K. Lipset; Norman Longmate MA(Oxon); John Lowry; Kenneth E. Lowther MA; Diana Lucas BA(Hons); Keith Lye BA, FRGS; Dr Peter Lyon; Dr Martin McCauley; Sean McConville BSc; D.F.M. McGregor BSc, PhD(Edin); Jean Macqueen PhD; William Baird MacQuitty MA(Hons), FRGS, FRPS; Jonathan Martin MA; Rev Canon E.L. Mascall DD; Christopher Maynard MSc, DTh; Professor A.J. Meadows; J.S.G. Miller MA, DPhil, BM, BCh; Alaric Millington BSc, DipEd, FIMA; Peter L. Moldon; Patrick Moore OBE; Robin Mowat MA, DPhil; J. Michael Mullin BSc; Alistair Munroe BSc, ARCS; Professor Jacob Needleman; Professor Donald M. Nicol MA, PhD; Gerald Norris; Caroline E. Oakman BA(Hons, Chinese); S. O'Connell MA(Cantab), MInstP; Michael Overman; Di Owen BSc; A.R.D. Pagden MA, FRHistS; Professor E.J. Pagel PhD; Carol Parker BA(Econ), MA(Internat. Aff.); Derek Parker; Julia Parker DFAstrolS; Dr Stanley Parker; Dr Colin Murray Parkes MD, FRC(Psych), DPM; Professor Geoffrey Parrinder MA, PhD, DD(London), DLitt(Lancaster); Moira Paterson; Walter C. Patterson MSc; Sir John H. Peel KCVO, MA, DM, FRCP, FRCS, FRCOG; D.J. Penn; Basil Peters MA. MInstP, FBIS; D.L. Phillips FRCR, MRCOG; B.T. Pickering PhD, DSc; John Picton; Susan Pinkus; Dr C.S. Pitcher MA, DM, FRCPath; Alfred Plaut FRCPsych; A.S. Playfair MRCS, LRCP, DObstRCOG; Dr Antony Polonsky; Joyce Pope BA; B.L. Potter NDA, MRAC, CertEd; Paulette Pratt; Antony Preston; Frank J. Pycroft; Margaret Quass; Dr John Reckless; Trevor Reese BA, PhD, FRHistS; Derek A. Reid BSc, PhD; Clyde Reynolds BSc; John Rivers; Peter Roberts; Colin A. Ronan MSc, FRAS; Professor Richard Rose BA(Johns Hopkins), DPhil(Oxon); Harold Rosenthal; T.G. Rosenthal MA(Cantab); Anne Ross MA, MA(Hons, Celtic Studies), PhD(Archaeol and Celtic Studies, Edin); Georgina Russell MA; Dr Charles Rycroft BA(Cantab), MB(London), FRCPsych; Susan Saunders MSc(Econ); Robert Schell PhD; Anil Seal MA, PhD(Cantab); Michael Sedgwick MA(Oxon); Martin Seymour-Smith BA(Oxon), MA(Oxon); Professor John Shearman; Dr Martin Sherwood; A.C. Simpson BSc; Nigel Sitwell; Dr Alan Sked; Julie and Kenneth Slavin FRGS, FRAI; Alec Xavier Snobel BSc(Econ); Terry Snow BA, ATCL; Rodney Steel; Charles S. Steinger MA, PhD; Geoffrey Stern BSc(Econ); Maryanne Stevens BA(Cantab), MA(London); John Stevenson DPhil, MA; J. Stidworthy MA; D. Michael Stoddart BSc, PhD; Bernard Stonehouse DPhil, MA, BSc, MInstBiol; Anthony Storr FRCP, FRCPsych; Richard Storry; Professor John Taylor; John W.R. Taylor FRHistS, MRAeS, FSLAET; R.B. Taylor BSc(Hons, Microbiol); J. David Thomas MA, PhD; Harvey Tilker PhD; Don Tills PhD, MPhil, MIBiol, FIMLS; Jon Tinker; M. Tregear MA; R.W. Trender; David

Trump MA, PhD, FSA; M.F. Tuke PhD; Christopher Tunney MA; Laurence Urdang Associates (authentication and fact check); Sally Walters BSc; Christopher Wardle; Dr D. Washbrook; David Watkins; George Watkins MSc; J.W.N. Watkins; Anthony J. Watts; Dr Geoff Watts; Melvyn Westlake; Anthony White MA(Oxon), MAPhil(Columbia); P.J.S. Whitmore MBE, PhD; Professor G.R. Wilkinson; Rev H.A. Williams CR; Christopher Wilson BA; Professor David M. Wilson; John B. Wilson BSc, PhD, FGS, FLS; Philip Windsor BA, DPhil(Oxon); Professor M.J. Wise; Roy Wolfe BSc(Econ), MSc; Dr David Woodings MA, MRCP, MRCPath; Bernard Yallop PhD, BSc, ARCS, FRAS; Professor John Yudkin MA, MD, PhD(Cantab), FRIC, FIBiol, FRCP.

The General Editor wishes particularly to thank the following for all their support:
Nicolas Bentley
Bill Borchard
Adrianne Bowles
Yves Boisseau
Irv Braun
Theo Bremer
the late Dr Jacob Bronowski
Sir Humphrey Browne
Barry and Helen Cayne
Peter Chubb
William Clark
Sanford and Dorothy Cobb
Alex and Jane Comfort
Jack and Sharlie Davison
Manfred Denneler
Stephen Elliott
Stephen Feldman
Orsola Fenghi
Dr Leo van Grunsven
Jan van Gulden
Graham Hearn
the late Raimund von Hofmansthal
Dr Antonio Houaiss
the late Sir Julian Huxley
Alan Isaacs
Julie Lansdowne
Andrew Leithead
Richard Levin
Oscar Lewenstein
The Rt Hon Selwyn Lloyd
Warren Lynch
Simon macLachlan
George Manina
Stuart Marks
Bruce Marshall
Francis Mildner
Bill and Christine Mitchell
Janice Mitchell
Patrick Moore
Mari Pijnenborg
the late Donna Dorita de Sa Putch
Tony Ruth
Dr Jonas Salk
Stanley Schindler
Guy Schoeller
Tony Schulte
Dr E. F. Schumacher
Christopher Scott
Anthony Storr
Hannu Tarmio
Ludovico Terzi
Ion Trewin
Egil Tveteras
Russ Voisin
Nat Wartels
Hiroshi Watanabe
Adrian Webster
Jeremy Westwood
Harry Williams
the dedicated staff of MB Encyclopaedias who created this Library and of MB Multimedia who made the IVR Artwork Bank.

The Physical Earth/Contents

7

Keystone

Lord Butler, Master of Trinity College,
Cambridge, knocks on the great door of
the college during his installation
ceremony on October 7, 1965

Preface

I do not think any other group of publishers could be credited with producing so comprehensive and modern an encyclopaedia as this. It is quite original in form and content. A fine team of writers has been enlisted to provide the contents. No library or place of reference would be complete without this modern encyclopaedia, which should also be a treasure in private hands.

The production of an encyclopaedia is often an example that a particular literary, scientific and philosophic civilization is thriving and groping towards further knowledge. This was certainly so when Diderot published his famous encyclopaedia in the eighteenth century. Since science and technology were then not so far developed, his is a very different production from this. It depended to a certain extent on contributions from Rousseau and Voltaire and its publication created a school of adherents known as the encyclopaedists.

In modern times excellent encyclopaedias have been produced, but I think there is none which has the wealth of illustrations which is such a feature of these volumes. I was particularly struck by the section on astronomy, where the illustrations are vivid and unusual. This is only one example of illustrations in the work being, I would almost say, staggering in their originality.

I think it is probable that many responsible schools will have sets, since the publishers have carefully related much of the contents of the encyclopaedia to school and college courses. Parents on occasion feel that it is necessary to supplement school teaching at home, and this encyclopaedia would be invaluable in replying to the queries of adolescents which parents often find awkward to answer. The "two-page-spread" system, where text and explanatory diagrams are integrated into attractive units which relate to one another, makes this encyclopaedia different from others and all the more easy to study.

The whole encyclopaedia will literally be a revelation in the sphere of human and humane knowledge.

Butler

Master of Trinity College,
Cambridge

The Structure of the Library

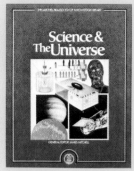

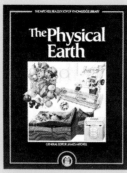

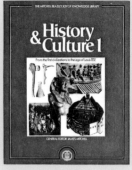

Science and The Universe

The growth of science
Mathematics
Atomic theory
Statics and dynamics
Heat, light and sound
Electricity
Chemistry
Techniques of astronomy
The Solar System
Stars and star maps
Galaxies
Man in space

The Physical Earth

Structure of the Earth
The Earth in perspective
Weather
Seas and oceans
Geology
Earth's resources
Agriculture
Cultivated plants
Flesh, fish and fowl

The Natural World

How life began
Plants
Animals
Insects
Fish
Amphibians and reptiles
Birds
Mammals
Prehistoric animals and
    plants
Animals and their habitats
Conservation

Man and Society

Evolution of man
How your body works
Illness and health
Mental health
Human development
Man and his gods
Communications
Politics
Law
Work and play
Economics

History and Culture

Volume 1 From the first
civilizations to the age of
Louis XIV

The art of prehistory
Classical Greece
India, China and Japan
Barbarian invasions
The crusades
Age of exploration
The Renaissance
The English revolution

The Physical Earth is a book of popular general knowledge about the earth. It is a self-contained book with its own index and its own internal system of cross-references to help you to build up a rounded picture of our world.

It is one volume in Mitchell Beazley's intended ten-volume library of individual books we have entitled *The Joy of Knowledge Library*—a library which, when complete, will form a comprehensive encyclopaedia.

For a new generation brought up with television, words alone are no longer enough—and so we intend to make the *Library* a new sort of pictorial encyclopaedia for a visually oriented age, a new "family bible" of knowledge which will find acceptance in every home.

Seven other colour volumes in the *Library* are planned to be *Science and The Universe, Man and Machines, Man and Society, History and Culture* (two volumes), *The Natural World*, and *The Modern World. The Modern World* will be arranged alphabetically: the other volumes will be organized by topic and will provide a comprehensive store of general knowledge.

The last two volumes in the *Library* will provide a different service. Split up for convenience into A-K and L-Z references, these volumes will be a fact index to the whole work. They will provide factual information of all kinds on peoples, places and things through approximately 25,000 mostly short entries listed in alphabetical order. The entries in the A-Z volumes also act as a comprehensive index to the other eight volumes, thus turning the whole *Library* into a rounded *Encyclopaedia*, which is not only a comprehensive guide to general knowledge in volumes 1–7 but which now also provides access to specific information as well in *The Modern World* and the fact index volumes 9 and 10.

Access to knowledge

Whether you are a systematic reader or an unrepentant browser, my aim as General Editor has been to assemble all the facts you really ought to know into a coherent and logical plan that makes it possible to build up a comprehensive general knowledge of the subject.

Depending on your needs or motives as a reader in search of knowledge, you can find things out from *The Physical Earth* in four or more ways: for example, you can simply browse pleasurably about in its pages haphazardly (and that's my way!) or you can browse in a more organized fashion if you use our "See Also" treasure hunt system of connections referring you from spread to spread. Or you can gather specific facts by using the index. Yet again, you can set yourself the solid task of finding out literally everything in the book in logical order by reading it from cover to cover: in this the Contents List (page 7) is there to guide you.

Our basic purpose in organizing the volumes in *The Joy of Knowledge Library* into two elements—the three volumes of A-Z factual information and the seven volumes of general knowledge—was functional. We devised it this way to make it easier to gather the two different sorts of information—simple facts and wider general knowledge, respectively—in appropriate ways.

The functions of an encyclopaedia

An encyclopaedia (the Greek word means "teaching in a circle" or, as we might say, the provision of a *rounded* picture of knowledge) has to perform these two distinct functions for two sorts of users, each seeking information of different sorts.

First, many readers want simple factual answers to simple factual questions, such as "What is beryl?" They may be intrigued to learn that it is a silicate mineral that forms as hexagonal crystals which may be white, blue, yellow, green or pink, and that it is already familiar to them—as aquamarine, emerald and morganite. Such facts are best supplied by a short entry and in the *Library* they will be found in the two A-Z *Fact Index* volumes.

But secondly, for the user looking for in-depth knowledge on a subject or on a series of subjects—such as "How does a farmer improve his breed of cattle?"—short alphabetical entries alone are inevitably bitty and disjointed. What do you look up first—"farming"? "breeding"? "cattle"? "stock improvement"? "genetics"?—and do you have to read all the entries or only some? You normally have to look up *lots* of entries in a purely alphabetical encyclopaedia to get a comprehensive answer to such wide-ranging questions. Yet comprehensive answers are what general knowledge is all about.

A long article or linked series of longer articles,

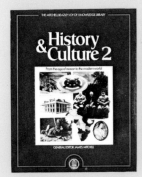

History and Culture

Volume 2 From the Age
of Reason to the
modern world

Neoclassicism
Colonizing Australasia
World War I
Ireland and independence
Twenties and the
  depression
World War II
Hollywood

Man and Machines

The growth of
  technology
Materials and techniques
Power
Machines
Transport
Weapons
Engineering
Communications
Industrial chemistry
Domestic engineering

The Modern World

Almanack
Countries of the world
Atlas
Gazetteer

Fact Index A-K

The first of two volumes
containing 25,000 mostly
short factual entries
on people, places and
things in A-Z order. The
Fact Index also acts as
an index to the eight
colour volumes. In
this volume, everything
from Aachen to Kyzyl.

Fact Index L-Z

The second of the A-Z
volumes that turn the
Library into a complete
encyclopaedia. Like the
first, it acts as an
index to the eight
colour volumes. In this
volume, everything from
Ernest Laas to Zyrardow.

organized by related subjects, is clearly much more
helpful to the person wanting such comprehensive answers.
That is why we have adopted a logical, so-called *thematic*
organisation of knowledge, with a clear system of
connections relating topics to one another, for teaching
general knowledge in *The Physical Earth* and the six
other general knowledge volumes in the *Library*.

The spread system

The basic unit of all the general knowledge books is the
"spread"—a nickname for the two-page units that
comprise the working contents of all these books. The
spread is the heart of our approach to explaining things.

Every spread in *The Physical Earth* tells a story—almost
always a self-contained story—a story on weather
forecasting, for example (pages 72 to 73) or on how man
explores the sea-bed (pages 90 to 91) or on how to improve
crops (pages 170 to 171) or on the origins of wine (pages
198 to 199). The spreads on these subjects all work to the
same discipline, which is to tell you all you need to know in
two facing pages of text and pictures. The discipline of
having to get in all the essential and relevant facts in this
comparatively short space actually makes for better
results—text that has to get to the point without any waffle,
pictures and diagrams that illustrate the essential points in
a clear and coherent fashion, captions that really work and
explain the point of the pictures.

The spread system is a strict discipline but once you get
used to it, I hope you'll ask yourself why you ever thought
general knowledge could be communicated in any other way.

The structure of the spread system will also, I hope,
prove reassuring when you venture out from the things you
do know about into the unknown areas you don't know,
but want to find out about. "Well, if they treat the story
of how earthquakes occur (pages 28 to 29) like that, then
they will probably treat the story of how caves are formed
(pages 110 to 111) in much the same way." There are many
virtues in being systematic. You will start to feel at home
in all sorts of unlikely areas of knowledge with the spread
system to guide you. The spreads are, in a sense, the
building blocks of knowledge. Like the various circuits
and components that go to make up a computer, they are
systematically "programmed" to help you to learn more

easily and to remember better. Each spread has a main
article of 850 words summarising the subject. The article
is illustrated by an average of ten pictures and diagrams,
the captions of which both complement *and* supplement
the information in the article (so please read the captions,
incidentally, or you may miss something!). Each spread,
too, has a "key" picture or diagram in the top right-hand
corner. The purpose of the key picture is twofold: it
summarises the story of the spread visually and it is
intended to act as a memory stimulator to help you to
recall all the integrated facts and pictures on a given subject.

Finally, each spread has a box of connections headed
"See Also" and, sometimes, "Read First". These are
cross-reference suggestions to other connecting spreads.
The "Read Firsts" normally appear only on spreads with
particularly complicated subjects and indicate that you
might like to learn to swim a little in the elementary
principles of a subject before being dropped in the deep end.

The "See Alsos" are the treasure hunt feature of *The Joy
of Knowledge* system and I hope you'll find them helpful
and, indeed, fun to use. They are also essential if you want
to build up a comprehensive general knowledge. If the
spreads are individual components, the "See Alsos" are the
circuit diagram that tells you how to fit them together into
a computer that stores all general knowledge.

Level of readership

The level for which we have created *The Joy of Knowledge
Library* is intended to be a universal one. Some aspects of
knowledge are more complicated than others and so
readers will find that the level varies in different parts of
the *Library* and indeed in different parts of this volume,
The Physical Earth. This is quite deliberate: *The Joy of
Knowledge Library* is a library for all the family.

Some younger people should be able to enjoy and to
absorb most of the spread in this volume on volcanoes,
for example, from as young as ten or eleven onwards—
but the level has been set primarily for adults and older
children who will need some basic knowledge to make sense
of the pages on continental drift or gravity and earth
magnetism, for example.

Whatever their level, the greatest and the bestselling
popular encyclopaedias of the past have always had one

Main text Here you will find an 850-word summary of the subject.

Connections "Read Firsts" and "See Alsos" direct you to spreads that supply essential background information about the subject.

Illustrations Cutaway artwork, diagrams, brilliant paintings or photographs that convey essential detail, re-create the reality of art or highlight contemporary living.

Annotation Hard-working labels that identify elements in an illustration or act as keys to descriptions contained in the captions.

A typical spread Text and pictures are integrated in the presentation of comprehensive general knowledge on the subject.

Captions Detailed information that supplements and complements the main text and describes the scene or object in the illustration.

Key The illustration and caption that sum up the theme of the spread and act as a recall system.

thing in common—simplicity. The ability to make even complicated subjects clear, to distil, to extract the simple principles from behind the complicated formulae, the gift of getting to the heart of things: these are the elements that make popular encyclopaedias really useful to the people who read them. I hope we have followed that precept throughout the *Library*: if so our level will be found to be truly universal.

Philosophy of the Library

The aim of *all* the books—general knowledge and *Fact Index* volumes—in the *Library* is to make knowledge more readily available to everyone, and to make it fun. This is not new in encyclopaedias. The great classics enlightened whole generations of readers with essential information, popularly presented and positively inspired. Equally, some works in the past seem to have been extensions of an educational system that believed that unless knowledge was painfully acquired it couldn't be good for you, would be inevitably superficial, and wouldn't stick. Many of us know in our own lives the boredom and disinterest generated by such an approach at school, and most of us have seen it too in certain types of adult books. Such an approach locks up knowledge, not liberates it.

The great educators have been the men and women who have enthralled their listeners or readers by the self-evident passion they themselves have felt for their subjects. Their joy is natural and infectious. We remember what they say and cherish it for ever. The philosophy of *The Joy of Knowledge Library* is one that precisely mirrors that enthusiasm. We aim to seduce you with our pictures, absorb you with our text, entertain you with the multitude of facts we have marshalled for your pleasure—yes, *pleasure*. Why not pleasure?

There are three uses of knowledge: education (things you ought to know because they are important); pleasure (things which are intriguing or entertaining in themselves); application (things we can do with our knowledge for the world at large).

As far as education is concerned there are certain elementary facts we need to learn in our schooldays. The *Library*, with its vast store of information, is primarily designed to have an educational function—to inform, to be a constant companion and to guide everyone through school and college.

But most facts, except to the student or specialist (and these books are not only for students and specialists, they are for everyone), aren't vital to know at all. You don't *need* to know them.

But discovering them can be a source of endless pleasure and delight, nonetheless, like learning the pleasures of food or wine or love or travel. Who wouldn't give a king's ransom to know when man really became man and stopped being an ape? Who wouldn't have loved to have spent a day at the feet of Leonardo or to have met the historical Jesus or to have been there when Stephenson's *Rocket* first moved? The excitement of discovering new things is like meeting new people—it is one of the great pleasures of life.

There is always the chance, too, that some of the things you find out in these pages may inspire you with a lifelong passion to apply your knowledge in an area which really interests you. My friend Patrick Moore, the astronomer, who first suggested we publish this *Library* and wrote much of the astronomy section in our volume on *Science and The Universe*, once told me that he became an astronomer through the thrill he experienced on first reading an encyclopaedia of astronomy called *The Splendour of the Heavens*, published when he was a boy. Revelation is the reward of encyclopaedists. Our job, my job, is to remind you always that the joy of knowledge knows no boundaries and can work miracles.

In an age when we are increasingly creators (and less creatures) of our world, the people who *know*, who have a sense of proportion, a sense of balance, above all perhaps a sense of insight (the inner as well as the outer eye) in the application of their knowledge, are the most valuable people on earth. They, and they alone, will have the capacity to save this earth as a happy and a habitable planet for all its creatures. For the true joy of knowledge lies not only in its acquisition and its enjoyment, but in its wise and living application in the service of the world.

Thus the Latin tag "Scientiam non dedit natura, semina scientiae nobis dedit" on the first page of this book. It translates as "Nature has given us not knowledge itself, but the seeds thereof." It is, in the end, up to each of us to make the most of what we find in these pages.

General Editor's Introduction
The Structure of this Book

The Physical Earth is a book of general knowledge containing all the information that we think is most interesting and relevant about the earth and man's use of it. In its 264 pages it covers everything from the structure of the earth to the question of whether, in the future, we will be able to produce enough food to feed ourselves. Our intention has been to present these facts in words and pictures in such a way that they make sense and tell a logical, comprehensible and coherent story rather than appear in a meaningless jumble.

For a long time in his history man did not understand the forces at work in his world. With greater understanding, came greater control of these forces. Above all, man acquired the technology that could in some part, at least, counteract the worst effects of flood and drought. And if there were still natural forces he could not control—like earthquakes—he could at least measure them and analyse them and thus take a step towards being able to predict them.

But step by step with that greater understanding and control came burgeoning populations that demanded greater and greater output from the soil. Again, science came to man's aid. New breeding methods, new equipment, vastly increased the output from each acre. For a considerable part of the world, however, there was—and still is— a catch: the rich life bolstered by new methods was available only to those who could afford it. Droughts and floods were still disasters for enormous numbers of people. And the rich world was beginning to turn its birthright into a polluted wasteland.

The Physical Earth devotes some forty pages to the earth's resources, man's exploitation of them, and the steps that are being taken to increase and improve them. It is a fascinating story.

Where to start
Before outlining the plan of the contents of *The Physical Earth* I'm going to assume for a moment that you are coming to this subject, just as I came to it when planning the book, as a "know-nothing" rather than as a "know-all". Knowing nothing, incidentally, can be a great advantage as a reader—or as an editor, as I discovered in making this book. If you know nothing, but want to

find things out, you ask difficult questions all the time. I spent much of my time as General Editor of this *Library* asking experts difficult questions and refusing to be fobbed off with complicated answers I couldn't understand. *The Physical Earth*, like every other book in this *Library*, has thus been through the sieve of my personal ignorance in its attempt to re-state things simply and understandably.

If you know nothing, my suggestion is that you start with William Nierenberg's introduction to the subject (pages 16 to 19). He places the study of geology in an historical setting and then goes on to discuss agriculture in a similar way. In that introduction you will also find a brief, but informed, summary of a subject that is rapidly becoming more and more important—oceanography. If, however, you prefer to plunge straight into the book, but don't have much basic knowledge, I suggest that you study eight spreads in the book before anything else (see panel on page 14). These spreads are the "Read First" spreads. They will give you the basic facts about the earth and agriculture and will provide you with a framework of essential information in its proper perspective. Once you have digested these spreads you can build up a more comprehensive general knowledge by proceeding to explore the rest of the book.

Plan of the book
Apart from the broad division into geology and agriculture, there are nine sections in *The Physical Earth*. The divisions between them are not marked in the text because we thought that that would spoil the continuity of the book. They are:
Structure of the earth
The planet on which we live was formed about 4,600 million years ago. The crust—it is, relatively, no thicker than an eggshell—can be observed directly, but for information about the interior scientists have to study the paths of earthquake waves. The heart of the earth is believed to be solid, even though the temperature is probably 3,000°C (5,400°F). The crust, though solid, is not stable: it is made up of huge plates that are still shifting about on the plastic layer underneath. This concept of continental drift is central to modern thinking about the earth and *The Physical Earth* discusses it in detail.

The Physical Earth, like most volumes in The Joy of Knowledge Library, tackles its subject topically on a two-page spread basis. Though the spreads are self-contained, you may find some of them easier to understand if you read certain basic spreads first. Those spreads are illustrated here. They are "scene-setters" that will give you an understanding of the fundamentals of technology. With them as background, the rest of the spreads in The Physical Earth can be more readily understood. The eight spreads are:

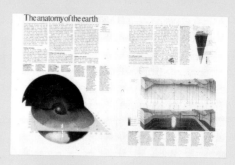

The anatomy of the earth

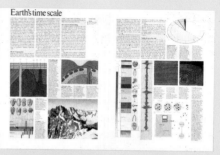

Earth's time scale

The earth in perspective

One of man's principal concerns has been to map the world he lives on. This section of The Physical Earth, after describing the techniques used in mapping, devotes 30 pages to physical maps of the world and to satellite photographs of the world from space.

Weather

Apart from being an ever-popular topic of conversation, the weather is a vital part of our lives. It affects the production of our food, our transport systems, our housing, our health and our leisure. Aristotle wrote what is probably the first scientific treatise on forecasting the weather, but it was not until the early 1800s that modern forecasting began; the international system of cloud classification, for example, dates from 1833. This section begins by discussing the nature of the atmosphere and large-scale weather systems before looking at forecasting and climate.

Seas and oceans

Life began in the sea; and increasingly the sea is becoming more important to life. It may well be that in the not-too-distant future much of our food will come from the world's oceans. That story is told later on in The Physical Earth; in this section we examine the constituents of the seas and oceans, the action of waves and tides, and the structure of the sea-bed. There are maps of the ocean floors, and two spreads look at how man has explored his underwater world.

Geology

Stalactites that grow only 2.5cm (1in) in 4,000 years; caves that are 1,174m (3,850ft) deep or 231km (144 miles) long; diamonds that weigh .60kg (1.3lb); waves that dislodge blocks of concrete weighing more than 1,000 tonnes—and the theory that the world was created at 9am on 23 October 4004BC: those are some of the statistics that the science of geology has produced. Archbishop Ussher, who so precisely worked out the biblical act of creation, according to the book of Genesis, started from the proposition that the Bible was in all respects literally "true". Modern geologists use radioactive decay to date the ages of the rocks they study. This section is concerned with the ground under our feet; with glaciers and rivers; with coastlines, mountains and deserts and the forces that shape them.

Earth's resources

Accurate measurement of existing resources, careful husbandry of them, and an ongoing search for as yet untapped sources of energy have assumed greater and greater importance in the last decade. Opinions differ, but most experts agree that if we go on using fuels and minerals at our present rate, available supplies will last only a generation or two. Even a doubling of known world reserves of oil would add only 10 years to the lifetime of all oil resources.

We are not only using up our reserves: we are also polluting our air, land, rivers, lakes and seas. There are, for example, nearly three million tonnes of oil floating on the oceans of the world every day. The Physical Earth looks at what man is doing to clean up his world.

Agriculture

The last hundred pages of The Physical Earth are devoted to the fruits of the earth. Beginning with a history of agriculture, the first section deals with farming and farm equipment, soil, irrigation, ways of improving and protecting crops, and plant breeding. It was relatively recently—about 10,000 years ago—that man first domesticated animals and discovered how to cultivate and harvest plants. Intensive cultivation dates from the fourteenth century, mechanization from the nineteenth. Today, intensive farming has combined with mechanization to produce the factory farm.

Techniques such as irrigation, which we may think of as relatively modern, are in fact ancient: more than 5,000 years ago farmers along the Nile irrigated their fields and in Iraq remnants of an irrigation canal 120m (395ft) wide and about 10m (33ft) deep have been traced for miles. Today, about 162 million hectares (400 million acres) of land are irrigated. Other techniques, like the use of chemical fertilizers and pesticides, are modern; synthetic organic pesticides date from the discovery of DDT in 1939. This section also discusses a farming technique that deliberately avoids such things as inorganic fertilizers and pesticides—the technique of organic farming, which relies on natural feeding of the soil.

Cultivated plants

Grains like wheat have been staple foods for thousands

Global tectonics

Gravity and the shape of the earth

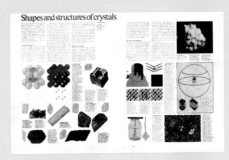

Shapes and structures of crystals

History of agriculture

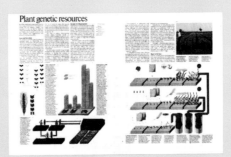

Plant genetic resources

Farm stock breeding and management

of years—wheat seeds have been found in the Tigris-Euphrates valley which date back to 7000BC. Barley, oats, rye and maize have also been staples for centuries and rice is said to provide food for more people than any other cereal. *The Physical Earth* shows where the world's grains are produced and how they are grown. It then looks in detail at two other staples: pulses and tubers.

In 15 following spreads, the section covers vegetables, fruits, beverages, spices, herbs, and plants that are used for their fibre or oil content. Here you will find the familiar plants of the vegetable garden and orchard, and exotics that are seen only in greengrocers' windows. There are some fascinating facts: the coffee trade is valued at more than $2,000 million a year; the tomato, which as a fruit comes from America, was originally grown in Europe as an ornamental plant; the Spaniards guarded the secret of cocoa for more than a century; and the Babylonians used to brew beer about 4000BC. The section ends with three spreads on trees and forestry.

Flesh, fish and fowl

The final section of *The Physical Earth* deals with the animals man uses for food. We are beginning to need new sources of protein: the section ends with a discussion of the new science of aquaculture—fish farming—and with an examination of new, non-animal proteins like TVP—textured vegetable protein—which are already finding their way into our hamburgers.

That, then, is the scope of this book: from the core of the earth to the problem of feeding the 6,500 million people who will cover its surface by the year 2000. I hope you will be as fascinated by reading it as I was by editing it.

The Physical Earth

Dr William A. Nierenberg,

Director, Scripps Institution of Oceanography

Until 1960, geology had, in a sense, no coherent history. It was simply a large collection of facts about the surface and near surface of the earth. But the ten years between 1960 and 1970 saw its transformation into a true science. At the heart of that transformation lay the hypothesis of sea-floor spreading and the theory of continental drift.

Historically there were earlier concepts that were used as tools in geological thought – for example, neptunism, plutonism, catastrophism and uniformitarianism. Of these, only the last two have survived as useful ideas. Catastrophism arose from the observation that at particular times in earth's history dramatic changes have occurred. They are largely unexplained in origin, but have given rise to interesting speculations. For example, there have been theories that the Cambrian revolution occurred at a time of closest approach of the moon to the earth – raising giant tides and causing great alterations in the ocean and the atmosphere. Fluctuations in the energy output of the sun – due either to internal variations or to periodic passage through absorbing cosmic dust clouds – are hypotheses put forward to explain climate changes and the massive disappearance of species. Other theories have been based on periodic increases in convection under the earth's mantle due to excess heat generated by radioactivity. However, it now appears likely that the "new geology" of sea-floor spreading and continental drift may provide a demonstrable basis for many of these changes.

Uniformitarianism has, however, survived the new geology very well. It is the concept that the geological forces we see operating about us today – erosion by stream, wind and wave; vulcanism; sedimentation, and so on – except for the catastrophic events – have operated as continuously in the geological past as they do today.

The transformation of geology between 1960 and 1970 need not have taken so long. Observations had been made that could have led more quickly to the sea-floor spreading hypothesis and the theory of continental drift. It was thought that erosion of the continents should, in a ten-million-year period, supply enough sediment to fill the oceans of the world – but manifestly had not. More subtle evidence in support of the theories came from the field of palaeomagnetism, which is concerned with the magnetic properties of rocks and with the history of the earth's magnetic field. But the concept of continental drift, which derived from sea-floor spreading, had to wait for the new science of marine geology. The probable reason for the delay was that when geology was confined to observations on land, the tremendous complexity of the twisting, warping, erosion and cutting that occurs offered so

much information that it obscured the ultimate simple truth, whereas the mechanics of the study of land underlying the oceans was (and is still) so complex and difficult that only the larger and simpler features could be observed.

Marine geology was accelerated by anti-submarine techniques developed during World War II, and from the outset the results were exciting. The study of underwater sedimentation, seismology, heat flow and magnetism combined to create a hypothesis that was handsomely verified by the first voyages of the *Glomar Challenger*, the best equipped oceanographic vessel in the world.

That story is told in detail in this volume of the encyclopedia. The sea-floor is analogous to a magnetic recording on an endless cassette. The tape can be reversed – and when it is, the present-day continents coalesce very nicely into a single continent about 200 million years ago. In particular, the South American and African plates join neatly, proving that the early theorists were right.

In a sense the new geology provides a dynamic picture. Corresponding questions can now be posed. What happens at the active continental margins where thick sediments pile up at the base of the continents and are transformed by temperature, pressure and erosion to what is observed at or near the surface? What is the origin and description of the circulation in the earth's upper mantle that gives the observed extrusions along the crests of the midocean ridges? Will the new theory help to locate useful thermal beds for the extraction of geothermal energy? Perhaps new light can be shed on the old question of the source of the earth's magnetism and its aperiodic reversals?

These and other important questions can be dealt with in greater confidence given the framework provided by the new geology.

The oceans and oceanography

Scientists can explain very little of what they observe of the oceans. There is a tradition that Aristotle was so frustrated by his inability to account theoretically for the tides he observed that he cast himself into the sea and drowned. We know that his knowledge of the oceans derived from observations of the Mediterranean – a very poor place to study tidal action because the restriction of the Straits of Gibraltar reduces the tidal variation to only a few centimetres in many areas.

Oceanography is now divided into a number of differently defined disciplines. The first is physical oceanography. It is concerned with the variations in temperature, salinity and density among the different waters and includes the

behaviour of the visible ocean currents of the world. But it also involves the study of the much slower motions of the bulk masses, including those near the bottom. Despite many years of work and measurement there is no thoroughly satisfactory explanation for most of the observed motions – so many factors are involved. In this sense it is akin to meteorology, which deals with another geophysical field and which continually advances but will never reach total understanding.

Another major field is biological oceanography. In today's meaning it could readopt its original name of ecology, because the first serious attempts to describe an ecological model dealt with the food chain in the ocean. Protein in the form of fish from the sea has always been important and now reaches nearly 80 million tonnes a year – close to the maximum expected yield – from the world's oceans. It is vital that the best possible understanding be reached of the interactions of oceanic life and the oceans. Unfortunately, the continental borders of the oceans, where the major fisheries lie, are also close to the densest concentrations of the world's population. The intentional discharge of wastes in the form of sewage and chemical wastes, and unintentional discharge such as the runoff of pesticides such as DDT from agriculture, pose threats to the fisheries that are becoming as important as those of overfishing.

One branch of physical oceanography that seems to be emerging as a separate discipline is the air-sea interaction which has become the meeting ground of the meteorologist and oceanographer. The earth has a continual history of severe climate changes, the best known and most recent being the ice ages. However, it is now clear that the oceans play a fundamental role in the shorter-range climate changes that have affected man's life on this planet in the last 2,000 years.

Man's commercial exploitation of the sea is at odds with his use of it for pleasure. The most violent conflict is between offshore oil production and onshore building, and the desire to preserve scenic beauty and activities such as swimming, angling, and boating. Some of man's most valuable assets are beaches and they are disappearing in many places of the world at an alarming rate. The principal reason is the damming of rivers for power, irrigation and flood control. The rivers are the principal source of replenishment for the sand lost from beaches.

Other important resources that can be obtained from the oceans are more speculative. The most interesting is tidal power, but it is inadequate when measured against global needs. More attractive as a long-term source is the energy available in the temperature difference between the upper 200–300m (650–1,000ft.) of the earth's oceans and the cold waters of the depths. This difference is about 15°C (59°F) for much of the world's oceans. With modern technology, proposals to tap that energy seem feasible and it has been estimated that this source could supply all the world's needs to beyond the beginning of the twenty-first century. Another attractive supply of energy resides in ocean waves. An idea of the magnitude of this source can be obtained from the fact that waves from an average sea, the width of a freighter, carry ten times more power than is needed to drive the ship.

The science of agriculture

Agriculture as a human activity is only about 10,000 years old. It is also a miracle of modern times. For many years predictions of calamity based on worldwide food shortages have been made but the combined actions of farmers, scientists, industry and government have averted crisis after crisis by achieving new levels of production at each epoch.

Once man had found it was more convenient to remain in one place and deliberately choose certain plants to cultivate, he then slowly improved the original choices by selective breeding of his plants and animals. (One of the techniques of modern agricultural science has been the patient "back-breeding" current highly hybridized grain plants to forms nearer the original species first used by man.) Even this first crude agriculture must have been economically more profitable than nomadic hunting and was the take-off point for man's rapid development. Storage of agricultural products smoothed the fluctuations in food supply and the economy of effort yielded extra time for development of other activities that could enrich man's life.

The earth recycled – the sea erodes the rocks to fragments and deposits them in sediments; from which new rocks are created.

A successfully developing agriculture depends on four qualities: selection and development of appropriate crops; land and fertilizers; water; and climate. In time man has learned how to manage the first three. The last is still elusive and potentially more dangerous because of the success with the first three. By genetic engineering, biologists are able to supply strains resistant to new pests and are capable of increasing yields in terms of both acreage and effort. Equally important in modern times has been the success in increasing the nutrient value of the crops.

From the very beginning the availability of water was a major factor in determining the type and quality of the local agriculture. In many areas irrigation had to be practised from the beginning. However, coupled with the problem of irrigation was the nature and suitability of the soil. Man soon learned that intensive cultivation exhausted the soil and this reduced the value of irrigation. Except for some particularly favourable areas such as the Nile valley, where the seasonal flooding and silting brought both water and nutrients, exhaustion of the soil resulted in wholesale displacement of populations – as happened with the Mayas. But in many regions man learned the practice of crop rotation and the enriching effect of certain nitrogen-fixing plants. Many cultures introduced the systematic recycling of nutrients back to the soil.

Man was able to cope using such simple methods because population was limited by disease. The great advances in medicine in the nineteenth century led to rapidly increasing populations demanding more and more food. The answer to this problem is the expansion of cultivation, the development of newer and better strains of plants – the so-called "green revolution" – and the introduction of more productive ani-mals. Yet for the breeding programme and the green revolution to work, large quantities of artificial fertilizers are required. Furthermore, for fullest efficiency only the optimum variety is used covering vast areas – hundreds of square metres in extent. This is the state of modern agriculture and is described as monoculture. In this state it is capable of feeding the world for generations to come. Arable land can be doubled. Water can be used more efficiently. More important, the protein supplement can displace the need for animal protein yielding a manifold increase in available grain. The technology can be applied to the underproducing nations if their political systems will allow it.

These advances are marginal at each stage. Even though modern agriculture is able to match the growing population, the methods open up new vulnerabilities. The first is the dependence on the gene pool of wild plants for the development of new strains. The monoculture technique makes the total harvest vulnerable to new pests – even a single one can cause havoc, spreading without the natural barriers of a heterogeneous culture. If they cannot be chemically or biologically controlled, new strains must be developed.

The one overriding concern today is the climate, in the understanding of which little progress has been made. Short-term climate changes can produce important changes in the world food supply and given the always marginal difference between need and production – a few months of reserve at best – genuine crises in food can be expected. In a sense, nothing has changed. In 10,000 years, despite the extraordinary changes in the technology of agriculture we, like the Mayas, live with the threat of food shortages – only the nature of the threat is different.

Anatomy of the earth

The earth is made up of several concentric shells, like the bulb of an onion. Each shell has its own particular chemical composition and physical properties. These shells are grouped into three main regions : the outermost is called the crust, which surrounds the mantle; the innermost is the core.

The solid crust on which we live is no thicker in relation to the earth than an egg shell, taking up only one-and-a-half per cent of the earth's volume. Scientists have been able to learn a great deal about the uppermost part of the crust by direct observation. Their knowledge of the earth's interior, on the other hand, comes from the study of earthquake wave paths.

Earthquake waves are bent, like light passing through a piece of glass, when they traverse rock boundaries with different densities. If the waves hit the boundary at a low angle, they are reflected instead. Waves from distant earthquakes emerge steeply through the crust while those from earthquakes nearby emerge at shallow angles. By knowing these angles, the velocities at which the waves emerge, their times of arrival and distances travelled, geophysicists have been able to compute the positions and densities of the earth's different shells.

Observing the earth's crust

The chemical composition of the crust [Key] and upper mantle is known from direct observation of rocks at or near the earth's surface [2]. Below the upper mantle, little is certain, although similarities may exist between iron and stony meteorites and the composition of the earth's deep interior.

The upper crust over continental areas is known as "sial" (from the first two letters of its most abundant elements, silicon and aluminium). Over oceanic areas, and underlying the continental sial, is the crust called "sima" (from silicon and magnesium, the most abundant elements found in it) [Key]. The sial has a density of $2.7g/cm^3$; it is lighter than the sima (density $2.9g/cm^3$), and lies above it to form the continents. The oceanic crust is made of sima with a thin veneer of sediments and lavas.

The crust is separated from the mantle by a sudden change of density (2.9 to $3.3g/cm^3$) which shows up as a good reflecting plane for earthquake waves. This plane is known as the Mohorovičić discontinuity (Moho for short) [2], after the Croatian who discovered it in 1909, and is taken to represent the base of the crust. The Moho is at an average depth of 35km (20 miles) under the continents and of a mere 10km (6 miles) below sea-level under the oceans and seas.

Beneath the crust

The upper mantle [1] consists of a thin rigid top layer extending from below the Moho to a depth of about 60–100km (40–60 miles), a pasty layer or asthenosphere down to about 200km (120 miles) and a thick bottom layer between 200 and 700km (120–430 miles). The uppermost layer together with the overlying crust forms the rigid lithosphere which is divided laterally into plates. These plates drift on the asthenosphere where pressure and temperature almost reach melting-point, leading to near-fluidity.

The upper mantle is separated from the lower mantle by another discontinuity, where the rock density again increases (3.3 to 4.3

CONNECTIONS

See also
28 Earthquakes
32 Gravity and the shape of the earth
24 Global tectonics

In other volumes
174 Science and The Universe
182 Science and The Universe

1 The earth's crust varies in thickness from 40km (25 miles) to 5km (3 miles) under the sea-floor. With the uppermost mantle it forms the rigid lithosphere [1] which overlies a plastic layer, the asthenosphere [2], on which it may drift sideways.

The upper mantle [3] goes down to 700km (430 miles) where it overlies the lower mantle [4]. The mantle is made of peridotite which is near melting-point in the asthenosphere. This at least is the explanation for the slowing-down of seismic waves at those depths and it fits the plate tectonics theory. The increase of density in the lower mantle is thought to be due to increased pressure and packing of the atoms, without a change of chemical composition. The mantle is separated from the outer core [6] by another seismic wave discontinuity, the Gutenberg discontinuity [5]. P wave velocity drops from 14km (9 miles) to 8km (5 miles) per second and S waves are not transmitted inside the outer core. These observations indicate that the outer core is in a liquid state. The density jumps from $5.5g/cm^3$ for the lower mantle to $10g/cm^3$ for the outer core where it increases downwards to 12 or $13g/cm^3$. Although the core is only 16% of the earth by volume, it represents 32% of its mass. The core is thought to consist of iron and nickel, a hypothesis that fits the data and is inspired by the iron-nickel meteorites which are thought to be the remnants of another planet. P waves show another discontinuity [7] and increase their speed in the centre of the earth or inner core [8], which is solid.

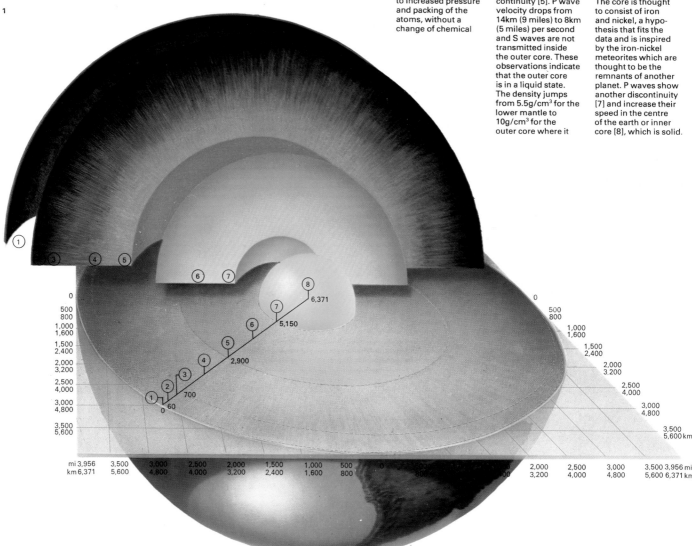

g/cm^3). Here, the composition is thought to be chiefly peridotite, plus minerals of higher density, the latter formed as a result of the crushing pressure of the rocks above.

Between the lower mantle and the core lies a further discontinuity at a depth of 2,900km (1,800 miles), at which the density increases from 5.5 to 10g/cm^3. This is the Gutenberg discontinuity, discovered in 1914. The core itself is divided at a depth of 5,150km (3,200 miles) into an outer and an inner zone thought to be composed of iron-nickel alloy. The outer zone is believed to be liquid because it stops S waves (shearing earthquake waves), while the inner zone is believed to be solid because P (compressional) waves travel slightly faster there. The density changes from 12.3 to about 13.3g./cm^3 at the boundary of outer and inner cores and increases to about 13.6g/cm^3 at the centre of the earth, 6, 371km (3,956 miles) down.

Meteorites reaching the earth's surface provide clues as to its composition. Such meteorites are made either of stone or else are made predominantly of iron. The propor-tion of stony to iron meteorites is more or less equal to the proportion of the mass of the mantle to the core of the earth. Meteorites probably represent the remains of another planet similar to the earth and which broke up at some unknown period [1].

Heat behaviour
The amount of heat reaching the surface from within the earth can be represented on a heat flow profile [2]. From the earth's surface, the temperature rises to about 375°C (710°F), increasing in the upper mantle to 800°C (1,480°F) at 50km (32 miles) and 1,800°C (3,300°F) at 1,000km (625 miles). The estimated temperature for the lower mantle is 2,250°C (4,600°F) at a depth of 2,000km (1,250 miles) and 2,500°C (4,600°F) at the mantlecore boundary – a depth of 2,900km (1,800 miles). The temperature at the centre of the earth is probably about 3,000°C (5,400°F).

Heat is transferred to the surface by convection and conduction. In the solid layers it is probably transferred by conduc-tion; in the liquid layers by convection.

The earth is com-posed of three main layers – the crust, mantle and core. The crust is subdivided into continental and oceanic material. The upper continental crust is mostly granite, rich in silicon and aluminium (*si*licon and a*l*uminium = sial). The oceanic crust is essentially basalt, rich in silicon and magnesium (*si*li-con and *m*agnesium = sima); sima also underlies continental sial. The sial conti-nents, of a lighter mat-erial than the sima, tend to float upon it like icebergs in the sea. The mantle con-sists of rock, rich in magnesium and iron silicates, and the dense core probably consists mainly of iron and nickel oxides in a molten condition.

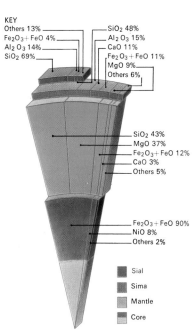

KEY
Others 13%
Fe$_2$O$_3$+ FeO 4%
Al$_2$O$_3$ 14%
SiO$_2$ 69%

SiO$_2$ 48%
Al$_2$O$_3$ 15%
CaO 11%
Fe$_2$O$_3$+ FeO 11%
MgO 9%
Others 6%

SiO$_2$ 43%
MgO 37%
Fe$_2$O$_3$+ FeO 12%
CaO 3%
Others 5%

Fe$_2$O$_3$+ FeO 90%
NiO 8%
Others 2%

Sial
Sima
Mantle
Core

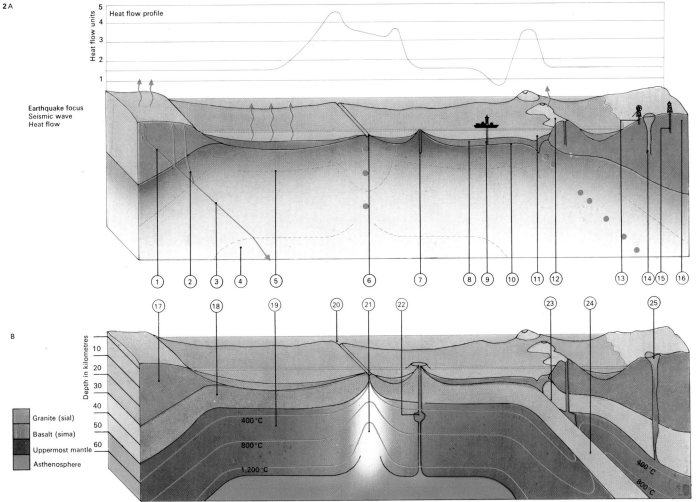

2 A

Heat flow profile

Heat flow units

Earthquake focus
Seismic wave
Heat flow

① ② ③ ④ ⑤   ⑥   ⑦   ⑧⑨ ⑩ ⑪ ⑫   ⑬ ⑭⑮ ⑯

⑰   ⑱   ⑲   ⑳ ㉑   ㉒   ㉓   ㉔   ㉕

B

Depth in kilometres

400°C
800°C
1,200°C

400°C
800°C

Granite (sial)
Basalt (sima)
Uppermost mantle
Asthenosphere

2 The mantle and crust are studied by various means [A]. Heat flow from the earth's interior is high over volcanic areas [7], island arcs [12] and along ocean ridges [6], and low along ocean trenches [11]. The zone of direct observation reaches its limits at [9] the deepest marine drill-hole, 1,300m (4,265ft) below the sea-bed, at [13] the deepest mine, 3,848m (12,600ft), and [15] the deepest land bore-hole, 9,583m (31,441ft). Earth-quake foci occur in particular planes.

Earthquake shocks generate seismic waves which travel faster the deeper they go through the crust [1, 2]. Beyond the Mohorovičić dis-continuity [5] lies the mantle, where the seismic waves enter a high-velocity zone [3], later slowing down in the under-lying low-velocity zone [4]. Marine sedi-ments [8] accumulate on the ocean floor above the oceanic crust [10]. The conti-nental crust [16] lies above the latter. At [14] is a peridotite massif, rich in iron and mag-nesium, intruded from the earth's interior. All these data are interpreted in [B]. The sea-floor crust [18] is composed of basalt, and the continental one [17] of granite over-lying a basaltic layer. The mantle [19] is made of peridotite.

High heat flow and low-intensity seisms are associated with oceanic ridges [20] where magma ascends from the mantle [21]. Sometimes magma is temporarily stored in a magma chamber [22] before it is released by eruption. At [24] the cold lithospheric plate is subducted along the Benioff zone [23], diving into the mantle beneath an adjacent plate and melting magma by friction. Peridotic intrusions [25] have risen through deep crustal cracks to the earth's surface.

The earth as a magnet

The earth has a strong magnetic field [Key]. A bar magnet suspended by a thread eventually comes to rest with one end pointing towards the earth's north magnetic pole, the other towards the south magnetic pole. It behaves similarly if another bar magnet, or large coil of wire with electricity flowing through it, is brought near.

Origin of the earth's magnetic field

As the earth spins on its axis the fluid layer of the outer core allows the mantle and solid crust to rotate relatively faster than the inner core. As a result, electrons in the core move relative to those in the mantle and crust. It is this electron movement that constitutes a natural dynamo and as a result produces a magnetic field, similar to that produced by an electric coil [1].

The magnetic axis of the earth is inclined slightly to the earth's geographical axis [Key] by about 11 degrees, and the magnetic poles do not coincide with the geographic north and south poles. The earth's magnetic axis is continually changing its angle in relation to the geographic axis, but over a long time –

some tens of thousands of years – an average relative position is established.

A compass needle points to a position some distance away from the geographical north and south poles. The difference, which is known as the declination [3], varies from one geographical location to the next. Small-scale variations in the earth's magnetism are probably caused by minor eddies or swirls in the outer core at the junction between the core and the mantle. Large bodies of magnetized rock and ore in the crust can have a similar effect.

The earth's magnetic field is distorted by electrically charged particles from the sun [5]. These particles flow in the upper atmosphere and create small variations of the magnetic field at ground level. Some variations are regular – such as the diurnal (night and day) variation – and some, such as magnetic storms, are occasional.

The earth's magnetic field in ancient times

The study of the magnetic field of the geological past is called palaeomagnetism. It relies on the fact that rocks may pick up a perma-

nent magnetization when they are formed, or when they remelt and cool at some later date. When rocks are heated they lose their magnetization, as an ordinary bar magnet does when heated. Rocks are remagnetized by the earth's field when they cool. This natural remanent magnetization, as it is called, lies parallel to the lines of the earth's magnetic field at the time of rock formation. Rocks magnetized in this way therefore carry a permanent record of the field, and can thus be used to study the geological history of the earth's magnetic field.

There are several ways in which magnetic "clues" to the earth's history can be deposited in the rocks. The technique of palaeomagnetic – literally "old magnetism" – investigation is to drill out a cylinder of rock and then measure its natural remanent magnetism. This gives the palaeomagnetic coordinates of the specimen, which allows its original position to be plotted. Magnetic coordinates, expressed in magnetic latitudes [4], are similar to geographical latitudes (but the pole considered is the magnetic pole instead of the rotation pole). Palaeomagnetic

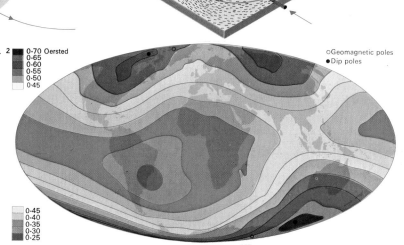

1 The magnetic field originating inside the earth makes up about 90 per cent of the field observed at ground level. The remainder is due to currents of charged particles coming from the sun and to the magnetism of rocks in the crust. The difference in rotation speed between the liquid outer core and the mantle creates a dynamo effect [A], which generates a field similar to that of a coil [B]. In reality the situation is more complex for it involves interaction between two types of magnetic fields and small variations may change the polarity of the earth's field. Irregularities in the magnetic field at the earth's surface are caused by minor eddies of the core liquid. Their displacement in time results in long-term variations of the geomagnetic field, causing gradual changes of direction for magnetic north in given locations.

Relative motion of core
Magnetic field
Magnetic moment
Iron filings
Mantle
Core

2 The intensity of the earth's magnetic field is strongest at the poles and weakest in the equatorial regions. If the field were purely that of a bar magnet in the centre of the earth and parallel to the spin axis, the lines of equal intensity would follow the lines of latitude and the magnetic poles would coincide with the geographic poles. In reality the "bar magnet" field is inclined at about 11° to the spin axis and so are its geomagnetic poles. Also the real field is not purely that of a bar magnet. The "dip poles", where the field direction is vertical (downwards at the north pole and upwards at the south dip pole), are themselves offset in respect to the geomagnetic poles – each by a different amount so that the S dip pole is not exactly opposite the N dip pole. The poles and the configuration of the field change slowly with time.

0·70 Oersted
0·65
0·60
0·55
0·50
0·45

○ Geomagnetic poles
● Dip poles

0·45
0·40
0·35
0·30
0·25

3 The declination is the angle between the direction of a magnetic compass and geographic north. The lines of force radiate from the southern dip pole [S] and converge towards the northern dip pole [N]. The arrows symbolize the direction of the magnetic north in 1955. Declination exists because the earth's field is not exactly like that of a bar magnet lined up along the axis of the earth. Account of this must be taken in navigation.

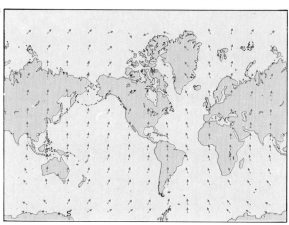

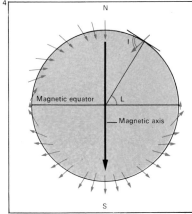

N
Magnetic equator
L
Magnetic axis
S

4 The dip or inclination [I] of the magnetic field at the surface of the earth is related to the magnetic latitude [L], measured relative to the magnetic axis. Assuming that the earth's field averages out like a bar magnet aligned along the geographical axis, this allows the calculation of ancient latitudes of a land using palaeomagnetic data. The inclination is measured with a special compass with a horizontal pivot but does not pose a navigation problem.

co-ordinates are respective to the apparent magnetic pole at the time the rock was magnetized. Evidence from this type of work reveals that the magnetic poles were not always in the position that they now occupy but have "wandered" over the years.

Polar wandering differs from continent to continent. But the poles for a specific time in geological history can be aligned through the various continents by envisaging the continents in different positions from those they now occupy. It is in this way that the progress of continental drift can be plotted. The results of this technique agree fairly well with other drift indicators such as sea-floor spreading and evidence of ancient climates shown by the rocks and their fossils. Palaeomagnetism is a powerful tool in continental drift research.

Some rocks formed over short time intervals show fossil magnetic polarities 180 degrees apart. This cannot be explained by a 180 degrees rotation of a continent because there was not enough time for that. Thus the earth's field must have undergone a switching of its magnetic polarity, as when the direction of the current in a coil is reversed [6]. This switching is known as a "reversal". Reversals mark the boundaries of periods of variable length throughout geological time when the magnetic field was of constant polarity. The dating of reversals (by studying the decay of radioactive isotopes in the rocks) provides the geologist with a palaeomagnetic time scale. This can be used to date other rocks by analysing their remanent magnetization. It was the comparison of this palaeomagnetic time scale with the "magnetic anomalies" on the sea floor that supported the sea-floor spreading hypothesis.

Magnetic and electrical prospecting

Many ore bodies and rocks rich in magnetic minerals have a local strong magnetic field [7]. This is utilized by prospectors using sensitive instruments to detect minerals of economic value. Another method utilizes natural electrical currents that are set up between the surface and an ore body by percolating ground water. Its interaction with the magnetic field can be measured and used to locate useful deposits.

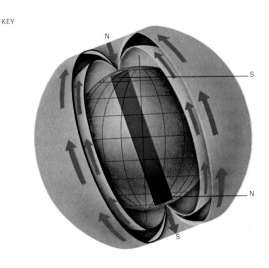

The earth's magnetic field is like that of a giant natural bar magnet placed inside the earth with its magnetic axis inclined at a small angle to the geographical axis. The poles of a compass needle are attracted by the magnetic poles of the earth and swing so that one end points to the north magnetic pole and the other to the south.

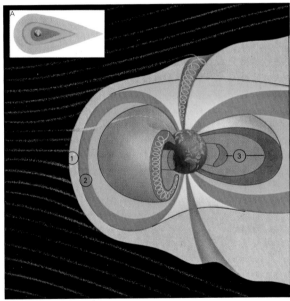

5 The magnetosphere is the region in which the earth's magnetic field can be detected. It would be symmetrical were it not for electrically charged particles streaming from the sun [A], which distort it to a teardrop shape. The particles meet the earth's magnetic field at the shock front [1]. Behind this is a region of turbulence and inside the turbulent region is the magnetopause [2], the boundary of the magnetic field. The Van Allen belts [3] are two zones of high radiation in the magnetopause. The inner belt consists of high-energy particles produced by cosmic rays and the outer belt of solar electrons.

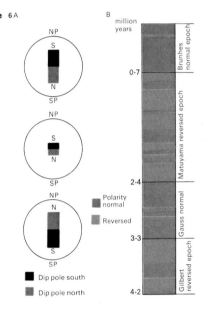

6 A

B
million years

Polarity normal
Reversed

■ Dip pole south
■ Dip pole north

6 Field reversals are the changes of polarity (north becoming south and vice versa) which occurred many times during geological history. The polarity of the earth's magnetic field does not flip over; its magnetic strength decreases to zero then slowly increases in the opposite direction [A]. Rocks "fossilize" the magnetic field when they are formed. If a sufficient number of rocks of different ages are dated, and their polarity measured, a worldwide magnetic time scale is obtained. The magnetic time scale featured here [B] shows the significant changes in magnetic polarity over the last 4.2 million years.

7 In mineral prospecting magnetometers can be used to detect variations in the earth's magnetic field due to ores: [1] regional magnetism of the country rocks; [2] background magnetism due to topsoils; [3] effects of deepseated ore bodies; [4] effects from near surface ores.

Gammas
20,000
10,000
2,500
0
−1,000

8 Millivolts
40
20
−20
−40
−60
−80
−100

8 Electrical prospecting for minerals makes use of natural ground currents [1] related to the magnetic field and influenced by ore bodies [2]. Two electrodes [3], placed in the ground at staked points [4], are connected to a millivoltmeter [5] and the voltages at the points are measured. Variations may indicate the presence of ores.

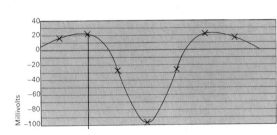

Global tectonics

The theory of plate tectonics was advanced in the late 1960s and has had a revolutionary effect on the earth sciences. It is a unifying, all-embracing theory, offering a plausible and logical explanation for many of the earth's varied structural and geophysical phenomena, ranging from mountain-building to earthquakes and continental drift.

Crustal plates

The theory envisages the crust of the earth together with the upper part of the mantle, which form the lithosphere, as consisting of rigid slabs or plates that are continuously moving their position in relation to each other [1]. Below the lithosphere is the asthenosphere which is thought to be plastic.

The plates are bounded by oceanic ridges, trenches and transform faults. Oceanic ridges are where two plates are moving apart leaving a gap which is continuously filled by magma (molten rock) rising from the asthenosphere. As the magma cools, new crust is created and becomes part of the moving plates. This is the phenomenon of sea-floor spreading. Spreading rates, though slow, are not negligible. The Atlantic is opening up by 2cm (0.75in) a year. The fastest rate is found at the East Pacific Rise, which creates 10cm (4in) of new crust every year – that is 1,000km (620 miles) in the short geological time of ten million years.

Trenches are formed where two plates converge. One of the plates slides steeply under the other [6] and enters the mantle. Thus trenches are areas where plate edges are destroyed. Since the volume of the earth does not change, the amount of crust created at the ridges is balanced by that destroyed at the trenches.

The leading edges of the colliding plates may be oceanic crust, such as in the Tonga-Kermadec trench north of New Zealand: or one may be oceanic (and will be the sinking plate) while the other is continental, such as at the Peru-Chile trench; or both plates may be continental, such as those of northern India and Tibet. In the last two instances the thick sedimentary covers are crumpled and injected with material melted by the heat generated by the collision, and mountains such as the Himalayas are created [7].

Transform faults form where two plates slide past one another [8]. They offset oceanic ridges and their continuation scars can be followed in places for thousands of kilometres [5]. They sometimes slice through continents, as does the famous San Andreas fault in the south-western United States.

The cause of plate movements

As early as 1927 the British geologist Arthur Holmes suggested that convection currents in the mantle could explain the continental drift theory. Convection currents are generated by heat differences – they can be observed in a saucepan of water placed over a fire. Global tectonics theory suggests that convection currents exist in the asthenosphere and perhaps in the lower mantle. They form convection cells that rise under the ridges and descend under the trenches. This theory is supported by measurements of the heat radiated from the earth which show high values along the ridges and low values along the trenches.

The amazing world pattern of ridges, trenches and faults was discovered in the 1940s and 1950s. Their distribution was seen

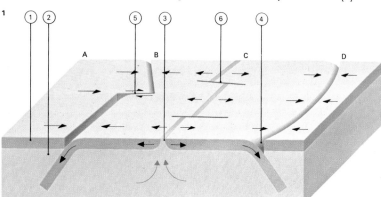

1 The plate tectonics theory envisages the earth's lithosphere [1] as a series of rigid but mobile slabs called plates [A, B, C, D]. The lithosphere floats on a plastic layer called the asthenosphere [2]. There are three types of boundaries possible. At the oceanic ridges [3], upwelling of mantle material occurs and new sea-floor is formed. A trench [4] is formed where one plate of oceanic crust slides beneath the other, which may be oceanic or continental. The third type of boundary is where two plates slide past one another, creating a transform fault [5, 6]. Transform faults link two segments of the same ridge [6], two trenches [5] or a ridge to a trench. Plates move from ridges and travel like conveyor belts towards trenches where they sink.

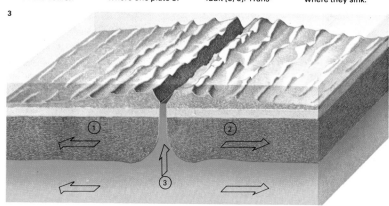

3 Oceanic ridges are found where two plates [1, 2] are moving away from each other. Magma from the mantle [3] continuously wells up from below. As the magma cools, it becomes part of the plates and it is in turn injected by fresh magma. The ridges are thus the newest part of the earth's crust, while the oldest oceanic crust is found where it plunges into the trenches. The dating of cores from the *Glomar Challenger* supports the theory.

4 A magnetic survey from a research ship [1] sailing back and forth over a mid-oceanic ridge gives readings [2, 3, 4] that indicate that the magnetism of the rocks of the sea-bed points alternately north and south in a series of bands parallel to the ridge [6]. The pattern of bands is identical at each side of the axis and corresponds to the pattern of reversals in the earth's magnetic field for the last few million years [5]. The rocks moving away from the axis carry a record of the earth's magnetic field.

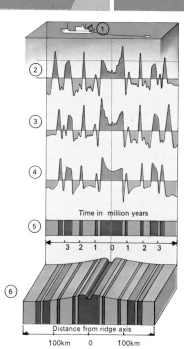

2 The birth and death of oceans is a continuous process. In [A] ocean 1 is growing by sea-floor spreading from a mid-ocean ridge while ocean 2 is closing because of the continents forcing the ocean floor down at the trenches. Ocean 3 is young and growing. In [B] ocean 1 has reached maturity, ocean 3 is still growing. Ocean 2 has disappeared with the joining of the continental masses. In [C] oceans 1 and 3 are declining while a new crack appears at 4. [D] shows ocean 1 still declining and ocean 4 growing. [E] shows a widening of the Red Sea and Gulf of Aden while [F] indicates that the Mediterranean Sea has been steadily shrinking.

to fit that of earthquakes and volcanoes. In 1962 the American geologist Harry Hess suggested that these features, as well as continental drift, could be explained by sea-floor spreading, but because of insufficient relevant data could not prove his point.

The proof of a theory

Mysterious zebra patterns of magnetic anomalies were mapped on the ocean floor. In 1963 two Cambridge University graduate students, Frederick Vine and Drummond Matthews, explained these in a way that strongly supported sea-floor spreading. They suggested that material welling up from the mantle along a ridge acquires a remanent magnetization as it cools which is parallel to the then prevailing magnetic field of the earth. The earth's magnetic field is known to have reversed its polarity many times during geological history. Assuming that the newly forming crust at oceanic ridges picked up the prevailing polarity signal of the earth's magnetic field, the result over a long period of time would be strips of ocean floor that were alternately normally and reversely mag-

netized. The correspondence between the patterns on the ocean floor and the known history of the earth's magnetic field is too remarkable to be due to chance [4].

By 1966 the hypothesis of sea-floor spreading had been further established by independent oceanographic data involving microfossils, sediment thickness (thicker on older crust where sediments had more time to accumulate), measures of heat flow from the earth's interior, and palaeomagnetic and seismological studies. The expression "global tectonics" came into use in 1968, to explain the links between spreading ridges, transform faults, sinking trenches, drifting continents and mountain-building. This new theory is undoubtedly the most significant put forward in earth sciences this century.

Also in 1968, the United States commissioned the deep-sea drilling ship *Glomar Challenger* [Key] for a major campaign of oceanographic exploration. Drilled cores have been collected from all the oceans and seas and the sea-floor has been directly dated, proving beyond doubt the validity of the global tectonics theory.

The *Glomar Challenger* is a purpose-built research vessel for the Deep Sea Drilling Project (DSDP). The vessel is equipped to drill and retrieve cores from the floors of the deep oceans.

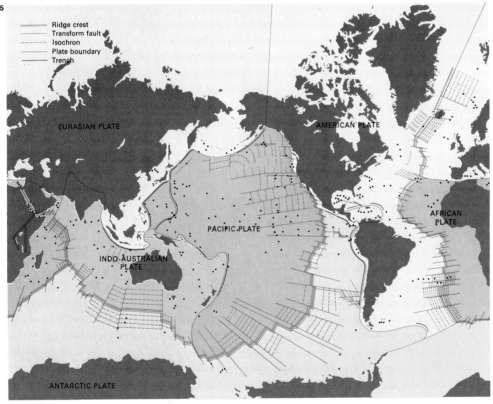

5 Ridge crest
Transform fault
Isochron
Plate boundary
Trench

EURASIAN PLATE

AMERICAN PLATE

AFRICAN PLATE

PACIFIC PLATE

INDO-AUSTRALIAN PLATE

ANTARCTIC PLATE

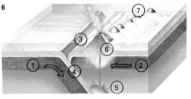

6 Destruction of the oceanic crust takes place where one plate [1] sinks beneath the other [2], forming a trench [3], while the dipping plane forms the sub-

duction zone [4]. Friction along this zone creates localized melting [5] and lava rises to the surface, creating volcanoes [6] forming island arcs [7].

7 Collision zones are where two plates each carrying a continental mass meet. When, in this zone, one of the plates is forced beneath the other, the buoyant

continental material thrusts upwards in a series of high overthrusts and folds, producing great mountain ranges. The Himalayas are the result of such forces.

5 The earth's outer shell is formed of six major mobile plates (the American, Eurasian, African, Indo-Australian, Pacific and Antarctic plates) separated by ridges, transform faults and trenches. These plates contain some smaller plates such as the Arabian and West Indian plates which "absorb" the geometrical discrepancies between the major plates by creating or destroying compensating amounts of crust. The dotted lines

parallel to the ridges are lines of equal age (isochrons) based on magnetic anomalies. The lines closest to the ridges are 10 million years old; each successive line is 10 million years older than its immediate neighbour towards the ridge. Fast rates of spreading are shown by widely spaced isochrons. The age of the crust has been directly verified at most of the DSDP drilling sites [black dots]. The African plate has no trenches

on its border; it is therefore growing in area. Its east-west growth is being compensated for by crust disappearing in the Tonga-Kermadec and Peru-Chile trenches, two and three plates away. Similarly the Antarctic plate is growing northwards, the compensation taking place in the Indonesian, North Pacific and Middle America trenches. Despite spreading along one of their margins, the Pacific and Indo-Australian plates are shrinking overall.

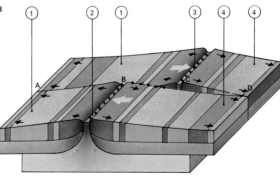

8

A   B   C   D

8 Transform faults separate two crustal plates [1, 4] where they move apart from one another. A transform fault links two segments of a ridge [2, 3] and the ridge offset gives an apparent motion [blue arrows] opposite to the real movement [black]. The transform fault is active only between the ridge crests [BC] where opposite crustal spreadings occur; it is only a dead scar along [AB] and [CD] where both sides move together.

Continents adrift

The idea that the continents were once joined together is by no means a new one: it was held by Francis Bacon (1561–1626) in 1620 and in 1658 R. P. Placet published a book in French whose title, in translation, means "The corruption of the great and little world, where it is shown that before the deluge, America was not separated from the other parts of the world". The first map showing the fit of the continents was published in 1858 by A. Snider-Pelligrini who based his theory on the similarity of the fossil plants which had been found in various parts of Europe and in North America.

Alfred Wegener's hypothesis

The man most closely associated with the theory of continental drift is Alfred Wegener (1880–1930) [2], a German meteorologist. An American, F. B. Taylor, had independently put forward the same ideas a few years before Wegener's first lectures on the subject in 1912. Like others before him, Wegener had been attracted to the idea of drifting continents because of the appearance of the land masses on the world map. He showed

great skill in reconstructing the ancient land masses in arguing his case and brought carefully assembled geological, geodetic, geophysical, palaeontological and palaeoclimatic evidence into his discursions. All of these facets of geology had to be considerably developed before they provided evidence of such a substantial nature that almost all geologists accepted the theory.

It was geologists in the Southern Hemisphere who collected evidence that strongly supported Wegener's own theories. Glaciation during Permo-Carboniferous times was more intense and widespread than that of the more recent Pleistocene ice age. Geological fieldwork revealed evidence of it in South America, Africa, Australia, India, Antarctica and Madagascar. Geologists examining deposits of till (sediment carried and deposited by a glacier) and fossil plants were able to correlate them between the continents. If these land masses had always been fixed in their present positions, it would mean that the ice stretched from the polar regions to the Equator – clearly a preposterous idea. Joining the areas together not only reconciles

the geological facts but also provides evidence of the land mass wandering over the South Pole. This, in turn, allows for an explanation of the other known climatic belts of the world during this period; for example, tropical conditions experienced in northern Europe during the Carboniferous period.

Corresponding structures and fossils

Work in India and in Australia has proved the existence of links between these continents; for example, the basins of Permian age in northwest Australia can be correlated with those of India, and features of eastern Australia match those of Antarctica.

The close links between the geology of West Africa and Brazil provide further proof of their former contact. There is a clear boundary between the 2,000 million-year old rocks of West Africa and the much younger geological province (about 400 million years old) to the east. The boundary between these two is near Accra in Ghana and it heads out into the Atlantic Ocean in a southwesterly direction. The continuation of this boundary in Brazil was found by a geological expedi-

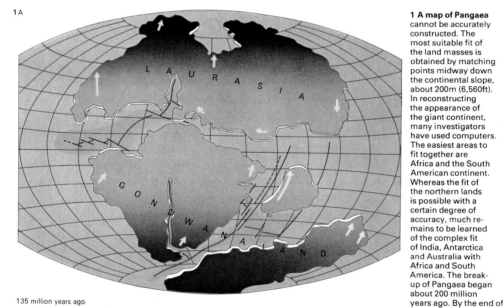

1 A

135 million years ago

B
65 million years ago

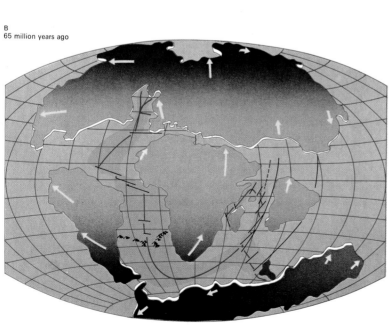

1 A map of Pangaea cannot be accurately constructed. The most suitable fit of the land masses is obtained by matching points midway down the continental slope, about 200m (6,560ft). In reconstructing the appearance of the giant continent, many investigators have used computers. The easiest areas to fit together are Africa and the South American continent. Whereas the fit of the northern lands is possible with a certain degree of accuracy, much remains to be learned of the complex fit of India, Antarctica and Australia with Africa and South America. The break-up of Pangaea began about 200 million years ago. By the end of the Jurassic, about 135 million years ago [A], the North Atlantic and Indian Oceans had become firmly established. The Tethys Sea was being diminished by the Asian land mass rotating in an anti-clockwise direction. South America had begun to move away from Africa to form the South Atlantic. By the end of the Cretaceous, *c*.65 million years ago [B], the South Atlantic had grown, Madagascar had parted from Africa and India had continued northwards. Antarctica was moving away from the central mass, while linked with Australia. The North Atlantic rift forked at the north, thus forming the island of Greenland.

2 Alfred Wegener was born in 1880. Trained as an astronomer, he became interested in meteorology and geophysics. He lectured on continental drift in 1912 and his first paper on the subject was published later the same year. In 1915, his classic exposition, *The Origin of Continents and Oceans*, appeared in print. He died while leading an expedition over the ice cap of Greenland in 1930.

3 Antarctica's tropical past is shown by the existence of coal seams up to 4m (13ft) thick. The study of changes of climate provides evidence confirming continental drift. In the Southern Hemisphere, rocks between 400 million and 180 million years old show marked similarities over now widely separated continents. Plant fossils in coal seams and layering of glacial deposits in one place match those in another, thus providing evidence for the theory.

tion exactly as was predicted, at São Luis.

The drawing together of the now widely separated northern lands (North America, Europe and Asia) has been a little more difficult, but there now exists overwhelming evidence for their once having formed part of a single continent, Laurasia [1]. Geologists have shown that the now widely separated Norwegian, Caledonian, Appalachian and east Greenland mountains were originally formed as a single chain.

Wegener paid considerable attention to the distribution of fossils. When his theory was first put forward, palaeontologists were still postulating land bridges to account for the distribution of some plants and animals in the fossil record. In many cases the land bridges would have to have covered an area equal to that of the continents they joined. It was long assumed that the land bridges disappeared by subsidence. The detailed study of the ocean floors in recent years has ruled out this idea. If, however, the idea of Gondwanaland – the old grouping of the continents of Africa, South America, Antarctica, India and Australia – is accepted [4], it becomes much

easier to explain the distribution of many animals and plants. For example, remains of the reptile *Mesosaurus*, which could not have swum an ocean, have been discovered only in western South Africa and in Brazil.

New developments
Recent investigations, especially the study of palaeomagnetism (the history of the earth's changing magnetic field [6]), have provided data which not only support continental drift but also locate the positions of the various land masses during past geological time. Perhaps the most important impetus to the theory of continental drift has come from the twin theories of sea-floor spreading and plate tectonics which have been so rapidly developed since the 1960s. One of the weakest points in Wegener's original arguments centred on the tremendous forces necessary to drive the continents apart. These new theories, which have been substantially proven, provide an explanation of the necessary motive power, but there is still much to learn about the break-up of Pangaea, the original continent [Key].

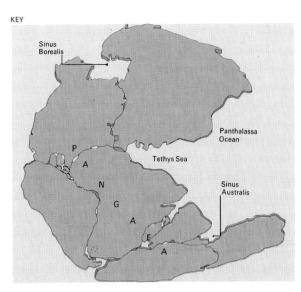

The theory of continental drift proposes a period when all the continents formed one land mass, called Pangaea. The initial break-up made a northern mass, Laurasia, and a southern one, Gondwanaland, called after a province in India.

4 The existence of Gondwanaland is confirmed in many different ways. Constant directions of ice flow in Permo-Carboniferous glaciations; structural trends traced and matched from continent to continent; and fossil distribution form part of the evidence. The distribution of *Mesosaurus*, *Lystrosaurus* and the fossil fern *Glossopteris* suggest the presence of a single land mass in Permian times. Palaeomagnetic techniques – which locate the magnetic pole of any stage in the past – give consistent results on each continent only when the continents are placed in the estimated configuration of Gondwanaland.

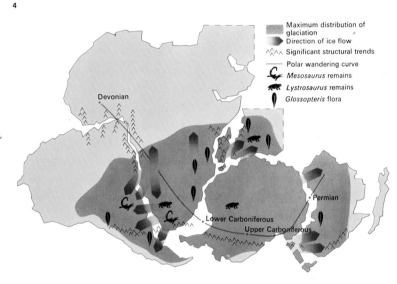

Maximum distribution of glaciation
Direction of ice flow
Significant structural trends
Polar wandering curve
Mesosaurus remains
Lystrosaurus remains
Glossopteris flora

Devonian
Lower Carboniferous
Upper Carboniferous
Permian

5 Fossil seed ferns (such as the tongue-like *Glossopteris*, shown here, and *Ganganop-teris*) in rocks of the Gondwana series in the southern continents evince support for continental drift theory. They reached their height in Permo-Carboniferous times and the complex nature of their species distribution can only be explained if the now widely separated locations had been in one land mass.

6 Palaeomagnetism – the study of the magnetization of rocks – has developed since the 1950s into a powerful tool for continental drift studies. Rocks record the earth's magnetic field at the time of their formation. Portable drills cut small cylindrical rock cores which are oriented relative to the present north and in a vertical position. In the laboratory a magnetometer determines their original north direction and their original latitude.

7   50 million years ahead

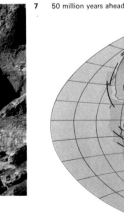

7 The continents are still drifting, and there is no reason to expect them to stop. The map shows how the world may look 50 million years from now if drift is maintained as predicted. The most striking change in the "new world" is the area of new land in the Caribbean, the splitting away from the USA of Baja California and the area west of the San Andreas fault line, the northward drift of Africa almost eliminating the Mediterranean, and the breaking away of that part of the continent east of the present-day rift valley. Australia has continued its journey northwards, but the great continent of Antarctica remains in its present southerly position.

27

Earthquakes

An earthquake at the earth's surface is the sudden release of energy in the form of vibrations and tremors caused by compressed or stretched rock snapping along a fault in the earth's surface. Rising lava under a volcano can also produce small tremors. It has been estimated that about a million earthquakes occur each year – but most of these are so minor that they pass unnoticed. Really violent earthquakes, which result in widespread destruction, occur about once every two weeks. Fortunately, most of these take place under the oceans, out of harm's way. It is not known what causes deep earthquakes – up to 700km (450 miles) below the surface

Waves and their measurement

Slippage along a fault is initially prevented by friction along the fault plane. This causes energy, which generates movement, to be stored up as elastic strain; a similar effect is created when a bow is drawn. Eventually the strain reaches a critical point, the friction is overcome and the rocks snap past each other releasing the stored-up energy in the form of earthquakes by vibrating back and forth.

Earthquakes can also occur when rock folds that can no longer support the elastic strain break to form a fault.

Seismic (earthquake) waves spread outwards in all directions from the focus – much as sound waves do when a gun is fired [Key, 4]. There are two main types of seismic wave: the compressional wave and the shear wave [2]. Compressional waves cause the rock particles through which they pass to shake back and forth in the direction of the wave. Shear waves make the particles vibrate at right-angles to the direction of their passage. Neither type of seismic wave physically moves the particles: instead it merely travels through them.

Compressional waves, which travel 1.7 times faster than shear waves, are the first ones to be distinguished at an earthquake recording station [3]. Consequently seismologists refer to them as primary (P) waves and to the shear waves as secondary (S) waves. A third wave type is recognized by seismologists – the long (L) or surface wave. It is L waves that produce the most violent shocks. The Richter scale is used to measure the magnitude of earthquakes. The scale of magnitudes is so arranged that each unit on the scale is equivalent to 30 times the energy released by the previous unit. A magnitude of 2 is hardly felt, while a magnitude of 7 is the lower limit of an earthquake that has a devastating effect over a large area.

Tsunamis – giant sea waves

Earthquakes are best known for the havoc they can wreak [1, 9]. The destruction may be the result of ground vibrations or giant tidal waves (tsunamis) generated by seismic disturbances on the sea floor. At sea tsunamis have wavelengths – the distance between one crest and the next – as great as 200km (120 miles). They can travel at speeds of 800km/h (500mph). When they reach a gently sloping shore they slow down and gather height. As the tsunami approaches, the sea withdraws, then rushes back in a series of giant waves that may travel far inland.

In 1755 the city of Lisbon was reduced to rubble in six minutes during one of the most devastating earthquakes ever recorded. The sea withdrew from the harbour and rushed

CONNECTIONS

See also
20 Anatomy of the earth
104 Folds and faults
30 Volcanoes

1 Severe surface disturbance ripped apart a road during the 1964 earthquake in Anchorage, Alaska, and minor subsidence left a crack 50cm (20in) wide. Ground waves such as those responsible for the break-up of this highway can disrupt underground services in cities and start fires. Broken water mains may hamper fire-fighting and encourage the spread of epidemics. The Alaskan "quake" affected only a sparsely populated area and 114 lives were lost – a small number considering the magnitude of the shock, which caused a permanent tilting of the land mass along the southern coast of Alaska.

2 Seismic waves are basically of two kinds. Primary (P) waves [A] are compressional and cause the particles of rock to vibrate backwards and forwards like a coil spring. Secondary (S) or shear waves [B] cause the particles to oscillate at right-angles to the wave direction like a vibrating guitar string. When P and S waves reach the surface they are converted into long (L) waves which either [C] travel along the surface vibrating horizontally at right-angles to the wave direction (known as Love waves) or travel like sea waves (Rayleigh waves). Some of the paths followed are shown in D.

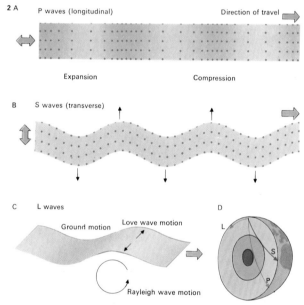

4 Seismic wave paths vary with the density of the rock, forming curving patterns as they move away from the focus [1]. Primary (P) waves can pass through gases, liquids and solids. The primary waves travel fastest, increasing their velocity as they pass through the mantle [2] but dropping in the outer core [3] and rising in the inner core [4] due to the conditions produced by pressure. Secondary (S) waves travel through solids only and do not penetrate into the dense molten core. As the waves travel down they meet concentric layers of increasing density which bend or refract the waves towards the surface so that these travel along curved paths. The region between 5 and 6 does not receive any direct waves. This area is known as the wave shadow zone. Seismic wave propagation has given invaluable information about the earth's interior.

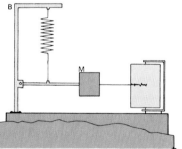

3 Seismographs are instruments that detect and record seismic waves of the three types (P, S and L). Most seismographs contain a sprung mass (M) which, when an earthquake passes, stays still while the rest of the instrument moves. Some seismographs detect horizontal motion [A] while others detect vertical motion [B]. The trace of the waves is recorded by a vibrating pen on a travelling strip of paper [C]. The time interval between the arrival of the P and S waves can be calculated and this interval, applied to a graph (see 5B), gives the distance between station and epicentre.

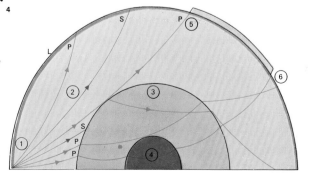

back as a 17m-high (55ft) *tsunami*, drowning hundreds. Smaller aftershocks caused landslides, fires broke out and by nightfall 60,000 people were dead. The shocks from this earthquake were felt over an area nearly 40 times as large as the United Kingdom.

Despite the innate destructive capacity of earthquakes it is possible in some circumstances to take precautions that minimize the hazards. Tall buildings can be constructed on reinforced concrete rafts that literally float during the passage of earthquake waves. Careful planning can ensure that streets are wide in relation to the height of buildings: many of the deaths caused during earthquakes are due to the collapse of tall buildings into narrow streets.

Control and prediction

Recent research indicates that it may now be possible to control earthquakes. In the mid-1960s, the dumping of water-based waste into a well in Denver, Colorado, set up a series of small earthquakes. Thus the idea was born that by drilling deep holes along a fault, and then pumping water down, it might

be possible to relieve strains in a series of small, non-destructive earthquakes instead of allowing them to build up until a major earthquake occurs [8].

Just before an earthquake the ground on either side of a fault suffers elastic deformation that can be measured by triangulation with a theodolite or laser beam. Tilt meters can also be used to discover how much warping of the ground has taken place. Monitoring of large areas has now been introduced. Using artificial satellites, information is transmitted from devices placed in the vicinity of major faults and radioed back to centres where it can be analysed. It is now possible to detect very small movements of the earth's surface and locate areas where strain is building up.

Another recently discovered method involves measuring the amount of water the rocks contain. Under strain, the pores in the rock enlarge, allowing more water to enter. Because of the importance of ground water in producing earthquakes, knowledge of the water level in wells in earthquake-prone areas is extremely valuable.

An earthquake takes place when two parts of the earth's surface move suddenly in relation to each other along a crack called a fault [1]. The point from which this movement originates is called the focus [2] and the point on the surface directly above this is called the epicentre [3]. Shock waves [4] travel outwards from the focus decreasing in intensity the farther they go. These shock waves travel more quickly as they pass through denser material at depth and so the direction of their travel [5] is curved as shown. On the surface the pattern of waves is similar to that of the isoseismal lines connecting points feeling equal shocks.

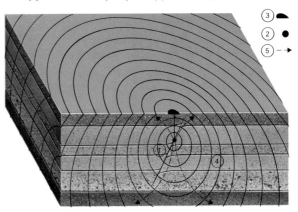

5 **The location of an epicentre** [1] is found by plotting its distance from three recording stations [2]. Each station notes the different arrival times of P and S waves and uses a graph [B] which allows the distance from the epicentre to be measured. The distance is then used as the radius of a circle round each station [A]. The epicentre is located at the intersection of these three circles.

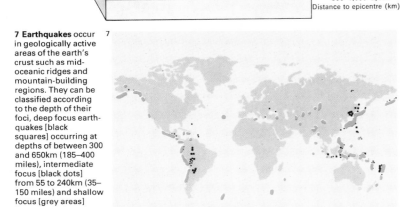

6 **Earthquake intensity** is based on a measure of the damage caused in populated areas. The most common intensity scale used today is the Wood-Neumann or modified Mercalli scale.

6 Modified Mercalli scale

1 Earthquake not felt, except by few.

2 Felt on upper floors by few at rest. Swinging of suspended objects.

3 Quite noticeable indoors, especially on upper floors. Standing automobiles may sway.

4 Felt indoors. Dishes and windows rattle, standing cars rock. Like heavy lorry hitting building.

5 Felt by nearly all, many wakened. Fragile objects broken, plaster cracked, trees and poles disturbed.

6 Felt by all, many run outdoors. Slight damage, heavy furniture moved, some fallen plaster.

7 People run outdoors. Average homes slightly damaged, substandard ones badly damaged. Noticed by car drivers.

8 Well-built structures slightly damaged, others badly damaged. Chimneys and monuments collapse. Car drivers disturbed.

9 Well-designed buildings badly damaged, substantial ones greatly damaged, shifted off foundations. Conspicuous ground cracks.

10 Well-built wood structures destroyed, masonry structures destroyed. Rails bent, ground cracked, landslides. Rivers overflow.

11 Few masonry structures left standing. Bridges and underground pipes destroyed. Broad cracks in ground. Earth slumps.

12 Damage total. Ground waves seem like sea waves. Line of sight disturbed. Objects thrown upwards in the air.

7 **Earthquakes** occur in geologically active areas of the earth's crust such as mid-oceanic ridges and mountain-building regions. They can be classified according to the depth of their foci, deep focus earthquakes [black squares] occurring at depths of between 300 and 650km (185–400 miles), intermediate focus [black dots] from 55 to 240km (35–150 miles) and shallow focus [grey areas] from the surface down to 55km (35 miles).

8 **Release of the pressure** that could cause a severe earthquake may be achieved by promoting a number of small "quakes" in the fault area. Researchers are investigating the possibility of minimizing the destructive effects of earthquakes in certain regions by regulating their occurrence. Many small earthquakes, for instance, may release as much energy as a single devastating one by lessening the strain built up over a period of time [A]. One method of achieving this may be to pump water to act as a lubricant [B]. A number of wells [1] may be set up along a fault line [2] in which stress has been detected. Large quantities of water from a reservoir [3] would then be pumped into the wells to lessen friction between rocks in the fault and allow them to slip smoothly in a series of gentle tremors. Another method of triggering off small "quakes" may be to explode nuclear devices along a fault plane.

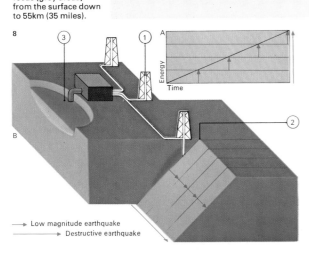

→ Low magnitude earthquake
→ Destructive earthquake

9 **The western Sicily earthquake** of 1968 completely destroyed many buildings in Gibellina and in nearby villages. Dazed survivors picked their way through piles of shattered masonry. Falling masonry is the greatest hazard to human life during a severe earthquake and most injuries are caused outside buildings. Disease often spreads among survivors after the breakdown of essential services. There were 224 deaths in the Sicilian "quake". Some of the most disastrous earthquakes since 1965 have occured in Peru in 1970 (66,794 deaths), in Guatemala in 1976 (more than 16,000 deaths), and in China, also in 1976 (estimates of up to 100,000 deaths).

Volcanoes

Volcanoes, the earth's most spectacular displays of energy, are responsible for forming large parts of the earth's crust. They give clues to the earth's history and evolution, and to the nature of the earth's interior. Soils formed by the weathering of volcanic rocks are so exceptionally fertile, that despite the danger large numbers of people often live in the shadow of volcanoes, and eruptions lead to major loss of life. The earth's upper mantle, under the crust, is nearly molten. A slight drop in pressure caused, say, by crustal plates drifting apart, completes the melting process. The molten rock (magma), being lighter than the surrounding rocks, then rises slowly to the surface, often along faults. A small increase in heat, will also melt the rock and it is believed that pockets of radioactive elements generate enough heat to form magma.

Along the mid-ocean ridges, where the crustal plates' drifting apart creates a drop of pressure, magma rises more or less continuously and cools to form new crust. Elsewhere it forms underground reservoirs, which, if they do not cool, can become unstable and produce eruptions. When this happens the flow is speeded up as the drop of pressure allows the gas dissolved in the magma to form bubbles [2]. Many of the gases, such as hydrogen sulphide and carbon monoxide, burn as they reach the air; this increases the temperature at the vent, making the lava even more fluid. If the lava is viscous, trapping the gases, they may escape explosively. The force of such explosions – and of normal eruptions – is increased when water seeps down to the magma and is turned into steam.

The volcanoes formed by escaping magma are characterized by their vent or crater at the summit. They often have side vents as well. Sometimes such large amounts of magma bubble out during an eruption that the chamber below the volcano is more or less emptied. The volcano then collapses into the void, forming a large, steep-sided depression known as a caldera.

Location of volcanoes

Volcanoes are found along the big tensional cracks of the earth's surface – the mid-ocean ridges and their continental continuations – and along the collision edges of crustal plates. The famous "ring of fire" encircling the Pacific is the boundary of the crustal plate that forms the Pacific.

The largest number of volcanoes is under the sea-floor, forming the abyssal hills. Most are probably extinct. They exist because the oceanic crust is very thin and easily pierced by the underlying magma. The Pacific alone is thought to have more than 10,000 volcanoes of above 1,000m (3,300ft) in height. The Hawaiian volcanoes are thought to be due to fixed "plumes" or "hot spots" in the mantle, which give rise to a string of volcanoes as the crust drifts slowly over them.

A few volcanoes that exist on land away from the plate boundaries are perhaps due to localized heating by radioactivity or to a hot spot in the mantle.

Apart from the uncounted abyssal volcanoes, there are about 500 active volcanoes, of which perhaps 20 or 30 erupt in any given year. Between eruptions a volcano is said to be dormant. An active volcano is one that has erupted in historic times. Volcanoes can, however, be dormant for periods longer than

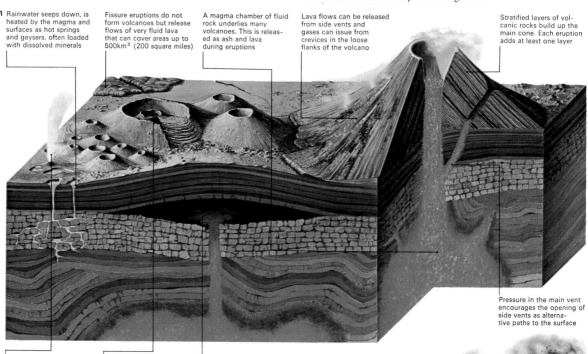

1 **Volcanoes** are fed by molten rock that rises from the earth's mantle. This material, called magma, may rise directly to the surface where it erupts, or be stored in a magma chamber that swells up like a balloon before erupting. The magma rises through a chimney and eventually reaches the surface at the vent. Matter spewed forth as lava or ejecta (bombs and ashes) builds up a volcanic cone or volcano. Vent explosions caused by expanding gases often form craters shaped as inverted cones. The magma does not always reach the surface and often cools at depth forming plutons (large bodies), laccoliths (lens-shaped structures), dykes (that cut through the strata) and sills (injected between two strata). Volcanic regions are also characterized by hot springs, gas vents and, in some areas, by geysers.

1 Rainwater seeps down, is heated by the magma and surfaces as hot springs and geysers, often loaded with dissolved minerals

Fissure eruptions do not form volcanoes but release flows of very fluid lava that can cover areas up to 500km² (200 square miles)

A magma chamber of fluid rock underlies many volcanoes. This is released as ash and lava during eruptions

Lava flows can be released from side vents and gases can issue from crevices in the loose flanks of the volcano

Stratified layers of volcanic rocks build up the main cone. Each eruption adds at least one layer

Pressure in the main vent encourages the opening of side vents as alternative paths to the surface

Geysers are intermittent fountains of water and steam created by the vaporizing of ground waters. They operate like giant safety valves.

Active or recent cones often form inside explosion craters or crater-shaped calderas due to the collapse of an empty magma chamber

A laccolith is a giant, lens-shaped intrusion that pushes up the strata above. It is fed from the magma chamber

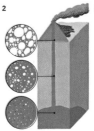

2 **As the pressure** in rising magma falls, dissolved gases are forced out of it; these form expanding bubbles and drive the magma out of the volcano.

3 **Active volcanoes** mostly occur at crustal plate boundaries. The principal zone of activity is found along a great arc around the Pacific, from Chile to the East Indies.

— Active volcanoes
•--Extinct volcanoes

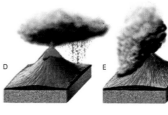

4 **Volcanic eruptions take various forms.** Fissure eruptions [A] release the most basic and runny lava. In Hawaiian eruptions [B] the lava is less fluid and produces a low cone. The Vulcanian type [C] is more violent and ejects solid lava. Strombolian eruptions [D] blow out incandescent material. In the Peléean type [E] a blocked vent is cleared explosively. A Plinian eruption [F] is a continuous gas blast that rises to immense heights.

historic times and "extinct" volcanoes sometimes come back to life, as Helgafell in Heimaey, Iceland did in 1973 [8]. The best-known new volcano on land is Paricutin, in Mexico, which appeared in a field in 1943. The map [3] shows active volcanoes and recent extinct volcanoes, many of which were seen erupting by prehistoric man, such as those in France. The geological record abounds in "fossil" volcanoes (not shown). For instance Scotland's capital, Edinburgh, was a volcano 325 million years ago.

Volcanic products and eruption types

Volcanoes emit gases, liquids and solids. The gases are mainly nitrogen, carbon dioxide, hydrogen chloride, water vapour, carbon monoxide and hydrogen sulphide. Liquid emissions, known as lavas, are either ropy *pahoehoe* [10] or clinker-like *aa* [9], depending on the temperature.

Fluid lava allows calm eruptions; more viscous lava, by preventing the escape of gas before it reaches high pressure, is accompanied by explosions; very viscous magma is thrown out as ash and rubble in huge explo-sions. Craters at rest are often filled by a lake; an eruption often creates a mud flow which is as destructive as, and even more lethal (owing to its speed) than, a lava flow.

Catastrophic eruptions

Volcanoes are in a sense safety valves in the earth's crust: the tighter the valve the greater the eruption will be. The 1815 eruption of Tambora in Indonesia ranks as the greatest volcanic disaster in history. Ten thousand people were killed outright during the eruption and 82,000 died later of disease and starvation. Again in Indonesia, the uninhabi-ted island of Krakatoa was blown to pieces [Key] in 1883, creating a tidal wave that killed 36,000 people. Evidence on the island of Thera [11] suggests that an even larger explosion occurred there around 1470 BC.

Eruptions cannot be prevented, but they can, in some cases, be predicted. This is done principally by monitoring the small earth-quakes created by rising magma, measuring the swelling of the ground with tiltmeters and watching variations in the output of gas and steam vents.

KEY

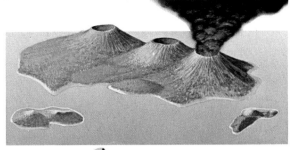

Krakatoa was a small volcanic island in the Sundra Straits of Indonesia that had been inactive since 1680. Two-thirds of the island were destroyed in an eruption in August 1883.

5 Hawaiian-type eruptions are characterized by lava flows consisting of basalt and they are often accompanied by fiery lava fountains which can sometimes be as tall as the Eiffel Tower – 300m (1,000ft). The Hawaiian eruption seen here shows how the incandescent lava from the fountain has filled the crater and has then overflown to form a lava flow that is partially chilled on its surface.

6 Lava formations include hornitos or spatter cones, small bursts [A] that form miniature volcanoes on lava flows [B]. Tree moulds [C] form where trees have burned away beneath the cooled lava [D]. Lava tubes occur when the surface flow cools [E] and the hot interior drains away [F].

7 Spindle bombs are the product of molten rock pulling apart after ejection. Larger blobs twist in flight and their drawn-out ends curl and give the bomb its characteristic lozenge shape.

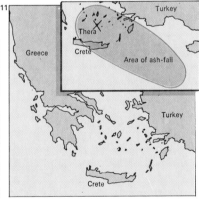

8 Ash-falls can cause more damage to property and agricultural land than lava flows because they cover greater areas. Volcanic ash is made of fine ejecta of less than 4mm (0.15 in) diameter and is produced by the cubic kilometre. Most is deposited within 10km (6 miles) of the volcano. The picture shows some houses of Heimaey, Iceland, buried by ash from the Helgafell eruption of 1973.

9 The texture of a lava flow depends on the temperature and the velocity of the flow during the eruption and on the composition of the lava. Geologists have borrowed two words from the Hawaiian language, *aa* and *pahoehoe*, to describe two typical surfaces. *Aa* lava has a very rough and clinker-like aspect and is formed by slowly moving or relatively cool out-pourings. Seven years after the solidi-fication of this flow, vegetation will gradually return.

10 *Pahoehoe* or ropy lava has a smooth but twisted surface. It is formed by fast-flowing fluid lava, which develops a plastic skin on its sur-face, by cooling. The skin is dragged into picturesque folds by the still liquid lava running beneath it.

11 In the late Bronze Age the island of Thera experienced an eruption that had a catastrophic effect on the people of Crete and may have been responsible for the fall of the Minoan civilization and the creation of the Atlantis legend.

31

Gravity and the shape of the earth

Gravity is the mutual attraction of two bodies and its strength depends on the mass of the bodies and their distance apart. The strength of the earth's gravitational field is therefore proportional to the earth's mass and decreases as the distance from the surface of the earth increases.

Gravity is responsible for nearly all major erosion which takes place. Rain falls under gravity; streams, rivers and glaciers move, and sediments compact under gravity.

Rotation, shape and sea-level
The earth's rotation creates a centrifugal force that is greater at the Equator than elsewhere, causing the earth to bulge out slightly at the Equator and to flatten at the poles [1]. This makes the earth's diameter at the Equator greater than that through the poles by about 41km (25 miles).

Mean sea-level is the average sea-level between tides and is taken as the base level when measuring altitudes. It is always perpendicular to the force of gravity. Such a surface, taken all over the earth is known as the "geoid", which is what the earth's true

shape is called [2]. The surface of the geoid is irregular because the gravity field varies locally and depends on the type of rocks in the crust. A large ore body or mountain chain will deflect a nearby plumbline away from the centre of the earth and thus the geoid's shape is obtained by directly measuring gravity on land [3] or its variation at sea, where wave motion precludes direct measurement. Perturbations of artificial satellite orbits are now extensively used for broad-scale geoid studies. The shape of the geoid is defined by its departure from a "reference ellipsoid" which fits most closely to the shape of the earth; in this case, the average level of the land and sea is taken as the norm. Mountains are then higher and sea-floors lower than the surface of this ellipsoid. (An ellipsoid is the regular geometric shape obtained by revolving an ellipse round one of its axes.)

The Trigonometrical Survey of India
During the Trigonometrical Survey of India, in the nineteenth century, it was found that some stations, whose positions were determined by astronomical surveying methods,

did not coincide with those determined by triangulation, which allowed for the extra sideways attraction of the Himalayas to the north. The Himalayas were exerting a smaller gravitational pull on the plumbline than expected and introducing discrepancies as large as 91.5m (300ft). As a result, both J. H. Pratt and G. B. Airy (1801-92) proposed that the continents consisted of lighter material floating on a denser substratum [5]. Pratt believed that the different heights of mountains were due to blocks of different densities floating at the same base level, while Airy thought the heights were due to blocks of different thickness and of the same density floating at different depths. Today, Airy's hypothesis is generally accepted.

Another way of putting Airy's hypothesis is that lighter crustal materials "float" in equilibrium on the denser but slightly plastic mantle, like a cork floating on water. If such a cork were covered with an extra load (such as a coin) it would sink a little until the extra immersed volume created enough buoyancy to compensate for the extra load. This is what is observed for the earth's crust: extra loads

CONNECTIONS

See also
20 Anatomy of the earth

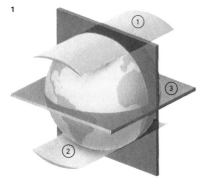

1 The earth's surface is not a true sphere but approximates to an ellipsoid whose equatorial diameter exceeds its polar diameter. The difference is about 41km (25 miles) and may be illustrated by forcing hypothetical wedges [1, 2] through the spaces over the poles; this would not be possible at the Equator [3]. This equatorial bulge is caused by the effect of centrifugal force as the earth rotates on its axis.

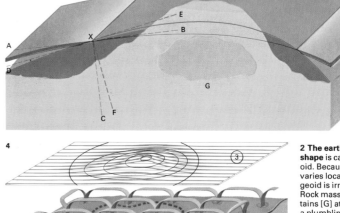

Reference ellipsoid
Geoid

2 The earth's true shape is called the geoid. Because gravity varies locally, the geoid is irregular. Rock masses in mountains [G] attract a plumbline and the assumed direction of the centre of gravity [XC] for the ellipsoid is deflected to the local direction of the centre of gravity [XF]. As XF is perpendicular to the true level [DE] based on the geoid, AB is only the assumed level based on the ellipsoid.

3 A gravimeter is a device used for measuring gravity at a point on the earth's surface by observing the extension of a weighted spring. A quartz spring [1] is housed in a partially evacuated chamber [2] which protects it from pressure changes. Levelling screws [3] keep the meter vertical and the movement of the spring is indicated by the position of the pointer [4] on a scale observed through an eyepiece [5].

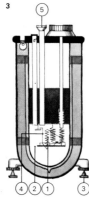

4 A negative gravity anomaly – a lower gravity reading than normal – is found over an intrusion of light rock near the surface. A less dense salt dome [1] rises through denser crustal rocks [2] disturbing the local gravitational field. Readings at regular stations on the surface can be plotted on a map [3] and these will show an area of low gravity above the dome. A dense metallic ore body would show a positive anomaly.

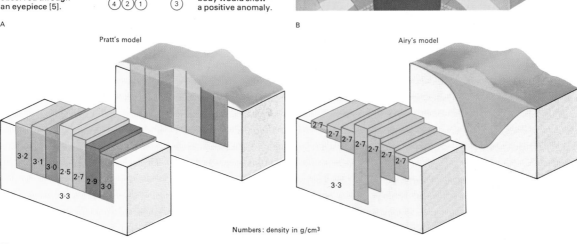

5A
Pratt's model

B
Airy's model

Numbers : density in g/cm³

5 In the 1850s Pratt and Airy proposed that the continents consisted of light material floating on a denser substratum. Whereas Pratt suggested that the different heights of mountains were due to blocks of different densities floating at the same base level [A], Airy considered that they were due to blocks of different heights of the same density floating at different depths [B]. Geophysical research has shown Airy's view to be the more likely.

cause downwarping of the crust [7]. When the load disappears, for example by the melting of ice or by erosion, the crust bobs up, regaining a new isostatic equilibrium.

Parts of Norway and Sweden are still rising, following the melting of the thick Pleistocene ice caps 10,000 years ago, and it is estimated that the crust there must rise a further 213m (700ft) before equilibrium is restored. However, as Scandinavia rises, the coastlines of The Netherlands and parts of Denmark are sinking [8] because mantle material which is flowing up under Scandinavia is drawn from beneath The Netherlands and parts of Denmark.

Isostatic equilibrium is maintained by variation in the depth of the earth's crust and scientists have shown that every mountain chain floats on a deep root in the mantle. Conversely, below the oceans only a thin layer of crust is found.

Anomalies in the gravitational field
Rock bodies, whose mass differs greatly from that of the surrounding area, cause small variations or anomalies in the local gravita-tional field, making it possible to detect them with sensitive gravimeters [Key]. Gravi-meters work on the principle that gravity changes will cause minute variations in a very fine quartz spring [3]. These instruments are sensitive enough to detect minute gravita-tional changes as small as one ten-millionth of a gramme.

Gravity surveys reveal that large salt domes near the surface (often associated with oil and gas) will show up as a negative ano-maly (that is a mass deficit, because salt is lighter than other rocks) [4], while a dense ore body will show up as a positive gravity anomaly (mass excess). Geologists exploring the terrain carry out surveys of the earth's gravity field and the observed values are cor-rected to eliminate the influence of latitude, height and the mass of material between the observation point and sea-level or the lowest level obtainable. This allowance is called the Bouguer correction and the resulting map is therefore known as a Bouguer anomaly map. Its value to industry lies in its use to deter-mine the possible positions and sizes of economic ores and oil reserves.

Underground rock formations can be detected by measuring the local variations in the pull of gra-vity on a delicate spring balance called a gravimeter. The lighter the material, the weaker the pull.

1 Normal gravity reading
2 Heavy igneous material near surface giving gravity high
3 Anticline giving gravity high
4 Rift valley where lighter surface material continues to a greater depth giving gravity low
5 Salt dome or upward emplacement of light material giving gravity low
6 Oceanic trough where lighter crustal material deep in mantle gives gravity low

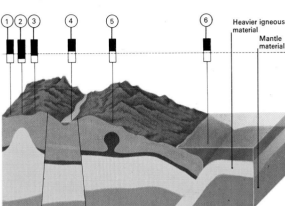

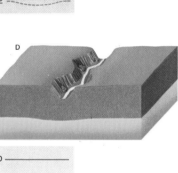

Heavier igneous material

Mantle material

6 A person sitting on a water bed down-warps the surface, causing fluid to flow from beneath him un-til equilibrium is reached and he no longer sinks. A sec-ond person also down-warps the surface and causes the fluid to flow from under her and push up under the first person thus causing him to rise slightly. This is analogous to the floating of mountains on the liquid mantle of the earth.

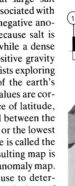

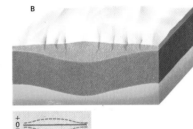

7 When the earth's surface acquires a heavy load, like an ice sheet, over a period of time the crust downwarps into the mantle. In A the crust is in equilib-rium with the mantle below. An ice sheet [B] is heavy and pro-duces a positive grav-ity anomaly. To com-pensate for this, the crust downwarps, giving a deficiency in mass and a nega-tive anomaly. The positive and negative anomalies cancel out and the crust remains in equilibrium. When the ice melts [C] the load and positive an-omaly are removed leaving a mass de-ficiency and a neg-ative anomaly. To re-store equilibrium [D] the land rises and rivers rejuvenate, cutting deep valleys.

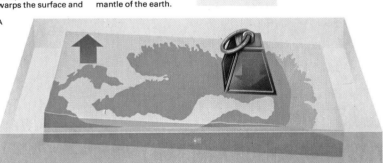

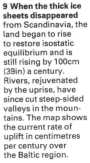

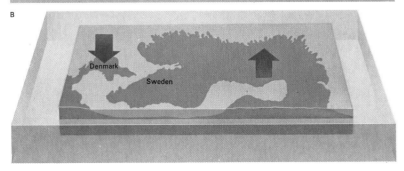

8A

B

Denmark

Sweden

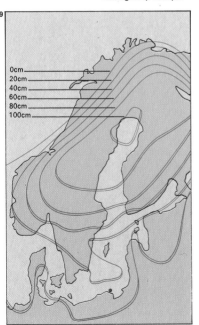

8 Glacial rebound is an example of iso-static activity. During the last ice age Scan-dinavia was weighed down by ice, causing the north of Europe to tilt [A]. After the ice melted the conti-nent returned to its original attitude [B], buoyed up on upper mantle material flow-ing from under the sinking areas to under areas of uplift.

9 When the thick ice sheets disappeared from Scandinavia, the land began to rise to restore isostatic equilibrium and is still rising by 100cm (39in) a century. Rivers, rejuvenated by the uprise, have since cut steep-sided valleys in the moun-tains. The map shows the current rate of uplift in centimetres per century over the Baltic region.

0cm
20cm
40cm
60cm
80cm
100cm

Mapping the earth

Throughout history man has recorded, analysed and communicated information in map form. The oldest map in existence is engraved on a Babylonian clay tablet dating from 3000 BC and like many surviving examples of early mapping it records land tenure. It was not until the fifth century BC, however, that Greek philosophy stimulated attempts to create a map of the world. Unfortunately these were based on philosophical theories rather than the geographical knowledge of the day. Nevertheless, in the following 600 years Greek scholars did develop a more scientific approach to cartography.

Early attempts at cartography
At the end of the first century AD Ptolemy of Alexandria compiled his *Geographia*. In it he discussed the problem of representing the spherical shape of the earth on a plane surface and also introduced the concepts of longitude and latitude [4].

After Ptolemy, cartography entered a period of decline until the Crusades and an expansion of trade revived interest. A cartographic renaissance came in the fifteenth century with the discovery and publication of Ptolemy's work, voyages of exploration like those of Vasco da Gama (*c.* 1469–1525) and Christopher Columbus (1451–1506) and the invention of printing and engraving. In the sixteenth century the work of map-publishing houses in Holland and France, and particularly that of Gerhardus Mercator (1512-94), founded modern map-making.

By the middle of the eighteenth century the French had initiated the first topographic survey. Many special-purpose or thematic maps have been produced since the nineteenth century. Their variety reflects the increasingly specialized demands of modern life, which require such aids as land-use maps, pilot charts and road maps.

Modern surveying
Small areas of the earth can be mapped by plane surveying but larger areas must be done by geodesy which takes into account the earth's curvature. A variety of instruments and techniques is used to determine position, height and extent of features – data essential to the cartographic process. Instruments such as graduated metal rods, chains, tapes and portable radar or radio transmitters are used for measuring distances; the theodolite [Key] is used for measuring angles. Using measured distances and angles further distances and angles are calculated by triangulation [1]. Heights are determined similarly [2].

Perhaps the most dramatic changes in cartography have come with the development of aerial survey techniques. Photographs from satellites or aircraft are used together with data from ground surveys to map accurately large areas of the earth. This technique, called photogrammetry, is particularly useful for mapping remote areas [3], and also for mapping the earth's natural resources using images produced by remote sensing equipment. Almost all topographic mapping today is done from aerial photographs and photomaps are made from the aerial photographs of an area.

Map projections
It is obviously impossible to represent accurately the surface of a sphere on a flat plane without distorting the relationships between

1 Triangulation is a method surveyors use to determine heights and distances from a base line [1–2], which is generally measured using a calibrated tape or a surveyor's chain. Angles from it – sometimes forming part of a map's grid – are measured by means of a theodolite [3]. Detailed surveying requires a fine network of triangles [B]. These may be simple triangles [4]; braced [5] or centred [6] quadrilaterals; double-centred figures [7]; or narrow quadrilaterals [8]. The type adopted depends on local conditions – braced quadrilaterals are preferred in hilly terrain and centred quadrilaterals are usually used in surveying flat regions.

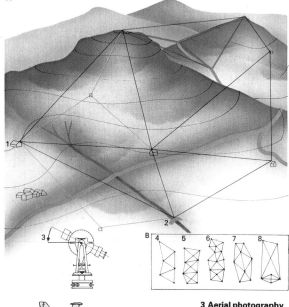

2 Height above sea-level is usually determined by means of a levelling instrument [1, 2, 3] and a measuring rod [4–10] with reference to a known height or bench mark [X]. Level 1 sights on the rod at 4 and then at 5. The instrument is moved to 2 to begin the second stage, sighting first on 5 and then on 6. Finally 10 will be reached using stage 3. Intermediate heights are determined by placing the rod at 8. The heights of the points 4–10 can be related to sea-level [Y] because of the height above sea-level of the bench mark [X] is already known.

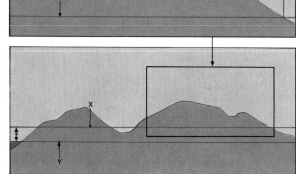

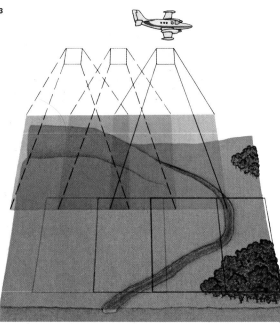

3 Aerial photography is one of the modern techniques that have helped the work of the cartographer. An aircraft flies over the area to be mapped, taking a continuous series of photographs. The area covered by each photograph overlaps by 60% the area of the previous one and so, after processing, any adjacent pair of photographs can be examined stereoscopically and the relief of the area studied directly in three dimensions. By means of an optical instrument the positions of corresponding points on each photograph can be compared and the height of that point calculated. Each "run" of photographs overlaps sideways by 10% the previous run, ensuring total coverage.

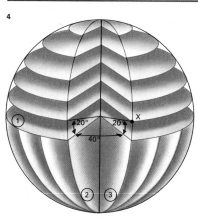

4 Any point on the surface of the earth can be located in terms of longitude and latitude – in degrees, minutes and seconds east or west of a prime meridian or line of longitude, and north or south of the Equator. The latitude of X (the angle between X, the centre of the earth and the plane of the Equator [1]) equals 20° while its longitude (the angle between the plane of the prime meridian [2] and that passing through X and the poles [3]) equals 40°.

features on that surface. A map projection is a device used to plot the earth's features with a minimum of such distortion. There are a number of different types of projection and the choice of one in preference to others depends on the purpose of the map.

If the map is to show relatively small areas of the world, as in a national topographic series that will be used by planners, engineers and the general public, then a projection must be selected that shows distance, angle and shape with the utmost accuracy. For this reason conformal projections are chosen. If on the other hand the map is to show distribution of, say, cultivated land throughout the world, then a projection that shows those areas at their correct relative size must be selected. Such projections are called equivalent or equal area projections.

Conformal projections are not used for world maps except in special circumstances because they exaggerate polar regions to an enormous extent. The Mercator projection, the best known example, is, however, invaluable to navigators as it shows all lines of constant direction as straight lines.

The first essential in making any map is to establish its purpose. The necessary data must then be assembled – this may be in the form of survey data, aerial photography, existing maps and written material – analysed, evaluated and edited before any drafting can begin. Many factors influence the presentation of information in map form, from the size of paper that can be handled in a cockpit or motor car to the visual preconceptions of the probable user. The cartographer has at his disposal all the techniques of graphic communication and the written word and he must carefully consider the possibilities they offer when designing the map. Various forms, symbols, lines and shading may be used. Contours are widely used to represent relief on maps [8] and provide accurate data; however, they give little visual impression of the appearance of the landscape. Cartographers often use layer-colouring for greater clarity and spot heights are marked where an accurate assessment of height or depth is desired, or when a particular crest or low point falls between the contour lines.

A theodolite is essentially a tripod-mounted telescope on a base plate that is marked in degrees, minutes and seconds to allow the surveyor to measure horizontal angles. He makes vertical readings from an upright plate at the side.

5 Map projections are mathematical constructions designed to maintain certain selected relationships of the earth's surface. Some projections are purely geometrical and may be thought of as projections of a transparent globe's parallels and meridians on to a cylinder, cone or plane. This illustration shows the construction of cylindrical projections [A] and how, by varying the point of projection, different types are produced: simple cylindrical [B], cylindrical stereographic [C] and cylindrical orthographic [D]. Mercator's [E] and Miller's projection [F] are both constructed mathematically.

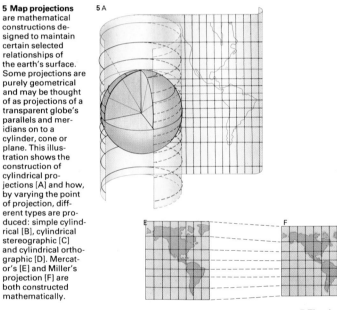
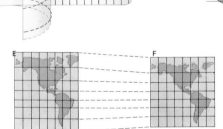

6 Azimuthal or zenithal projections are those produced on plane surfaces [A]. Angles measured from the centre (the point of contact with the globe) are correct. However, distortion of shape and area increases with distance from the centre. The gnomonic projection [B] shows all great circles (circumferences of planes through the centre of the earth) as straight lines. As these are the shortest distances between two points the projection is of importance in navigation. Lambert's azimuthal equal area projection [C] combines usefully the properties of azimuthal projections and equivalence.

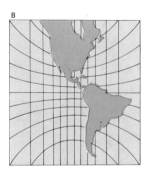
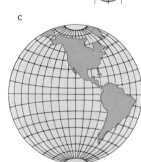

7 The simple conic projection is constructed from a cone tangential to the globe [A]. Scale along the parallel in contact with the globe only – the standard parallel – is correct. Projections constructed from a secant cone [B] have two standard parallels and because scale error increases away from them more of the projection is nearer the correct scale. The polyconic projection [C], mathematically constructed to have all parallels standard, is very accurate over small areas and is therefore used for topographic series. Alber's equal area projection [D] is a modification of the conic projection with two standard parallels [red lines].

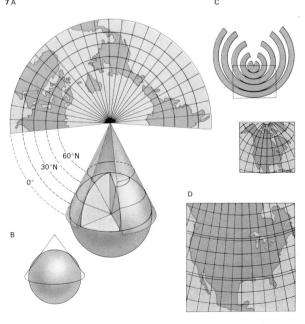

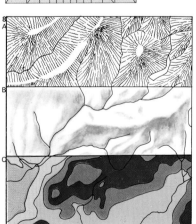

8 Height and slope can be represented on a map in many ways, including hachuring [A] in which fine lines follow the direction of greatest slope. This method can give an excellent impression of the landscape but may obscure other information. Hill shading [B], the representation of a landscape illuminated from one direction, is used alone or with colours. Contours [C] can be separated by colour and intermediate heights given as spot heights.

The face of the earth

The earth is the third nearest planet to the sun. It is the heaviest of the stony planets (the gas giants such as Jupiter are heavier) and the densest of all planets. Its orbit, 150 million kilometres (93 million miles) away from the sun, ensures that the planet is neither scorching nor freezing and the presence of water and an atmosphere reduce the temperature extremes, allowing the evolution of life.

From a distance the earth looks one of the most interesting objects in the Solar System. This is largely due to the variable cloud cover which does not prevent a distant observer from seeing the lands and oceans on the surface. Astronaut Neil Armstrong (1930–) said, during the Apollo 11 flight in 1969, that the earth "looks like a beautiful jewel in space". The earth is much brighter than the moon, reflecting about 40 per cent of the light falling on it, compared to the moon's seven per cent. From Mercury, Venus and Mars the earth would look to the naked eye like a brilliant bluish star but from Jupiter and the more distant planets a telescope would be required as the earth would be hidden by the glare of the sun.

1 This sunset seen from a manned satellite in space shows the diffraction of sunlight by the earth's atmosphere. Only the lower few tens of kilometres are dense enough to produce this effect.

2 A "full earth" was a spectacle unseen by man before the Apollo flights. North and South America and Africa can be seen here but the cloud cover tends to obscure the shapes of continents.

3 These spectacular cloud vortices, photographed from Skylab in 1973, are the result of air being drawn into the pronounced low-pressure area in the lee of Guadalupe Island, off the coast of Baja California, Mexico.

The World

Land Features

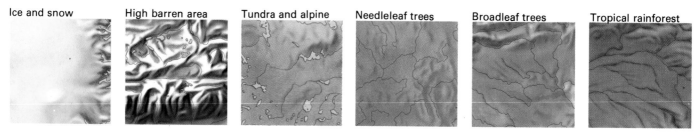

Ice and snow   High barren area   Tundra and alpine   Needleleaf trees   Broadleaf trees   Tropical rainforest

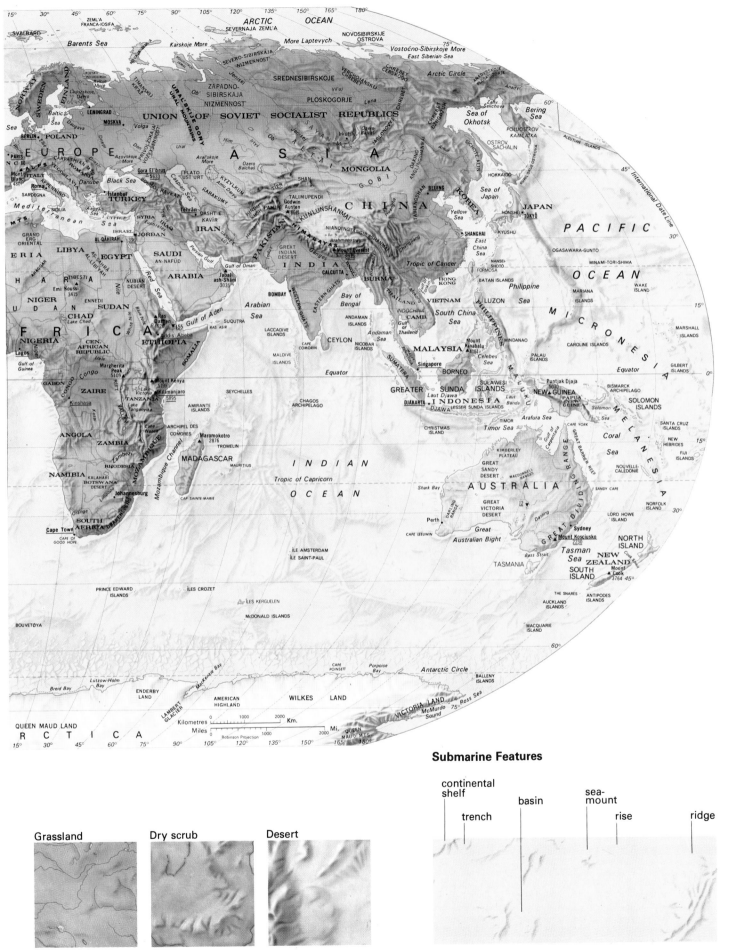

Grassland

Dry scrub

Desert

Submarine Features

continental shelf

trench

basin

sea-mount

rise

ridge

39

Europe and North Africa

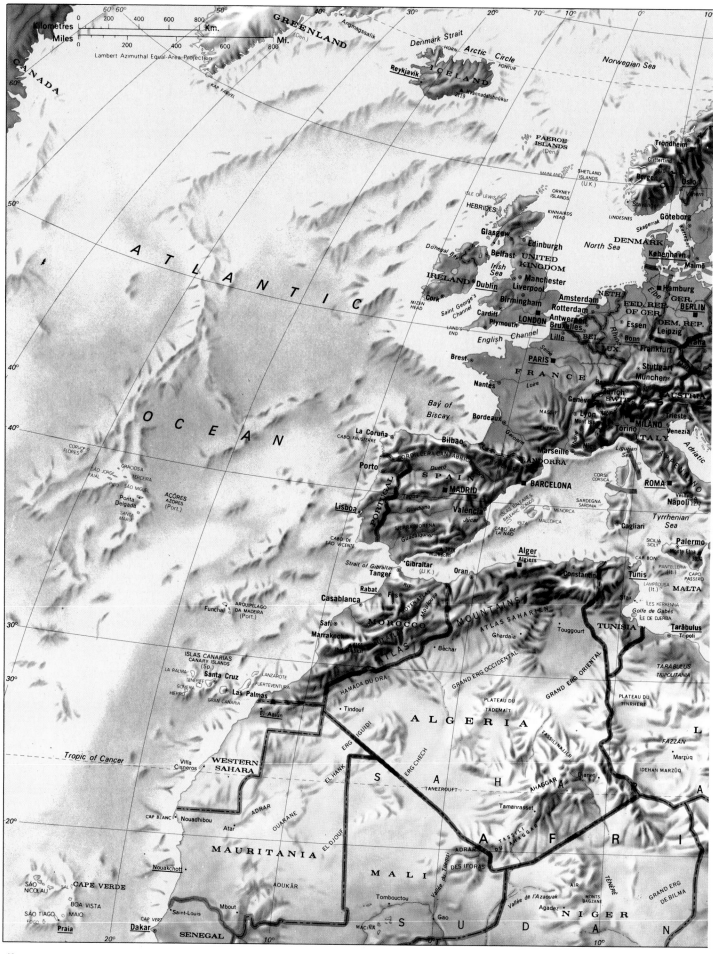

NORDKAPP
Hammerfest
MYS
KANIN NOS
OSTROV
KOLGUJEV
Narjan-Mar
Arctic Circle
Vorkuta
Obskaja Guba
Jenisej
ASIA
60°
Murmansk
Pečora
Usa
1894
Gora Narodnaja
ZAPADNO- SIBIRSKAJA
KOLSKIJ
POLUOSTROV
KOLA PENINSULA
White Sea
Beloje More
Mezen'
TIMANSKIJ KRJAŽ
Ob'
Kazym
Put
Tura
NIZMENNOST'
Kiruna
Oulu
FINLAND
Onežskoje
Ozero
Severnaja Dvina
Suchona
Kotlas
URAL'SKIJE GORY
URAL MOUNTAINS
Kama
Čusovaja
Chanty-
Mansijsk
Konda
Tavda
Om'
Helsinki
Ladožskoje
Ozero
Lake Ladoga
Vologda
Nižnij Tagil
Irtyš
Tobol'sk
Ittyš
Om
Stockholm
Tallinn
Gulf of Finland
LENINGRAD
Rybinskoje
Vodochranilišče
Perm'
SVERDLOVSK
T'umen'
Omsk
Čudskoje
Ozero
Jaroslavl'
Gor'kovskoje
Vodochranilišče
Kirov
Volga
Čel'abinsk
SOVIET
Rižskij
Zaliv
Rīga
Ozero
Il'men'
Kalinin
Gor'kij
Kazan'
Ufa
Magnitogorsk
Celinograd
50°
Klaipėda
Daugava
Velikaja
MOSKVA
MOSCOW
Oka
Sura
Kujbyševskoje
Vodochranilišče
Kujbyšev
REPUBLICS
Orsk
TURGAJSKAJA
STRANA
Ozero
Tengiz
Kalininград
Vilnius
Nemunas
Smolensk
Tula
Penza
UNION
Orenburg
STOLOVAJA STRANA
POLAND
Gdańsk
Minsk
Dnepr
SREDNERUSSKAJA
VOZVYŠENNOST'
Voronež
Saratov
Ural
Novokazalinsk
Warta
Warszawa
Bug
Kursk
PE
Don
Volgogradskoje
Vodochranilišče
Emba
Aral'skoje
More
Aral Sea
Syrdarja
Katowice
Kraków
L'vov
Dnestr
Kijev
Char'kov
SOCIALIST
VOLGOGRAD
Volga
PRIKASPIJSKAJA NIZMENNOST'
PLATO UST' URT
KYZYLKUM
CZECHOSLOVAKIA
Graz
VIENNA
WIEN
BUDAPEST
Cluj
ROMANIA
Dnepropetrovsk
Doneck
Zaporožje
Astrachan'
Amudarja
Nukus
KARAKUMY
Zagreb
HUNGARY
Sava
Rostov-na-Donu
Ciml'anskoje
Vodochranilišče
Zaliv Kara-
Bogaz-Gol
Krasnovodsk
Beograd
Bucureşti
Kišin'ov
ODESSA
Azovskoje
More
Krasnodar
Caspian Sea
BAKU
YUGOSLAVIA
DINARA
Danube
Varna
Sevastopol'
Gora El'brus
Groznyj
 KOPPER DAGH
Sofija
STARA PLANINA
Black
Sea
Batumi
Ašchabad
Tirané
ALBANIA
BULGARIA
İSTANBUL
Samsun
TBILISI
Krasnovodsk
Bari
Ölimbos
Marmara
Denizi
Kızıl Irmak
Jerevan
Mashhad
Thessaloniki
Kırıkkale
Sakarya
TURKEY
Euphrates
Van Gölü
5184
Tabrīz
Daryācheh-ye
Rezā'īyeh
Rasht
GREECE
Eskişehir
Ankara
3916
Tigris
ALBORZ
Aegean
Sea
İzmir
Tuz Gölü
Ereyan Dağı
Hamadān
Daryācheh-ye Namak
TEHRĀN
İmir
KIKLADHES
Adana
AS MINOR
Al-Mawşil
Daryācheh-ye
DASHT-E KAVIR
AFG.
Athinai
TOROS DAĞ
Halab
Hamadān
Esfahān
IONIOS NISOI
RÓDHOS
İskenderun Körfezi
Levkosia
SYRIA
BĀDIYAT ASH-SHĀM
MESOPOTAMIA
Baghdād
DASHT-E LUT
Iráklion
CYPRUS
LEBANON
Bayrūt
Dimashq
IRAQ
Tigris
Yazd
KRITI CRETE
ISRAEL
Ammān
Euphrates
Al-Başrah
Kermān
Tel Aviv-Yafo
AL-WIDYĀN
Ābādān
Shīrāz
Banghāzi
Bengasi
Tubruq
Al-Iskandariyah
Alexandria
Bur Sa'īd
Ghazzah
Dead Sea
396
JORDAN
AL-HAMAD
Al-Kuwayt
KUWAIT
Khalīj Surt
BĀRQAH
CYRENAICA
MUNKHAFAD
AL-QAŢŢĀRAH
AL-QĀHIRAH
CAIRO
SHIBH
JAZĪRAT
Jabal al-
2580 Lawz
Qanat
as-Suways
NEUTRAL
ZONE
BAHRAIN
OMAN
42▼
▼133
AŞ-ŞAHRĀ'
Al-
Manāmah
Strait of Hormuz
LIBYA
AS-ŞAHRĀ
AL-GHARBĪYAH
SĪNĀ'
Al-Hufūf
QATAR
Gulf of Oman
SARIR NERASTRO
EGYPT
Asyūţ
AL-HIJĀZ
AN-NAFŪD
AD-DAHNĀ
UNITED ARAB
EMIRATES
Jabal
ash-Shām
3035
AL-KUFRAH
HADABAT AL-JILF
Aswān
Nile
Tropic of Cancer
Al-Madinah
Ar-Riyāḍ
SAUDI
ARABIAN
SARIR TIBESTI
Lake
Nasser
Jabal al-
Uwaynāt 1934
NUBIAN
DESERT
Makkah
NAFŪD
AD-DAHY
ARABIA
AR-RUB' AL-KHĀLĪ
BODELE
Tarso Tieroko
3315
TIBESTI
Émi Koussi
3415
CHAD
ENNEDI
SUDAN
Nile
'Atbarah
Wadi al-Malik
Būr Sūdān
PENINSULA
OMAN
Umm
Durmān
Al-Khurtum
Khartoum
ETHIOPIA
Asmera
Mesewa
Şan'ā'
DJAHLAK
ARCHIPELAGO
YEMEN
P.D.R.
OF
YEMEN
Ghubbat al-Qamar
Arabian Sea
RA'S
FARTAK
Copyright © by Rand M°Nally & Co.
B -519394-764-3-3-3-6

41

Southern Africa

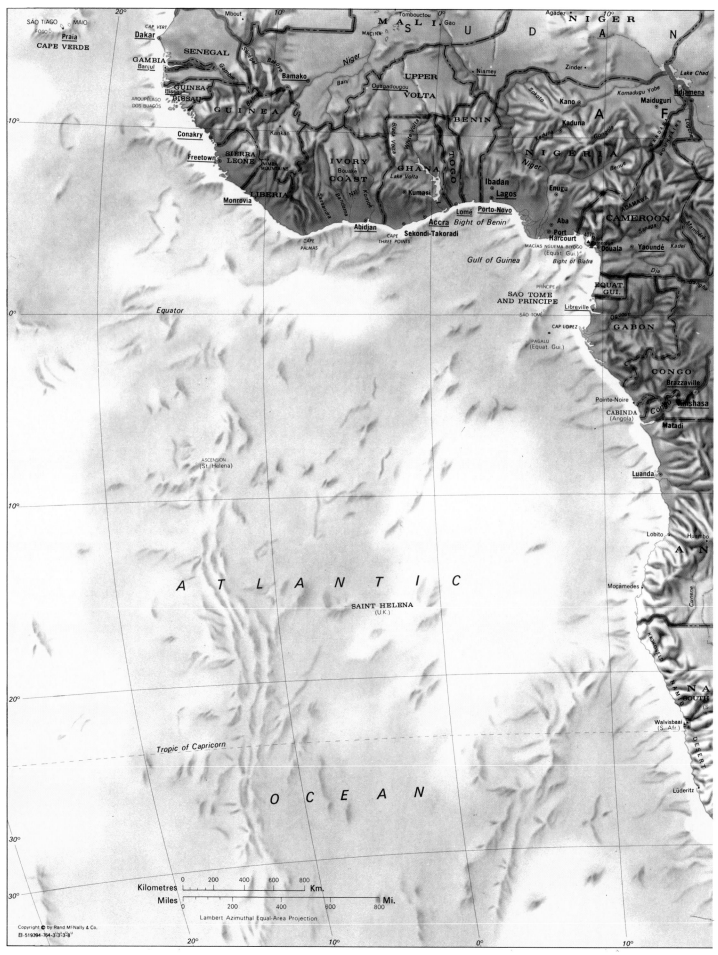

Earth panorama: Europe

Europe is the second smallest continent. It is bounded by the Arctic Ocean to the north, the Atlantic Ocean to the west and the Mediterranean and Black Seas to the south. It merges into Asia to the east and the conventional boundary from south to north follows the Caucasus mountains, the Caspian Sea and the Ural mountains.

A line following the northern edge of the Pyrenees mountains, the Rhône valley and the northern edge of the Alps and Carpathian mountains separates northern from southern Europe. Northern Europe thus consists of large sedimentary plains, a Precambrian shield and worn-down Palaeozoic highlands. Southern Europe is characterized by Cenozoic mountains (Alps, Pyrenees, Carpathians) surrounding restricted basins.

Apart from a small subarctic fringe in the extreme north, most of Europe is in the temperate zone. The distance from the Atlantic Ocean and the situation of the mountains create a climatic subdivision of Europe into marine areas to the west, Mediterranean areas to the south and continental areas to the east.

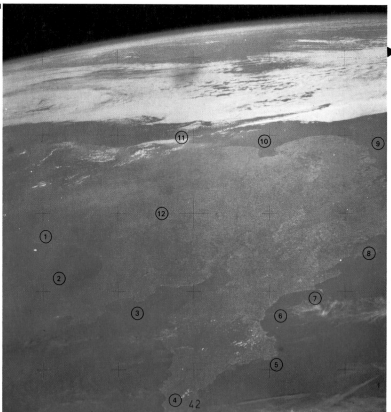

1 Great Britain and Ireland are islands of the European continental shelf and they were part of the mainland during the recent Ice Age. The Irish seashore is indistinct north of Anglesey [1] but Cardigan Bay [2], the Bristol Channel [3], Cornwall [4] and Start Point [5] are clearly visible. All this part of Britain consists of ancient Precambrian and Palaeozoic rocks. East of a line from Lyme Bay [6] to Grimsby [11] the rocks are Mesozoic and Cenozoic. North of Derby [12] the rocks are again Palaeozoic. Other features include Portland Bill [7], the Isle of Wight [8], Orford Ness [9] and the Wash [10].

2 The Dutch coast, from the Schelde and Rhine estuaries [1] to the Frisian islands [2], is seen here with the cities of The Hague [3], Rotterdam [4] and Amsterdam [5] as well as the IJsselmeer [6].

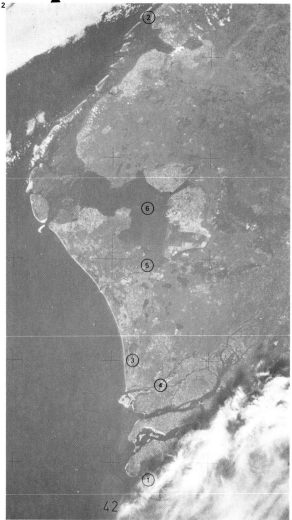

3 The Alps are the highest mountains in western Europe, extending 1,000km (620 miles) from the Mediterranean to Vienna. They are the western limb of a much larger system of mountains which extends to Indonesia through the Balkans and the Himalayas. The highest summit is Mont Blanc [1] at 4,807m (15,771ft). It is part of the inner granite core which has been thrust up in places and uncovered by erosion. Lake Geneva [2], which divides the upper Rhône valley [3] from the lower [4], is in a depression between the Alps and the Jura mountains [5] which were folded, but less severely, as a consequence of the Alpine upheaval. The lakes of Neuchâtel [6] and Thun [7] can also be seen.

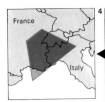

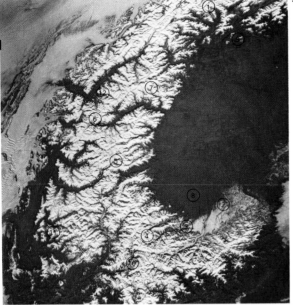

**4 The Western
Alps** extend from
the Mediterranean
coast [lower right]
to the Adula massif
[1]. They enclose
the Po valley. The
following massifs can
be seen here: Argen-
tera [2], Monte Viso
[3], Mont Pelvoux [4].
Vanoise [5], Mont
Blanc [6] and Monte
Rosa [7]. The major
rivers on the Italian
side are the Po [8]
and the Adda [9],
draining through
Lake Como; and the
Durance [10], Isère
[11] and Rhône [12]
on the French side.

**5 Part of the south
coast of France**
from the Vaccares
marshes [1] to
Toulon [6] is shown
here. The main
Rhône outlet [2] is
seen near the huge
dock area of the
modern harbour of

seille-Fos [3].
Port-de-Bouc docks
on the sea side of
the canal link the
Berre lake [4] with
the Mediterranean.
The artificial break-
water [5] of the New
Marseille-Fos [3].
Port-de-Bouc docks

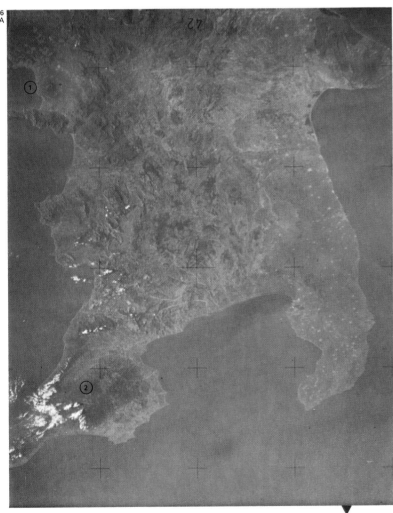

**6 The famous boot-
shape** of southern
Italy [A] appears far
more squat here be-
cause of the camera
angle. The western
side of the penin-
sula right down to
the "toe" of Cala-
bria has a pro-
nounced relief due
to the Apennine

range. The Bay of
Naples [1] is limited
to the north by the
island of Ischia and
to the south by
Capri. Just inland
of the bottom of the
bay the volcano
Mt Vesuvius is
conspicuous, as is
Botte Donato moun-
tain [2]. The "in-

step" between the toe
and the heel is the
Gulf of Taranto;
the heel is term-
inated by Cape Santa
Maria di Leuca.
Between the heel and
the spur of Gargano
(which has some
lakes on its north
side) is a dry lime-
stone area, Puglia.

The eastern coast
of Sicily [B] is
seen in this infra-red
photograph. Mt Etna,
the highest volcano
in Europe, is still
active as is evidenced
by the thin plume of
smoke rising from
its crater. Etna's
height of 3,340m
(10,960ft) is approxi-

mate because it
changes at each erup-
tion. Recent lava
flows appear as
black in contrast
to the older red
ones. The numerous
small "warts" on
the flank of the vol-
cano are cinder
cones built by side
vents. The town of

Catania nestles at
the foot of Mt Etna
by the sea. Beyond
it is a cultivated
plain with fields
of various colours
and a meandering
river. The town of
Augusta, enclosed
by a breakwater,
can be seen at
bottom left.

Earth panorama: Africa

Africa is the third largest continent and is devoid of any peripheral island arcs. It is entirely surrounded by mid-oceanic ridges (one of them, in the Red Sea, coming right up to its shores) except to the north where it abuts the Mediterranean and the alpine system. The Maghreb (Morocco, Algeria and Tunisia) is the only geologically recent province and it is separated from the Precambrian basement and shield forming the rest of the continent by a big fault running from Agadir to Gabes.

The rolling basement to the south of this fault line forms great basins (Niger, Chad, Congo, Kalahari) surrounded by highlands that dominate the coasts. More than half of the continent lacks drainage towards the sea and the large rivers (Nile, Congo, Niger and Zambezi) have difficult paths to the sea.

The climate is zoned, with a central equatorial band grading the north and the south into tropical lands with a marked dry season, the length of which increases polewards until the desert areas. South Africa has a warm maritime climate, whereas the north coast has a Mediterranean climate.

CONNECTIONS

See also
40 Europe and North Africa
42 Southern Africa

1 Where the Nile flows, the land is lush and green; where its waters do not reach there is desert. The ribbon oasis along the Nile valley to the south (right) of Cairo [1] spreads out into a rich alluvial delta. The river branches into two, the Rosetta Nile [2] and the Damietta Nile [3]. The front of the delta is marked by large lagoons and infertile desert-sand. Between Alexandria [4] and Cairo there are large fields and modern irrigation projects. The Suez Canal runs from Port Said [5] to Suez [6]. A narrow strip of vegetation links the Nile with Ismailia [7].

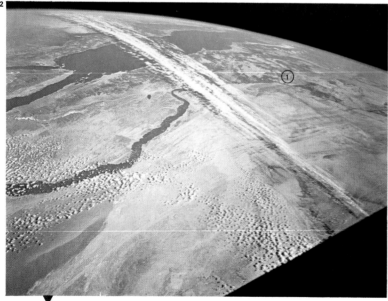

2 The ribbon oasis of the Nile divides the Libyan desert in the foreground from the Arabian desert in the background, beyond which the Red Sea, the gulfs of Suez and Aqaba and the Sinai peninsula can be seen. The Nile's yearly flood used to bring about 55 million tonnes of new fertile silt but much of this is now stopped by the Aswan Dam [1] and it is silting up the artificial reservoir, Lake Nasser. The lake loses water through evaporation, a situation aggravated by a plague of water hyacinths. The reduced flow has increased the salinity of the eastern Mediterranean and seriously affected its plankton population. The local sardine fishing industry has suffered as a result.

3 The Arabian peninsula [1] has moved and rotated away from Africa, opening up the Gulf of Aden [2] and the Red Sea [3]. The Bab el Mandeb strait [4] is a triple junction of three spreading axes: the Gulf of Aden which links up with the Carlsberg Ridge in the Indian Ocean, the Red Sea axis which extends to the Dead Sea, and the Afar Triangle [5] linking up with the East African Rift.

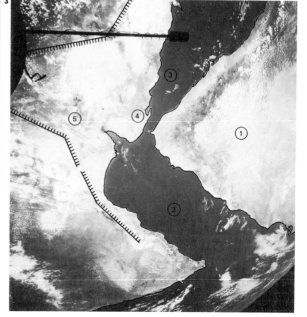

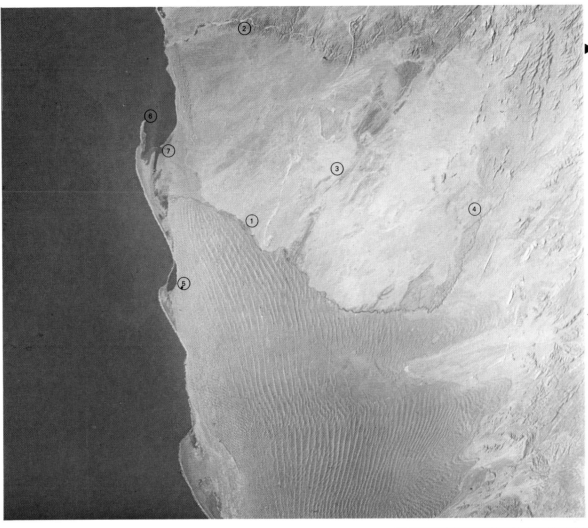

4

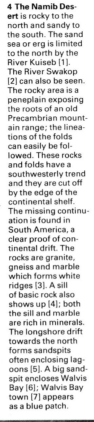

4 The Namib Desert is rocky to the north and sandy to the south. The sand sea or erg is limited to the north by the River Kuiseb [1]. The River Swakop [2] can also be seen. The rocky area is a peneplain exposing the roots of an old Precambrian mountain range; the lineations of the folds can easily be followed. These rocks and folds have a southwesterly trend and they are cut off by the edge of the continental shelf. The missing continuation is found in South America, a clear proof of continental drift. The rocks are granite, gneiss and marble which forms white ridges [3]. A sill of basic rock also shows up [4]; both the sill and marble are rich in minerals. The longshore drift towards the north forms sandspits often enclosing lagoons [5]. A big sandspit encloses Walvis Bay [6]; Walvis Bay town [7] appears as a blue patch.

5 Lake Chad lies across the borders of Chad, Niger, Nigeria and Cameroon. It is the centre for an inland drainage system and has no outlet to the sea. Its main tributary is the 1200km (750 miles) Chari flowing from the south. The intermittent wadi Bahr el Ghazal drains the rare rains from the Saharan north. Because of the high evaporation and the marked seasonal variation of the inflow, the area of the lake varies and therefore precise contours are rarely shown on maps. It is a shallow lake whose average level is dropping at a rate of 1.25cm (0.5in) a year.

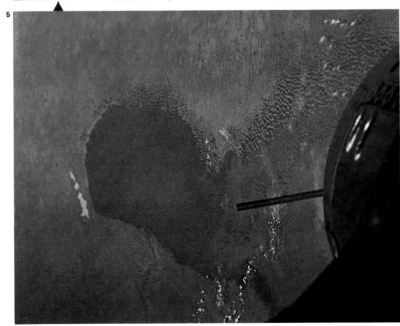

5

6

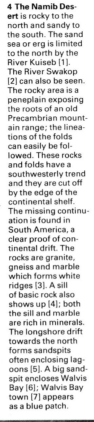

6 This is the bare African shield in southwest Africa – the roots of Precambrian mountains which have been worn down to a peneplain. The area consists mainly of gneiss. The conspicuous circular patches are plugs of granite known as inselbergs. Two intermittent wadis, [1], [2] can be seen. The coastline is underlined by a narrow strip of blue sea; the rest of the sea is clouded over because of the cold Benguela current, which promotes cloud formation.

Northern Asia

ATLANTIC OCEAN

ARCTIC

GREENLAND
(Den)

Greenland Sea

JAN MAYEN (Nor.)

SVALBARD (Nor.)
SPITSBERGEN
NORDAUSTLANDET

ZEML'A FRANCA-IOSIFA
FRANZ JOSEF LAND

Barents Sea

NOVAJA ZEMLA

Karskoje More
Kara Sea

Dikson

Denmark Strait

ICELAND
2119 Hvannadalshnúkur
Reykjavík

FAEROE ISLANDS (Den)

Norwegian Sea

Arctic Circle

HEBRIDES

IRELAND
Dublin
Irish Sea
Glasgow
Edinburgh
KINNAIRDS HEAD
Manchester
UNITED KINGDOM
Birmingham
Cardiff
LONDON
North Sea

ORKNEY ISLANDS
SHETLAND ISLANDS

NORWAY
Stavanger
Bergen
Oslo
SWEDEN
Trondheim

Murmansk

POLUOSTROV KANIN
MYS KANIN NOS

Beloje More
White Sea

Archangel'sk

OSTROV KOLGUEV

OSTROV VAJGAC
POLUOSTROV JAMAL

Narjan-Mar

Vorkuta

Salechard

Ob'

BELGIUM
Bruxelles
Amsterdam
NETHERLANDS
DENMARK
København
Malmö

Gulf of Bothnia

FINLAND

Helsinki
Tallinn
Gulf of Finland

Baltic Sea

GOTLAND
ÖLAND

Mezen'
Severnaja Dvina
Kotlas
Suchona

TIMANSKIJ KRJAZ
Pečora
Usa

ZAPADNO-

SIBIRSKAJA

Kazym

Chanty-Mansijsk

Vach

Tym

FRANCE
LUX
Köln
Essen
Bonn
Frankfurt
Leipzig
BERLIN
Hamburg
BRD GER.
DDR GER.

MÜNCHEN

POLAND
Gdańsk
Kaliningrad
Klaipeda
Riga
Warra

Vilnius

Minsk

Smolensk
Kalinin

Kurgan
Ladozskoje Ozero
Lake Ladoga
LENINGRAD

Onezskoje Ozero

SEVERNYJE UVALY

Kirov
Perm'
Kama
Nižnij Tagil

URAL'SKIJE GORY
URAL MOUNTAINS

SVERDLOVSK

T'umen'

Tobol'sk

NIZMENNOST' OB'

SOV

Konda

Tura

Irtyš

Omsk
Novosibirsk

CZECHOSLOVAKIA
Praha
WIEN
WARSZAWA
Kraków
Katowice
L'vov

HUNGARY
BUDAPEST
YUGOSLAVIA
Beograd

ROMANIA
Cluj
București

Gomel'
Desna
Kijev
Kursk

Char'kov

MOSKVA
Moscow
Tula
Oka

Gor'kij
Jaroslavl'
Volga

Gor'kovskoje Vodochranilišče

Rybinskoje Vodochranilišče
Vologda

E U R O P E

Penza

VOLZ.SENNOST'
VOLZ.SENNOST'

Kazan'
Kujbyševskoje Vodochranilišče
Ufa
Kujbyšev

PRIVOLZSKAJA

Čel'abinsk
Tobol
Išim
Om'

Ozero Cany

Barnaul

Adriatic Sea

BULGARIA
Sofia
Varna

GREECE

Sevastopol'
Black Sea

İSTANBUL

Marmara

Kišin'ov
ODESSA
Zaporožje
Dnepropetrovsk
Azovskoje More
Doneck
Rostov-na-Donu
Krasnodar

Voronež
Don

Saratov

VOLGOGRAD
Volgogradskoje Vodochranilišče
Ciml'anskoje Vodochranilišče

Orenburg
Ural

Orsk

Magnitogorsk

TURGAJSKAJA
STOLOVAJA STRANA

NIZMENNOST'

Emba

Celinograd
Ozero Tengiz
KAZACHSKIJ
Karaganda

MELKOSOPOČNIK

Ust'-Kamenogorsk

Semipalatinsk

Ozero Zajsan

GREECE
Aegean Sea

İzmir
Eskişehir
Ankara
TURKEY
ASIA MINOR
TOROS DAĞLARI
Adana
CYPRUS
Nicosia
İskenderun Körfezi

Samsun
Kizil Irmak

5633 ELBRUS
Batumi
Groznyj
TBILISI
BOL.

KAVKAZ
Jerevan
BAKU

Caspian Sea

Astrachan'
Volga

PRIKASPIJSKAJA NIZMENNOST'

PLATO UST'URT

Aralskoje More
Aral Sea

Novokazalinsk

Syrdarja

Nukus
KYZYLKUM

Balchaš

Ozero Balchaš

Džezkazgan

Ozero Alakol'

Mediterranean Sea
Halab
SYRIA
LEBANON
Bayrüt
ISRAEL
Tel Aviv-Yafo
Dimashq
Amman
JORDAN
Jerushalayim
Baghdad
IRAQ

Tabriz
Rasht
TEHRAN
Hamadan
Al-Mawşil
MESOPOTAMIA

DASHT-E KAVIR

Zaliv Kara-Bogaz-Gol

Krasnovodsk

KARAKUMY

Amudarja
Aşchabad

Samarkand
Buchara

Mashhad

KOPPEH DAGH

Dušanbe
PAMIR

Frunze
Ozero Issyk-Kul'

Taškent
Fergana

Alma-Ata

T'EN

Yining

TAKLA MAKAN
TALIMUPENDI

Red Sea
SAUDI ARABIA
EGYPT
Jabal al'Uwal

Al-Başrah
Euphrates
Tigris

IRAN
Eşfahān
Qom

HINDU KUSH

Kabul

Kilometres
0 200 400 600 Km.
Miles
0 200 400 600 Mi.
Lambert Azimuthal Equal Area Projection

Copyright © by Rand McNally & Co.
B-515200-764

48

North Pole

OCEAN

UNITED STATES   Nome

POINT HOPE
Bering   Strait
MYS DEŽNEVA

Chukchi
Sea

OSTROV VRANGEL'A
ČUKOTSKIJ POLUOSTROV

Proliv Longa

UNIMAK ISLAND

NUNIVAK ISLAND

Unalaska

ALEUTIAN ISLANDS

Anadyrskij Zaliv

Uel'kal'

Amparcik

Anadyr

MYS NAVARIN

SAINT LAWRENCE ISLAND

PRIBILOF ISLANDS

Bering
Sea

Markovo
Velikaja

ANADYRSKOJE PLOSKOGORJE

SEVERNAJA ZEML'A
NORTH LAND

NOVOSIBIRSKIJE OSTROVA
NEW SIBERIAN ISLANDS
OSTROVA NOVAJA SIBIR

OSTROV KOTEL'NYJ

Vostočno-Sibirskoje More
East Siberian Sea

OSTROV AJON

Arctic Circle

Penžina

Omolon

Kavača

MYS OLJUTORSKIJ

ATTU

OSTROV KARAGINSKIJ

MYS CELJUSKIN

POLUOSTROV TAJMYR

Ozero Tajmyr

SEVERO-SIBIRSKAJA NIZMENNOST

More Laptevych
Laptev Sea

OSTROV BOL-SOJ BEGIČEV

Nordvik

Tiksi

Olenjok

Karačá

JUKAGIRSKOJE PLOSKOGORJE

Srednekolymsk

Kolyma

Jansk

Zaliv Šelichova

POLUOSTROV KAMCATKA KAMCHATKA PENINSULA

KOMANDORSKIJE OSTROVA

Dudinka   Noril'sk
Igarka

GORY PUTORANA

SREDNESIBIRSKOJE PLOSKOGORJE

Cheta

Chatanga

Kotuj

Jessej

Anabar

Muna

Indigirka

Jana

CHREBET ČERSKOGO

Indigirka

CHREBET SETTE-DABAN

Ojmjakon

Palana

SREDINNYJ CHREBET

Petropavlovsk-Kamčatskij

MYS LOPATKA

Jenisej

Nižnaja

Tunguska

Ziganska

Vil'uj

Lena

Jakutsk

ALDANSKOJE NAGORJE

Aldan

Tommot

STANOVOJ CHREBET

Ochotsk

Ajan

Sea of Okhotsk

MYS JELIZAVETY

OSTROV SACHALIN SAKHALIN

KURIL'SKIJE OSTROVA KURIL ISLANDS

Podkamennaja Tunguska

Cuna

Lenak

Vitim

STANOVOJE NAGORJE

Bratskoje Vodochranilišče

Angara

Skovorodino

Amur

Heilongjiang

Komsomol'sk na-Amure

Sočí

Nikolajevsk

Ocha

SANTARSKIJE OSTROVA

Aleksandrovsk-Sachalinskij

SICHOTE-ALINJ

Tatarskij Proliv

MYS TERPENIJA

La Pérouse Strait

Asahikawa

HOKKAIDO

Angara

Kan

Tomsk

Krasnojarsk

Kemerovo

Novokuzneck

Abakan

VOSTOČNYJ SAJAN

Oka

Angara

Ozero Bajkal

Irkutsk

Ulan-Ude

JABLONOVYJ CHREBET

Chita

BORSČOVOČNYJ CHREBET

Argun

CHREBET

Eergu-nahe

Hailar

Hulunchi

Buir Nuur

DAXINGANLINGSHANGMAI

Chabarovsk

Bureja

Jamusi

Ozero Chanka

Vladivostok

Sapporo

Hakodate

Tsugaru

Aomori

Sea of Japan

HONSHÚ

Niigata

Sendai

Qiqihaer

HANGGUANGCALLING

Haerbin

Changchun

NORTH KOREA

ZAPADNYJ SAJAN

Kyzyl

Jenisej

4506 Gora Belucha

ALTAI

Uvs-Nuur

Chirgis Nuur

CHANGAJN

Chövsgöl Nuur

Selenga

Orchon

Kerulen

Xihe

Liaohe

Fushun

SHENYANG MUKDEN

Andong

Laodongwan

P'yongyang

SÓUL

SOUTH KOREA

Tongjosón-Man

OKI GUNTO

Nagoya

Kyóto

OSAKA

TOKYO

Char Us Nuur

Dzavchan

Nuruu

MONGOLIA

GOBI

Ulaanbaatar

Zhangjiakou

Huhehaote

BEIJING PEKING

TIANJIN TIENTSIN

Bohai

Luda Dairen

CHENGSHANJIAO

Korea Bay

Taegu

Pusan

Hiroshima

SHIKOKU

Fukuoka

JAPAN

Korea Strait

KYUSHU

Bungo-suido

Kagoshima

AERJINSHANMAI

S SHAN

Wulumuni

Luobubo Lop Nor

QILIANSHANMAI

CHAIDAMUPENDI

Qinghai

CHINA

TATHANGSHAN

Taiyuan

Yinchuan

Huanghe

Huanghe

Ruoshui

Lanzhou

Qinghai

Zhengzhou

Yinghe

Nanjing

Grand Canal

Yunhe

Ji'nan

Qingdao Tsingtao

Yellow Sea

Mokp'o

CEJU DO (S. Kor)

Hangzhou

SHANGHAI

EAST CHINA SEA

AMAMI-O-SHIMA

Magadan

SOCIALIST

REPUBLICS

IET

SOVIET

A S I A

Southern Asia

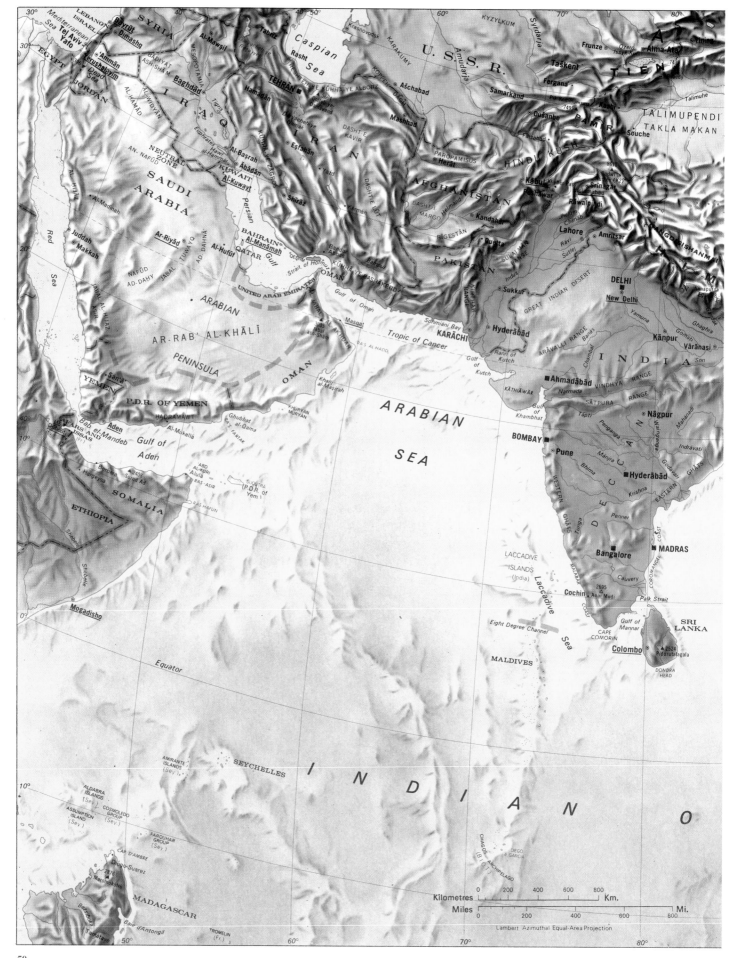

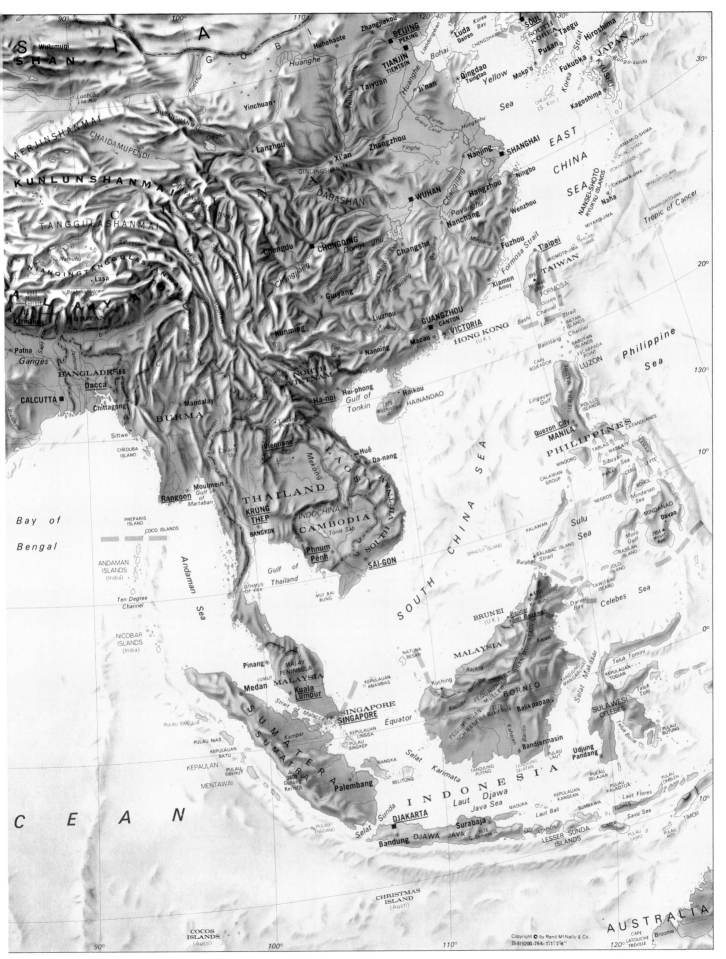

S • Wulumuqi

SHAN

A   S   I   A

AERJINSHANMAI

CHAIDAMUPENDI

KUNLUNSHANMAI

TANGGULASHANMAI

Luobubo Lop Nor

Qinghai

• Lanzhou

GOBI

Huhehaote

Zhangjiakou   BEIJING PEKING

• Taiyuan

Yinchuan •

Xi'an •   Zhengzhou

Huangjiakou

Liaodongwan   Korea Bay

Luda Dairen

CHENGSHANTOU   SOUTH KOREA   SOUL

Pusan

Taegu

Hiroshima

JAPAN

SHIKOKU

30°

Yellow Sea

Qingdao Tsingtao

Mokp'o   Korea Strait

Fukuoka   KYUSHU

Kagoshima

Bungo-suido

Bay of Bengal

Bay of Bengal

51

Earth panorama: Asia

Asia is the largest of the seven continents. The eastern and southeastern parts of the continent are fringed by a series of island arcs which have frequent earthquakes and many volcanoes: Indonesia, the Philippines, Ryukyu, Japan and the Kurils.

The Himalayas are the seam welding India and Asia proper. Similarly the Arabian peninsula is geologically part of Africa, not of Asia. The recent mountains (Caucasus, Zagros, Himalayas, Tien Shan, Altai) sometimes enclose highland plateaus such as Anatolia, the Iranian plateau and Tibet. These fringing mountains surround the geological heartland of Asia – the Siberian shield, which also extends under Mongolia and much of China.

Asia has four broad climatic domains – Mediterranean, desert, continental and monsoon. The Mediterranean zone is limited to a narrow fringe in Turkey and the Middle East. The continental zone comprises Siberia, Mongolia and Tibet and is characterized by very harsh winters. The desert climate is found from Arabia to Pakistan. Monsoon Asia extends from India to Japan.

CONNECTIONS

See also
48 Northern Asia
50 Southern Asia

1 The Red Sea [1] is a giant rift in the earth's surface that separates Asia from Africa. It is a cleft formed by sea-floor spreading. To the north it is split in two. The Gulf of Suez [2] follows the same line and at its inner end the Suez Canal and the Great Bitter Lake can be seen. The Gulf of Aqaba [3] is the southern end of another big rift that can be followed through the Dead Sea [4] and Lake Tiberias [5]. The Sinai is the peninsula between the gulfs of Suez and Aqaba. The dark mountains in the south are Precambrian terrains that can be traced into Egypt and Arabia.

3 The high mountain ranges of the Himalayas rise above the Indo-Gangetic plain [1]. Katmandu [2] is the capital of Nepal. Annapurna (8,078 m [26,503 ft]) is visible [3] above the deep valley of the River Gandak. The border with Tibet passes at the head of the valley along the narrow snowy "connection". The big central valley in Tibet is that of the upper Brahmaputra. This river runs across to the horizon on the left and then crosses the Himalayas to join the Indo-Gangetic plain. Lhasa is at [4]. Mt Everest [5] is the highest mountain in the world, reaching 8,848m (29,030ft).

2 Lakes Neyriz [1] and Tashk [2] in the Zagros Mountains of Iran are normally dry except following rare rainfalls or the spring thaw in the mountains. The lakes are fed by the River Kur [3]. The thick salt deposits are conspicuous. The terrains to the top of the picture have been pushed over those to the bottom in a thrust line [WX], while to the left there is an ordinary fault [YZ]. An eroded anticline [4] and a dome [5] are visible. The dark patches [6] are extinct volcanoes.

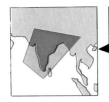

4 This Gemini 9 picture of the Indian subcontinent, taken from an altitude of 740km (460 miles), has a peculiar perspective due to a wide-angle lens. The indentation at the north end of the west coast is the Gulf of Cambay. The Western Ghats are to the left and the Deccan plateau basalts are the dark areas to the upper left. The shallowness of the strait separating Sri Lanka from the mainland is apparent. The conspicuous river and delta just to the north is the Coleroon, whereas, to the north of the Bay of Bengal, the huge delta of the River Ganges can be seen. The Himalayas are hidden on the horizon by clouds.

6 Japan is an archipelago bounding the shallow epicontinental Sea of Japan on its eastern side. On Japan's Pacific coast the sea-bed plunges down to the Ryukyu and Japan trenches, part of the major Pacific trench system; thus the Japanese islands are part of the same tectonic system as other island arcs such as the West Indies and Indonesia. In this photograph of the island of Kyushu evidence of tectonic activity can be seen in the plume of smoke rising from the volcano on Sakura-Jima [1]. Aso-san [2] is the world's largest active volcanic crater.

5 The complicated coastline of southeast China is due to the flooding of a worn-down peneplain by the sea, following subsidence of the land in the last few million years. The remaining knolls are predominantly made of granite and lavas. Particularly conspicuous is the island of Hainan, the land mass to the right is Taiwan. The deeper-blue current flowing north through the strait is an arm of the "black current" or Kuroshio, which warms local climates.

53

Australia and Antarctica

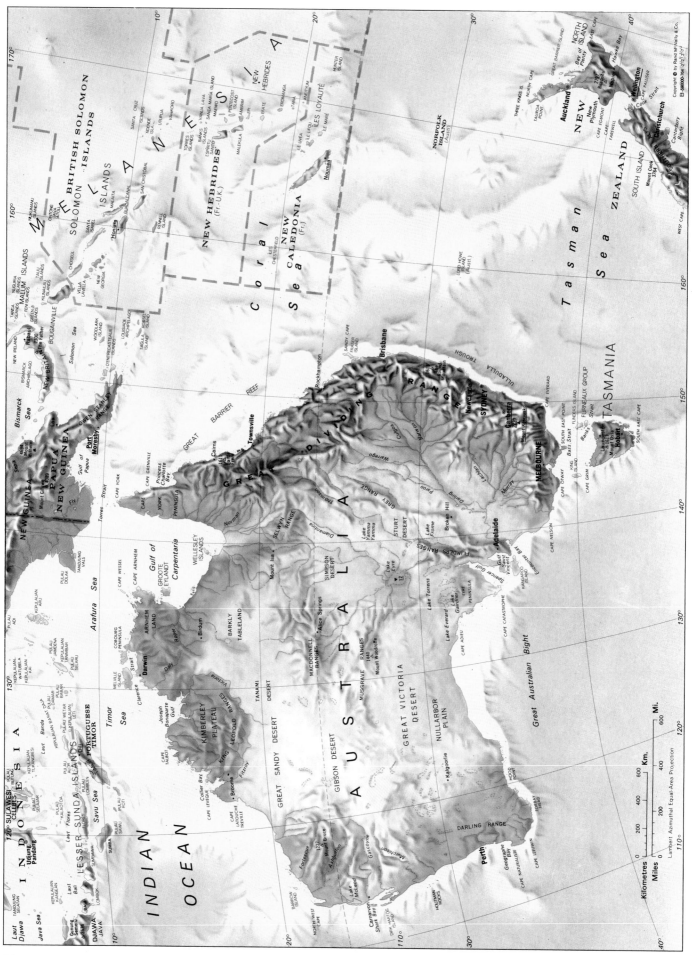

Scots firm completes contract in Antarctica

PERRY GOURLEY

A Scottish engineering firm has just completed work on a project 8,000 miles away in Antarctica.

AJ Engineering & Construction Services of Forres won the contract to create a steel platform which will be installed in Bird Island, South Georgia. The project, which has taken three months to complete, involved fabricating some 30 individual pieces which are now being shipped to the island where the platform will be used as part of a modernisation project at the research station.

Jazmin Kellas, AJ Engineering's quality manager, said the island's unique bio-diversity means the company had to carefully consider how to package the items as cardboard can harbour foreign objects such as seeds.

"The island has been largely untouched by the outside world and the researchers there understandably want to keep it that way. Bubble wrap has been one of the solutions we have used," he said.

25. 11. 2017

There was little let-up along the way. Water was sprayed

the gulag conditions. Sin her release, she has conti

adverse effects from a minor stroke last year as he provided the lithe, swinging foundation for Steve Morse's keening blues guitar and Don Airey's keyboard arpeggios on All I Got Is You.

Airey was given ample latitude throughout and indulged in a Scottish-flavoured solo incorporating a bagpipe effect burst of Amazing Grace, Loch Lomond and I Belong to Glasgow which eventually led into the apocalyptic march of Perfect Strangers.

The group collectively made a meal of the groovy Hush and a more lumbering Black Night but there was no denying that behemoth of a riff from Smoke

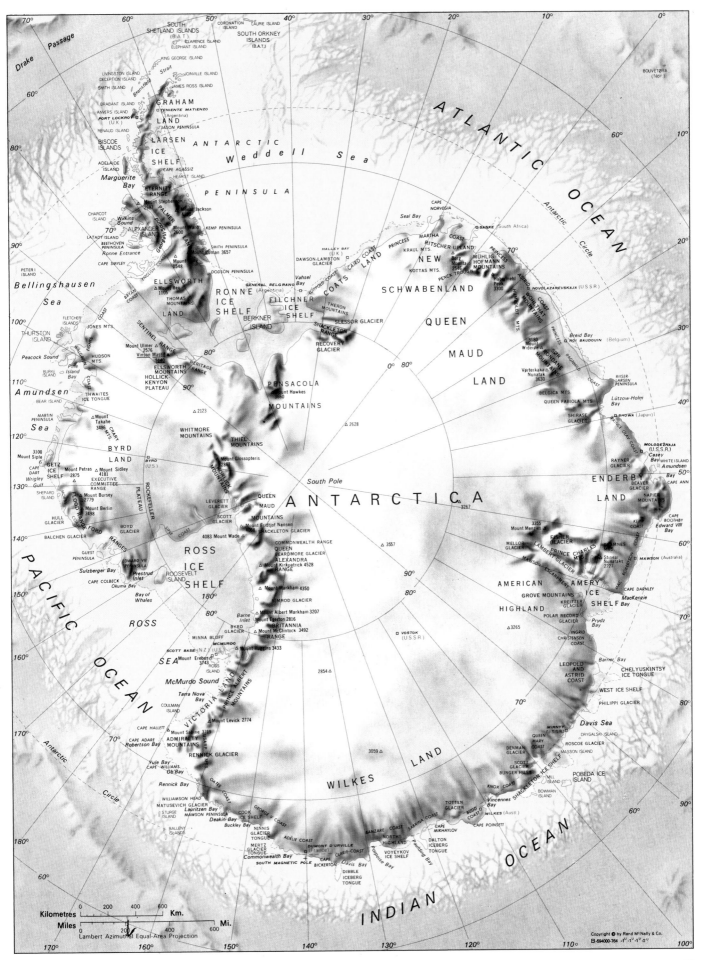

DRAKE PASSAGE
SOUTH SHETLAND ISLANDS (B.A.T.)
CORONATION ISLAND
LAURIE ISLAND
SOUTH ORKNEY ISLANDS (B.A.T.)
Clarence Island
Elephant Island
King George Island
Livingston Island
Deception Island
Smith Island
Brabant Island
Anvers Island
Port Lockroy (U.K.)
Renaud Island
Biscoe Islands
Adelaide Island
Marguerite Bay
Charcot Island
Alexander Island
Latady Island
Beethoven Peninsula
Cape Smyley
Peter I Island
Bellingshausen Sea
Thurston Island
Fletcher Islands
Peacock Sound
Burke Island
Pine Island Bay
Amundsen Sea
Bear Island
Martin Peninsula
Cape Dart
Getz Ice Shelf
Mount Siple 3100
Wrigley Gulf
Shepard Island
Hull Glacier
Balchen Glacier
Guest Peninsula
Sulzberger Bay
Cape Colbeck
Okuma Bay
Bay of Whales

ATLANTIC OCEAN
BOUVETØYA (Nor.)
Antarctic Circle

GRAHAM LAND
Teniente Matienzo (Argentina)
Jason Peninsula
LARSEN ICE SHELF
ANTARCTIC
Weddell Sea
Cape Agassiz
Hearst Island
ETERNITY RANGE
Mount Stephenson 2987
PALMER LAND
Mount Jackson
PENINSULA
Mount Ward 2600
KEMP PENINSULA
Smith Peninsula
Mount Coman 3657
DODSON PENINSULA
Mount 3548

Seal Bay
MARTHA COAST
RITSCHER UPLAND
SANAE (South Africa)
PRINCESS
CAPE NORVEGIA
KRAUL MTS.
MÜHLIG HOFMANN MOUNTAINS
HALLEY BAY (U.K.)
DAWSON-LAMBTON GLACIER
CAIRD COAST
COATS LAND
KOTTAS MTS.
PENCK TROUGH
NEW SCHWABENLAND
Habermehl Peak 3300
NOVOLAZAREVSKAJA (U.S.S.R.)
QUEEN MAUD LAND
Mount Widerøe 3180
Breid Bay
ROI BAUDOUIN (Belgium)
Vörterkaka Nunatak 3630
RIISER-LARSEN PENINSULA
BELGICA MTS.
QUEEN FABIOLA MTS.
Lützow-Holm Bay
SHIRASE GLACIER
SHOWA (Japan)

ELLSWORTH LAND
Mount Rex 1105
THOMAS MOUNTAINS
BRYAN COAST
General Belgrano (Argentina)
RONNE ICE SHELF
BERKNER ISLAND
FILCHNER ICE SHELF
Vahsel Bay
THERON MOUNTAINS
Slessor Glacier
SHACKLETON RANGE
RECOVERY GLACIER

SENTINEL RANGE
Mount Ulmer 2576
Vinson Massif 5140
HERITAGE RANGE
ELLSWORTH MOUNTAINS
HOLLICK-KENYON PLATEAU
Ronne Entrance
ENGLISH COAST
SIPLE COAST
WALGREEN COAST
HUDSON MTS.
JONES MTS.

PENSACOLA MOUNTAINS
3660
2123

BYRD LAND
Mount Takahe 3486
CRARY MTS.
WHITMORE MOUNTAINS
THIEL MOUNTAINS
2628
3267
MOLODEŽNAJA (U.S.S.R.)
RAYNER GLACIER
Casey Bay
WHITE ISLAND
Amundsen Bay
CAPE ANN
ENDERBY LAND
BEAVER GLACIER
NAPIER MTS.

Mount Petras 2875
Mount Sidley 4181
EXECUTIVE COMMITTEE RANGE
BYRD (U.S.)
QUEEN MAUD MOUNTAINS
Mount Glossopteris 2867
EXECUTIVE COMMITTEE
Mount Bursey 2779
Mount Berlin 3498
FLOOD RANGE
EDWARD VII PENINSULA
ROCKEFELLER PLATEAU
FORD RANGES
BOYD COAST
LEVERETT GLACIER
SCOTT GLACIER
Mount Fridtjof Nansen
SHACKLETON GLACIER
4083 Mount Wade
3355
Mount Menzies
3557
AMERICAN HIGHLAND
GROVE MOUNTAINS
MELLOR GLACIER
PRINCE CHARLES MTS.
LAMBERT GLACIER
FISHER GLACIER
MAWSON ESCARPMENT
BARNES
Stinear Nunataks 2727
MAWSON (Australia)
CAPE DARNLEY
AMERY ICE SHELF
MacKenzie Bay

ROSS ICE SHELF
ROOSEVELT ISLAND
Prestrud Inlet
COMMONWEALTH RANGE
QUEEN ALEXANDRA RANGE
BEARDMORE GLACIER
Mount Kirkpatrick 4528
3265
VOSTOK (U.S.S.R.)
POLAR RECORD GLACIER
Prydz Bay
Ingrid CHRISTENSEN COAST

PACIFIC OCEAN
Bay of Whales
ROSS SEA
Mount Markham 4350
Mount Albert Markham 3207
NIMROD GLACIER
Mount Egerton 2816
BRITANNIA RANGE
Mount McClintock 3492
Barne Inlet
BYRD GLACIER
MINNA BLUFF
2854
3059
LEOPOLD AND ASTRID COAST
Barrier Bay
CHELYUSKINTSY ICE TONGUE
WEST ICE SHELF
PHILIPPI GLACIER

SCOTT BASE (N.Z.) (U.S.)
McMURDO
Mount Huggins 3433
Mount Erebus 3743
ROSS ISLAND
McMurdo Sound
Terra Nova Bay
Coulman Island
Cape Hallett
Cape Adare
Robertson Bay
ADMIRALTY MOUNTAINS
Mount Levick 2774
Mount Sabine 3719
VICTORIA LAND
PRINCE ALBERT MOUNTAINS
OATES COAST
RENNICK GLACIER
Yule Bay
Cape Williams
Ob'Bay
Rennick Bay
WILLIAMSON HEAD
MATUSEVICH GLACIER
Lauritzen Bay
MAWSON PENINSULA
Deakin Bay
Sturge Island
Balleny Islands
NINNIS GLACIER TONGUE
MERTZ GLACIER TONGUE
Buckley Bay
Commonwealth Bay
SOUTH MAGNETIC POLE
DUMONT D'URVILLE (France)
Cape Bickerton
ADÉLIE COAST
CLARIE COAST
Davis Bay
DIBBLE ICEBERG TONGUE
VOYEYKOV ICE SHELF

ANTARCTICA
South Pole
WILKES LAND
GEORGE V COAST
NORTHS HIGHLAND
BANZARE COAST
SABRINA COAST
Cape Mikhaylov
Cape Poinsett
BUDD COAST
KNOX COAST
TOTTEN GLACIER
Vincennes Bay
SCOTT GLACIER
BUNGER HILLS
DENMAN GLACIER
Queen Mary Coast
MIRNYY (U.S.S.R.)
Davis Sea
DRYGALSKI ISLAND
ROSCOE GLACIER
MASSON ISLAND
SHACKLETON ICE SHELF
MILL ISLAND
BOWMAN ISLAND
POBEDA ICE ISLAND
DALTON ICEBERG TONGUE
Porpoise Bay
Paulding Bay

INDIAN OCEAN
Antarctic Circle

Kilometres 0 200 400 600 Km.
Miles 0 200 400 600 Mi.
Lambert Azimuthal Equal-Area Projection

Copyright © by Rand McNally & Co.
B-594000-764 -1°-1°-1°-2°

55

Earth panorama: the Pacific

The Pacific is the largest of all the oceans and has an area of 165 million square kilometres (64 million square miles). It is roughly circular and is bounded on three sides by Australia, Asia and America. It has wide contact south of Australia with the Indian Ocean, a limited contact with the same ocean through the Indonesian archipelago and a smaller contact with the Atlantic Ocean (through Drake Passage). To the north it has a very narrow passage through Bering Strait into the Arctic Ocean.

The hydrography of the Pacific is relatively simple. In the Northern Hemisphere there is a clockwise-current loop, driven by the northeast trade winds towards the Philippines and curving up towards Japan before carrying on towards Alaska and looping down past California back to its departure in the North Equatorial Current. The new sea-floor created by sea-floor spreading is compensated by sea-floor "sinking" into the trenches that extend from New Zealand round to Alaska and from Central America to Chile. Active volcanoes associated with these trenches circle the Pacific.

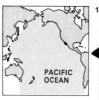

1 The oceanic hemisphere could well be another name for the expanse of the Pacific Ocean. Despite its many islands such as New Zealand, Papua and New Guinea, Borneo, Sakhalin and Japan, the proportion of dry land within the area is extremely small. The Pacific is the still-shrinking remains of the original world ocean Panthalassa that surrounded the dry land before it broke up into the continents we know today.

them cannot grow in depths of more than 45m (150ft). The early mariners thought that atolls were created by divine providence as convenient shelters for seafarers. Some naturalists believed that they were founded on shallow crater rims. In 1837 Charles Darwin proposed that they were once volcanic islands that sank by subsidence of the seafloor as coral growth towards the surface kept pace.

2 Many of the tropical Pacific Islands, for instance the Tuamotus, are coral atolls – ring-shaped islands surrounding a shallow lagoon. The atolls rise from great depths, yet the corals that built

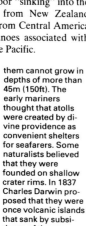

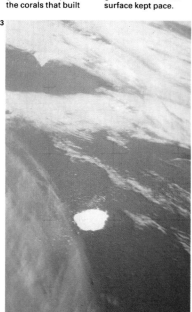

3 Icebergs shed from the huge Antarctic ice sheet drift in a northeasterly direction and are a danger to shipping. In the Pacific they can drift as far north as 41°S before they melt away. Some of them are over 600m (2,000ft) thick.

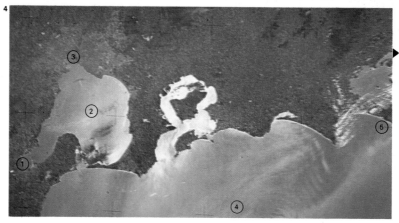

4 This section of the south coast of the state of Victoria, Australia, extends from Geelong [1] to Wilson's Promontory [5]. Bass Strait [4] separates mainland Australia from Tasmania. Melbourne [3] is at the head of Port Phillip Bay [2].

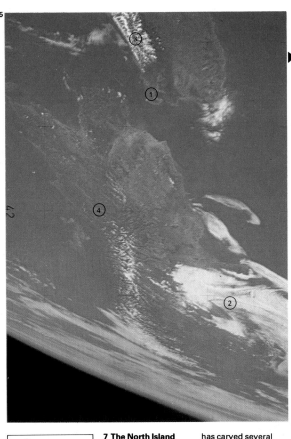

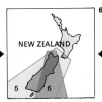

5 Cook Strait, named after James Cook the explorer, separates the North and South Islands of New Zealand. Wellington, the capital, can be seen [1] and also Christchurch [2]. The Tararua Range [3] and the Southern Alps [4] are covered in snow. The curvature of the earth is visible to the south.

6 The South Island of New Zealand consists of Cenozoic mountains, the Southern Alps, which were uplifted by the collision of the Pacific crustal plate to the east [left] and the Australian plate to the west. This Skylab 4 photograph was taken over Cook Strait [1] looking south, with Christchurch [2] visible.

7 The North Island of New Zealand is a volcanic area. Mount Egmont at 2,520m (8,260ft) is a symmetrical volcano although it carries several secondary cones on its flanks. Water erosion has carved several radial gullies round the craters. The limit between the volcanic scoriae and rock forming the volcano and the agricultural plain that surrounds it is distinct.

NEW ZEALAND

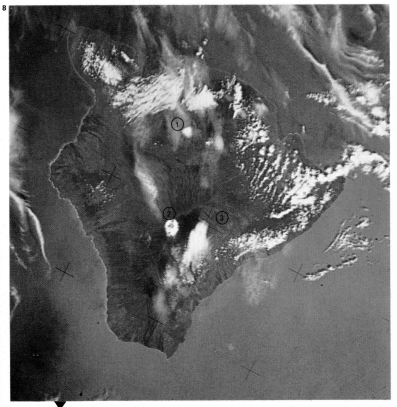

HAWAIIAN ISLANDS

8 Hawaii is the largest island in the Hawaiian archipelago. It was formed by the coalescence of two large volcanoes – Mauna Kea [1], 4,200m (13,796ft), and Mauna Loa [2], 4,160m (13,680ft). With its foot resting on the sea-floor 5,500m (18,000ft) below sea-level, Mauna Kea is the biggest mountain in the world, with a greater base-to-summit difference than even Mt Everest. It is a dormant, perhaps extinct, volcano, but Mauna Loa is one of the most active volcanoes in the world. Its most active vent, Kilauea [3], is on its southeast flank. The outer rim of the Kilauea caldera (collapse crater) is 13km (8 miles) in circumference. The caldera floor has an inner crater called Halemaumau.

North America

A search for life near Earth's molten core

By Alister Doyle
IN OSLO

Scientists will start drilling off Japan this month to explore a uncharted realm deep below the seabed, which they believe to be the hottest place to support life.

The drilling under the Nankai Trough in the Pacific Ocean will, it is hoped, reveal clues on everything from the origin of life on Earth to the formation process of oil and gas.

Previously, microbes have been found living at 121°C around a volcanic vent on the seabed in the Pacific Ocean off the United States.

Scientists will now drill into rocks, heated by Earth's molten core, that are believed to be as hot as 130°C, said Kai-Uwe Hinrichs, of the University of Bremen, who led the proposal for the mission.

He believes life could exist. "We've been surprised in these systems before," he said. REUTERS

3 days
by AIR
from
£299pp

London
Weekend Break

ATLANTIC OCEAN

PACIFIC OCEAN

GULF OF MEXICO

CARIBBEAN SEA

WEST INDIES

BAHAMAS

CUBA

JAMAICA

MEXICO

BAJA CALIFORNIA

SIERRA MADRE OCCIDENTAL

SIERRA MADRE ORIENTAL

SIERRA MADRE DEL SUR

ROCKY MOUNTAINS

APPALACHIAN MOUNTAINS

UNITED STATES

GREAT PLAINS

SOUTH AMERICA

COLOMBIA

VENEZUELA

PANAMA

COSTA RICA

NICARAGUA

HONDURAS

GUATEMALA

EL SALVADOR

BELIZE (U.K.)

CAYMAN ISLANDS (U.K.)

DOMINICAN REPUBLIC

HAITI

PUERTO RICO (U.S.)

NETHERLANDS ANTILLES

NEW YORK
WASHINGTON
BOSTON
PHILADELPHIA
MONTRÉAL
TORONTO
CHICAGO
DETROIT
CLEVELAND
ST. LOUIS
KANSAS City
DALLAS
LOS ANGELES
SAN FRANCISCO
SAN DIEGO
Phoenix
Denver
CIUDAD DE MÉXICO
Guadalajara
Monterrey
CARACAS
BOGOTÁ

Tropic of Cancer

Lambert Azimuthal Equal-Area Projection

Copyright © by Rand McNally & Co.
B-520000- 764

Kilometres
Miles

Km.
Mi.

59

Earth panorama: North America

North America extends from 15° to 83° latitude north, from the Isthmus of Tehuantepec in Mexico to the Arctic. Nearly all types of climate are found in this great geographical region, from the polar climate in the north, through the Subarctic tundra and conifer forest climates, the temperate climates, the high-altitude climates of the Rockies and the Sierras, the tropical deserts of Arizona, New Mexico and northern Mexico to the tropical climates of Florida, the Gulf Coast and southern Mexico.

The core of the continent consists of a Precambrian basement of granite and gneiss. This is overlaid by a horizontal sedimentary cover in the Middle West Plains and reaches the surface north of the Great Lakes and the St Lawrence to form the Canadian Shield. To the west of this basement, geologically recent foldings have uplifted the Rocky Mountains which run from Alaska to eastern Mexico. To the west of these mountains lie even more recent ranges. They are still the site of faulting, folding and volcanic eruptions. To the east of the basement is an ancient and highly eroded range, the Appalachians.

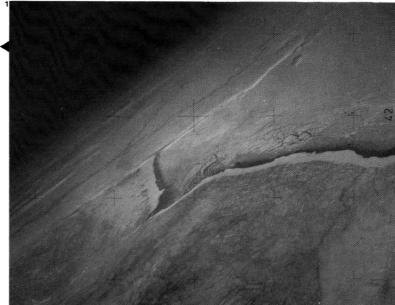

1 Hudson Bay is a large inland sea of 1,230,250km² (475,000 sq miles) which is open to navigation for only three months a year because of ice The area shown here is the Ontario and Manitoba shoreline. Hudson Bay is a shallow sea underlaid by the North American continental shield. Like the Baltic, this sea has filled a depression made by the weight of the Ice Age ice sheet and the sea-bed is now slowly rising.

2 The Great Lakes, the largest body of fresh water in the world, occupy depressions carved by the Quaternary ice sheet. The St Lawrence Waterway allows large ships to reach Duluth on Lake Superior.

3 The Gulf of St Lawrence is icebound in winter. The elongated island is Anticosti. To the north is the mainland of Quebec and the rounded coastline to the south is the Gaspé peninsula.

4 This Skylab view of Chesapeake Bay shows the cities of Washington [below] and Baltimore [above]. The Potomac River flowing between Washington and Alexandria (on the south bank) can be seen, at the bottom. The beltways around Washington and Baltimore and the interstate 95 freeway joining the two cities are conspicuous, as is the bridge of US Route 50 across Chesapeake Bay. The US Naval Academy at Annapolis is to the south of the Washington side of the bridge. The tunnel across Baltimore's harbour can be guessed from its aerial accessways. The patterns of murkiness in the bay help in the study of sedimentation and circulation.

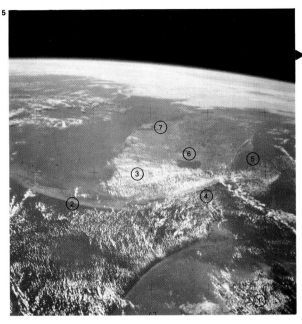

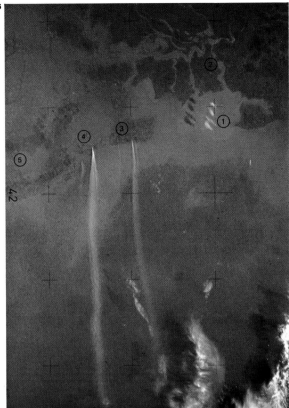

5 The Straits of Florida, through which the Gulf Stream flows, appear as a dark blue zone between the Bahama bank (bottom right) and the Florida peninsula. The Bahamas, of which only Andros [1] can be seen here, and the Florida Keys [2] are built of coral and algal reefs. The Everglades [3], Miami [4], Cape Canaveral and the John F. Kennedy Space Center [5], Lake Okeechobee [6] and Tampa Bay [7] can also be seen.

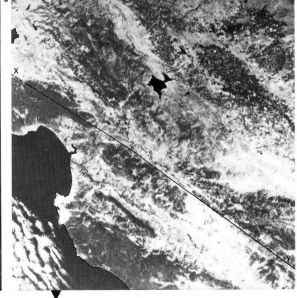

6 This swampy coast of Louisiana consists of Atchafalaya Bay [1], Atchafalaya River [2], which is a secondary effluent of the Mississippi, small muddy islands off the delta such as Isles Dernieres and Marsh Island [3] in front of Vermilion Bay [4]. White Lake [5] is another conspicuous feature. The two smoke plumes are oil well fires. These extend 320km (200 miles) over the Gulf.

7 Baja California peninsula was part of the Mexican mainland before it drifted 480km (300 miles) in a northwesterly direction, opening up the Gulf of California. This sliding motion is also shearing California along the San Andreas Fault which starts near the mouth of the Colorado River [1]. The large amount of sediment carried by this river is shown by the discoloration of the water. The islands of Angel de la Guarda [2] and Tiburón [3] are clearly seen and so is that of Cedros [4] off Sebastián Vizcaíno Bay. At the head of this bay are two lagoons, the largest of which is Scammon's Lagoon [5] to which the California grey whale migrates each year to mate and calve.

8 The San Andreas Fault is a huge break in the earth's crust running 435km (270 miles) from the top of the Gulf of California to a point north of San Francisco. Its movement caused the 1906 San Francisco earthquake. It can be seen in this Skylab picture [X Y] running parallel and to the east of the cultivated Salinas River valley. To the west Monterey Bay can be clearly seen. The large mottled expanse to the east is the Central Valley.

South America

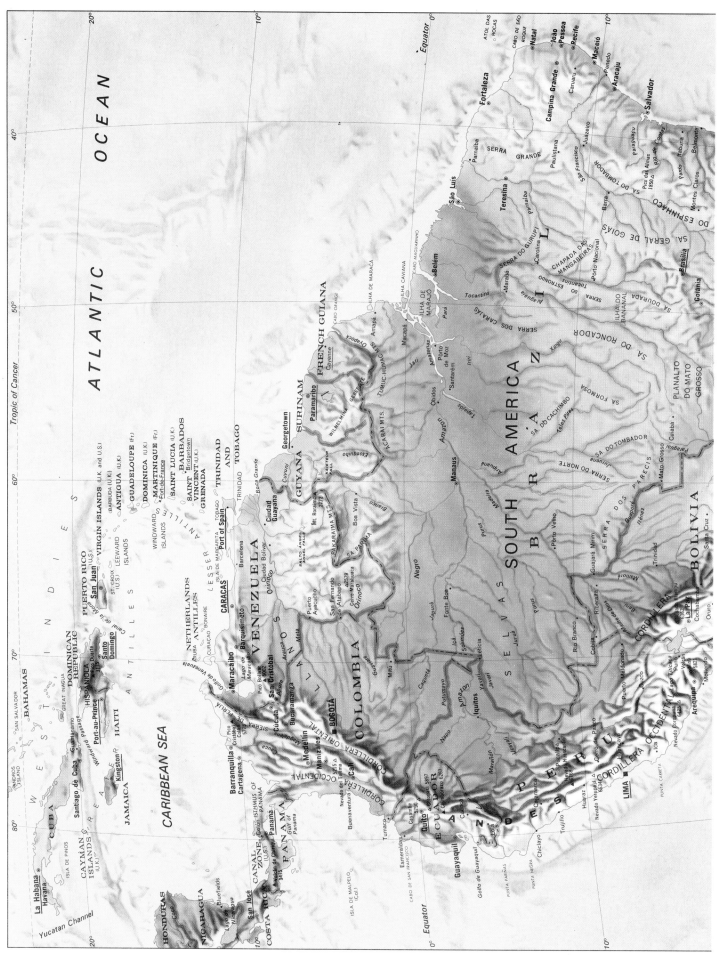

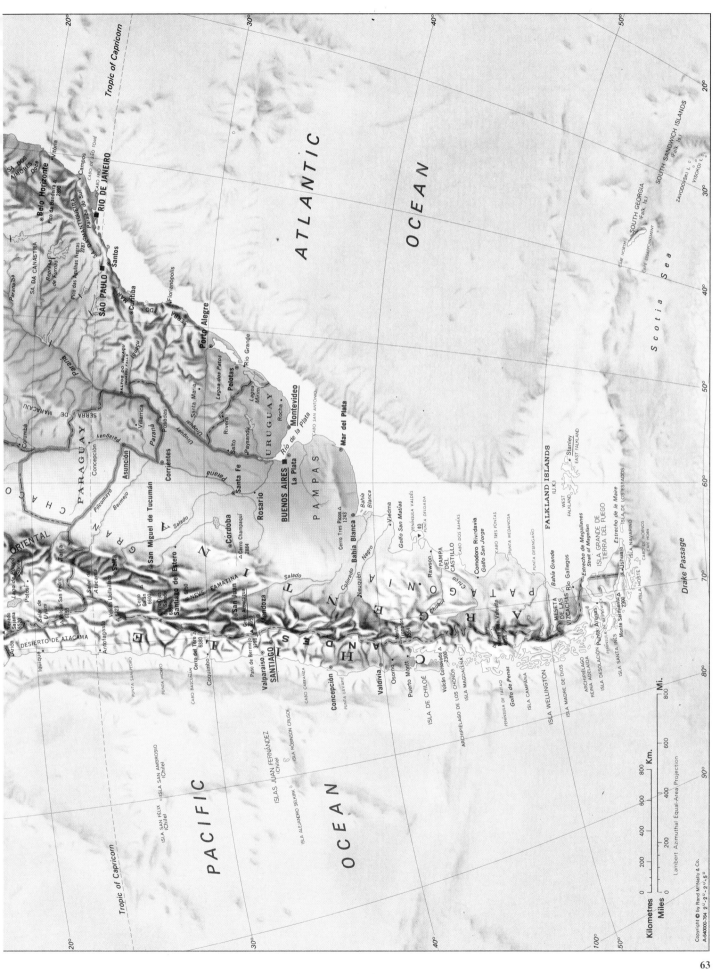

PACIFIC

OCEAN

ATLANTIC

OCEAN

Tropic of Capricorn

Tropic of Capricorn

SA DOS AMORES
Doce

Belo Horizonte
Pico da Bandeira 2890
SA DA CANASTRA
Represa de Furnas
Paranaíba
SERRA DA MANTIQUEIRA 2787
Pico das Agulhas Negras 2787
Paraíba do Sul
RIO DE JANEIRO
Vitória
Campos
CABO FRIO
CABO DE SÃO TOMÉ

Santos
SÃO PAULO
Curitiba
SERRA DO MAR

Corumbá
PANTANAL
SERRA DE MARACAJÚ

Florianópolis

ORIENTAL
Lago de Poopó
Sucre
Potosí
Tarija

PARAGUAY
Concepción
Asunción
GRAN CHACO
Pilcomayo
Bermejo

Porto Alegre

Rio Grande
Lagoa dos Patos
Pelotas
Santa Maria
Lagoa Mirim

Rivera
Salto
Paysandú
URUGUAY
Rocha
Montevideo
CABO SAN ANTONIO
Río de la Plata
Mar del Plata

Posadas
Corrientes
Paraná
Villarrica
Itaipú
Uruguay

Salado
Santa Fe
Rosario
BUENOS AIRES
La Plata

Arica
DESIERTO DE ATACAMA
Iquique
Antofagasta
Salar de Atacama
Salar de Uyuni
Volcán Llullaillaco
Salta
Cerro Galán 6600
Volcán Ojos del Salado 6885
San Miguel de Tucumán
Santiago del Estero
FAMATINA
San Juan
Cerro Aconcagua 6959
Cerro del Toro 6380
Paso de Bermejo 3798
Valparaíso
SANTIAGO
Córdoba
Cerro Champaquí 2884

Coquimbo
CABO BASCUÑÁN
CABO CARRANZA
PUNTA LAVAPIE
ISLA ROBINSON CRUSOE
ISLA ALEJANDRO SELKIRK
ISLAS JUAN FERNÁNDEZ (Chile)
ISLA SAN FÉLIX ISLA SAN AMBROSIO (Chile)

Mendoza
Neuquén
Colorado
Negro
Salado
PAMPAS
Cerro Tres Picos 1243
Bahía Blanca
Bahía Blanca

Concepción
Valdivia
Osorno
Puerto Montt
Volcán Corcovado 2300
ISLA DE CHILOÉ
ARCHIPIÉLAGO DE LOS CHONOS
ISLA MAGDALENA
PENÍNSULA DE TAITAO
Golfo de Penas
ISLA CAMPANA
ISLA WELLINGTON
ISLA MADRE DE DIOS
ARCHIPIÉLAGO REINA ADELAIDA
ISLA SANTA INÉS
Monte San Valentín 4058
Monte Tronador 3554

Viedma
Golfo San Matías
PENÍNSULA VALDÉS
Río
PUNTA DELGADA
Rawson
PAMPA DEL CASTILLO
Chubut
CABO DOS BAHÍAS
Comodoro Rivadavia
Golfo San Jorge
CABO TRES PUNTAS
PUNTA MEDANOSA
PUNTA DESENGAÑO
MESETA DE LAS VIZCACHAS
Santa Cruz
Río Gallegos
Bahía Grande
Estrecho de Magallanes
Strait of Magellan
Punta Arenas
PENÍNSULA DE BRUNSWICK
ISLA GRANDE DE TIERRA DEL FUEGO
Ushuaia
ISLA HOSTE
ISLA NAVARINO
CABO DE HORNOS
ISLA DE LOS ESTADOS
Estrecho de le Maire

FALKLAND ISLANDS (U.K.)
WEST FALKLAND
EAST FALKLAND
Stanley

SOUTH GEORGIA (Falk. Is.)
SOUTH SANDWICH ISLANDS (Falk. Is.)
ZAVODOVSKI I.
VISOKOI I.

Scotia Sea

Drake Passage

Kilometres
Miles
Mi.
Km.
0   200   400   600   800
0   200   400   600   800
Copyright © by Rand McNally & Co.
A-540000-764
Lambert Azimuthal Equal-Area Projection

Earth panorama: South America

The structure of South America is in many ways comparable to that of North America. High recent mountains – the Andes – follow the Pacific coast and old and worn highlands are found to the east: the Guiana Highlands and the Brazilian plateau. The Andes and the eastern highlands define vast alluvial basins that are drained by large rivers such as the Orinoco, Amazon, Tocantins, São Francisco Uruguay and Paraná. More than 90 per cent of the continent's drainage is towards the Atlantic Ocean; in terms of water flow the imbalance is even higher because the Andean coast receives very little rain between 5° and 35° of latitude south.

The Andes are a young and still extremely active chain of mountains. Their geological crumpling is the result of the Andes area being squeezed between the American and the East Pacific (Antarctic) plates, which are moving towards one another.

The climate is equatorial in the north and in the Amazon basin. It is tropical south of the Amazon basin and temperate south of southern Brazil. The Andean mountains south of Ecuador have a dry, cold climate.

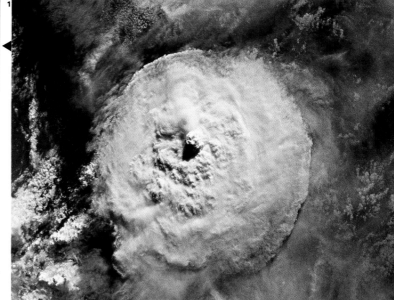

1 Hurricanes are prominent weather features when seen from space, as in this Apollo photograph. The Caribbean island arc experiences about a dozen hurricanes a year (the word is derived from the name of the native Mayan god of the big wind, Hunraken). Hurricanes are tropical depressions with extremely steep pressure gradients. They often originate in the Atlantic and travel westwards towards the American coast.

2 The Gulf of Venezuela lies between the peninsulas of Guajira [1] and Paraguaná [2]. The town of Maracaibo is on the channel leading from the gulf to Lake Maracaibo [3].

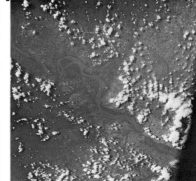

3 The Orinoco is the largest river in Venezuela. It meanders its way to the coast. Not far from its spring it links up with the Rio Negro by a natural canal, the Casiquiare.

4 The large body of reddish water seen here is the Rio de la Plata, between Uruguay and Argentina, which flows into the South Atlantic. The red plume is probably sediment moving seawards. Montevideo is the lighter area surrounding the deep bay where the coastline changes direction. To the west, the River Santa Lucía enters the Rio de la Plata and it is the major drainage for the area. The small island at its mouth is Isla del Tigre. The white beaches and sand dunes are visible along the coast. Major thoroughfares and residential areas are seen. Green and grey rectangular patterns are fields and show local types of agriculture.

5 Taken high over the Andes, looking south, this photograph reveals their basic shape and structure. The Pacific Ocean [1] bathes the feet of the Cordillera Occidental [2], some summits of which are snow-covered. This chain is made up of Mesozoic sediments and has numerous volcanoes, some of them active, located along fault lines that run parallel to the axis. East of the Cordillera Occidental is a central zone. To the north are some high folded mountains, the Cordillera Central [3]. To the south they dip under a debris-filled highland plain, the Altiplano [4], which is a graben (depression bounded by faults). Lake Titicaca [5] drains into the salt lake Poopó (not shown), which in turn drains into the salt pans of Coipasa and Uyuni [6]. The divide between the waters draining into the Altiplano and into the Amazon basin is distinct.

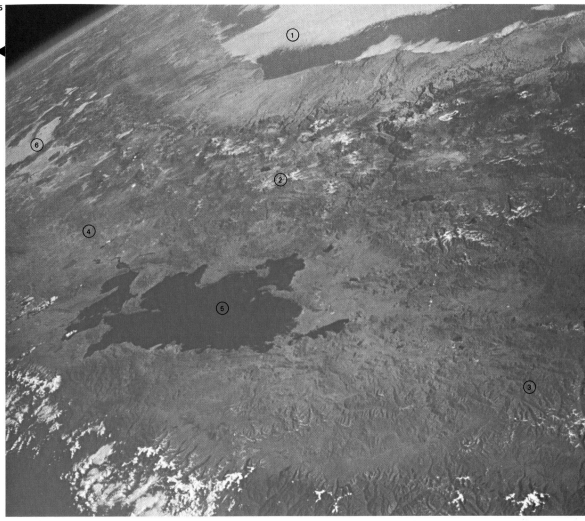

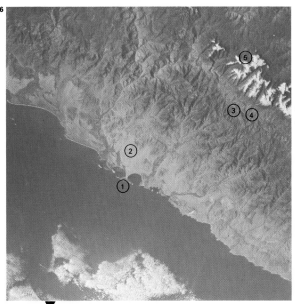

6 The direction of the sand bars [1] of this stretch of the Peruvian coast shows the northward Humboldt current's drift. The current's coldness prevents rainfall over the coast and the light areas [2] are deserts. Parallel to the coast run the Cordillera Negra [3] of volcanic origin and the snow-covered Cordillera Blanca [5]. The town of Yungay [4], in the Rio Santa valley was wiped out in 1970 by a landslide triggered by an earthquake; about 25,000 people were killed.

7 An interesting pattern of valleys is displayed in this area of the Andes between Chile and Argentina. They were carved by glaciers leaving the valley floors covered in moraine debris.

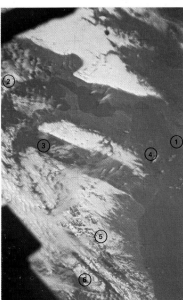

8 The eastern half of the Magellan Strait extends from Cape Vírgenes [1] past the town of Punta Arenas [2] and Useless Bay [3] to Cape Froward [4]. Tierra del Fuego [5] is separated from Navarino Island [6] by Beagle Channel.

The atmosphere

The origin of the atmosphere was no doubt closely associated with the origin of the earth. When the earth was still a molten ball, it was probably surrounded by a large atmosphere of cosmic gases, including hydrogen, that were gradually lost into space. As the earth began to develop a solid crust over a molten core, gases such as carbon dioxide, nitrogen and water vapour were slowly released to form an atmosphere with a composition not unlike the present emanations from volcanoes. Further cooling probably led to massive precipitation of water vapour so that today it occupies less than four per cent by volume of the atmosphere. At a much later stage, the oxygen content of the atmosphere was caused by green plants releasing oxygen as a result of combining water and carbon dioxide to form carbohydrates [Key].

Heated from below
Up to a height of about 50km (31 miles) the composition of the atmosphere [1] is remarkably homogeneous, comprising a mixture of gases each with their own physical properties. Carbon dioxide, water vapour and ozone, although only small constituents of the atmosphere, play vital roles in absorption of solar and terrestrial radiation, thus allowing life on earth. Due to the action of gravity, this homogeneous mixture of gases is compressed [2] giving the highest values of density and pressure near the earth's surface; average surface density is 1.2kg/m³ and average surface pressure is 1,013 millibars (mb) (roughly 1kg/cm² or 14lb per square inch). At a height of 16km (9 miles), pressure falls to 100mb and the density is less than 11 per cent of the density at sea-level.

The constituent gases of the atmosphere largely allow the sun's radiation to pass without interception. Fortunately the small amount of ozone, concentrated most strongly at 24km (15 miles) height, but in significant amounts up to 50km (31 miles), filters out most of the ultra-violet rays harmful to life on earth. If all the ozone were brought down to sea-level, it would form a layer only 0.25cm (0.1in) thick. After scattering, reflection and some absorption in the lower, denser layers of the atmosphere, only about 46 per cent of the solar radiation reaching the upper atmosphere is absorbed by the solid earth's surface as heat. This input of energy raises the earth's surface to a mean temperature of 14°C (57°F). Because this is lower than the 5,700°C (10,290°F) of the sun's surface, the earth radiates energy of much longer wavelengths (infra-red or heat rays) than solar radiation and these longer waves are absorbed by the carbon dioxide, water vapour and clouds in the lower atmosphere.

This means that the atmosphere is directly heated from below, not from above as one would perhaps expect. Just as the earth radiates heat, so does the atmosphere – upwards to be lost to space and downwards to be reabsorbed by the earth. The net effect of these exchanges [3] is that together they lose as much heat to space as they gain from solar radiation, thus maintaining a balance.

Temperature distribution
In the bottom 80 per cent (in mass) of the atmosphere, temperature falls with height in accord with the heating from below [4]. This layer of the atmosphere, 8km (5 miles) deep in polar regions and 16–19km (about 11

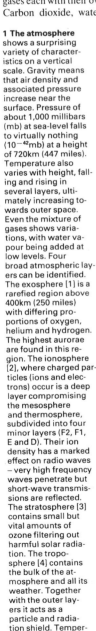

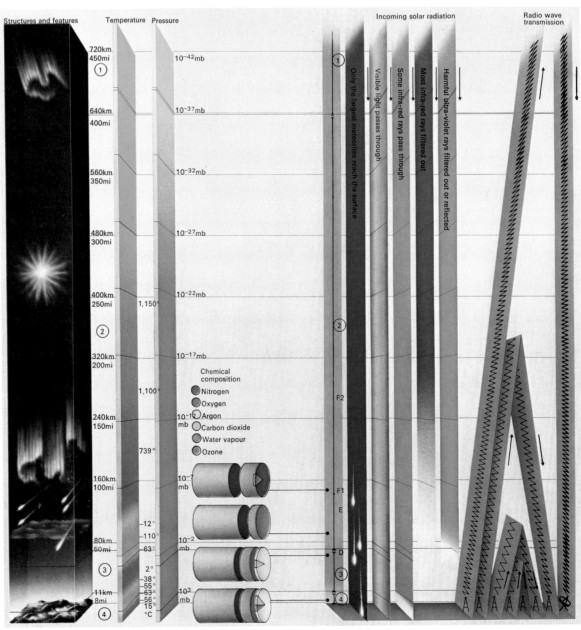

1 The atmosphere shows a surprising variety of characteristics on a vertical scale. Gravity means that air density and associated pressure increase near the surface. Pressure of about 1,000 millibars (mb) at sea-level falls to virtually nothing (10⁻⁴²mb) at a height of 720km (447 miles). Temperature also varies with height, falling and rising in several layers, ultimately increasing towards outer space. Even the mixture of gases shows variations, with water vapour being added at low levels. Four broad atmospheric layers can be identified. The exosphere [1] is a rarefied region above 400km (250 miles) with differing proportions of oxygen, helium and hydrogen. The highest aurorae are found in this region. The ionosphere [2], where charged particles (ions and electrons) occur is a deep layer comprising the mesosphere and thermosphere, subdivided into four minor layers (F2, F1, E and D). Their ion density has a marked effect on radio waves – very high frequency waves penetrate but short-wave transmissions are reflected. The stratosphere [3] contains small but vital amounts of ozone filtering out harmful solar radiation. The troposphere [4] contains the bulk of the atmosphere and all its weather. Together with the outer layers it acts as a particle and radiation shield. Temperatures decrease to its upper boundary.

Structures and features

Temperature Pressure

Incoming solar radiation

Radio wave transmission

720km 450mi — 10⁻⁴²mb

640km 400mi — 10⁻³⁷mb

560km 350mi — 10⁻³²mb

480km 300mi — 10⁻²⁷mb

400km 250mi 1,150° 10⁻²²mb

320km 200mi 1,100° 10⁻¹⁷mb

Chemical composition
Nitrogen
Oxygen
Argon
Carbon dioxide
Water vapour
Ozone

240km 150mi 739° 10⁻¹²mb

160km 100mi 10⁻⁷mb

−12°
−110°
80km 50mi −63° 10⁻²mb

2°
−38°
−55°
11km 8mi −63° 10³mb
−56°
15°
°C

Only the largest meteorites reach the surface

Visible light passes through

Some infra-red rays pass through

Most infra-red rays filtered out

Harmful ultra-violet rays filtered out or reflected

F2

F1
E

D

miles) deep over the equatorial regions, is known as the troposphere. It is characterized by wind speeds increasing with height, lots of moisture at low levels and appreciable vertical air movement, and it is generally the source of all the "weather" we experience. The tropopause marks the boundary between the troposphere and the stratosphere.

The temperature is virtually constant throughout the lower stratosphere but this layer has strong air circulation patterns and high wind speeds in the jet streams which are used (when they blow in the right direction) by airliners. In the upper stratosphere, above about 25km (15 miles), temperature gradually increases with height to a broad maximum at the stratopause. Above the stratopause, in the mesosphere, the temperature begins to decline sharply with increasing height, to a minimum at about 85km (52 miles). Above this level, which is called the mesopause, is the thermosphere where temperature is believed to increase to the thermopause at 400km (250 miles). Beyond, in the exosphere, the pressure drops to virtually a vacuum – equivalent to that of the sun's

outer atmosphere in which the earth orbits.

Within the troposphere another type of heat balance operates. More radiant heat is received than lost in tropical latitudes and the converse is true in polar latitudes. This broad temperature gradient from equator to pole generates a pressure gradient in the same direction; warm air moves down the gradient, reducing temperature extremes by cooling the tropics and warming the polar areas [3C].

Humidity of the atmosphere
The water content of the atmosphere is primarily in vapour form. Humidity decreases with height [5] because water enters the atmosphere by evaporation from the earth's surface. The driest parts of the lower atmosphere are over the subtropical deserts, the wettest are over the equatorial and summer monsoon regions, especially ocean surfaces. Water is constantly being cycled between the earth and the atmosphere. The amount in the atmosphere at any one time is only a fraction of one per cent of the total water in the planet, but it provides enough rainfall to sustain life on earth.

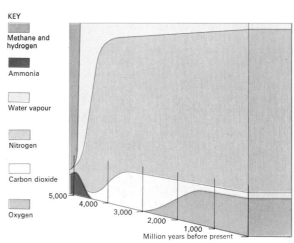

KEY
Methane and hydrogen
Ammonia
Water vapour
Nitrogen
Carbon dioxide
Oxygen

Important changes have occurred in the earth's atmosphere since it formed 4,600 million years ago as hydrogen, methane and ammonia. Most of the primitive hydrogen was lost to outer space and large quantities of steam and other gases were produced. This led to an atmosphere consisting mainly of nitrogen, water, sulphur dioxide and carbon dioxide. Photosynthesizing algae appeared 3,500 million years ago to produce free oxygen, and resulting ozone made up an ultra-violet shield, permitting life to spread on land.

2 Air is easily compressed, so the atmosphere becomes "squashed" by the effect of gravity. This results in the bulk (80%) of the atmosphere being in the troposphere, occupying a volume of about 6×10^9 cu km. As air density decreases with altitude, the very much smaller amounts of air present in the stratosphere (19%) and the ionosphere and above (1%) occupy a greater and greater volume.

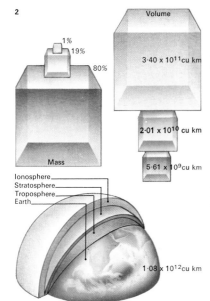

3 Temperatures in the atmosphere and on earth result mainly from a balance of radiation inputs and outputs. Average annual solar radiation reaching the earth, measured in kilolangleys (one calorie absorbed per sq cm) is highest in hot desert areas [A]. Comparison with the average annual long-wave radiation back from the earth's surface [B] shows an overall surplus radiation for nearly all latitudes but this is absorbed in the atmosphere and then lost in space, ensuring an overall balance. The extreme imbalance of incoming radiation between equatorial and polar latitudes is somewhat equalized through heat transfers by atmosphere and oceans [C]. This balancing transfer between surplus and deficit radiation is greatest in middle latitudes where most cyclones and anticyclones occur, shown at a latitude of 40° on the chart.

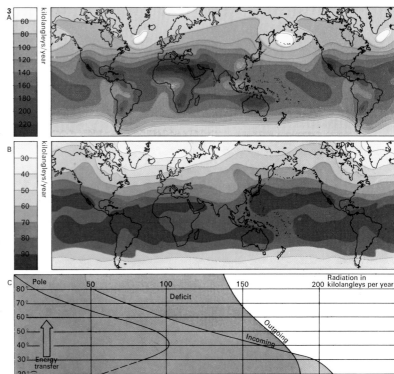

4 Atmospheric temperatures tend to decrease evenly with increase in height and latitude up to a level called the tropopause at a height of about 9km (5.5 miles) at the poles rising to 18km (11miles) in tropics.

5 Humidity falls with height in the troposphere. Warm air can hold more water vapour than cold air and therefore the warmer mid-latitude atmosphere holds more water vapour than the colder air over the Antarctic region.

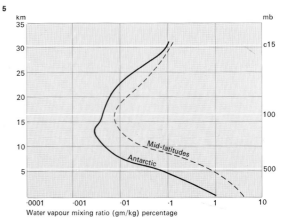

Winds and weather systems

Wind is the movement of air, and large-scale air movements, both horizontal and vertical, are important in shaping weather and climate. The chief forces affecting horizontal air movements are pressure gradients, the Coriolis effect and friction.

Pressure gradients are caused by the unequal heating of the atmosphere by the sun [1]. Warm equatorial air is lighter and, therefore, has a lower pressure than cold, dense, polar air. The strength of air movement from high- to low-pressure areas – known as the pressure gradient – is proportional to the difference in pressure.

The Coriolis effect, caused by the earth's rotation, deflects winds to the right in the Northern Hemisphere [3] and to the left in the Southern. As a result, winds do not flow directly from the point of highest pressure to the lowest. Instead, winds approaching a low-pressure system are deflected round it rather than flowing directly into it. This creates air systems, with high or low-pressure, in which winds circulate round the centre. Horizontal air movements are important around cyclonic (low-pressure) and anticyclonic

(high-pressure) systems. Horizontal and vertical movements combine to create a pattern of prevailing winds.

Along the Equator is a region called the doldrums, where the sun's heat warms the rising air. This air eventually spreads out and flows north and south away from the Equator. It finally sinks at about 30°N and 30°S, creating subtropical high-pressure belts, from which trade winds flow back towards the Equator and westerlies towards the mid-latitudes of the earth.

Cyclones and anticyclones

Along the polar front in the Northern Hemisphere, the warm air of the westerlies meets the polar easterlies. Waves, or bulges, develop along the polar front, some of which grow quickly in size [4]. Warm air flows into the bulge and cold air flows in behind it.

The warm, light air rises above the cold air along the warm front. Behind, the cold air forces its way under the warm air along the cold front. Gradually, the cold front catches up the warm front and the warm air is pushed above the cold in an occlusion. In cyclones in

the Northern Hemisphere, the air circulates in an anticlockwise direction (clockwise in the Southern). Along the warm front, a broad belt of cloud forms, bringing rain and sometimes thunderstorms. The cold front usually has a much narrower belt of clouds. Clouds and rain normally persist for some time along occluded fronts.

The circulation of air in anticyclones is the reverse of cyclones, being clockwise in the Northern Hemisphere and anticlockwise in the Southern. Many anticyclones are formed in warm, subtropical regions by sinking air. In winter anticyclones form over continental interiors in temperate latitudes as a result of the cooling of air.

How monsoons occur

Monsoons [2] are seasonal reversals of wind direction. The most celebrated monsoon occurs in India, where the generally northerly winds of winter are replaced by generally southerly winds in summer. The summer winds contain a lot of water vapour, which falls in heavy rain storms.

Another reversal of winds on a local scale

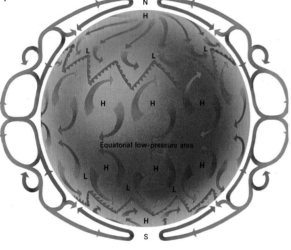

Warm air
Cool air
Cold air

▲▲▲▲ Warm front
▲▲▲▲ Cold front

H = High pressure
L = Low pressure

1 The earth's atmosphere acts as a giant heat engine. The temperature differences between the poles and the Equator provide the thermal energy to drive atmospheric circulation, both horizontal and verti-

cal. In general, warm air at the Equator rises and moves towards the poles at high levels and cold polar air moves towards the Equator at low levels to replace it. The pattern of prevailing winds is

complicated by the rotation of the earth, (which causes the Coriolis effect), by cells of high-pressure and low-pressure systems (depressions) and by the distribution of land and sea.

2 World winds in July and January form a pattern. Patterns at a low level are influenced by cells of low pressure, into which air flows, and high-pressure cells from which air flows outwards. If the earth did not rotate, winds would

blow directly from high-pressure cells to low-pressure cells. But the Coriolis effect causes winds to be deflected to the right in the Northern Hemisphere and to the left in the Southern Hemisphere. Wind patterns are remarkably constant between

summer and winter west of Africa. But, in the east, variations are caused by monsoons (reversals of wind flows). Monsoons arise from the unequal heating of land and sea. For example, dry winds blow outwards in winter across India from

the cold high-pressure system over southern Siberia. In summer, the land heats quickly and a low-pressure system develops over northwestern India. Moist, southeasterly trade winds are drawn into this system, bringing heavy rainfall.

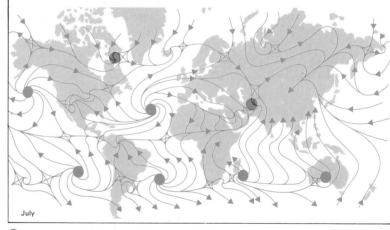

July

● Low-pressure areas
● High-pressure areas

3 A weather chart shows a "low" or depression to the south of Iceland and a "high" or anticyclone over southern Portugal and Spain. The isobars join points with equal atmospheric pressure. The values of the isobars are in millibars (1,000 millibars is the equivalent of about 750.1mm [29.53in] of mercury). Because winds are deflected, they circulate in an anticlockwise direction round a "low" and clockwise round a "high" in the Northern Hemisphere.

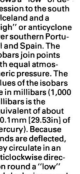

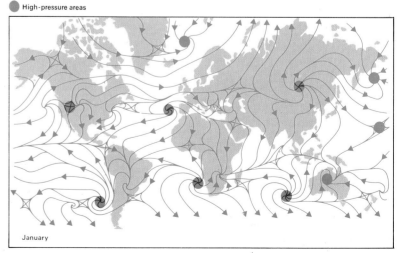

January

occurs with land and sea breezes. Sea breezes spring up on warm days along sea and lake shores when a pressure gradient is established between the rapidly heating land and the less rapidly heating water. As a result, winds blow on to land. At night, the land cools faster than the water, so a reverse gradient reverses the wind.

Thunderstorms, hurricanes and tornadoes

The most common storms are thunderstorms [6]. About 45,000 occur every day, in both temperate and tropical regions and prerequisites for their formation are strong, rising air currents. As the air rises, it is cooled and latent heat is released as condensation occurs. The release of heat provides energy that intensifies the upsurgence of air and the development of the storm. The condensation causes cumulo-nimbus clouds to rise sometimes more than 4,570m (15,000ft) from their base to their top. These clouds bring with them rain and hail and, sometimes, thunder and lightning.

Hurricanes [5], also called typhoons or tropical cyclones, form over warm oceans.

They have fast spiralling winds which may reach 240–320km/h (150–200mph). The calm centre, or eye, contains warm subsiding air. The eye may be between 6.5 and 48km (4 and 30 miles) across. The hurricane itself may have a diameter of 480km (300 miles). The warmth of the air in the eye contributes to low air pressure at the surface. Warm, moist air spirals upwards around the eye. Condensation creates cumulo-nimbus clouds and releases latent heat which further increases the upward spiral of air. Hurricanes are especially destructive along coastlines where storm waves and torrential rain cause destruction through flooding.

Tornadoes [Key] are violent whirlwinds, but they cover a far smaller area than hurricanes. A tornado forms when a downward growth starts from a cumulo-nimbus cloud. When the funnel-shaped extension of the cloud reaches the ground, it may be between 50–500m (164–1,640ft) wide. It crosses land at speeds of 32–65km/h (20–40mph) and usually dies out after 32km (20 miles), although a few are known to have travelled as far as 480km (300 miles).

Hundreds of tornadoes strike the United States each year, especially in the Midwest. They may last for several hours, travelling up to 480km (300 miles) and causing great damage. At the centre winds may reach 644km/h (400mph).

4

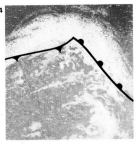

Fronts form in temperate latitudes where a cold air mass meets a warm air mass.

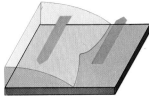

The air masses spiral round a bulge causing cold and warm fronts to develop.

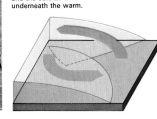

The warm air rises above the cold front and the cold air slides underneath the warm.

Eventually, the cold air areas merge, and the warm air is lifted up or occluded.

4 A front is a narrow band of changing weather lying between two air masses of different temperatures and humidity. When the two air masses meet, each pushes against the other to form a cold, warm or occluded front.

5

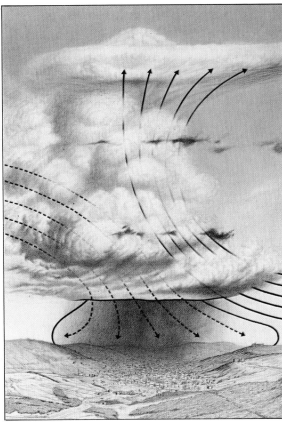

Hurricane winds

Prevailing winds

5 Hurricanes consist of a huge swirl of clouds rotating round a calm centre – the eye – where warm air is sucked down. Hurricanes may be 400km (250 miles) in diameter and they extend through the troposphere, which is about 15-20km (9-12 miles) thick. Clouds, mainly cumulonimbus, are arranged in bands round the eye, the tallest forming the wall of the eye. Cirrus clouds usually cap the hurricane.

6 A storm cloud or cumulo-nimbus has developed along a cold front. These clouds occur when the air mass is unstable over a great vertical distance. Air moves upwards in a convection current and cooling causes condensation. Flat anvil-shaped cloud heads mark the level where stability is re-established. Cumulo-nimbi are formed along fronts or in overheated areas. In depressions, a line of cumulo-nimbus marks the front and thunderstorms and violent squalls occur.

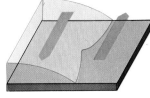

Weather

"Weather" in anybody's language means rain and sunshine, heat and cold, clouds and wind. Humidity and visibility might be added to the list. In fact, if not in terminology, this layman's catalogue comprises the six elements which, for the meteorologist, make weather: in his language they are air temperature, barometric pressure, wind velocity, humidity, clouds and precipitation.

Cloud formation

Clouds are made up of millions of very small drops of liquid water or ice crystals that are too light to fall out of the atmosphere. The cloud particles form from air that contains water vapour when the temperature falls to a critical level called the dew-point. These liquid droplets may then freeze into an ice crystal. But before either water or ice particles can form, two things must happen. First, the moist air must rise, reducing its pressure and giving up its heat to the surrounding atmosphere. Second, dust particles must already be present on which the cooled vapour can condense to form droplets or ice crystals. These tiny particles are respectively

called condensation nuclei and ice nuclei.

The formation of a cloud does not necessarily mean that it will precipitate. Condensation cannot create droplets or crystals that would survive the fall to the ground. They would evaporate even if they were large enough to overcome the force of the rising air. Two other mechanisms, the Bergeron or ice-crystal process, and the coalescence process, account for precipitation-sized particles. In clouds that contain both ice crystals and droplets of supercooled water (water at a temperature less than 0°C), the droplets evaporate and the vapour condenses on to the ice crystal. Thus the crystals grow at the expense of the droplets until they are large enough to fall out of the cloud. If they melt on the way down (as frequently happens), rainfall is observed at ground level. If the cloud contains no ice crystals, precipitation particles grow by the coalescence of different-sized droplets as they fall through the cloud. The larger a drop becomes, the more efficient it is at collecting smaller ones and the greater its chance of reaching the ground.

The two basic shapes of clouds – in layers

or in heaps – are caused by the two different ways in which air can move upwards. When air rises slowly over large areas at rates of a few centimetres a second, layer or stratified clouds are formed. This frequently occurs in cyclones, particularly in warm sectors and at warm fronts. Rapidly rising air (several metres per second) occurs in convection currents which are usually only a few hundred metres across near the ground. These currents widen with altitude but the resultant heaped or cumuliform clouds are rarely more than a few kilometres across. If the atmosphere is unstable they may grow into very large cumulo-nimbus clouds.

The easiest way to identify a cloud [1] is by its shape and height above ground. This was recognized by Luke Howard, a London chemist, in 1833 when he presented his first cloud classification. This still forms the basis of the World Meteorological Organization's International Classification of ten cloud types, which fall into three families according to their height. The highest clouds – about 8–10km (5–6 miles) – made of ice, are called cirrus, cirro-stratus and cirro-cumulus; the

1 The different cloud types are best illustrated within the context of the familiar mid-latitude frontal depression. Most of the major types occur within such cyclones. Here a schematic, generalized Northern Hemisphere depression is

viewed from the south as it moves from west [left] to east [right]. It is in a mature state, prior to the occlusion stage, and both warm [1] and cold [2] fronts are clearly visible. Over the warm front, which may have a slope

ranging from 1/100 to 1/350, the air rises massively and slowly over the great depth of the atmosphere. This results in a fairly complete suite of layer-type clouds ranging from cirrus [3] and alto-cumulus [4] to nimbo-stratus [5]. The precipitation

area often associated with such cloud types, and especially with nimbo-stratus, usually lies ahead of the surface warm front and roughly parallel to it [6]. Turbulence may cause some clouds to rise and produce heavy convective rainfall, as well as

the generally lighter and more widespread classical warm front rainfall. Stratus often occupies the warm sector, but a marked change occurs at the cold front. Here the wind veers (blowing in a more clockwise direction) and cum-

ulus clouds [7] are often found in the cold air behind the front. At the front itself the atmosphere is often quite unstable and cumulus clouds grow into cumulo-nimbus formations [8]. The canopy of cirrus clouds – of all types – may

extend over the whole depression and is often juxtaposed with the anvil shape of the nimbus. These cloud changes are accompanied by pressure, wind temperature and humidity changes as the fronts pass the individual observer on the ground.

middle clouds – 3–8km (2–5 miles) – of water and ice, are called alto-cumulus and alto-stratus; and the low clouds – below 3km (2 miles) – usually made of water, are called stratus, strato-cumulus and nimbo-stratus. The two remaining types are cumulus and cumulo-nimbus. There are, however, many variations on these ten types. Rarely is a cloud seen in isolation or conforming exactly to its textbook form; clouds of different types occur together and, as a result many hybrid forms are to be found.

Sun, wind and humidity

Long periods of sunshine are, of course, marked by clear skies, which usually result from sinking air in anticyclones. The longest periods of sunshine occur in the polar summer when the sun never sets, but the highest intensities and temperatures occur in the main deserts of the world which lie roughly at the latitudes 30°N and S. The daily maximum temperature in these areas may be more than 35°C (95°F), falling to below freezing at night.

Wind speed and direction at low levels are affected by friction between the air and the ground and by local topography. Friction means the wind speeds near the ground are generally less than at high levels and it also accounts for the generally higher speeds over water as compared with the rougher land surface. Air flow is often channelled in both valleys and urban areas.

There are several ways of expressing the humidity of the air but relative humidity is the most widely used. This is the percentage of water vapour actually held in a given volume of air relative to the amount that the air could hold if saturated at the same temperature. In middle latitudes the daily values usually lie between 60 and 80 per cent, but they can range from 8 to 100 per cent.

Fog, a modern menace

Visibility has assumed a great importance in our communications-conscious world. Fog [3], which is cloudy air at ground level, presents dangers to aircraft, ships and motor vehicles alike. It can also affect the man in the street; if fog becomes contaminated with pollutants it may become lethal smog.

The structure of a hurricane may be difficult to discern at ground level but from a satellite in orbit the pattern of air movements involved can be clearly seen.

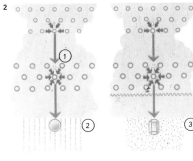
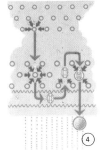

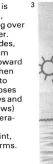

3 Advection fog is caused by warm, damp air blowing over cold land or water. In warmer latitudes, heat rises to warm the air above (upward blue arrows). When the air passes into cooler areas, it loses heat (blue arrows and upward red arrows) and as the temperature of the air reaches dew point, fog gradually forms.

4 Radiation fog occurs when air is cooled to its dew-point by contact with land that has itself been cooled by long-wave radiation loss [long brown arrows]. As the ground cools, surrounding air transfers heat by conduction [short brown arrows]. The cooling of the air is shown by the grey arrows.

2 Repeated coalescence of droplets [1] forms drops [2] too large to float on air currents. Ice crystals collect in hexagonal patterns [3] then agglomerate into snowflakes. Water can freeze round an ice embryo [4] to form hail.

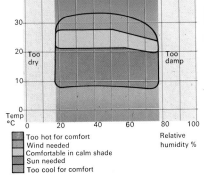

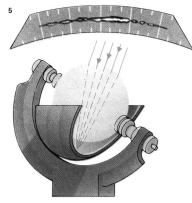

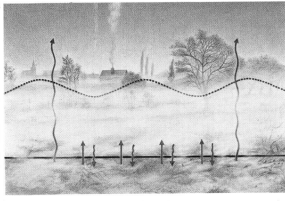

5 A Campbell-Stokes recorder registers the duration of sunshine. A glass ball focuses the sun's rays on to a specially prepared piece of card on which a trace is burnt as long as the sun shines brightly. The instrument must be orientated to the noonday sun at an angle determined from the declination of the sun. Three sets of grooves in the bowl behind the glass sphere will accommodate the different cards which are used for summer, winter and the equinoxes.

6 Human beings can tolerate only certain ranges of temperature and humidity. Even within these ranges, other elements such as sun and wind are needed to produce comfortable conditions in which people can live and work.

7 The power of the wind to erode and transport is well seen in this photograph of an approaching dust storm in the Midwest of the United States of America. These storms occur only in arid areas where the soil is loose.

Too hot for comfort
Wind needed
Comfortable in calm shade
Sun needed
Too cool for comfort

Forecasting

Day-to-day weather depends on the movements of huge air masses, which take their characteristics of temperature and humidity from the land or water surface beneath them and shift slowly over the surface of the earth. Some are virtually static, providing steady weather conditions for days or weeks on end in their area of origin. These produce, for example, the constant, predictable weather of tropical deserts and oceans and the heartlands of the great continents. Other air masses are affected by the earth's rotation and, as a result, they move and swirl rapidly, interacting in different ways with neighbouring masses. These provide the changeable weather of temperate latitudes, which is much more difficult to predict accurately.

Factors that influence the weather
To predict the weather over a particular area, the forecaster must first know the pattern of air masses that overlies it at any given time. Then he must try to predict how the pattern will change during the period of forecast – usually the next few hours or days – drawing on his experience of the ways in which similar patterns have changed in the past.

Weather forecasting originated in the observations of farmers and sailors, whose especial interest caused them to watch the weather closely and discover the patterns underlying it. Even in temperate regions this is not as difficult as it may seem at first. For example, much of western Europe's weather depends on a west-to-east procession of cyclones or depressions, and the passage of "fronts" – planes of contact between neighbouring air masses of differing temperature and humidity. Fronts that bring the worst weather lie generally in the southern half of the depressions. An observer who sees the barometer falling and notices a change of wind (often veering to the southwest accompanied by a thickening and lowering of cloud from high cirrus to cirro-stratus, alto-stratus and nimbo-stratus), is keeping track of the movement of a depression and warm front. He can predict fairly accurately the sequence of weather that will arise from it, and even the speed with which the changes will take place. Similarly, a rising or high and steady barometer, with clear skies and light breezes, usually means that an anticyclone has formed. This often brings clear, steady weather for several days until the next depression moves in to replace it.

The forecaster at work
The professional forecaster begins his work by preparing a synoptic chart [3], that is, an accurate map of the weather prevailing at the time over a large area surrounding his position. In this he is helped by observations from many surface stations, which come to him through teleprinter and radio networks.

There are more than 8,000 surface stations providing this service round the world; they include mountain outposts, ships at sea, polar bases and automatic (unmanned) units which record the weather and send information out at regular intervals.

In his synoptic chart the forecaster plots pressure, wind, temperature, cloud types, humidities and pressure tendency, and makes a note of past and present weather. This enables him to draw in isobars (lines connecting points of equal atmospheric pressure) and the position of fronts. Knowing the weather

CONNECTIONS

See also
68 Winds and weather systems
70 Weather
66 The atmosphere
74 Climates

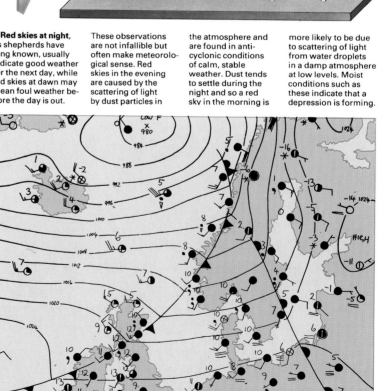

1 Red skies at night, as shepherds have long known, usually indicate good weather for the next day, while red skies at dawn may mean foul weather before the day is out.

These observations are not infallible but often make meteorological sense. Red skies in the evening are caused by the scattering of light by dust particles in the atmosphere and are found in anticyclonic conditions of calm, stable weather. Dust tends to settle during the night and so a red sky in the morning is more likely to be due to scattering of light from water droplets in a damp atmosphere at low levels. Moist conditions such as these indicate that a depression is forming.

2 Analysis and forecasting require simultaneous, standard and regular observations at many stations. The international, standardised weather station includes a variety of equipment. The Stevenson screen [1] is a box that shelters thermometers and other instruments from sunlight; it contains wet- and dry-bulb thermometers and recording instruments. Open land [2] allows the state of the ground to be assessed. A grass minimum thermometer [3] records the lowest ground temperature during the past 24 hours. The anemometer and arrow [4] show wind speed and direction. The Campbell-Stokes sunshine recorder or radiometer [5] records hours of sunshine. Radiosonde balloon [6] and theodolite [7] show wind speed and direction, and other data from high altitude. A rain-gauge [8] records the amount of precipitation. The weather office also has a barometer and barograph, to record atmospheric pressure.

3 The synoptic chart is compiled from data collected over a wide area. The circles indicate weather stations and the symbols show weather conditions at a particular time – the synoptic hour. The cloud cover symbols show the fraction, in eighths, of the observable sky at each station that is covered by clouds. The numbers by the symbols on the map indicate the air temperature in degrees centigrade, and wind direction is shown by an arrow from the appropriate direction and the speed in knots by the fletches. From pressure and wind readings, the analyst draws in a series of isobars (lines connecting points of equal pressure) which provide a full picture of the horizontal flow of surface winds.

CLOUD
- ○ 0
- ◍ 1 or less
- ◒ 2
- ◔ 3
- ◑ 4
- ◕ 5
- ◖ 6
- ◕ 7 or more
- ● 8
- ⊗ Sky obscured
- ⊠ Doubtful data

WEATHER
- ≡ Mist
- Fog
- ▪ Drizzle
- Rain and drizzle
- ● Rain
- ✳ Rain and snow
- ✳ Snow
- ▽ Rain shower
- Rain and snow shower
- ▽ Snow shower
- Hail shower
- ⚡ Thunderstorm

WIND
- ○ Calm
- ○⌐ 1-2
- ○⌐ 3-7
- ○⌐ 8-12
- ○ 13-17
- ▲▲▲▲ Cold front
- ━━━ Warm front

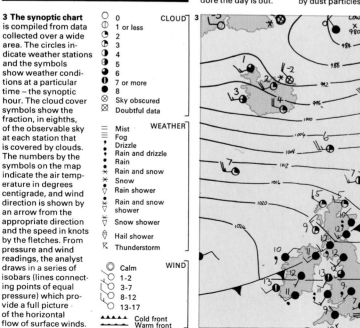

picture, and the rate at which it is changing, he is then able to predict what the weather will be like at any point on his map in the near future. His work is nowadays made easier by upper-air observations (taken by weather balloons with radiosonde attachment) and photographs from weather satellites [6] that give him an astronaut's eye-view of patterns. Much of the forecaster's work of plotting and analysis has been automated and mathematical analysis plays an increasing role in forecasting as more accurate data from all levels of the atmosphere become available.

Short-term and long-term forecasting

Short-term forecasting is still of great importance to farmers and sailors, and the safety of many millions of airline passengers depends on it each year. There is also an increasing demand for long-term forecasting, covering periods of from five days to six months ahead. Different techniques of analysis are required for this kind of forecasting. In areas of the world where the climate varies little from one year to another, comparatively simple statistical methods are used to relate the

character of one season to the next as a basis for prediction. More variable climates demand more sophisticated methods and detailed research into the nature, origins and movements of air masses. Recently, analysis of relationships between the atmosphere and the ocean have been of great value to long-range forecasters. Britain's predominantly maritime climate depends on air masses that have passed over the ocean, and anomalously cold or warm patches of ocean can have marked effects on weather to come.

The first attempt to co-ordinate meteorological observations on an international basis was not made until 1853, when the major maritime nations formulated a system of weather observations over the oceans to help navigation. In 1878 the International Meteorological Organization (IMO) was set up to keep a constant watch on the weather. International co-operation continued to improve during the years that followed, and in 1951 the IMO was reorganized to become the World Meteorological Organization (WMO), which was recognized by the United Nations.

KEY

Television brings the daily weather forecast into viewers' homes. From internationally collected data, the pre-senter prepares a simplified synoptic chart. On this he explains the pressure situation, discusses the position and movement of fronts and predicts the kind of weather to be expected during the next few hours.

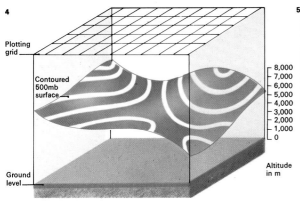

4 Basic mathematical models are designed to forecast the altitude of a pressure surface, usually about 500 or 600 millibars (mb). From them, air flow at that level can be defined and the future position of cyclones and anticyclones plotted. Pressure surfaces are contoured in the same way as ground maps: surface heights are plotted by calculation over a grid. This method is suitable for large-scale forecast-ing, usually on a 250km (155 mile) grid. A two-level model may be used, on which two pressure surfaces – eg the 500mb and 1,000mb surfaces – are plotted at the intersections of the same grid.

5 Accurate weather forecasting involves the processing of large quantities of data that flow into the weather centres several times each day. The introduction of computers and plotting machines allows forecasters to handle the data more rapidly. Recent developments in plotting machinery display processed data in map form using a line printer or plotting table. Here both types of output are shown; isobars can be interpolated from the pattern of figures [left], or the isobars can be computed and drawn in automatically [right]. These techniques can produce maps like this one in seconds rather than hours.

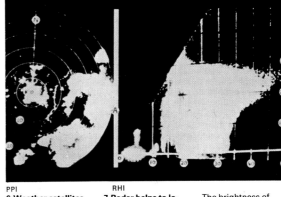

PPI   RHI

6 Weather satellites have provided photographs of clouds since 1960, with steady improvement in photographic quality and coverage. A sequence of photographs showing developing cloud patterns greatly helps the work of the forecaster.

7 Radar helps to locate rain clouds. Signals are reflected by raindrops and ice particles, registering on the Plan Position Indicator (PPI) showing the pattern over an area, or on the Range Height Indicator (RHI) showing the vertical distribution.

The brightness of the echo shows the intensity of precipitation. Rangemarkers (radiating from the centre of the PPI and bottom left of the RHI) are in miles while heights (on the right of the RHI) are shown in thousands of feet.

Climates

The climate of an area is its characteristic weather considered over a long period. Climate depends first on latitude, which determines whether an area is hot or cold and how strongly marked are its seasons. It depends also on the moving air masses that prevail in the area; these may be purely local in origin or they may have moved into the area from several hundred kilometres away, bringing cooler or warmer, wetter or drier conditions with them. Climate is also influenced by the relative distribution of land and sea, high ground and low, and the presence nearby of forests, lakes, valleys, glaciers and many other physical factors.

On a world scale, macroclimate is defined primarily in terms of temperature and rainfall and the world can be divided into large climatic zones on this basis. On a smaller scale, humidity, wind strength, aspect in relation to the sun and other local features determine local climate. On an even smaller scale, microclimate refers to the conditions in a particular woodland or under a particular stone.

The range of climates can be broadly grouped under three headings, according to latitude. Tropical climates are hot and dominated by equatorial air masses throughout the year. Temperate climates of the mid-latitude zone are variable, dominated alternately by subtropical and sub-polar air masses, and usually seasonal. Polar climates of high latitudes are uniformly cold, under the continuous control of sub-polar and polar air masses and strongly seasonal.

Characteristics of tropical climates

Because of constant daily sunshine, equatorial and tropical regions are warm throughout the year and the moving air masses that affect them are also warm [6, 7]. The wettest regions lie in a belt of shallow depressions and convection formed where the trade winds meet. This belt shifts north and south seasonally on either side of the Equator, but temperatures vary only slightly and rainfall is fairly steady throughout the year. Monsoon climates of India, South-East Asia and China occur where seasonal winds blow from almost opposite directions; warm, moist winds alternate with warm, dry ones, giving cloudy, wet "summers" and drier "winters"

Dry, tropical climates occur in broad zones on either side of the Equator between latitudes 15° and 30°. These are anticyclonic areas of warm, dry air, where cloudless skies bring strong sunshine and little rain except in rare torrential thunderstorms.

Temperate climates and their features

The middle latitudes of both hemispheres are battlegrounds where warm subtropical and cool sub-polar air masses jostle for position. The day-to-day battle-lines are the warm and cold fronts of the weather charts, which tend to occur along broad frontal zones. On the equatorial side of these zones warm air is present for most of the time. The zones shift north and south with the seasons so that an area such as the south of France may bask reliably in subtropical air throughout the summer but suffer occasional draughts of cold sub-polar air in winter. On western flanks of the continents in the warmer zone the air tends to be dry, bringing hot, dry summers and mild, damper winters – the "Mediterranean" climate of California, southwestern Australia and the eastern

Tropical rainy climates ☐ Dry climates ☐ Temperate climates ☐ Humid cold climates ☐ Polar climates

1 Climate on a global scale presents a bewildering variety and has provided would-be classifiers with a challenge for more than a century. The most generally accepted classifications reflect the close links that exist between vegetation and climate. A system in wide use today is that of W. Köppen (1846–1940), a German biologist who devoted most of his life to climatic problems and modified his own system many times before he was satisfied with it. The Köppen system of classification recognizes five major climatic categories, each quite distinct. These are equatorial and tropical rain climates, dry climates, temperate climates of the (mainly) broad-leaf forest zone, humid cold climates and polar climates. Each category is defined in terms of temperature and some in terms of rainfall too. Köppen also devised additional symbols for times of year in which most rain falls and other climatic qualities that affect the growth of vegetation. The map locates the hot, wet tropical rain forests of South America, Africa and the Far East. Farther away from the Equator the world's great deserts, dominated by the Sahara, straddle subtropical latitudes and the edges of the tropics, as a result of the stable high pressures there. The deserts are more evident in the Northern than in the Southern Hemisphere, mainly because of the extensive southern oceans. Nearer the poles, a mosaic pattern of mid-latitude climates occurs. This is less complex over the huge continental areas of Siberia and North America, particularly in tundra and boreal regions. The extreme polar and highland climates (dissimilar but with common features) occupy smaller areas.

Mediterranean itself. Eastern flanks of the continents draw moist, unstable air from over the sea; they tend to be warm all year, with frequent thunderstorms in summer.

In higher latitudes, farther from tropical influences, cool sub-polar air masses prevail. A procession of cyclones or depressions, swinging eastwards round the earth, brings moist maritime air to the western flanks of North America and Europe. Britain and western Canada stand in prevailing southwesterlies, giving mild, cloudy and damp conditions in winter and summer alike. Alternating air masses from the eastern continents bring cold, clear winter and hot, dry summer weather, and air from the north is usually cold and crisp. Central and eastern regions of the continents tend to be drier, with colder winters and hotter summers.

The cold, dry polar climates
Nearer the poles are climatic regions controlled by polar air masses [2]. Despite brief, sunny summers they tend to be cold and dry throughout the year. The broad boreal zone is forested, the tundra zone supports shrubs,

rough grassland and mosses. The true polar climate, which covers the northern fringes of Canada, Europe and Asia and the whole of the Antarctic continent, is generally too cold and dry to support any but the most meagre and hardy vegetation. The coldest regions of the Northern Hemisphere lie in the heartlands of northern Canada and northeastern Siberia, where winter temperatures fall well below −30°C (−22°F). On the high polar plateau of Antarctica summer temperatures are about −30°C (−22°F), while winter temperatures average −70°C (−94°F) or lower.

Geological evidence suggests that the modern situation in which different parts of the globe have certain well-defined climatic patterns is unusual. In past ages climates tended to be fairly even over most of the earth's surface. For example, the Permian period 280 million years ago was characterized by extensive desert areas over most of the continents and Jurassic sediments 195 million years old show evidence of warm, wet conditions in most places. The present pattern may be due, in part, to the fact that the earth is still recovering from the last Ice Age.

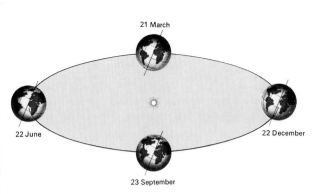

21 March

22 June

23 September

22 December

The seasons are primarily controlled by the rotation of the earth round the sun and the inclination of the earth's axis to the plane of rotation. Inclination of 23.5° means that the sun is directly over the Tropic of Cancer (23.5°N) on 22 June and over the Tropic of Capricorn (23.5°S) on 22 December. At the equinoxes, 21 March and 23 September, the sun is over the Equator. The sun's apparent movement is accompanied by a similar shift in belts of pressure and wind.

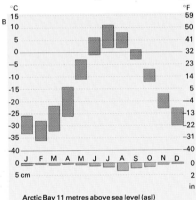

°C / °F

Arctic Bay 11 metres above sea level (asl)

2 Polar climates, as in Arctic Bay [A], are very cold and dry. Only three months of summer are frost-free [B]. There is little precipitation but much surface water due to poor drainage and low evaporation.

3 A continental climate is found in Calgary, Canada. Temperatures there are high in summer and low in winter; annual precipitation is low. This type of climate supports grasslands and cereal crops.

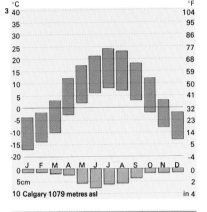

10 Calgary 1079 metres asl

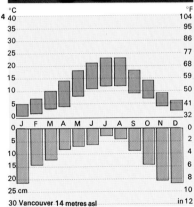

30 Vancouver 14 metres asl

4 The Canadian city of Vancouver has a mid-latitude maritime climate, with warm summers and mild winters. Monthly temperature ranges are greater in summer, precipitation is highest in winter.

5 A Mediterranean climate is typified by Rome [A] and its environs [B], with hot, dry summers and warm, moist winters. Similar latitudes, of western USA, South America and Australia, enjoy similar climates.

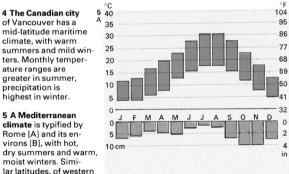

Rome 115 metres asl

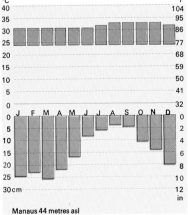

6 Hot deserts, near Timbuktu [A] for example, show little annual variation of temperature but extreme monthly variation [B]. The scarce rainfall occurs as a result of convectional summer storms.

7 An equatorial climate, that of Manaus in central Amazonia for example, is typified by high, constant temperature throughout the year and very heavy rainfall. There is no dry season, although the level of rainfall may vary from time to time.

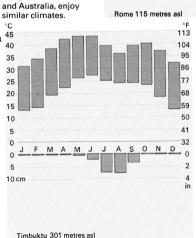

Timbuktu 301 metres asl

Manaus 44 metres asl

The sea and seawater

Photographs of the earth from space suggest that "Ocean" would be a more suitable name for our planet, because the oceans cover 70.8 per cent of the earth's surface [Key]. There are three major oceans, the Pacific, the Atlantic and the Indian, but the waters of the Arctic and Antarctic are also described as oceans. These five oceans are not separate areas of water but form one continuous oceanic mass. The boundaries between them are arbitrary lines drawn for convenience.

The study of oceanography
The vast areas of interconnected oceans contain 97.2 per cent of the world's total water supply. The study of oceans, including their biology, chemistry, geology and physics, has become a matter of urgency, because man's future on earth may depend on his knowledge of the ocean's potential resources of food, minerals and power.

The most obvious resource of the oceans is the water itself. But seawater is salty, containing sodium chloride (common salt), which makes it unsuitable for drinking or farming. One kilogramme (2.2lb) of sea-

water contains about 35g (1.2oz) of dissolved material, of which chlorine and sodium together make up nearly 30g (1oz) or about 85 per cent of the total.

Seawater is a highly complex substance in which 73 of the 93 natural chemical elements are present in measurable or detectable amounts [1]. Apart from chlorine and sodium seawater contains appreciable amounts of sulphate, magnesium, potassium and calcium, which together add up to over 13 per cent of the total. The remainder, less than one per cent, is made up of bicarbonate, bromide, boric acid, strontium, fluoride, silicon and trace elements. Because the volume of the oceans is so great there are substantial amounts of some trace elements. Seawater contains more gold, for example, than there is on land, although in a very low concentration of four-millionths of one part per million [3].

Also present in seawater are dissolved gases from the atmosphere, including nitrogen, oxygen and carbon dioxide. Of these, oxygen is vital to marine organisms. The amount of oxygen in seawater varies

according to temperature. Cold water can contain more oxygen than warm water. But cold water in the ocean deeps, which has been out of contact with the atmosphere for a long period, usually contains a much smaller amount of oxygen than surface water.

Other chemicals in seawater that are important to marine life include calcium, silicon and phosphates, all of which are used by marine creatures to form shells and skeletons. For cell and tissue building, marine organisms extract such chemicals as phosphates, certain nitrogen compounds, iron and silicon. The chief constituents of seawater – chlorine, sodium, magnesium and sulphur – are hardly used by marine organisms.

The salinity of the oceans
The volume of dissolved salts in seawater is called the salinity of the sample. The average salinity of seawater ranges between 33 and 37 parts of dissolved material per 1,000 parts of water. Oceanographers usually express these figures as 33 parts per thousand (33‰) to 37‰. The salinity of ocean water varies with local conditions [6]. Large rivers or

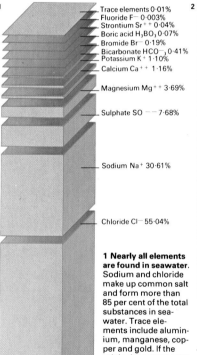

1 Nearly all elements are found in seawater. Sodium and chloride make up common salt and form more than 85 per cent of the total substances in seawater. Trace elements include aluminium, manganese, copper and gold. If the salt in the oceans were precipitated, it would cover the earth's land areas with a layer 153m (520ft) thick.

Trace elements 0·01%
Fluoride F— 0·003%
Strontium Sr++ 0·04%
Boric acid H₃BO₃ 0·07%
Bromide Br— 0·19%
Bicarbonate HCO—₃ 0·41%
Potassium K+ 1·10%
Calcium Ca++ 1·16%
Magnesium Mg++ 3·69%
Sulphate SO—— 7·68%
Sodium Na+ 30·61%
Chloride Cl— 55·04%

2 The salt in the world's salt mines was formed either from ocean water or from saline water in inland seas. The salt layers accumulated over extremely long periods in basins where evaporation caused the salt to precipitate from the water. The salt mines of Wieliczka, Poland, contain a layer of salt 366m (1,200ft) thick, while those of Texas are up to 3,658m (12,000ft) thick. A

column of seawater about 305m (1,000ft) high would precipitate only about 4.6m (15ft), so very thick salt deposits are difficult to explain. Illustration 4 depicts the creation of salt and other saline rock deposits.

3 One of the elements in seawater is gold. Although gold forms only 0.000004 part per million, the total amount of gold in ocean water, represented by the large cube, is about 6×10⁹ kg, about 100

times more than all the gold in man's possession, which amounts to 6×10⁷kg, represented by the small cube. The Germans once tried to extract oceanic gold to pay war debts, but it was too expensive.

4 The formation of saline rocks, called evaporites, takes place in dry, arid regions. Seawater flows into the gulfs where it tends to evaporate, concentrating the dissolved salts and making the water very saline. Finally, the salts are precipitated out when the solution becomes too concentrated to hold them. When subsidence occurs at the same time as the precipitation of the salts very thick deposits are formed. The sequence [A–E] is explained in illustration 5.

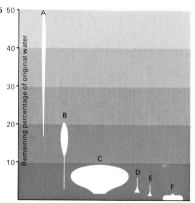

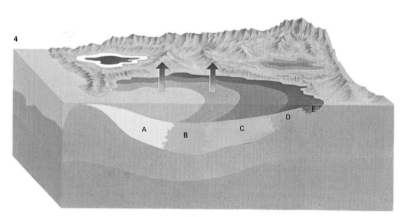

5 When half a seawater sample has evaporated, calcium and magnesium carbonates [A] precipitate out. They are completely removed from the brine when it is 15% of its original volume. At about 20% calcium sulphate [B] starts precipitating

followed by sodium chloride or common salt [C], other sulphates [D], rare salts of magnesium, potassium, sodium and borates and fluorides [E]. Finally magnesium and potassium chloride precipitate [F]. White areas show amounts of salts.

Shrinking Arctic ice sheets 'now the new normal'

By Louis Ashworth

Global warming means dramatically shrinking Arctic ice sheets are now the "new normal", say scientists who warn that this year's ice coverage is the smallest ever recorded.

The changes have dramatic effects on both the polar region and the planet as a whole, according to the experts at the US National Oceanic and Atmospheric Administration (NOAA). Its annual report showed warmer air temperatures, declining levels of sea ice and above-average sea temperatures.

The figures from this year showed slightly less warming in many areas than 2016, which was a record hot year. But ice coverage was even lower.

"What happens in the Arctic doesn't stay in the Arctic," said Timothy Gallaudet, the NOAA's acting administrator, "it affects the rest of the planet."

Looking at longer-term trends, researchers found the region is warming at double the speed of the rest of the planet and is growing warmer at a rate unprecedented in modern times.

The impact of environmental change in the world's northernmost region was brought into stark relief last week after a video showing an emaciated polar bear struggling to walk went viral.

Professor Chris Stokes from Durham University, who studies Arctic glaciers, said the report's findings were expected, but deeply worrying.

"This latest report will come as no surprise to scientists who have been observing changes in the Arctic over the last few decades," he said.

Back in 1985, 45 per cent of Arctic sea ice was over a year old. Today, this proportion is just 21 per cent – it has more than halved in just over 30 years.

According to NOAA, "sea ice more than four years old has nearly disappeared". Professor Stokes added: "If anyone ever asks me about the impact of climate change I always point them to the Arctic – the changes there are so rapid and so obvious that we ignore them at our peril."

{i} The melting of ice sheets can **exacerbate global warming.** When the reflective sea ice melts, it is replaced by a much darker ocean surface that absorbs and stores more heat.

On thin ice: A polar bear in the Canadian Arctic Archipelago

DAVID GOLDMAN/AP

14.12.2017

Christmas can be the hardest time of all. There's often nothing to enjoy, no one to share it with. Where Christmas should be, there's just

...ace for someone who's homeless

...arity for homeless people. We're ...melessness and changing lives, and for ...f we can welcome homeless people with ...l and good company, it can be their first ...melessness for good.

...uests as possible to Crisis at Christmas, ...d we need it urgently.

...e most vulnerable the situation is ...s and evictions, and a lack of affordable ...y people isolated and in danger. ...t to be ready to welcome as many guests ...why your gift is so important.

...ng a place at Crisis at Christmas for ...ss today. One place costs just £26.08; ...£260.80 would give ten people a great ...ards leaving homelessness for good.

...an for Crisis at Christmas 2017...

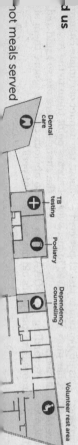

...d us

...ot meals served

Dental care

TB testing

Podiatry

Dependency counselling

Volunteer rest area

Please reserve a place for someone who's homeless at Crisis at Christmas today.

Dave's story is 100 per cent genuine, but his name has been changed and a model photographed to protect his identity. Photo: Ali Tollervey.

A whole new life for Dave

After a troubled childhood, Dave was determined to make his way in life – but when the jobs ran out, he found himself without a home.

Life on the streets was filled with danger. "Some people kick you for the fun of it," Dave says. "You're sleeping on a bench, some people have said I hope he dies."

At Crisis at Christmas, Dave found safety, good food and company. As well as having the chance to shower and get fresh clothes and walking boots, Dave was able to talk to Crisis about his longer-term problems and what to do next.

"Crisis are trying to find me a place," Dave says, "and I'm looking at work..."

What better present could you give someone this Christmas?

Reserve a place for someone who's homeless – for just £26.08 – and you'll be providing more than just a hot meal. You'll also provide a hot shower, clean clothes, a health check, plus access to Crisis' year-round services for training and support to leave homelessness in the year ahead.

Crisis at Christmas must open on 22 December.

melting ice reduce salinity, for example, whereas it is increased in areas with little rainfall and high evaporation. The Baltic Sea, which receives large quantities of fresh water from rivers and melting snow, has a low salinity of 7.2°/oo. The highest salinity of any ocean is in the Red Sea where it reaches as much as 41°/oo.

To produce fresh water from seawater the dissolved salts must be separated out. This desalination can be carried out by electrical, chemical and change of phase processes. Change of phase processes involve changing the water into steam and distilling it, or changing it into ice, a process that also expels the salt. Eskimos have used sea ice as a source of fresh water for hundreds of years [8] and primitive coastal tribes still take salt from the sea by damming water in pools and letting it gradually evaporate in the sun.

Density, light and sound
The density of seawater is an important factor in causing ocean currents and is related to the interaction of salinity and temperature [7]. The temperature of surface water varies

between −2°C and 29°C (28°F and 85°F). Ice will begin to form if the temperature drops below −2°C (28°F).

The properties of light passing through seawater determine the colour of the oceans. Radiation at the red or longwave end of the visible spectrum is absorbed near the surface of the water [9] while the shorter wavelengths (blue) are scattered, giving the sea its blue colour. The depth to which light can penetrate is important to marine life. In clear water light may penetrate to 110m (360ft) whereas in muddy coastal waters it may penetrate to only 15m (50ft).

Water is a good conductor of sound, which travels at about 1,507m (4,954ft) per second through seawater, compared with 331m (1,087ft) per second through air. Echo-sounding is based on the measurement of the time taken for sound to travel from a ship to the sea floor and back again. However, temperature and pressure both affect the speed of sound, causing the speed to vary by about 100m (328ft) per second and creating phenomena such as sound "shadow" zones and slow velocity zones [10].

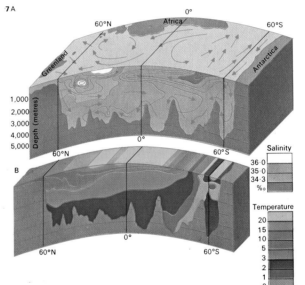

Pacific 165,063,000km²
Atlantic 84,133,000km²
Indian 65,522,000km²
Antarctic 32,248,000km²
Arctic 14,090,000 km²
Land 148,900,000km²

The oceans cover about 70 per cent of the earth's surface. No other planet in the Solar System has as much water. The five oceans are connected together and can be thought of as one large oceanic mass. Oceanography is the study of this great area of water.

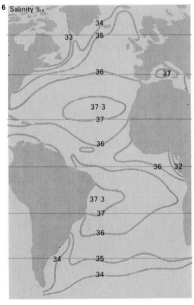

6 In an isohaline map of the Atlantic Ocean the lines join the places of equal salinity. The range of salinity in most of the ocean is between 33 and 37 parts per thousand. The map shows that salinity in the tropics, where evaporation is considerable, is relatively high. In the almost enclosed Mediterranean Sea it is higher still. In the Arctic, however, the salinity is lowered by melting ice and rainfall. In Hudson Bay the salinity falls considerably below the normal ocean values because of the inflow of fresh water from rivers. In the tropics, large rivers, such as the Amazon, reduce the salinity locally.

7 The physical properties of seawater, including the salinity [A] and temperature [B], are relatively constant at depth compared with the variations produced by local conditions at the surface. The surface salinity varies greatly with inflows of fresh water and variations in the rate of evaporation but it remains quite constant at depth as mixing of water by deep currents is very slow. The same is true of the temperature, which remains fairly constant at depth despite the climatic variation at the surface. In the Atlantic Ocean, as illustrated, a larger body of cold water exists at the south than the north and this has some effect on the temperature range.

Salinity
36·0
35·0
34·3
‰

Temperature
20
15
10
5
3
2
1
0
°C

8 One of the ways of removing salt from seawater is the direct freezing method. This happens naturally when the temperature of seawater with a salinity of 35 parts per 1,000 falls below −2°C (29°F). This freezing results in the formation of surface ice that contains little if any salt. Eskimos and other peoples living in polar climates have long used sea ice as a source of fresh water. Such freezing leads to an increase in the salinity of the water beneath the ice.

9 Seawater reduces the intensity of sunlight (attenuates it) selectively according to its wavelength. Attenuation is minimum for blue and maximum for red and infra-red and is caused by absorption and the scattering of light in all directions. The blue wavelengths, being less absorbed, are more scattered. As a result clear seawater looks blue. Impurities in seawater, such as organic life and silt, especially around coastlines, greatly increase the attenuation. The diagram shows the attenuation in pure seawater and depicts the attenuation for different light wavelengths, the bottom of each band being the point at which only one per cent of the light intensity at the surface remains.

10 The velocity of sound in seawater is more than 4.5 times as great as it is in air but it can vary with pressure, salinity and temperature. Sound waves passing through water in which these vary are refracted, or bent like light passing through a lens. Such refraction can take place at a thermocline – the boundary between warm surface water and cold water at depth. Submarines and other naval vessels can make use of this effect to hide the sound of their passage from an enemy.

Ocean currents

No part of the ocean is completely still, although, in the ocean depths, the movement of water is often extremely slow. Exploration of the deeper parts of the oceans has revealed the existence of marine life. If the water were not in motion, the oxygen – upon which life depends – would soon be used up and not replaced. Life would therefore be impossible. The discovery that all ocean water moves is of great significance. It was once thought that dangerous radioactive wastes could be dumped in sealed containers in the ocean depths. If the containers were to corrode, the radioactive substances would be released into the water and gradually circulate around the globe, poisoning marine life.

Causes of ocean currents
Surface currents in the oceans have been recorded since ancient times and were used by early navigators. In 1947, Thor Heyerdahl sailed on his raft, the *Kon-Tiki*, from Peru to the Tuamotu Islands east of Tahiti in 101 days. The journey, of nearly 7,000km (4,300 miles), was powered partly by the wind, but the raft was mainly carried by the Peru Cur-

rent and the South Equatorial Current.
Prevailing winds sweep surface water along to form drift currents. These surface currents do not conform precisely with the direction of the prevailing wind because of the Coriolis effect [Key] caused by the rotation of the earth. This effect, which increases away from the Equator, makes currents in the Northern Hemisphere veer to the right of the wind direction and currents in the Southern Hemisphere veer to the left. The result is a general clockwise circulation of water in the Northern Hemisphere and an anticlockwise circulation in the Southern Hemisphere [5].

Other factors affecting currents are the configuration of the ocean bed and the shapes of land masses. For example, in the Atlantic Ocean, the North Equatorial Current flows towards the West Indies. Most of this current is channelled into the Gulf of Mexico where it veers northeastwards, bursting into the Atlantic between Florida and Cuba as the Gulf Stream. (The term "stream" is used for currents with fairly clear boundaries.) This current, properly known as

the North Atlantic Drift once it leaves the American coast, then flows at four to five knots in a northeasterly direction. However, even this marked current is confined to waters near the surface. At a depth of about 350m (1,150ft), its effect is hardly noticeable. In the late 1950s, another large current was discovered flowing under the Gulf Stream in the opposite direction.

Variations in density
The causes of currents that are not powered by winds are related to the density of ocean water, which varies according to temperature and salinity. Heating at the Equator causes the water to become less dense. Cooling round the poles has the opposite effect. Salinity is affected by the inflow of fresh water from rivers, melting ice and rainfall, and by evaporation. For example, a high rate of evaporation in the Mediterranean Sea increases the salinity and therefore the density of the water [2]. As a result, currents of less dense (less salt) water flow into the Mediterranean from the Atlantic and the Black Sea. Smaller counter-currents of a

CONNECTIONS

See also
90 Oceanographic exploration
80 Waves and tides
76 The sea and seawater

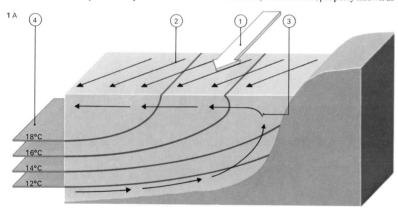

1 Upwelling [A] occurs when a longshore wind [1] pushes surface water away from a coast at an angle [2], allowing sub-surface water to rise [3]. This slow motion can best be seen as temperature gradients [4] as the deeper water is colder. Sub-surface water often contains many nutrients and so areas where upwelling occurs are often exceptionally rich fishing grounds, such as off the west coast of South America [B].

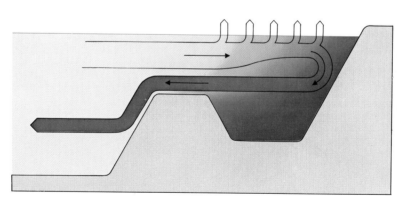

2 The water of the Mediterranean is constantly evaporating from its surface causing its salinity to increase. A current of normal salinity flows in and the excess salt is carried out by a deep-water current.

3 In the Baltic, a surface current of low salinity flows outwards. The overall salinity of the Baltic is maintained by a small under-current bringing in as much salt as the amount that is carried away by the outflow.

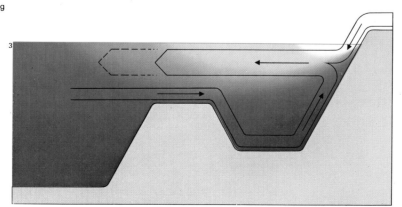

4 Current meters are the most accurate instruments used to measure and record the direction and rate of flow of ocean currents. The meters are sometimes attached to buoys or they may be anchored to the sea-bed and left to measure the current over a period of time. They usually contain some kind of propeller that is turned by the moving water and a vane connected to a compass, which orients the meter so that it always faces the current.

higher density or salinity flow outwards beneath these currents so that the salt content of the basin remains constant.

One of the simplest ways of measuring the speed and direction of surface currents is to record the movement of floating objects such as icebergs or wreckage. Ships record the flow of currents, sometimes by trailing a drift buoy and noting its movements. Current meters [4] of many kinds are also used.

Effects of ocean currents

One of the most important effects of ocean currents is that they mix ocean water and so affect directly the fertility of the sea. Mixing is especially important when sub-surface water is mixed with surface water. The upwelling [1] of sub-surface water may be caused by strong coastal winds that push the surface water outwards, allowing sub-surface water to rise up. Such upwelling occurs off the coasts of Peru, California and Mauritania. Sub-surface water rich in nutrients (notably phosphorus and silicon) rises to the surface, stimulating the growth of plankton which provides food for great

shoals of fish, such as Peruvian anchovies. The anchovies are adversely affected by another current: when the winds fail, about the end of December, disaster occurs in the form of a warm current called El Niño which flows into the area, killing the cold-water plants and animals.

Water has a high heat capacity and can retain heat two and a half times as readily as land. The heat of the sun absorbed by water around the Equator is transported north and south by currents. Part of the North Atlantic Drift flows past Norway warming offshore winds and giving northwest Europe a winter temperature that is 11C° (20F°) above the average for those latitudes [6]. The northward-flowing Peru and Benguela currents have a reverse effect, bringing cooler weather to the western coasts of South America and southern Africa. In such ways, currents have a profound effect on climate. Currents from polar regions can also create hazards for shipping. The Labrador and East Greenland currents carry icebergs and pack ice into shipping lanes and fog often occurs where cold and warm currents meet.

Surface currents are caused largely by prevailing winds. The Coriolis effect results in the deflection of currents to the right of the wind direction in the Northern Hemisphere.

In the same manner, the surface motion drives the sub-surface layer at an angle to it, and so on. Each layer moves at a slower speed than the one above it and at a greater

angle from the wind. The spiral created has the overall effect of moving the water mass above the depth of frictional resistance at an angle of about 90° from the wind direction.

Warm currents →

1 North Pacific
2 Alaska
3 Kuro Shio
4 Gulf Stream
5 North Equatorial
6 South Equatorial
7 Counter Equatorial
8 Brazil
9 Indian Counter Equatorial
10 Equatorial
11 East Australian

Cold currents →

12 California
13 Oya Shio
14 Canaries
15 Peru
16 Benguela
17 West Wind Drift
18 West Australian

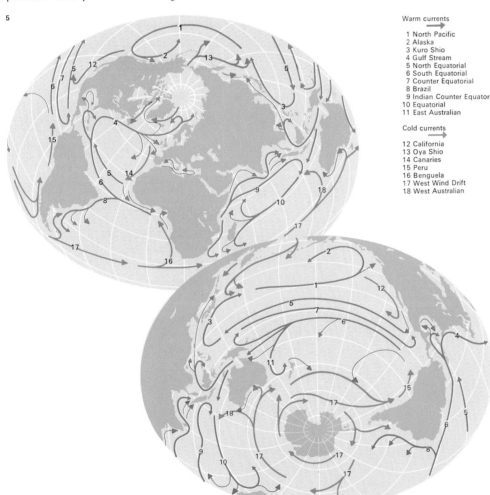

6 Climate is profoundly affected by ocean currents. Where winds blow off the warm sea rather than the cold land the North Atlantic Drift can bring mild weather to some European coasts. New York City [A] lies at a latitude only 160km (100 miles) north of Lisbon [B]. While New York has average January temperatures of −1°C (31°F), Lisbon has a sunny average of 10°C (50°F) at the same time.

5 The surface currents of the world circulate in a clockwise direction in the Northern Hemisphere and in an anticlockwise direction in the Southern Hemisphere. These circulatory systems are called gyres. There are two large clockwise gyres in the Northern Hemisphere (in the North Atlantic and in the North Pacific) and three anticlockwise gyres in the Southern Hemisphere (in the South Atlantic, the South Pacific and the Indian Ocean). Beneath the surface are undercurrents whose direction may be opposite to those at the surface. Beneath the northeastward flowing Gulf Stream off the eastern USA lies a large, cold current flowing south from the Arctic. The Gulf Stream finally splits in the North Atlantic, branching past eastern Greenland, northern Europe and southern Europe, while part of the current returns southwards to complete the gyre. Surface cold currents in the Northern Hemisphere generally flow southwards. In the Southern Hemisphere, cold water circulates around Antarctica, while offshoots flow northwards. The warm currents are very strong in tropical and sub-tropical regions. They include the Equatorial and Indian currents.

Waves and tides

Waves and tides are the most familiar features of oceans and seas. Sometimes, the energies of waves, tides and high winds combine with devastating effect. In January 1953, a high spring tide, storm waves and winds of 185km/h (115mph) combined to raise the level of the North Sea by 3m (10ft) higher than usual. This "surge" in the sea caused extensive flooding in eastern England, but in The Netherlands, 4.3 per cent of the entire country was inundated, about 30,000 houses were destroyed or damaged by the waters and 1,800 people died.

Waves and wave movements
Some wave motion occurs at great depth along the boundary of two opposing currents. But most waves are caused by the wind blowing over an open stretch of water. This area where the wind blows is known as the "fetch". Waves there are confused and irregular and are referred to as a "sea". As they propagate beyond the fetch they combine into more orderly waves forming a swell which travels for large distances beyond the fetch. Waves are movements of oscillation –

that is, the shape of the wave moves across the water, but the water particles rotate in a circular orbit with hardly any lateral movement [Key]. As a result, if there is no wind or current, a corked bottle bobs up and down in the waves, but is more or less stationary.

Waves have two basic dimensions [1]. Wave height is the vertical distance between the crest and the trough. Wave length is the distance between two crests. At sea, waves seldom exceed 12m (39ft) in height, although one 34m (112ft) high was observed in the Pacific [3] in 1933. Such a wave requires a long fetch measuring thousands of miles and high-speed winds. Wave motion continues for some way beneath the surface, but the rotating orbits diminish and become negligible at a depth of about half the the wave length; this is known as the wave base.

Waves that break along a seashore may have been generated by storms in mid-ocean or by local winds. As waves approach shallow water [1], which is defined as a depth of half a wave length, their character changes. As a wave "feels" the bottom, it gradually slows down and the crests tend to crowd together.

When the water in front of a wave is insufficient to fill the wave form, the rotating orbit, and hence the wave, breaks. There are two main kinds of breakers. Spilling breakers occur on gently sloping beaches, when the crests spill over to form a mass of surf. Plunging breakers occur on steeper slopes.

Tsunamis – tidal waves
Tsunamis [4] are sometimes called tidal waves but they have no connection with tides. Tsunamis are caused mainly by earthquakes, but also by submarine landslides and volcanic eruptions. At sea, the height of the wave is seldom more than 60–90cm (2–3ft). But the wave length may be hundreds of miles long and tsunamis travel at hundreds of kilometres per hour because of their long wave length. For example, an earthquake in the Aleutian Trench in the far north of the Pacific in 1946 triggered off a tsunami which devastated Honolulu. The tsunami took 4 hours 34 minutes to reach Honolulu, a distance of more than 3,220km (2,000 miles) – a speed of about 700 km/h (438mph). Waves more than 15m (50ft) high struck Honolulu,

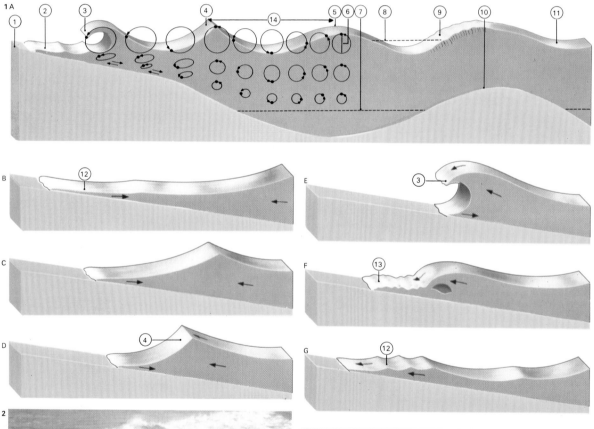

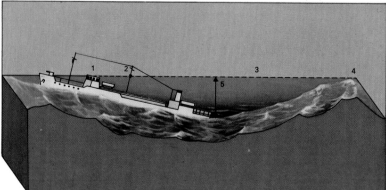

1 **Waves have length and height** [A]. The wave length [14] is the distance between one crest [5] and another, in this case, a peaking wave [4]. Between crests is a trough [11]. The wave height [6] is the distance between the crest and the trough. If wave action ceased, the water would settle at the "still water level" [8]. Wave action extends to the wave base [7]. Wave distortion is caused by frictional drag on the bottom. If waves pass over a sand bar [10], a spilling breaker [9] may form. Sometimes, waves in shallow water move the whole body of the water forward in translation waves [2] towards the shore [1]. In the development of a breaking wave, B shows backwash [12]. C shows the advance of the next wave which peaks [4] in D and then becomes a plunging breaker [3] in E. F and G shows swash [13] rushing up the beach after the wave has broken. Backwash then begins the cycle again.

2 **Surfing** is a popular sport in many countries, such as Australia. Surf forms as the crest of a wave breaks over.

3 **The tallest recorded wave** in open sea was 34m (112ft) high. An officer on the USS *Ramapo*'s bridge [1] in the Pacific in 1933 saw the crest of a wave [4] in line with the horizon [3] and the crow's nest [2], enabling him to work out the wave height [5].

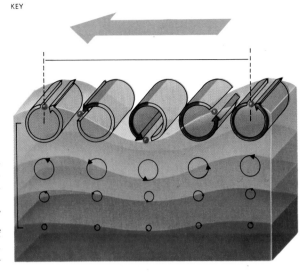

causing $25 million damage and killing 173 people. Although the height of the crest is low at sea, tsunamis have immense energy, which, as they lose speed in shallower water, is converted into an increase in height. The waves, on reaching the shore, may be 38m (125ft) or more high.

Most destructive tsunamis occur in the Pacific [5], but they have been recorded in the Atlantic. A tsunami battered Lisbon shortly after the earthquake of 1755. It was later felt in the West Indies in the form of a destructive wave 4–6m (13–20ft) high. Other tidal waves can be due to the surge of water when the barometric pressure is exceptionally low, such as in a hurricane.

Tides and their causes

Tides are alternate rises and falls of the sea's surface, caused chiefly by the gravitational pull of the moon and the sun [6]. The tidal effect of the sun is only 46.6 per cent that of the moon. Tides are also affected by the shapes of ocean basins and land masses. The moon's gravitational pull makes the waters of the earth bulge outwards when the moon is

overhead at any meridian. Another bulge occurs on the opposite side of the earth at the same time. Because the moon orbits the earth once every 24 hours 50 minutes, it causes two high tides and two low tides in that period.

Spring tides [7] occur when the earth, moon and sun are in a straight line. The combined gravitational attraction makes high tides higher and low tides lower, giving a high tidal range. Neap tides, which have the lowest tidal range, occur when the sun, earth and moon form a right-angle.

In the open sea, the tidal range is no more than a few feet and in enclosed basins, such as the Mediterranean, it is little more than 30cm (12in). However, in shallow seas, it may be more than 6m (20ft) and in tidal estuaries 12–15m (40–50ft). The highest tidal range recorded is about 16m (53ft) in the Bay of Fundy in eastern Canada. In some estuaries, including those of Hangchow Bay in China and the Severn in England, tidal bores occur. Bores are bodies of water with a wall-like front that surge up rivers. They form because estuaries act as funnels, leading to a rise in the height of the water as it flows upstream.

Most waves are generated by the wind. As a wave travels in deep water, however, the water particles do not move up and down but rotate in circular orbits. As depth increases, the rotations of the water particles diminish rapidly. This is why submarines escape the effects of severe storms.

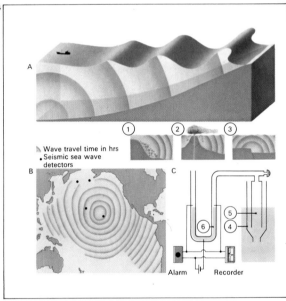

4

A

↘ Wave travel time in hrs
• Seismic sea wave detectors

B

C

Alarm   Recorder

4 Tsunamis [A], caused by landslides [1], volcanoes [2] or earthquakes [3], reach great heights near land. Pacific warning stations [B] use detectors [C]. A container [4], half in seawater, has air in a tube [5]. When a wave increases the pressure, mercury is forced around [6], closing an electrical circuit and setting off an alarm.

5 Tremendous damage is caused by tsunamis. They occur mainly in the Pacific Ocean.

5

6

B

→ Average gravitational pull
→ Actual gravitational pull
→ Tide generating force

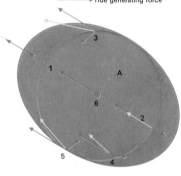

6 Water on the earth [A] is attracted to the moon [B] on the near side but pushed away on the far side. This causes tidal bulges at the nearest [1] and farthest [2] points, tidal flows at [3] and [4] and low tides at median points [5]. The force at work [red arrows] is the difference between the moon's actual gravitational pull and its average pull at the earth's centre [6] where it is exactly balanced by centrifugal force.

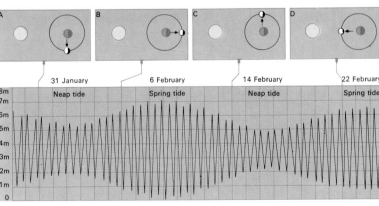

7 A   B   C   D

31 January   6 February   14 February   22 February

Neap tide   Spring tide   Neap tide   Spring tide

8m
7m
6m
5m
4m
3m
2m
1m
0

8
A

7 The height of tides varies according to the positions of the sun and moon in relation to the Earth. When the moon is in its first quarter [A], and again when it is in its last quarter [C], the moon, earth and sun form a right-angle. The gravita-tional forces are therefore opposed, causing only a small difference between high and low tide. Such tides are called neap tides. However, the moon, earth and sun form a straight line at full moon [B] and at new moon [D]. The high tides then become higher and the low tides lower. These are called spring tides. Because of friction and inertia, both spring and neap tides come about two days after the moon's phases. The graph shows the tidal range over the period of a month.

8 At low tide [A] the sea recedes from Mont St Michel, off the northern coast of Britanny, making it part of the mainland. At high tide [B], the sea surrounds it and it becomes an island. A similar feature in Cornwall, England, is St Michael's Mt.

The sea-bed

The floor of the deep ocean has always fascinated man. Pluto's legendary lost continent of Atlantis, "beyond the Pillars of Hercules", has never lost its hold on the imagination although there is no geological evidence to support the belief that it lay south of the Azores, on the mid-Atlantic Ridge.

Early knowledge of the sea-bed was restricted to depth soundings, taken by lead and line, of the areas around the known islands. Magellan tried – and, of course, failed – to reach the bottom of the Pacific with 370 m (1,200 ft) of rope. The first true oceanic sounding was made by James Clark Ross in 1840, when he measured a depth of nearly 3,700m (12,140ft) with a line.

Probing the sea-floor

The epic voyage of HMS *Challenger* which led to the first true oceanic depth survey was made between 1872 and 1876. The *Challenger* expedition used soundings weights with tube-like cups to obtain a sample of the material forming the sea-floor. Thus, when Jules Verne (1828–1905) wrote *Twenty Thousand Leagues under the Sea* (1870),

man first developed a systematic knowledge of the deep sediments. Their classification, which was developed by the *Challenger* expedition's geologist John Murray (1841–1914), and others who studied the samples after the ship's return, has been improved but never completely discounted.

Even so, the widely prevalent idea that the ocean floor consisted of a sandy waste extending for thousands of miles, dotted with a few islands and occupied by exotic fish, was gradually discarded. The early samples of sediment obtained by *Challenger*, and the soundings obtained on the mid-Atlantic which actually mapped the mid-Atlantic Ridge, did not lead to a full understanding of the variation in the extent and thickness of the sediments. They did not supply any scientific reasons for this variation nor was it appreciated that mid-ocean ridges ran through all the world's major oceans.

During this century, improved coring devices greatly enlarged the collective knowledge of oceanic sediments.

The actual topography of the deep ocean has been revealed by echo-sounding using

sonic or ultra-sonic signals. Scientists can calculate the depth of the water by noting how much time passed between sending the signals and receiving the echo. Since the 1940s seismic methods have also been used; they have shown that the ocean floor is made up of hills, volcanic mountains, island complexes – of which the islands are only the visible tips – and of huge and complex submarine mountain chains, with median rift valleys, faults and numerous flanking ridges.

The continental shelf

If one walks down a pebbly beach, the pebbles usually give way to sand, which continues out to sea. This is the continental shelf [1, 2] which may be covered by relatively coarse sediments or muds and silts. It is inhabited by seaweeds of many kinds, as well as numerous animals: corals, sea anemones and other coelenterates, many species of burrowing worms and minute colonial rock-encrusting animals (Bryozoa), and clams, mussels, oysters, scallops and other molluscs. There are sea-urchins, starfish, brittle stars, sea-cucumbers and sponges. Bottom-

CONNECTIONS

See also
90 Oceanographic exploration
84 The Atlantic Ocean
86 The Pacific Ocean
88 The Indian Ocean and the polar seas
24 Global tectonics
92 Man under the sea
134 Mineral resources of the sea
122 Coastlines

In other volumes
120 The Natural World

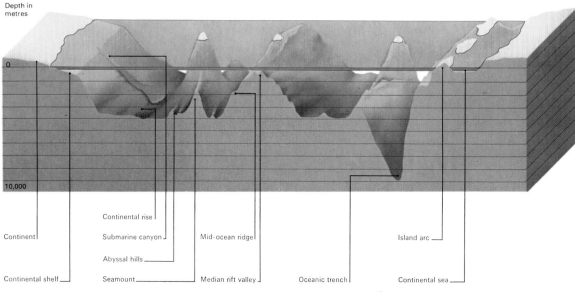

1 **The sea-bed** consists of different zones, the shallowest of which is the continental shelf that lies between the coast and the 200m (656ft) depth contour. The shelf area occupies 7.5% of the sea-floor and corresponds to the submerged portion of the continental crust. Beyond, the downward slope increases abruptly to form the continental slope, which occupies some 8.5% of the sea-floor. This area may be dissected by submarine canyons. The continental slope meets the abyssal basins at a more gentle slope called the continental rise. The basins lie at depths of 4,000m (13,200ft) and show many mountain ranges and hills.

Depth in metres

Continent
Continental shelf
Continental rise
Submarine canyon
Abyssal hills
Seamount
Mid-ocean ridge
Median rift valley
Oceanic trench
Island arc
Continental sea

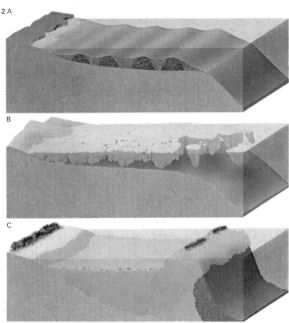

2 **Continental shelves** are the regions immediately off the land masses. There are several different sorts. Off Europe and North America the shelf [A] has a gentle relief, often with sandy ridges and barriers. In high latitudes, floating ice wears the shelf smooth [B] and in clear tropical seas a smooth shelf may be rimmed with a coral barrier like the Great Barrier Reef off eastern Australia [C], leaving an inner lagoonal area "dammed" by the reef.

3 **Submarine canyons** like the 1.5km (5,000ft) deep gorge off Monterey, California [B], are found on the continental slopes [A]. They can be caused by river erosion before the

land was submerged by the sea or by turbidity currents. Mud and sediment-laden water often pour out of major estuaries scouring gorges out of slope rock and sediment. These canyons compare with the Grand Canyon [C].

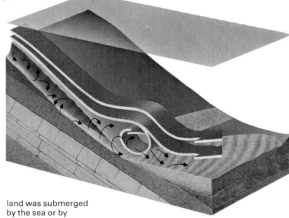

dwelling fish such as plaice and flounder also live on the continental shelf.

This is the region of sand banks and sand waves (underwater dunes). In the North Sea, sand waves are formed as masses of loose sand move around like sand dunes in the desert, propelled by currents. Here, too, vast oil and gas reservoirs, sometimes associated, as in the Gulf of Mexico, with salt domes, are found in the continental rocks deep beneath the surface sand.

At the edge of the continental shelf, at about 200m (656ft) depth, the sea-floor begins to dip markedly down: this is the upper boundary of the continental slope. It is dissected in places by submarine canyons [1, 3], in which underwater avalanches, known as turbidity currents, carry mud, pebbles and sand far out to sea and deposit them at the foot of the slope, on the continental rise [1], at depths of about 2,000m (6,560ft). Life is much more scarce on the continental slope and rise: large free-swimming molluscs – octopuses, cuttle fish and large squid – brittle stars, worms and strange fish are among the most common species.

The continental rise leads to the abyssal plains – vast empty basins occupied by few, even stranger fish, worms, brittle stars, and deep-sea free-swimming molluscs, with no plant life to speak of. From these plains rise huge mountain ranges [1], the mid-ocean ridges, from 4,000m (13,200ft) deep to some 1,000m (3,300ft) below the surface with occasional peaks reaching the surface as islands. Seamounts [5] also rise from the abyssal plains, sometimes part of island chains like the Hawaii-Emperor chain in the Pacific, but often isolated. They are nearly all volcanic and may be crowned with coral, formed when they were near the surface.

Sea-bed maps
On the six pages following, the floors of the five oceans are mapped. The projections used were chosen to maximize coverage of sea areas relative to land and the colours reflect those thought to exist on the sea-bed. Continental shelves are shown in the greyish-green of the terrigenous oozes, while the calcareous oozes of the deeper areas are shown in pale greys and buff.

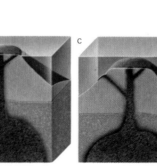

The existence of manganese nodules on the deep ocean floor was discovered by the *Challenger* expedition. These potato-sized and -shaped nodules form on the ocean bed by processes that are not fully understood, although a great deal of research is going on to find out how they grow. They consist of a rock nucleus surrounded by concentric layers of metal oxide and are of potential economic importance because they contain enough copper, cobalt and nickel, as well as manganese, to last the world for many thousands of years.

4 Basalt is the rock most commonly found on the sea-floor. It is a lava that forms the bulk of the seamounts and ocean ridges and it underlays the marine sediments in the abyssal plains. Submarine eruptions produce lumps of basalt "frozen" into pillow shapes (pillow lavas). The submarine basalt shown here is seen through a microscope. It shows small crystals, glassy patches and gas bubbles now filled with green mica.

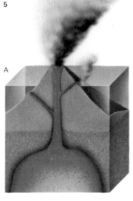

5 Seamounts are submarine mountains rising at least 1,000m (3,300ft) above their surroundings; they are nearly always volcanoes. Some seamounts called tablemounts or guyots have flat tops at depths down to 2,500m (1.5 miles). These tops are often too large to be explained as ancient craters filled to the rim by sediments. Thus it was proposed that guyots were volcanoes above sea-level [A] which after extinction were worn flat by waves [B] and which then sunk as the sea-level rose or as the sea-bed subsided [C]. The has been confirmed by the composition of beach pebbles.

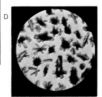

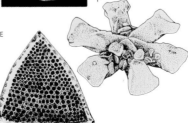

7 Constituents of marine sediments include micro-organisms and new (authigenic) minerals formed on the sea-bed, as well as clays. Radiolarians, such as *Calocycletta virginis* [A] are single-celled animals with siliceous skeletons. Foraminifera such as *Globigerina nepenthes* [B] and *Globigerinoides ruber* [C] are single-celled animals with calcareous shells; together with radiolarians they can be used for dating the sediments. Siliceous diatoms [E] and the tiny calcareous plates from flagellates known as coccoliths [F] are also common. Philipsite [D] is a typical authigenic mineral of the deep sea.

Terrigenous deposits
Red clay
Globigerina ooze
Pteropod ooze
Diatom ooze
Radiolarian ooze

6 Deep-sea sediments are related to surface water temperature, depth and distance from land. Terrigenous deposits consist of mineral particles derived from the weathering of land rocks. They are carried out to sea by rivers and winds and are found near the coasts. The deep sea floor is often blanketed by ooze which is formed by the endless "snowfall" of the shells or skeletons of countless tiny planktonic animals and algae. Globigerina and radiolarian ooze are made from the remains of single-celled animals with calcareous and siliceous skeletons respectively. Occasionally the shells of pteropods, small swimming molluscs, form deposits of pteropod ooze. In cold seas, silica-shelled microscopic algae, the diatoms, thrive and form diatom ooze. In areas away from land and where planktonic life is scarce, atmospheric dust settles very slowly as abyssal red clays.

The Atlantic Ocean

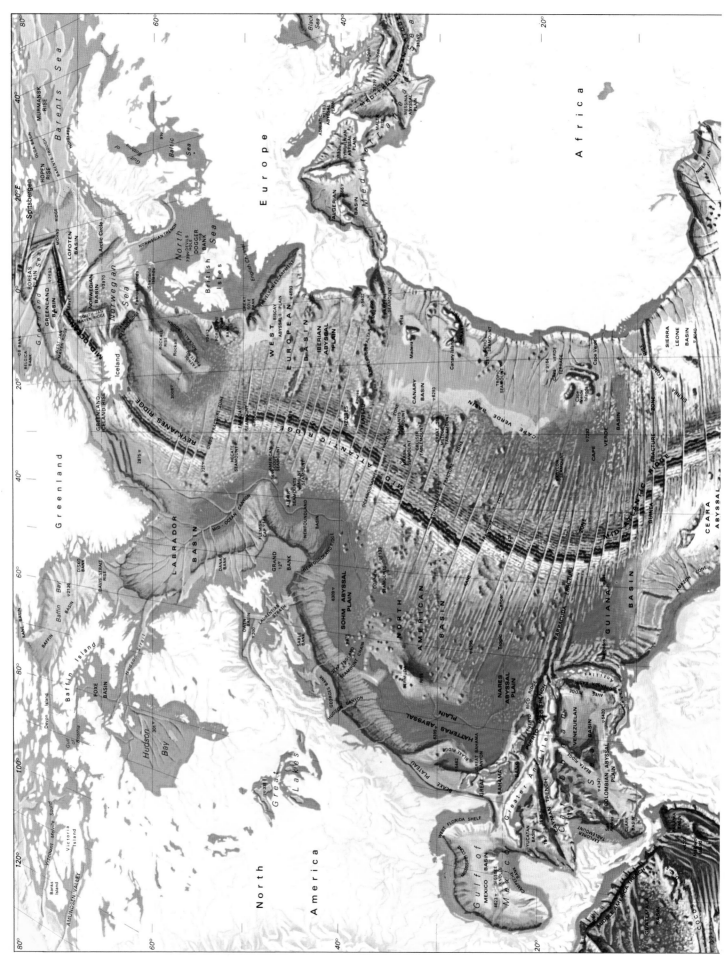

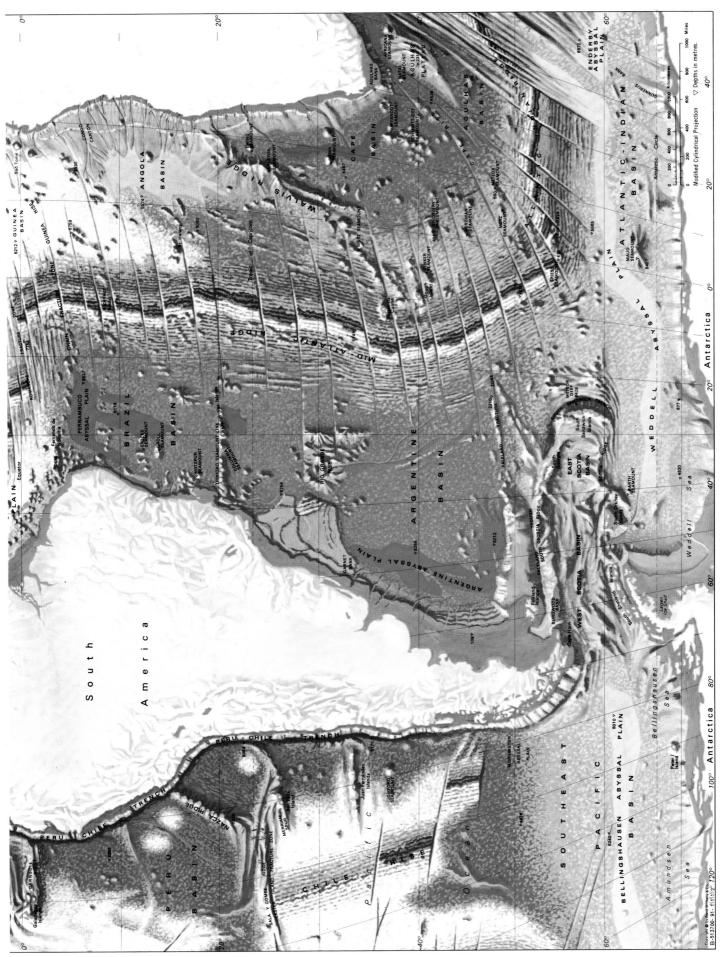

South America

South

America

PLAIN

Equator

São Tomé

0°

CONGO CANYON

ANGOLA BASIN

GUINEA BASIN

GUINEA RISE

St. Helena

WALVIS RIDGE

MID-ATLANTIC RIDGE

BRAZIL BASIN

PERNAMBUCO ABYSSAL PLAIN

CHAIN FRACTURE ZONE

ROMANCHE

FRACTURE ZONE

Fernando de Noronha

Rocas

Trinidade I.

ATLANTIC

CAPE BASIN

AGULHAS BANK

AGULHAS PLATEAU

Tristan da Cunha

Gough Island

Bouvet Island

CAPE AGULHAS BASIN

MID ATLANTIC RIDGE

ATLANTIC INDIAN BASIN

ATLANTIC - INDIAN ABYSSAL PLAIN

ENDERBY ABYSSAL PLAIN

Antarctic Circle

GUNNERUS BANK

40°

20°

0°

20°

40°

60°

ARGENTINE BASIN

RIO GRANDE RISE

ARGENTINE ABYSSAL PLAIN

Falkland Islands

FALKLAND PLATEAU

SOUTH GEORGIA RIDGE

FALKLAND TROUGH

BURDWOOD BANK

Cape Horn

South Georgia

SANDWICH TRENCH

South Sandwich Islands

SOUTH SANDWICH TRENCH

EAST SCOTIA BASIN

WEST SCOTIA BASIN

South Orkney Islands

South Shetland Islands

Larsen Ice Shelf

WEDDELL ABYSSAL PLAIN

Weddell Sea

Antarctica

PERU - CHILE TRENCH

PERU - CHILE TRENCH

NAZCA RIDGE

PERU BASIN

PERU CHILE TRENCH

Galapagos Islands

CARNEGIE RIDGE

SALA Y GOMEZ RIDGE

EASTER ISLAND FRACTURE ZONE

Juan Fernandez Islands

San Felix Island

MERRIAM SPUR

CHILE RISE

Pacific Ocean

MORNINGTON ABYSSAL PLAIN

GIFFORD SEAMOUNT

SOUTHEAST PACIFIC

Peter I Island

BELLINGSHAUSEN ABYSSAL PLAIN

Bellingshausen Sea

Amundsen Sea

Antarctica

SOUTHEAST PACIFIC BASIN

80°

100°

120°

40°

60°

Miles

Kilometers

Modified Cylindrical Projection

▽ Depths in metres.

85

The Pacific Ocean

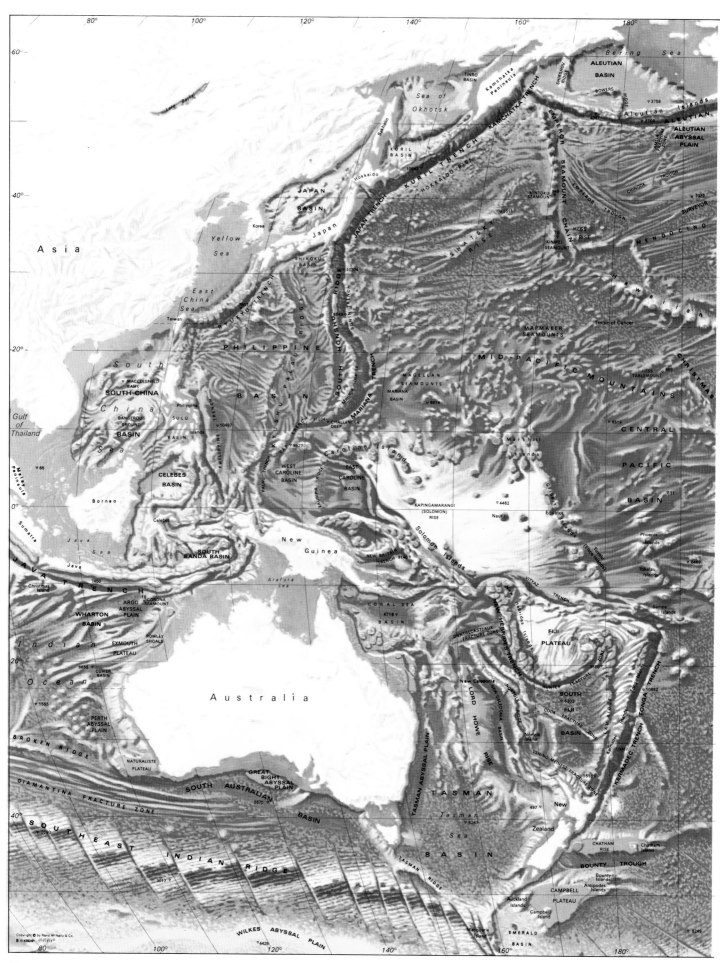

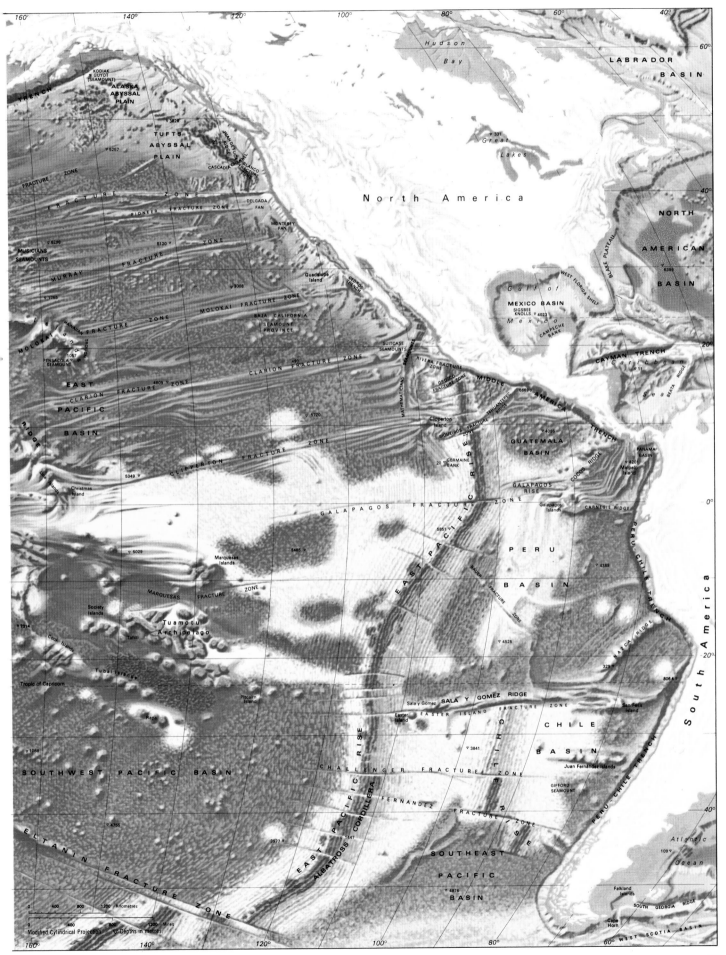

LABRADOR
BASIN

Hudson
Bay

KODIAK
GUYOT
(SEAMOUNT)
ALASKA
ABYSSAL
PLAIN

TRENCH

▽ 3826

TUFTS
ABYSSAL
PLAIN

▽ 5257

JUAN DE FUCA RIDGE

CASCADIA

BLANCO FRACTURE ZONE

FRACTURE ZONE

FRACTURE ZONE

PIONEER FRACTURE ZONE

DELGADA
FAN

MONTEREY
FAN

FRACTURE ZONE

▽ 6298

MUSICIANS
SEAMOUNTS

MURRAY

FRACTURE

▽ 1765

HAWAIIAN FRACTURE ZONE

MOLOKAI FRACTURE ZONE

ZONE

▽ 3008

Guadelupe
Island

PEDRO
TRENCH

331
Great
Lakes

North America

NORTH

AMERICAN

BASIN

▽ 6399

Gulf of
MEXICO BASIN
SIGSBEE
KNOLLS ▽ 4023
Mexico

WEST FLORIDA SHELF

BLAKE PLATEAU

BAJA CALIFORNIA
SEAMOUNT
PROVINCE

CAMPECHE
BANK

MOLOKAI ISLANDS

PENSACOLA
SEAMOUNT

HAWAIIAN TROUGH

1057

EAST

CLARION

PACIFIC

BASIN

RIDGE

Line Islands

Christmas
Island

▽ 5349

CLARION FRACTURE ZONE

4809 ▽

490

CLARION FRACTURE ZONE

▽ 5720

CLIPPERTON FRACTURE ZONE

SUITCASE
SEAMOUNTS

MATHEMATICIANS

RIVERA FRACTURE
ZONE

OROZCO
FRACTURE ZONE

Clipperton
Island

SIQUEIROS FRACTURE ZONE

TEHUANTEPEC RIDGE

20 ▽

GERMAINE
BANK

MIDDLE

AMERICA

TRENCH

6669

▽ 4086

GUATEMALA
BASIN

CAYMAN TRENCH
Caribbean

11

BEATA RIDGE

Sea

EAST PACIFIC RISE

COCOS RIDGE

GALAPAGOS
RISE

PANAMA
BASIN
▽ 3201
Malpelo
Island

GALAPAGOS FRACTURE ZONE

▽ 5029

Marquesas
Islands

▽ 5485

5851

PERU

BASIN

Galapagos
Islands

CARNEGIE RIDGE

▽ 4389

MARQUESAS FRACTURE ZONE

▽ 7314

Society
Islands

Cook Islands

Tahiti

Tuamotu
Archipelago

Tubai Islands

BAUER
FRACTURE
ZONE

▽ 4525

NAZCA RIDGE

PERU-CHILE TRENCH

South America

Tropic of Capricorn

Rapa

▽ 1068

Pitcairn
Island

SOUTHWEST PACIFIC BASIN

329 ▽

806 A ▽

Sala y Gomez

SALA Y GOMEZ RIDGE

Easter
Island

EASTER ISLAND FRACTURE ZONE

San Felix
Island

CHILE

BASIN

EAST PACIFIC RISE

▽ 3841

CHILE RISE

Juan Fernández Islands

▽ 4765

CHALLENGER FRACTURE ZONE

EAST PACIFIC RISE

ALBATROSS CORDILLERA

FERNANDEZ

FRACTURE ZONE

GIFFORD
SEAMOUNT

3977 ▽

▽ 1447

SOUTHEAST

PACIFIC

BASIN

▽ 4876

PERU-CHILE TRENCH

Atlantic

109 ▽

Ocean

Falkland
Islands

ELTANIN FRACTURE ZONE

0   400   800   1200 Kilometres

0   400   800   1200 Miles

Modified Cylindrical Projection   ▽ Depths in metres

SOUTH GEORGIA RIDGE

Cape
Horn

WEST SCOTIA BASIN

160°   140°   120°   100°   80°   60°

87

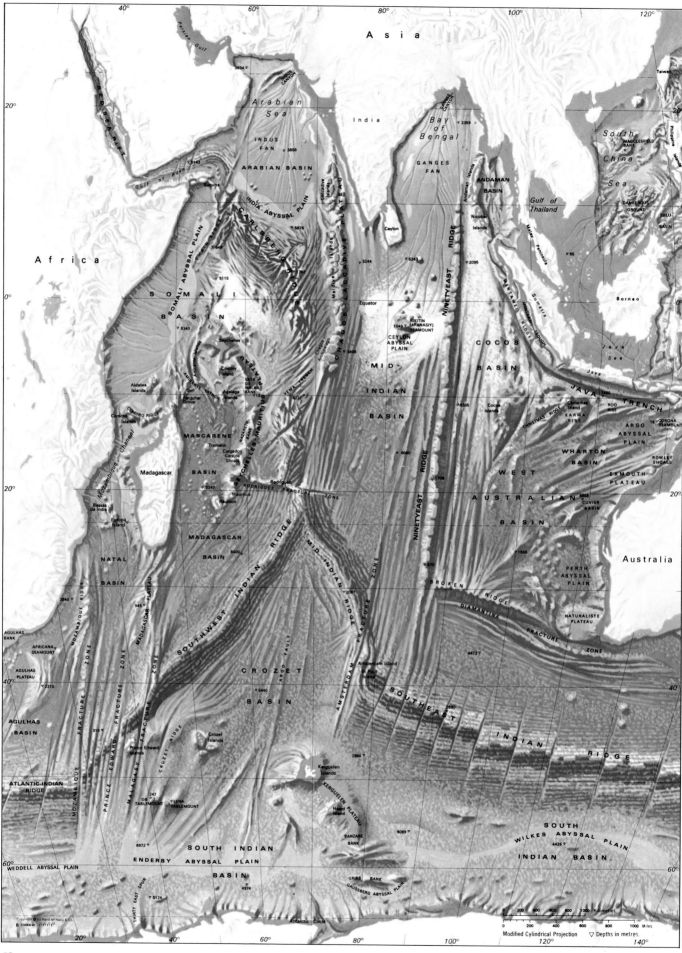

The Indian Ocean and the polar seas

A s i a

RED SEA

Arabian
Sea

India

Bay
of
Bengal

South
China
Sea

Taiwan

Philippine
Islands

Gulf of Aden

Gulf of
Thailand

Gulf of

INDUS
FAN

MACCLESFIELD
BANK

ARABIAN BASIN

Socotra

INDUS CANYON

GANGES
FAN

ANDAMAN
BASIN

DANGEROUS
GROUND

SULU
BASIN

INDIA ABYSSAL PLAIN

Andaman Islands

Nicobar
Islands

Africa

CARLSBERG RIDGE

Maldive Islands

Ceylon

Malay Peninsula

Sumatra

Borneo

SOMALI

BASIN

Seychelles

Equator

NIKITIN
(AFANASIY)
SEAMOUNT

CEYLON
ABYSSAL
PLAIN

MID-

INDIAN

BASIN

COCOS

BASIN

Java
Sea

Java

JAVA TRENCH

Cocos
Islands

Christmas
Island

ROO
RISE

KARMA
RISE

CORONA
SEAMOUNT

ARGO
ABYSSAL
PLAIN

Aldabra
Islands

COMORO RIDGE

Comoro
Islands

Coetivy
Island

Agalega
Islands

MASCARENE

Tromelin

Cargados
Carajos
Shoals

Rodriguez

SEYCHELLES MAURITIUS PLATEAU

VEMA TRENCH

WHARTON

BASIN

WEST

ROWLEY
SHOALS

EXMOUTH
PLATEAU

Madagascar

BASIN

Mauritius

Réunion

RODRIGUEZ FRACTURE ZONE

AUSTRALIAN

CUVIER
BASIN

PERTH
ABYSSAL
PLAIN

Australia

Bassas
da India

Europa
Island

MADAGASCAR

BASIN

MID-INDIAN RIDGE

SOUTHWEST INDIAN RIDGE

NINETYEAST RIDGE

BASIN

BROKEN

RIDGE

DIAMANTINA

NATURALISTE
PLATEAU

NATAL

BASIN

MOZAMBIQUE RIDGE

MADAGASCAR PLATEAU

MADAGASCAR FRACTURE ZONE

AMSTERDAM FRACTURE ZONE

FRACTURE ZONE

FRACTURE

ZONE

AGULHAS
BANK

AFRICANA
SEAMOUNT

CROZET

ARGO FAULT

Amsterdam Island

St Paul
Island

SOUTHEAST

AGULHAS
PLATEAU

BASIN

INDIAN

AGULHAS

BASIN

PRINCE EDWARD FRACTURE ZONE

MALAGASY FRACTURE ZONE

CROZET RIDGE

Crozet
Islands

Prince Edward
Islands

Kerguelen
Islands

KERGUELEN PLATEAU

RIDGE

ATLANTIC-INDIAN
RIDGE

MOZAMBIQUE

OB
TABLEMOUNT

LENA
TABLEMOUNT

Heard
Island

BANZARE
BANK

SOUTH
WILKES ABYSSAL PLAIN

INDIAN BASIN

WEDDELL ABYSSAL PLAIN

THIRTY EAST SPUR

SOUTH INDIAN

ENDERBY ABYSSAL PLAIN

BASIN

GRIBB BANK

GAUSSBERG ABYSSAL PLAIN

Antarctic Circle

Modified Cylindrical Projection

▽ Depths in metres.

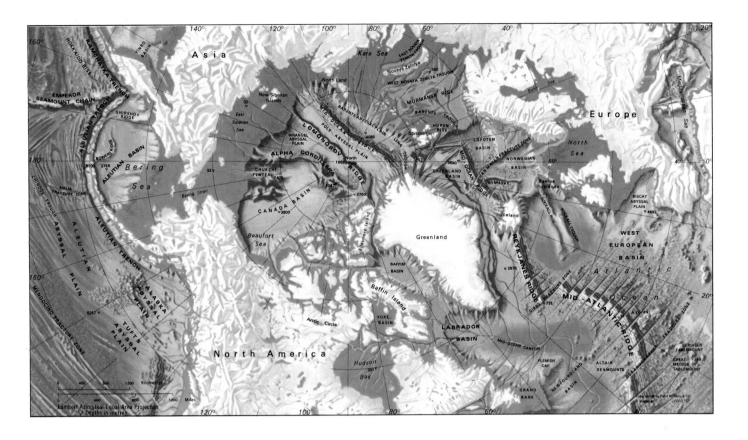

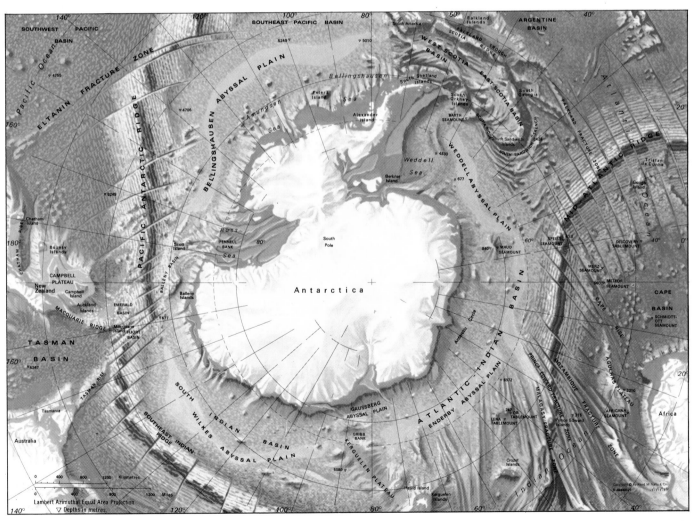

Oceanographic exploration

The topography of the ocean floor is hardly more familiar to man than that of the hidden side of the moon. Much remains to be discovered about the oceanic two-thirds of our planet and oceanographic research is not just a matter of scientific curiosity – man is only just beginning to appreciate and to exploit the vast untapped food, mineral and energy resources of the sea.

Early observations
The earliest scientific observations of the sea were recorded by Aristotle (384–322 BC) who described 180 marine animal species. Little further progress was made until the age of discovery during the fifteenth and sixteenth centuries which advanced the geographical knowledge of the seas and their currents. In 1670 the Irishman Robert Boyle (1627–91) published his *Observations and Experiments on the Saltiness of the Sea* in which he correctly deduced that the salt is derived from the weathering of the land. He introduced the silver nitrate test to measure the chloride content of seawater, a method which is still used today.

The Italian count Luigi Marsigli (1658–1730), a contemporary of Boyle, deserves to be called the first oceanographer because he studied the whole realm of the sea, from its flora and fauna to its currents. He invented the propeller current meter and discovered the deep counter-current in the Bosporus, correctly attributing it to differences in salinity between the Black Sea and the Mediterranean. (The Black Sea receives more water from rivers than it loses through evaporation, hence the main surface current flows towards the Mediterranean.)

In the eighteenth century the American Benjamin Franklin (1706–90) published a chart of the Gulf Stream which successfully shortened the passage time of the mail packets between North America and England. James Cook (1728–79), during his famous Pacific voyages from 1768 to 1779, finally exploded the geographical myth of a great southern continent in the South Pacific where he took soundings down to 1,243m (4,078ft) which hinted at even greater depths.

At the very beginning of the nineteenth century the German Alexander von Humboldt (1769–1859) described the cold current that flows up the Andean coast to the Galapagos Islands. In 1835 the English naturalist Charles Darwin (1809–82) visited these islands during his round-the-world voyage in HMS *Beagle*. Darwin made many sea-related observations, including some on plankton (the microscopic life forms of the sea). He also proposed a theory on the origins of coral reefs which after more than a century of heated debate was finally proved correct by the well-drilling on Eniwetok atoll in 1952. In the middle of the nineteenth century, Lieutenant Matthew Fontaine Maury of the US Navy (1806–73) published *Wind and Current Charts*, the first pilot charts, compiled from the data in ships' log books.

The *Challenger* expedition
The findings of the early scientists and the many unanswered questions they posed induced the British Government in 1872 to finance a scientific expedition to circumnavigate the world, for which the Royal Navy provided a ship – HMS *Challenger*.

Challenger, under the scientific direction

1 **The natural history laboratory** aboard HMS *Challenger* was fitted for the study of marine, bird and island life. The first true oceanographic ship, *Challenger* was a converted man-of-war and this laboratory was installed on her gun deck and lit by a gunport.

2 **Dredging** on the *Challenger* expedition was a slow, tedious operation and was nicknamed drudging by the crew who failed to share the enthusiasm of the scientists. The dredge was streamed out at the end of a warp and was made to sink by the attachment of a sliding weight.

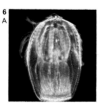

3 **Instruments used on the *Challenger*** included a current drag [A], which was lowered to a pre-determined depth, the drift being shown at the surface by a buoy. The Baillie sounding machine [B] measured depth and collected samples of sediment. The slip water bottle [C] was used for sampling the bottom water for in-board analysis.

4 **This dredge** was used on the *Challenger* for collecting biological specimens from the sea-floor.

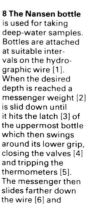

5 **The first world map of marine sediments** was provided by the *Challenger* expedition. Vast areas of the sea-floor were found to consist of dead shells of single-celled animals (mainly Foraminifera and radiolarians) and algae (diatoms). Foraminifera, such as *Globigerina digitata* [A], have calcareous shells whereas radiolarians, such as *Panartus tetrathalamus* [B], have siliceous shells. The *Challenger* discovered 3,508 new species of radiolarians.

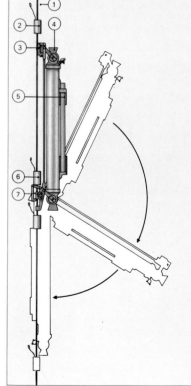

8 **The Nansen bottle** is used for taking deep-water samples. Bottles are attached at suitable intervals on the hydrographic wire [1]. When the desired depth is reached a messenger weight [2] is slid down until it hits the latch [3] of the uppermost bottle which then swings around its lower grip, closing the valves [4] and tripping the thermometers [5]. The messenger then slides farther down the wire [6] and trips the lever [7], releasing a new messenger to operate the next bottle down.

6 **The *Challenger* expedition** also studied the species and individuals forming plankton, such as the comb jelly [A], a ctenophore and the copepod [B], a crustacean.

7 **The plankton net** is a cone of fine muslin kept open by a hoop and weighted by a sinker on its bridle. The small end is tied to the neck of a jar in which the plankton is collected.

of Charles Wyville Thomson (1830–82), sailed 69,000 nautical miles between 1872 and 1876. Among other observations, ocean depths and surface currents were measured, water samples taken for analysis [3] and sea life was dredged or trawled [2, 4]. The ship was also fitted with laboratories [1]. The *Challenger* expedition outlined the broad features of the sea-floor and the nature of its sediments, and discovered some 4,417 new species of animals and plants. It also showed that life existed in the most extreme depths. Thus modern oceanography began.

Modern oceanography

Many expeditions were to follow using improved techniques and instruments. The Norwegian polar explorer Fridtjof Nansen (1861–1930) invented the deep-water sampling bottle [8] that bears his name. It can carry thermometers for determining the temperature at depth [9]. Plankton nets [7] were perfected and many ingenious instruments, such as the bathythermograph [10], were invented. A breakthrough came with the discovery of echo-sounding [13] just

before World War I, but it was only perfected for use at great depths after World War II.

A new era in marine geosciences was introduced in 1961 by the Mohole project on the US drilling ship *CUSS I* (from the initials of the participating oil companies). The aim to drill down to the earth's mantle was not achieved, but the accumulated experience was put to use in the American Deep Sea Drilling Project that started in 1968. This project confirmed the sea-floor spreading theory, which states that new sea-floor is generated along the mid-oceanic ridges.

The emphasis in oceanography has changed since the early days when the main object was to collect samples of sediment, water and marine life. While this is still part of the research, the aims are increasingly to find new food and physical resources (minerals and energy), to control pollution and to conserve the biological resources of the sea. Oceanography has become an advanced science involving satellite navigation, expensive ships [Key] with on-board computers and laboratories and backed up by sophisticated shore-based facilities.

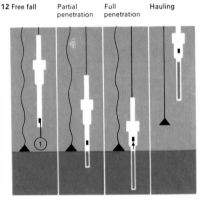

Modern research vessels such as the AGOR–class R/V *David Starr Jordan* of the Scripps Institution of Oceanography, USA, are designed for versatility. This ship is 80m (254 ft) long and carries four laboratories. She has two winches and two cranes for lowering scientific gear. A central well through the hull allows further use as a drilling platform. Propulsion is provided by two cycloid-al propellers, which give total manoeuvrability in all directions. The ship has an automatic satellite navigation system which provides a digital display of both the latitude and longitude.

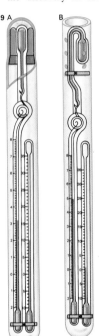

9 Deep-sea reversing thermometers (DSRT) are fixed on Nansen bottles. The protected thermometer [A] is insulated from pressure and records the temperature when the bottle is tripped. The unprotected thermometer [B] is similar but the hydrostatic pressure squeezes the reservoir so that the reading is a function of both temperature and depth. The small auxiliary thermometers indicate the temperature at the time of the reading and this must be introduced as a correction.

10 The bathythermograph simultaneously records temperature and depth down to 300m (1,000ft) but it lacks the precision and the depth range of the DSRT.

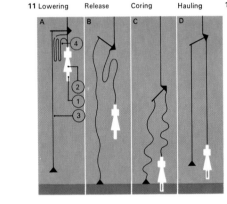

11 The gravity corer, which takes samples of the sea-floor sediments, consists of a metal tube [1] with a lead weight [2] and a tripping device [3]. The instrument is lowered in the sea [A] with the lower part of the wire coiled [4]. When the tripping device touches the bottom [B] the coil is released and the core barrel falls and punches into the sediments [C]. It is then brought back to the surface [D].

13 Echo-sounding is a method of measuring depth from the speed of sound in seawater – The lapse is timed between sending a sound signal and receiving its echo which has bounced off the bottom. In A both the sound receiver [1] and the transmitter [2] are on the ship, and the water depth is measured. In B a transmitter [3], weighted by a sinker [4] sends signals direct to a single receiver and bounces others off the bottom to calculate the depth.

12 The piston corer is an improvement on the gravity corer. Its penetration is increased by an internal piston [1] which sucks the corer deeper into the sediment. Piston cores can exceed 20m (60ft) in length whereas gravity cores seldom exceed 2m (6ft).

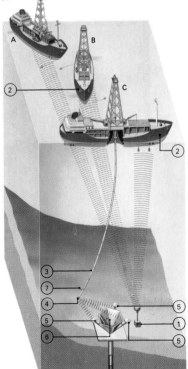

14 The American deep-sea drilling vessel *Glomar Challenger* maintains its station by dynamic positioning; the drift of the ship is computed relative to a sonar beacon [1] and is automatically corrected [A, B, C] by side thrusters [2] and the main propeller. After a worn drill bit at the end of the drill string [3] is renewed, the string is guided back to the hole by fitting the core barrel with a sonar device [4], which determines its position relative to three sonar reflectors [5] placed around the re-entry funnel [6]. The drill is then guided into the funnel by a sideways jet [7].

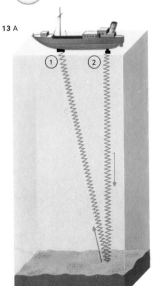

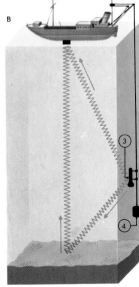

Man under the sea

For centuries man has striven to conquer the world beneath the sea. Even in the fourth century BC Alexander the Great (356–323 BC) was lowered into the sea in a large glass barrel and he used divers in military operations such as the siege of Tyre (334 BC).

Early diving apparatus
The earliest reliable diving bells, which were open at the base and supplied with air by a hose from the surface, date from the sixteenth century. In 1663–4 one such bell was used to recover 53 cannons from the Swedish galleon *Vasa*. The first practical bell holding more than one diver was built by Edmond Halley (1656–1742) [1] in 1690, and bells are still used for harbour construction and salvage. The familiar "hard hat" diving suit introduced by Augustus Siebe in 1837 is still used extensively in underwater engineering down to 60m (200ft) but it is the aqualung [3], developed by Jacques Yves Cousteau and Emile Gagnan in 1943, that gives the diver greatest mobility.

The air supplied to divers, whether from a pump or from aqualung tanks, must be at the same pressure as the surrounding water so that the diver's body is not crushed. At a depth of less than 10m (33ft) water pressure equals that of the atmosphere (1.03kg/sq cm [14.7lb/sq in]); each 10m from the surface increases pressure by one atmosphere. The result of breathing air at higher than normal pressure is that nitrogen (which forms 80 per cent of air) becomes highly concentrated in the blood and tissue fluids. This dissolved nitrogen can turn back into a gas in the organs and blood-stream if pressure is lowered too suddenly, leading to decompression sickness, usually called "the bends". If a diver rises from below 14m (45ft) too quickly, the nitrogen in his blood is not expelled in the normal way through his lungs; and the bubbles formed in his system prevent the proper circulation of his blood. To avoid the bends, divers are raised in stages, stopping for a set period at predetermined depths as they ascend. If for any reason this has not been possible the diver can be put into a decompression chamber [5]. This subjects him to the same pressure at which he was working under water and then slowly returns him to normal atmospheric pressure.

At depths greater than 40m (130ft) dissolved nitrogen can produce narcosis, a state in which a diver becomes so confused or euphoric that he may even remove his air supply. Narcosis can be avoided by using a mixture of oxygen and helium, but the mixture alters the diver's voice, making his speech almost unintelligible, and causes him to lose body heat rapidly – a hazard in cold waters unless he wears a heated suit.

The deepest dive made at sea to date using self-contained underwater breathing apparatus (SCUBA) gear was 133m (437ft), accomplished in 1968. Much deeper dives have been simulated in compression chambers, in 1970 two Royal Navy divers went to an equivalent of 457m (1,500ft) for ten hours. The "dive" and subsequent decompression took 15 days to complete.

Saturation diving
The disadvantages of the need for decompression after each dive are being overcome by saturation diving techniques. Twenty-four hours' exposure to nitrogen-

CONNECTIONS

See also
90 Oceanographic exploration

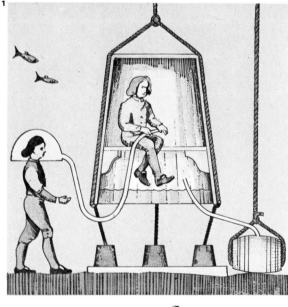

1 Edmond Halley's diving bell was 2.4m (8ft) high and 1.5m (5ft) wide at the base. It was wooden with glass portholes and was weighted with lead. Air was supplied from one of two lead-lined barrels. When the first barrel was exhausted it was pulled back to the surface to be refilled.

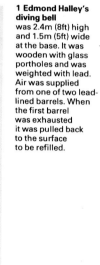

2 The standard diving suit consists of a heavy metal helmet with breastplate, tough watertight diving dress, heavily weighted boots and a flexible tube carrying air pumped from the surface.

3 The aqualung (or SCUBA) diver wears a rubber suit. Compressed air carried in tanks is delivered at ambient water pressure (equal to that of the surrounding water) via a demand valve.

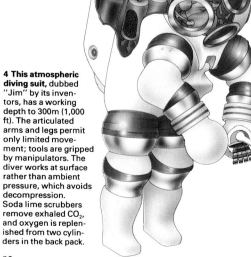

4 This atmospheric diving suit, dubbed "Jim" by its inventors, has a working depth to 300m (1,000 ft). The articulated arms and legs permit only limited movement; tools are gripped by manipulators. The diver works at surface rather than ambient pressure, which avoids decompression. Soda lime scrubbers remove exhaled CO_2, and oxygen is replenished from two cylinders in the back pack.

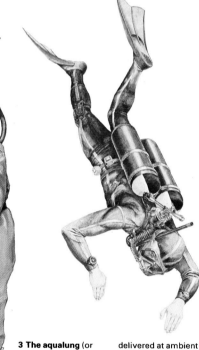

5 A submerged decompression chamber can be used to treat an injured diver or one who is suffering from the bends. He can be admitted to the main chamber through an air lock from a portable pressure vessel [left], which has its own bottled air supply. The rate of decompression to ordinary pressure can then be carefully controlled.

free artificial air (such as a mixture of oxygen and helium) under pressure causes a diver to become "saturated" at that pressure. He can remain under pressure for several weeks, greatly increasing his working capacity, after which only one decompression is necessary. Divers working under saturation conditions live in a large deck decompression chamber; they then transfer under pressure into a smaller chamber from which they work. On their return the transfer is reversed.

Underwater habitats such as Conshelf, Tektite [6] and Sealab, are variations on the saturation diving system. The living chamber lies on the sea-bed and divers enter and leave through an entry trunk. These habitats are used mainly for scientific research in depths down to 100m (328ft).

Submersibles for industry and research

Cornelius van Drebble built one of the earliest submersibles in 1620. Powered by 12 oarsmen, it travelled 5m (16ft) below the surface of the Thames. Subsequent development of small submarines was largely directed towards military objectives. Only in

the 1960s was much attention paid to the development of submersibles for scientific research or underwater engineering [Key]. Since 1960 more than 50 submersibles have been built with depth ranges from 100m to 2,000m (329–6,560ft) and displacements of between five and 100 tonnes.

The present generation of working submersibles, used in biological and geological research, are mostly in the 10–20 tonne range and carry a pilot and one or two observers who enter the submersible on the mother ship. The interior is at atmospheric pressure throughout the dive.

Since 1973 submersibles have been used increasingly by the offshore oil and natural gas industry for pipeline inspection, repairs and platform site surveys. The serious quest to probe great depths began in 1930 when Otis Barton and William Beebe descended to 425m (1,400ft) off Bermuda in a steel pressure sphere, or bathysphere, lowered on a cable from a ship. On 23 January 1960 Jacques Piccard and Donald Walsh descended 10,917m (35,820ft) to the bottom of the Challenger Deep in the Mariana Trench.

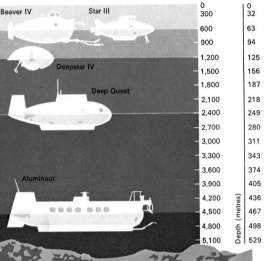

Depth (metres)	Pressure (kg/cm²)
0	0
300	32
600	63
900	94
1,200	125
1,500	156
1,800	187
2,100	218
2,400	249
2,700	280
3,000	311
3,300	343
3,600	374
3,900	405
4,200	436
4,500	467
4,800	498
5,100	529

Beaver IV   Star III   Deepstar IV   Deep Quest   Aluminaut

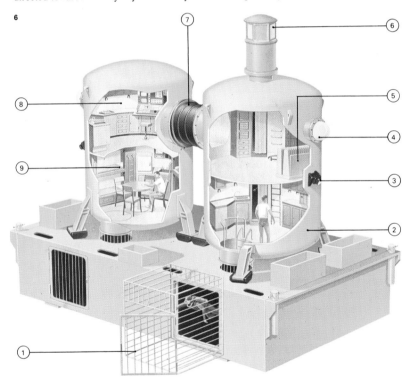

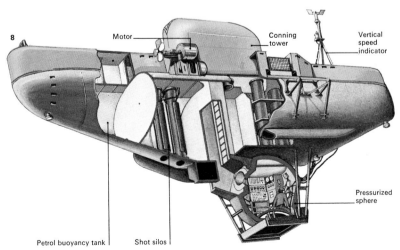

1 Shark cage protecting the entrance
2 Wet room
3 External light
4 Observation port
5 Air conditioning and purification equipment
6 Beacon light
7 Connecting passage
8 Control and communication room
9 Living quarters

6 Tektite was designed to study the reactions of scientists working underwater for long periods under saturated diving conditions. It has four chambers and accommodates four or five people.

7 The submersible Pisces III is 5.8m (19ft) long, weighs 10.8 tonnes and has a maximum operating depth of 1,100m (3,600ft). It is launched from an A-frame at the stern of the mother ship.

8 The bathyscaphe FRNS 3 consists of a pressure sphere fitted with an entrance hatch and a conical Plexiglas window. Entry is through an air lock. The buoyancy tanks are in compartments built of light sheet metal and filled with petrol for buoyancy. To descend, the remaining tanks and air lock are flooded. Electric motors provide lateral movement at depth and lead shot ballast is jettisoned for ascent.

Motor   Conning tower   Vertical speed indicator

Petrol buoyancy tank   Shot silos   Pressurized sphere

9 The submersible VOL L-1 operates down to 365m (1,200ft). The pilot, diving supervisor and observer travel in the forward compartment at atmospheric pressure while the two divers in the lockout are pressurized to the working depth.

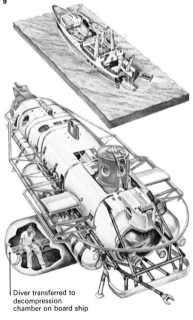

Diver transferred to decompression chamber on board ship

Shapes and structures of crystals

Crystals are solids that have their atoms, ions or molecules arranged in mathematically regular patterns. Because most solid matter is crystalline, this ordered construction gives such substances important properties that are not possessed by liquids or gases.

Ordinary table salt and sugar are perhaps the most obviously crystalline substances in common use, but even substances like clay and steel are crystalline. It is often difficult to recognize a single crystal in nature because the basic regularity of its true, individual form is usually hidden by the aggregation of several small crystals.

The science of crystallography began with the study of "well formed" crystals. The Danish physician Nicolaus Steno (1638–86) discovered in 1669 that the angles between the faces of different quartz crystals were constant; in 1783 the Frenchman Jean-Baptiste Romé de l'Isle (1736–90) established that the angles between a crystal's faces were characteristic of the substance of which it is formed. Another Frenchman, the abbot René Just Häuy (1743–1822), explained the constancy of the angles between

the faces by the stacking of tiny unit blocks known today as unit cells. He also described the seven basic crystalline systems [2–8] and the principles of their symmetry. What transformed crystallography from a side branch of mineralogy into an essential branch of physics was the discovery, in 1912, of the internal structure of crystals through the phenomenon of X-ray diffraction [9] by Max von Laue (1879–1960), a German physicist, and, jointly, by the British physicist, William Braggg (1862–1942), and his Australian son Lawrence (1890–1971).

Crystal lattices and crystalline systems

The rows of particles (atoms, ions or molecules) in a crystal form a lattice. The simplest three-dimensional particle structure that can re-create the crystal by repetition is the unit cell. The unit cell is represented by a geometrical solid, the corners of which are centred on particles. These corner particles must all be the same type; the other types (if they exist) are contained within the cell. For example, the unit cell of a crystal of halite (common salt or sodium chloride) can be

considered as a cube with eight chlorine (Cl) ions at the corners [1B]. This cube contains one sodium (Na) ion at its centre. Each of the corner Cl ions, being also shared by seven other contiguous cubes, has only one-eighth of its volume within the unit cell under consideration, so that the unit cell has one Na ion and the equivalent of one Cl ion, which together are equal to the compound's chemical formula – NaCl.

The external shape of well-developed crystals reflects precisely the symmetry of the unit cell. The unit cell of sodium chloride is a cube; the crystal shape of halite is therefore also a cube, or another closely related form such as an octahedron.

The elements of symmetry that can be found in crystals are axes, planes and centres. A symmetry axis is such that an object (including a crystal) rotated around it by a given angle will produce a configuration identical to the original one. The number of such rotations to obtain a full 360° turn is the order of the axis. Crystals can have two-, three-, four- and six-fold (order) axes. Planes of symmetry are like mirrors; a centre of sym-

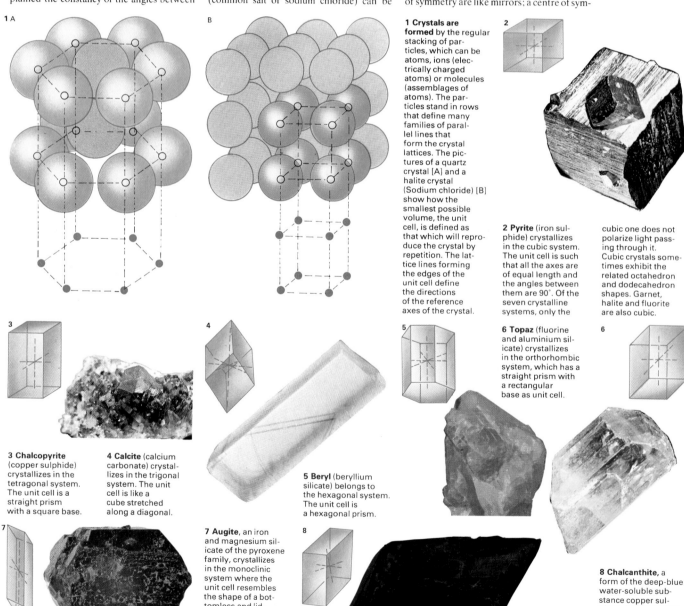

1 Crystals are formed by the regular stacking of particles, which can be atoms, ions (electrically charged atoms) or molecules (assemblages of atoms). The particles stand in rows that define many families of parallel lines that form the crystal lattices. The pictures of a quartz crystal [A] and a halite crystal (Sodium chloride) [B] show how the smallest possible volume, the unit cell, is defined as that which will reproduce the crystal by repetition. The lattice lines forming the edges of the unit cell define the directions of the reference axes of the crystal.

2 Pyrite (iron sulphide) crystallizes in the cubic system. The unit cell is such that all the axes are of equal length and the angles between them are 90°. Of the seven crystalline systems, only the cubic one does not polarize light passing through it. Cubic crystals sometimes exhibit the related octahedron and dodecahedron shapes. Garnet, halite and fluorite are also cubic.

3 Chalcopyrite (copper sulphide) crystallizes in the tetragonal system. The unit cell is a straight prism with a square base.

4 Calcite (calcium carbonate) crystallizes in the trigonal system. The unit cell is like a cube stretched along a diagonal.

5 Beryl (beryllium silicate) belongs to the hexagonal system. The unit cell is a hexagonal prism.

6 Topaz (fluorine and aluminium silicate) crystallizes in the orthorhombic system, which has a straight prism with a rectangular base as unit cell.

7 Augite, an iron and magnesium silicate of the pyroxene family, crystallizes in the monoclinic system where the unit cell resembles the shape of a bottomless and lidless rectangular crate that has been pushed sideways.

8 Chalcanthite, a form of the deep-blue water-soluble substance copper sulphate, belongs to the triclinic system, which has a unit cell with no right-angles.

metry is such that any feature is repeated upside down at an equal distance from the centre and on the opposite side of it. Thirty-two combinations of symmetry elements are found in crystals; these 32 classes are grouped into seven systems [2–8].

X-raying crystal lattices
The shape and size of the unit cells, and the positions of the particles within them, are determined by using X-rays. These rays have a very short wavelength that is about the same as the spacings between the lattice planes of crystals. They are consequently diffracted by these planes. Bragg's law relates the X-ray wavelength, the spacing between a given set of parallel lattice planes and the angle of incidence [9B]. Thus by using X-rays of known wavelength and measuring angles of incidence where diffraction occurs it is possible to calculate the distance between the lattice planes. Since atoms, ions or molecules are assumed to touch each other, these distances also give the diameters of the particles.

X-ray diffraction methods can use either a single rotating crystal [9] or powdered crystals, some of which will be at the correct angles for diffraction. The diffracted rays either expose a photographic film or are measured by a Geiger or scintillation counter.

The properties of crystals
The size and configuration of crystals in a metal affect its mechanical properties: stress behaviour, fatigue and resistance depend on the crystals. Impurities in the crystals, in concentrations of only a few parts per thousand million, account for the semiconducting properties of elements such as silicon and germanium. The magnetic properties of many materials are influenced by the internal disposition and shapes of crystals. Some crystals respond to vibrations by generating electricity (piezo-electricity) and this is the principle behind record-player needles; the converse is used in ultrasonic transducers and radio- or clock-tuning crystals. Transparent crystals of all the seven crystalline systems, except the cubic will rotate the plane of polarization of polarized light [12]. Polaroid polarizing sheets or glasses are actually composed of very small crystals.

KEY

Crystal shapes, when apparent, are related to the underlying lattice forming them, but the same lattice can produce different shapes. The dogtooth calcite shown here has the same lattice as the Iceland-spar calcite shown in illustration 4. Both types of crystals have the same internal symmetry.

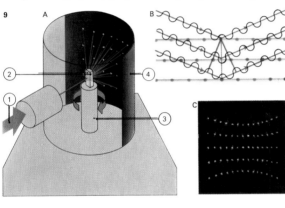
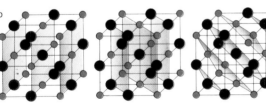
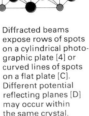

9 X-rays are scattered in selective directions by crystal lattices and are used to study the spacings and positions of the particles within the crystals. In the assembly shown [A] the source

[1] generates single-wavelength X-rays to hit a crystal [2] on a revolving axis [3]. Diffraction occurs in accordance with Bragg's law, when X-rays reflected by parallel lattice planes are in phase [B].

Diffracted beams expose rows of spots on a cylindrical photographic plate [4] or curved lines of spots on a flat plate [C]. Different potential reflecting planes [D] may occur within the same crystal.

10 Stereographic projection is a means of representing a three-dimensional crystal as a two-dimensional figure. The mathematics are quite complex but in theory the crystal is placed at the centre of a sphere and perpendicular lines are drawn from each face to meet the surface of the sphere [A]. These points on the northern hemisphere are then connected to the south pole and the points at which the connecting lines cut the equatorial plane are noted. The pattern of the points on this plane is the stereographic projection of the crystal [C]. In any crystal of a particular substance the angles between corresponding faces are always equal no matter how distorted the crystal may be, and so the stereographic projection will always be the same. Hence the two distorted crystals of quartz [B], despite the differences in size of their corresponding faces, will give the same projection.

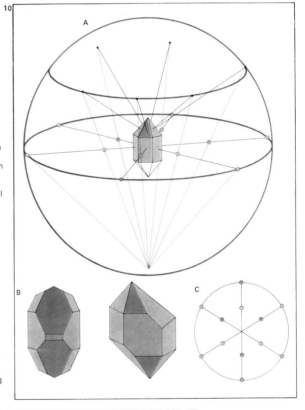

11 Crystals grow by precipitation out of a solution or a cooling melt. The atoms or ions coalesce into tiny "seeds" around which further particles build up the lattice layers. If alum powder is dissolved in hot water with a drop or two of sulphuric acid and placed in a jar as shown, alum crystals will grow as the solution cools. Slower cooling gives larger crystals. A similar experiment can be done by melting and then cooling sulphur powder.

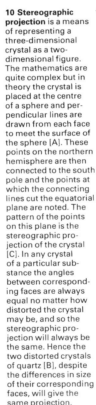

12 The colours of mineral crystals in a thin section of rock viewed through a polarizing microscope can help to identify the minerals in that rock section. Polarized light (light vibrating in one plane) passing through the microscope and through the thin section is distorted by the internal structure of the crystals and these distortions give rise to the colours observed. In this example the large yellow crystals are of pyroxine while the small grey ones are of felspar.

Earth's minerals

A rock is not a homogeneous mass of material with a constant chemical composition throughout its bulk. If a rock is examined closely it is seen to be made of many components, each quite different from the others and usually forming discrete crystals. These individual components are the minerals.

The economic minerals such as precious stones and the ores of useful metals are those that usually come to mind when minerals are mentioned, but these actually represent a small part of the mineral kingdom. The largest components are the rock-forming minerals, the building blocks from which the earth's crust and all its rocks are constructed. These can be so attractive, with their great variety of crystal shapes and range of colours, that finding and collecting minerals has been a popular hobby for thousands of years.

The constituents of the rocks

Minerals are defined as naturally occurring inorganic substances made up of one or more elements. Most have a constant chemical composition and are usually crystalline [1]. There are exceptions to this, however. Some minerals, such as opal [1B], are non-crystalline and in others the chemical composition varies: in olivines and pyroxenes, for example, the proportions of magnesium and iron atoms are not constant.

The quantities of the earth's elements are reflected in their relative abundance in minerals. The element oxygen is most abundant, thus a large number of minerals contain oxygen. Haematite [7] is an oxide of iron. Silicon is the next most abundant, so silica, the oxide of silicon, plays a large part in the composition of minerals. Quartz [2, 3] can be pure silica and is an extremely common mineral but the silica is more often found in combination with other elements to produce the numerous rock-forming minerals known as silicates. Olivine, for instance, is a common silicate mineral in igneous rocks where no pure quartz is present and consists of silica combined with varying amounts of iron and magnesium. Other groups of minerals found are compounds of sulphur such as anhydrite, gypsum [13] and galena [9] and carbonates such as calcite and malachite [1C, 12]. Occasionally, as with native copper and gold

[8], a mineral has only one element.

Many minerals crystallize from the molten state. The overall composition of magma, the molten material from which igneous rocks are formed, is fairly constant but when it starts to solidify the minerals that crystallize out vary greatly from place to place. The resulting rock has a composition very different from that of another rock formed elsewhere from the same magma.

The formation of minerals

Olivines have a high melting-point so they tend to crystallize first from a cooling magma. Once crystallized they may sink to the bottom of the magma chamber and leave the rest of the liquid deficient in iron and magnesium. Other minerals such as felspar [3, 14B], in which sodium, calcium, potassium and aluminium combine with what silica is left, then crystallize out, leaving a cooling magma with yet another composition that solidifies into even more minerals.

Some minerals are formed in sedimentary environments. For example, when seawater evaporates in restricted basins the salts dis-

3 The last minerals to crystallize in a rock are crowded by the others and are xenomorphic – unable to assume their normal external crystalline form. They have irregular shapes

although the internal crystalline lattice is retained. This section of granite contains xenomorphic quartz crystals [shown grey] surrounded by automorphic felspar crystals.

1 A crystal is a solid in which the molecules, atoms or ions are arranged along regular and repetitive lattices, its external shape reflecting its internal symmetry. Not all crystals

are minerals – sugar [A] is a crystalline organic substance. Opal [B] is an "amorphous" mineral without a crystalline internal structure. Calcite [C], as most minerals, is crystalline.

2 Where a mineral such as quartz grows unimpeded it shows its typical "automorphic" shape, with plane faces that reflect the internal lattice structure of its atoms.

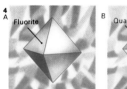

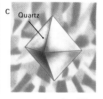

4 Pseudomorphism is the replacement of one mineral with an external crystal shape by another that is in every way identical in shape and volume, but not in internal crystalline lattice.

The quartz shown is forming a pseudomorph of a fluorite crystal that was dissolved. The quartz would normally form a six-sided prism but has filled the octahedral shape of the fluorite.

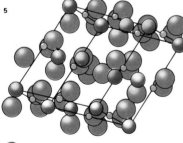
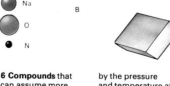

○ Ca A
○ O
○ C

● Na B
○ O
● N

5 Minerals of different chemical composition may possess identical lattice structures and will have similarly shaped crystal faces under normal circumstances. This is called isomorphism and is shown for calcite

[A] and soda-nitre [B] crystals. Because of a near similarity in size the ions of isomorphic minerals can readily substitute for each other in the structure of a crystal if they are also chemically similar.

6 Compounds that can assume more than one crystalline structure are said to be polymorphous and they form different kinds of minerals. The type of crystal lattice in which the compound will crystallize is determined

by the pressure and temperature at the time of formation. Kyanite, sillimanite and andalusite are three different crystal types and minerals of the same silicate compound that is common in metamorphic rocks.

Pressure in kilobars
Kyanite
Sillimanite
Andalusite
200 400 600 800
Temperature °C

● Oxygen ● Aluminium

is the feature that gives many gems their special quality. The transparent red ruby [13] and the blue sapphire [15] (both of which are forms of a normally dull, grey or colourless mineral called corundum), the green emerald, a form of the mineral beryl and the yellow topaz are all admired because of their pure tints. Opaque or cloudy gems such as opals depend entirely upon their colour to make them attractive.

Specific gravity is the weight of the mineral compared with the weight of an equal amount of pure water. Diamond, for example, has a specific gravity of 3.52 which means that it weighs 3.52 times an equal amount of pure water, whereas amber has a specific gravity of 1.07. The weight of a diamond is usually expressed in carats – one carat is equivalent to 200 milligrammes.

Hardness ensures durability and accordingly the most valuable gems are stones that will wear for a long time. Hardness is measured on the Moh's scale, which consists of numbers from one to ten, indicating the relative hardness of substances. The diamond has a hardness of ten on this scale and is by far the hardest of all natural substances. It is about 90 times harder than corundum, which rates as nine on the scale. Some gems are quite soft and are valued for other properties.

Polishing and cutting of gems

The beauty of gems is greatly enhanced by skilful cutting and polishing [4], for this removes the surface flaws and heightens the colour or brilliance of a stone. The oldest form in which gems were cut was a rounded shape called *cabochon*, a French word for head. The *cabochon* is used for stones showing the effects of chatoyancy (cat's-eyes) and asterism (star-stones), which are caused by reflections from inclusions.

Faceting – first started by Indian cutters polishing small facets on diamonds – soon became applied to other stones. Thus evolved the brilliant, step and mixed cuts, which depend on various facets being ground and polished in symmetrical arrangements on the stones. The facets on a diamond are cut and polished in one operation but other precious stones have their facets first ground and then subsequently highly polished.

KEY

The largest cut diamond in the world is the Star of Africa, weighing a little over 530 carats. It came from the biggest diamond ever found, the Cullinan. This stone, found in 1905 in the Premier mine, South Africa, was named after Thomas Cullinan, chairman of the mining company. It weighed 3,106 metric carats (0.60kg [1.3lb]) but was cut into two large stones, seven medium, and 96 smaller stones. The largest of these, the Star of Africa, is now among the British Crown Jewels in the Tower of London. Another diamond in this collection is the Indian diamond Koh-i-noor (Mountain of Light), which was given to Queen Victoria in 1850 by the East India Company.

6
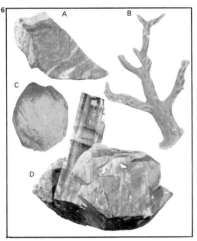

6 Most gems are minerals. Lapis lazuli [A] is the name given to a rock rich in lazurite. Tourmaline [D] is a complex boro-silicate. Organic gems include coral [B] (the skeletons of coral polyps) and amber [C] which is a fossil resin.

7A

B

7 Garnet [A], the birthstone for January, symbolizes faithfulness. Garnets are formed from silica and two metals [B]. Those with aluminium and magnesium are the prized ruby-red pyrope.

8A
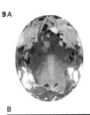
B

8 Amethyst [A], the birthstone for February, symbolizes sincerity. Amethysts are a form of transparent quartz [B], with a violet or purple colour. They are mined in USSR and South America.

9A

B

9 Aquamarine [A], one of the March birthstones, symbolizes courage. Aquamarines are a blue or blue-green variety of the mineral beryl [B]. The best are mined in Brazil and the Urals.

10A
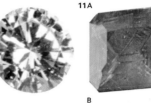
B

10 Diamond [A], the April birthstone, symbolizes innocence. Diamonds are pure crystalized carbon [B], the hardest natural substance. South Africa is the main source. The most prized are colourless.

11A
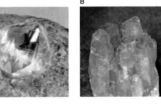
B

11 Emerald [A], the birthstone for May, represents love. Emerald is a gem-quality, rare green variety of the mineral beryl [B]. The best emeralds occur in Colombia in South America.

12A

B

12 Pearl [A], one of the June birthstones, symbolizes health. Pearls are organic gems produced mainly by oysters [B] from nacre, the iridescent substance forming the inner layer of the shell. Pearls are prized for their lustre.

13A
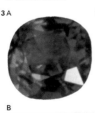
B

13 Ruby [A], the July birthstone, represents contentment. Ruby is a red variety of the hard grey or colourless aluminium oxide mineral, corundum [B]. The finest rubies, from Burma, are coloured a deep bluish-red by a chromium impurity.

14A

B

14 Peridot [A] is one of the birthstones for August. It symbolizes married happiness. Peridots are a transparent green variety of the mineral olivine [B], which is a magnesium-iron silicate, finest come from Burma and Thailand.

15A
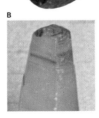
B

15 Sapphire [A], the September birthstone, represents clear thinking. Sapphire, like ruby, is a variety of the mineral corundum [B]. It occurs in many colours but the most valued are blue. The best kinds come from Burma and Thailand.

16A
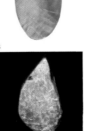
B
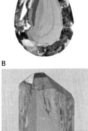

16 Opal [A] is the October birthstone. It symbolizes hope. Opals are a form of hydrated silica [B]. The most prized specimens are the so-called black opals found in Australia, which show flashes of several iridescent colours.

17A

B

17 Topaz [A], the birthstone for November, symbolizes fidelity. The most prized varieties are yellow. Topaz is a mineral compound of aluminium, silica and fluorine [B]. It is found mainly in Brazil, USSR and the United States.

18A
B
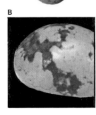

18 Turquoise [A], the birthstone for December, represents prosperity. Turquoise is a hydrous copper-aluminium phosphate [B] sometimes containing iron. The most prized colour is sky blue and comes from Iran. Its name means Turkish stone.

99

The rock cycle and igneous rocks

The rocks on the earth's surface can be divided into three kinds: igneous, metamorphic and sedimentary. Igneous rocks have formed by the cooling of molten magma [3]. Metamorphic rocks have formed by the baking or the compression of older rocks. New crystals have grown in the rock and because these were under pressure they grew in only one direction and are thus aligned. Sedimentary rocks are composed of the weathered or eroded fragments of older rocks or of the remains of living organisms.

Each of these three classes formed under very different conditions from the others. Sedimentary rocks formed on the surface of the earth under extremely low pressures, metamorphic rocks formed below the surface where both the temperature and pressure are high, and intrusive igneous rocks formed, again beneath the earth's surface, but where the temperatures are even higher.

Rock cycle

The rock cycle [2] is the relationship between these three types of rocks. The first part of the rock cycle takes place on the earth's sur-face. This is the erosion and weathering of older rocks to soil and sand and the transportation of the resulting sediment by rivers down to the sea. Nearly all the sediment produced, whether on the land or along the coast, is eventually transported to deep basins under the sea. In these areas great thicknesses of sediment accumulate. For instance, the Mississippi has been pouring sediment into the Gulf of Mexico at the rate of approximately 500 million tonnes a year for the last 150 million years. The pile of sediment is now 12km (7 miles) thick.

Formation of rocks from sediments

The water circulating through the sand deposits iron oxide, silica or lime between the grains and this "cements" the loose sand into sandstone. Mud is squeezed by the weight of the sediment above until all the water is pressed out and it becomes shale. This process of changing sediment into rock is called lithification.

Most great thicknesses of sedimentary rocks accumulate in long, narrow depressions on the sea-floor called geosynclines. These depressions are caused by descending convection currents which, over a period of millions of years, carry the crust of the earth down into the earth's interior where both the pressure and temperature are high. The sedimentary rock in the depression is carried down with the crust. It is folded and squeezed and heated up to between 200°C (392°F) and 500°C (932°F). This changes the sedimentary rock to a metamorphic rock.

The movement of the earth's crust may carry the rock as much as 700km (454 miles) below the surface. Here the temperature and pressure will be even higher and the rock will begin to melt. Molten rock is lighter than solid rock and it will begin to rise up through the overlying rock towards the surface. If it reaches the surface as a lava flow it will immediately be ready for weathering and erosion and the start of a new cycle. More often the molten rock solidifies underground and then all the rock above it must be eroded away before it can begin the cycle again.

Although the complete cycle is from sedimentary to metamorphic to igneous, many rocks short-cut the cycle, usually by

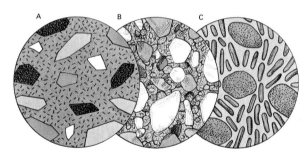

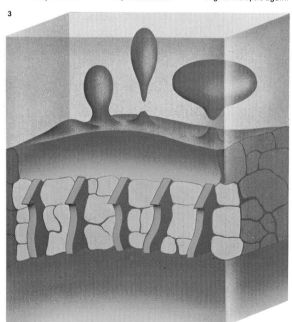

1 Textures can be used to identify rock types. Igneous rocks [A] show well-developed crystals, sedimentary [B] contain older pieces and metamorphic [C] show the stresses under which they formed.

2 The rock cycle is the slow change from one rock type to another. Erosion produces sediments which harden to form sedimentary rocks. If these are deeply buried the temperature and pressure turn them into metamorphic rocks. Intense heat at great depths melts metamorphic rocks. This rock may be pushed up to the surface where it cools to form igneous rocks. There erosion begins the cycle again.

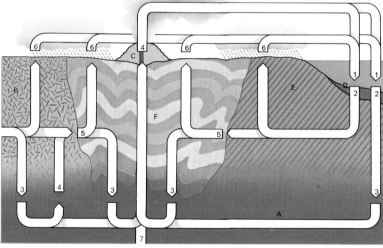

A Molten matter
B Intrusive igneous rocks
C Extrusive igneous rocks
D Sediments
E Sedimentary rocks
F Metamorphic rocks

1 Deposition
2 Lithification
3 Fusion
4 Solidification
5 Metamorphism
6 Erosion
7 Emplacement of new material from earth's interior

3 Granite is the most abundant igneous rock. It is formed by the partial melting of older deeply buried rocks. Initially, the molten liquid stays between the remaining grains but later it migrates to form small pods which in turn collect together into layers. Because the liquid is lighter than the surrounding rock it rises upwards and intrudes the rock above, forming large masses called batholiths.

	Granite	Basalt
Silica	70·8%	49·0%
Alumina	14·6%	18·2%
Ferric oxide	1·6%	3·2%
Ferrous oxide	1·8%	6·0%
Magnesia	0·9%	7·6%
Titanium oxide	0·4%	1·0%
Calcium oxide	2·0%	11·2%
Sodium oxide	3·5%	2·6%
Potassium oxide	4·2%	0·9%

4 Basalt and granite are the two commonest rocks found on the earth's surface. Both rocks are composed mostly of the elements silicon and oxygen. These are combined with minor amounts of other elements into natural chemical compounds called minerals. Basalts occur either as lava flows or dikes, while granites occur as batholiths.

missing out the igneous or metamorphic stages. For instance, a sediment may be lithified to a sandstone but then may be uplifted out of the sea and eroded.

Igneous rocks
Igneous rocks [5] are divided into extrusive and intrusive rocks. Extrusive rocks are those that were ejected by volcanoes and cooled as lavas on the surface of the earth. Intrusive rocks are those that solidified beneath the earth's surface. The grain or crystal size of a rock depends on how fast it cooled; coarse-grained rocks are the result of slow cooling which has given crystals time to grow to sizes greater than two millimetres in length. Rocks cool slowly when deep in the earth's crust and coarseness is characteristic of intrusive rocks. Fine-grained rocks have cooled rapidly either on or near the earth's surface; most extrusive rocks are fine grained although some are cooled so rapidly that no crystals have time to grow and obsidian is formed.

Igneous rocks are classified by the amount of silica they contain [6], and the size of the grains. The chemical composition [4]

and the silica content in particular depends on the origin of the magma from which the rock was made. The magma may have resulted from the partial melting of the rocks beneath the earth's crust or from the melting of the crust itself as part of the rock cycle. Magma from the crust contains more silica than that from below it and produces light coloured rocks, whereas the magma from below the crust gives dark-coloured rocks.

The partial melting of the rocks beneath the earth's crust produces basalts (fine-grained extrusive lavas), dolerite (medium-grained intrusive rock) and gabbro (coarse-grained intrusive rock) [7]. Basalts form the floors to the oceans and occur extensively in Iceland and in some continental areas. Dolerites are found in thin extensive sheets called dikes and sills [5] injected in or between the sedimentary rock layers. Gabbro occurs in large layered intrusions which were the source of the dolerite and basalt.

The melting of rocks which were once sediments on the crust produces granites [3] and andesites. Granites occur in very large intrusions called batholiths [5].

KEY

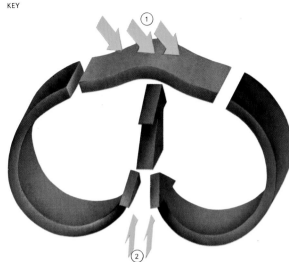

The energy to power the rock cycle is derived from the heat of the sun [1], which indirectly breaks down existing rocks to sediments, and the heat from the earth's interior [2], due to radioactivity which melts existing material to give igneous rock and also causes the movements of the earth's crust.

5 Many different shapes of igneous rocks exist. A neck [1] is a circular vertical feed channel of a volcano. A stock [2] is a large mass of rock which solidified at great depth. A batholith [3] is a large body of granite with no detect-able bottom. A laccolith [4, 7] is a dome-shaped mass which has arched up the rock above. A dike [5] is a vertical sheet-like mass of rock and sill [6] is a horizontal sheet-shaped body of rock. A lopolith [8] is a saucer-shaped mass of rock.

6 Igneous rocks comprise varying amounts of minerals in which the quantity of silica (SiO₂) determines the acidity of the rock and thus its classification. This proportion of silica determines the type

Intermediate (Diorite)

Acidic (Granite)

% of minerals
100
80
60
40
20
0

Olivine
(Iron and magnesium minerals)
Pyroxene

Sodium-calcium felspars

Quartz (silica)
Potassium felspars
White mica
Dark mica

Ultrabasic (Dunite)   Basic (Gabbro)

and proportion of the minerals present. Thin rock sections examined by polarized light reveal the individual minerals in distinctive colours, helping to identify them and to classify the rock.

7 Intrusive rocks can often be identified with the naked eye. Granite [A] contains a great deal of free silica in the form of quartz, giving the rock a light colour. Diorite [B] is darker, having less quartz and a quantity of dark minerals. These dark minerals, such as olivine and pyroxine, are more common in gabbro [C]. The light-coloured minerals in this are felspars. Ultrabasic rocks such as dunite [D] consist almost entirely of dark ferro-magnesium minerals.

Sedimentary and metamorphic rocks

The rocks of the earth's surface are of three sorts: igneous (formed from molten magma), sedimentary and metamorphic [Key]. Sedimentary rocks are formed from chemicals, organic materials and fragments produced by the weathering and erosion of older rocks; metamorphic rocks by the heating under pressure of older rocks.

Sedimentary rocks are divided into three types [1]. The first, which is called clastic, is formed by fragments of older rocks; the second is organic, composed of the remains of animals or plants; and the third is chemical, produced by the precipitation of minerals and salts from water. Streams, moving ice and waves, break up older rocks into fragments, some large, like stones or boulders, some about 1mm (0.04in) across, like sand, and some too fine to see, which form mud. Most of these are carried down to the sea by rivers and deposited in deltas or farther out in the sea-bed. Stones too large to be moved by the water remain near the heads of streams or on beaches and are eventually cemented together to form a rock known as conglomerate. Sand is deposited near the coast or on the continental shelf and eventually forms sandstone. Sands are also deposited by wind in desert environments. Mud is often carried far from the shore to become clay or shale.

Plant and animal origins

Organic sedimentary rocks may be made of plant remains, like coal, or from the hard parts of animals; many limestones are made of fossil shells and corals which have extracted lime from seawater to make their skeletons and, dying, have left their remains on the sea-floor. In time the movement of the sea wears the shells into fragments. Over a period of millions of years after burial, the fragments are compacted by weight and cemented to one another by various processes to give limestone. This is called lithification. Accumulation of lime is at present taking place in the Bahamas and the Persian Gulf, but in the past warm seas were much more extensive and limestone was produced over large areas. Chalk is made of countless small shells so minute that they can be seen only with a microscope. Seawater contains large amounts of salts and, if it is evaporated, they are precipitated out as lime is precipitated in a kettle. In tropical areas where hot, dry winds blow over shallow seas, much of the water is evaporated and lime forms on the sea-floor, hardening to a fine-grained limestone [4]. If there is a partially enclosed basin, then not only lime but salts such as gypsum will be precipitated.

Sedimentary rocks are important because they provide oil, natural gas, coal and building stone. They are of great interest, too, because they were formed at the earth's surface and provide much evidence of its nature many millions of years ago. Fine red sandstone [3], for example, indicates the former presence of deserts. The study of ancient environments through the analysis of present-day rocks is called stratigraphy.

New rocks from old

Metamorphic rocks are usually much harder than sedimentary rocks. Some are formed by deep burial, others by the heat of igneous intrusions. Their grains are all interlocking crystals and many, such as slates, schists and gneisses, split easily along certain planes.

1A

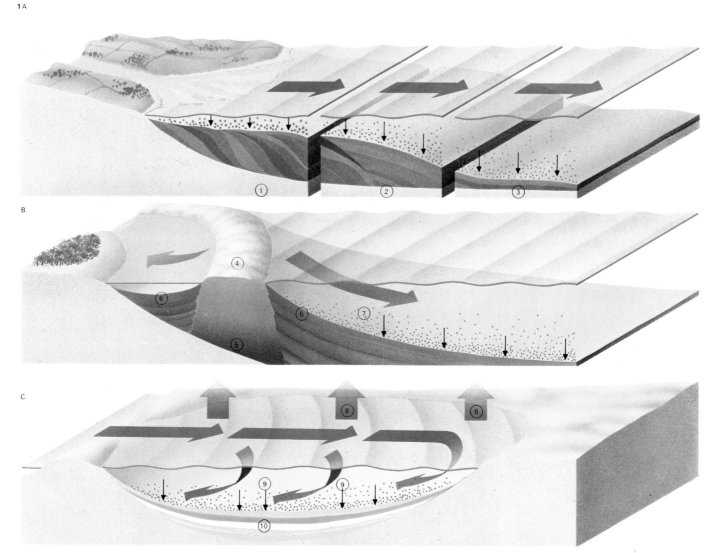

1 The three types of sedimentary rocks are clastic [A], produced by erosion from older rocks; organic [B], formed by the decomposition of living matter; and chemical [C], formed from salts deposited by evaporation. [A] Erosion produces sand and mud which is brought down to the sea by rivers, deposited in deltas [1] and on the sea-floor where it hardens into rock [2]. Large grains are deposited inshore, fine mud [3] farther away – a common origin for clastic rocks. [B] On a coral reef, living coral [4] is found only near the surface. It rests on hundreds of metres of dead coral [5] and on both sides are piles of broken pieces of coral eroded by the waves [6]. The shells of crustaceans also accumulate on the sea-floor [7]. All will form reef limestone, a typical organic rock. [C] Where a partially enclosed basin is found in the tropics the seawater is evaporated [8] and the salts in it deposited [9] forming chemical rocks [10].

Others, such as quartzite and marble, are compact rocks which break in any direction.

When sedimentary rocks are intruded by a molten mass of magma they are altered. This process is a form of metamorphism known as thermal or contact metamorphism. Small intrusions such as dikes and sills merely bake a thin skin of rock and make it harder; large intrusions, such as the Dartmoor (southwest England) granite, alter the rock for several miles around. A large intrusion may heat the rock to 700°C (1,292°F) and take more than a million years to cool, giving time for new minerals to form.

The rocks surrounding an igneous intrusion can be divided into zones, depending on how much they have been altered. Shales will have been changed to slates on the outside and near the intrusion new minerals, such as andalusite, will occur in the slate. Next to the intrusion, hornfels, a hard rock, will form.

Alteration of large areas
Regional (or dynamic) metamorphism [5] occurs where large areas of rock have been buried sufficiently deep for the increase in temperature and pressure to alter the rocks. The pressure increase is caused by the weight of rocks above and the increase in temperature by the earth's interior heat. Slates are formed by both regional and thermal metamorphism but schists and gneisses are found only in regional metamorphism. Regionally metamorphosed rocks outcrop over a large part of the earth's surface, where old mountain ranges have been eroded away, leaving on the surface rocks that were once deeply buried. Examples are the Canadian Shield and parts of Scotland and Sweden.

A third, rare type of metamorphism is dislocation metamorphism, caused by large areas of rock moving past one another. The pressure shatters the rock and the friction is so great that the rock is partially melted, producing a rock called mylonite. It occurs only in narrow strips, of which the Lizard, in Cornwall, England, is an example. Unlike sedimentary rocks, metamorphic rocks are of no great use to mankind, containing no oil and few useful minerals, and are of little use in building, apart from slate for roofing and marble used in decorative work.

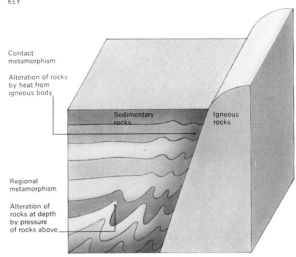

Contact metamorphism

Alteration of rocks by heat from igneous body

Sedimentary rocks

Igneous rocks

Regional metamorphism

Alteration of rocks at depth by pressure of rocks above

The three groups of rock are igneous, sedimentary and metamorphic. They are seen together where igneous rock has intruded sedimentary rocks. Its heat has caused the sedimentary rock to be thermally metamorphosed. Deeply buried sedimentary rock is regionally metamorphosed.

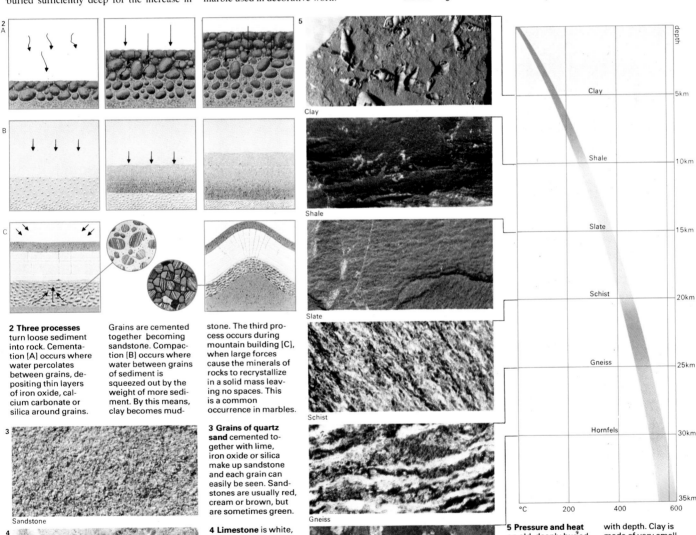

2 Three processes turn loose sediment into rock. Cementation [A] occurs where water percolates between grains, depositing thin layers of iron oxide, calcium carbonate or silica around grains.

Grains are cemented together becoming sandstone. Compaction [B] occurs where water between grains of sediment is squeezed out by the weight of more sediment. By this means, clay becomes mudstone. The third process occurs during mountain building [C], when large forces cause the minerals of rocks to recrystallize in a solid mass leaving no spaces. This is a common occurrence in marbles.

3 Grains of quartz sand cemented together with lime, iron oxide or silica make up sandstone and each grain can easily be seen. Sandstones are usually red, cream or brown, but are sometimes green.

Sandstone

4 Limestone is white, grey or cream-coloured rock, often containing many fossils. Limestones are made of calcium carbonate and are formed by the partial evaporation of seawater or from broken shells which are often preserved as fossils.

Limestone

Clay

Shale

Slate

Schist

Gneiss

Hornfels

depth

5km

10km

15km

20km

25km

30km

35km

°C 200 400 600

5 Pressure and heat on old, deeply buried rocks produces regional metamorphism. The temperature and pressure increase with depth of burial, causing new minerals to grow in the rock, and the size of the mineral crystals increases with depth. Clay is made of very small crystals, but gneiss has 2cm-long (0.75in) crystals. Minerals grow in the direction of least pressure, which means they are aligned and will split easily one way. Hornfels does not show lineations.

Folds and faults

The earth's mountains and valleys are formed by folds and faults in its ceaselessly changing crust. Folds are rock waves and faults are cracks, and both are caused by the intense pressures of continental drift. They are of major significance to industrial geologists because they often form structural traps for valuable mineral deposits.

How folds and faults form

Folds and faults are usually well developed in both sedimentary and volcanic rocks. They may also form in plutonic rocks such as granite and gabbro. Correct interpretation of their structure is essential in mining. Recumbent folding and reverse faulting can cause beds of coal, for example, to be repeated vertically, while normal faulting can cause a horizontal gap. A coal seam may thus be passed through several times in drilling, or alternatively missed altogether.

Faults that develop above an intrusive granite allow mineralizing fluids to pass into the overlying rocks and there deposit minerals such as lead, tin, zinc and copper ores. Similarly, faults that do not reach the surface

may form channels up which oil and gas can rise. In downward folds, where porous sandbeds overlie impermeable clays and shale, collections of water form which can produce artesian springs.

Movement of the massive plates of rock that compose the earth's crust produces intense pressure at the margins of the plates. Where two plates converge, these sometimes throw the rocks up into highly folded and faulted mountain chains. At other plate edges, stretching pulls the rocks apart and forms long depressions bounded by faults, such as the rift valleys of East Africa.

Folds vary greatly in size, from a few millimetres to hundreds of kilometres across. Downward or basin-shaped folds are called synclines and upward folds are called anticlines [Key]. Synclinoria and anticlinoria are the names given to large synclines or anticlines that have smaller folds on their limbs. The Weald of southeast England and parts of the Paris basin, for example, are anticlinoria.

Folds that form at the same time as deposition are known as supratenuous folds. These occur when material that compacts at

different rates is deposited at the same time in the same area, as when sand is deposited around coral. Domes are folds in which the beds dip outwards, whereas basins are formed when the beds dip inwards [3].

Classification of folds

There are three main kinds of folds. First true, or flexure, folding forms by the compression of competent (strong) rocks. This may grade into the second type, flow folding, in areas where incompetent (weak) rocks occur [4]. The incompetent rocks behave like a thick paste; they cannot easily transmit pressure and many minor folds usually form. Third, shear folding [5] may occur in brittle rocks by the formation of minute cleavage-like fractures in which thin slices of rock are able to move in relation to each other like a pack of cards pushed in from the side. Except where cut by a fault, all folds eventually die out by closure, the shape of the fold resembling a half basin or dome.

Simple folds usually occur in young rocks like those of the Tertiary and Quaternary eras. Complex folds are found in older rocks

CONNECTIONS

See also
106 Life and death of mountains
28 Earthquakes

1 Compression creates folds in the earth's crust. First, a simple fold may be created, probably a symmetric anticline [A]. But if there is a continuation of pressure, the fold

may become uneven and develop into an asymmetric anticline [B]. At a later stage, a recumbent fold may develop [C]. The anticline is then lying on top of the syncline and

the layers of rock on one side of the anticline are inverted. If pressure continues to be exerted, these layers will thin and eventually break to produce an overthrust

fold [D]. When these layers disappear due to stretching and fracture, a nappe is formed [E]. Over a long period this nappe may be pushed out many kilometres from its original position.

2 A symmetric anticline [1] and syncline [2] have limbs that dip at similar angles on either side of the axial plane of the fold. The position of the axial plane of

an asymmetric anticline [3] and of a syncline [4] may be more difficult to establish. Where compression produces a reverse fault [5] one side of the fault (in this case

the left) overrides the horizontal strata on the other side. In the case of a monocline fold [6], rock strata lying at two levels may be separated by a limb that is relatively steep.

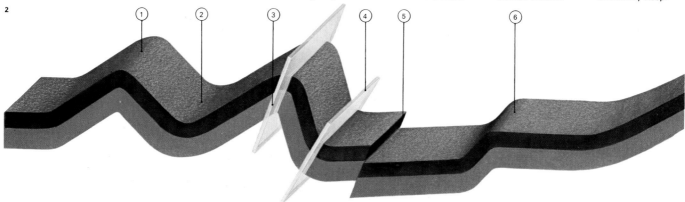

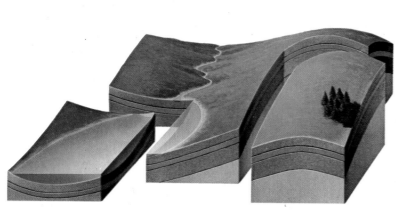

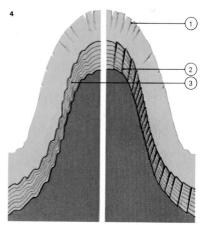

3 Domes and basins are folds that are about as wide as they are long. They are due to complex compressions of the crust. Isolated domes can be due to the subterranean rise of magma or rock-salt.

4 Beds can be competent or incompetent according to their reaction to folding. Competent beds bend and crack [1] without much flowage while incompetent beds shear [2] or form shearing microfolds [3] that alter the thickness of the bed.

which have been exposed to earth movements for a longer period and which have often been deeply buried within the earth's crust. Very old rocks, such as the Precambrian of Norway, have been refolded many times with the development of structures such as boudinage, mullion and cleavage [6, 7]. These ancient rocks have also been considerably altered by heat and pressure from igneous intrusions and deep burial. Platy minerals, such as micas, then develop parallel to each other and the rock tends to split easily along thin planes. Rocks with this property are known as schists.

With increasing distance from the source of pressure that causes folding, the folds gradually die out both in horizontal and vertical directions. This is well displayed in the Alps where the folds become less complex to the north and also to the west.

Faults and refaulting

When rocks can no longer bend under pressure they crack and a fault is formed [Key]. If the rocks are pulled apart a normal fault forms [10A], while if they are compressed reverse and thrust faults form [10B]. Due to movement along the fault plane, grooves and scratches are ground out on adjacent walls. These scratches allow geologists to measure the relative lateral and vertical movements of faults, and tell, for example, whether the movement was linear or rotational. Faults, which are often associated with earthquakes, are well expressed at the surface as fault scarps and rift valleys such as the San Andreas fault and the Rhine rift valley.

As they are produced by the same pressures, faults are frequently associated with folded areas. Sometimes the surface strata may crack into a complex mosaic of blocks by renewed movement along an existing buried fault. Reactivation of such a buried fault is believed to have been responsible for the disastrous 1966 Tashkent earthquake. Refaulting occurs in many areas where new and different stress fields are superposed upon ancient ones. Some regions have been refaulted and refolded several times, as in the complex Precambrian areas of Finland and Canada.

KEY

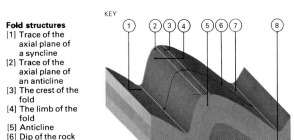

Fold structures
[1] Trace of the axial plane of a syncline
[2] Trace of the axial plane of an anticline
[3] The crest of the fold
[4] The limb of the fold
[5] Anticline
[6] Dip of the rock strata
[7] Trough between folds
[8] Syncline

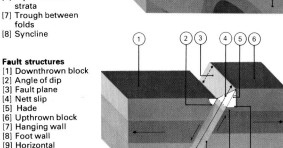

Fault structures
[1] Downthrown block
[2] Angle of dip
[3] Fault plane
[4] Nett slip
[5] Hade
[6] Upthrown block
[7] Hanging wall
[8] Foot wall
[9] Horizontal dip slip (known as heave)
[10] Vertical slip (throw)

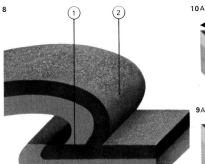

5 Rocks can break instead of bending. Coarse-grained rocks cleave on planes perpendicular to bedding. When the space between fractures exceeds a few centimetres, they are called joints.

Fractures form at the tops of anticlines where weak beds are pulled apart during folding. Finer-grained rocks split by close-spaced faults into slices parallel to the pressure direction.

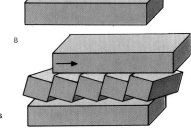

6 Strong rock in a competent bed between two incompetent beds [A] is sometimes subjected to stretching force [B]. The competent bed then deforms and breaks up into flattened rods which are called boudins [C]. The incompetent beds on either side flow into the spaces between the boudins.

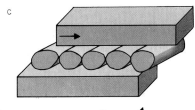

7 Shearing of a competent bed between two incompetent beds [A] may break up the competent bed [B] which is ground into rods called mullions [C]. Geologists use boudinage and mullion structures to tell them what kind of forces have been at work and in what directions they have been acting.

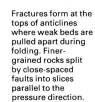

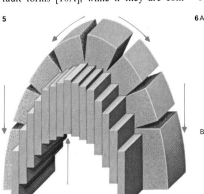

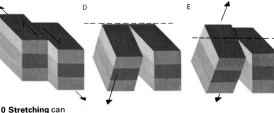

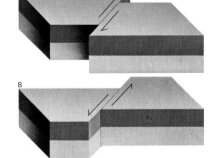

8 Lateral pressure, when applied to a recumbent fold, can produce a low-angle crack [1] along which the overturned limb of the fold [2] may slide. This type of crack in a rock structure is called a thrust fault.

9 Vertical displacement may be more or less equal in some types of block faulting. Where one block is lower relative to those on either side, a graben [A] is formed and where one is raised, a horst [B] is formed.

10 Stretching can produce a normal fault [A], compression a reversed fault [B], shearing a strike-slip fault [C], slumping a hinge fault [D] and twisting a rotational fault [E].

11 Strike-slip faults that move to the right [A] are dextral faults while movement to the left [B] produces sinistral faults. Transform faults are a special kind of strike-slip fault linking major earth structures.

Life and death of mountains

A true mountain is more than just a piece of high ground. It also has underneath it specific geological formations such as strongly folded and faulted rocks, ancient volcanic deposits or large igneous masses such as granite intrusion. Conversely, layers of sedimentary rocks that barely rise from the horizontal should not strictly be described as mountains.

Types of mountains

There are four main types of mountains: fold, block, dome and volcanic. Fold mountains can vary widely in complexity but still conform to the basic type. The Alps [10], Carpathians and Himalayas form the world's largest fold mountain chain. The rocks have been compressed and crumpled in extremely complex ways, with intrusions of molten rock, widespread metamorphism (changes in the rocks) and faulting. The numerous earthquakes in Turkey and Iran indicate that mountains there are still moving.

Block mountains are large-scale faulted structures. Internally a block mountain is usually highly folded and faulted and may have been created either by a deep fault or by an exceptionally large horst (block of raised strata) which was then shaped by erosion. Many block mountains – for example, in the Basin and Range province of Nevada – rise abruptly above the surrounding lowland.

Domes are formed by the lifting of strata, as when granitic magma is intruded. As the lifting increases the surface is worn away by erosion and underlying granite is exposed. When such domes are large and high they constitute true dome mountains, such as the Black Hills of Dakota.

Volcanic mountains differ from others in that they grow visibly during eruptions [4]. When their last growth was recent, their shape is relatively unaffected by erosion. As more eruptions take place on the same site, so the successive outflows of ash and lava increase the height of the volcano. Volcanic mountains in continental interiors are comparatively rare. They are mostly submarine or island features and can form island arcs several thousands of kilometres long; one such arc is the Aleutian Island chain.

Fold mountains are by far the most important because they form very large ranges thousands of kilometres long. Fold mountains are often associated with block-faulted mountains and with volcanic mountains because the forces causing the folding of the rocks are the same that produce faulting and promote vulcanism. The mechanism of the large horizontal compressions leading to the crumpling and folding of the sedimentary cover of the earth's crust was poorly understood before the plate tectonics theory.

According to plate tectonics theory, fold mountains are formed by the movements and collisions of large plates that make up the earth's crust. These plates are usually enormous and may underlie and carry whole continents. When two plates collide the more resistant one slides beneath the other, squeezing upwards the sediment deposited in the geosyncline [6], or trough, between them. The great folds formed in the compressed sediment eventually break out above the surrounding region as mountains.

If the initial clash involves a fast-moving continental plate, the folds may be thrown even higher forming much larger mountain ranges. A continental plate thrust under

CONNECTIONS

See also
104 Folds and faults
24 Global tectonics
114 Land sculptured by rivers
116 Rivers of ice
30 Volcanoes

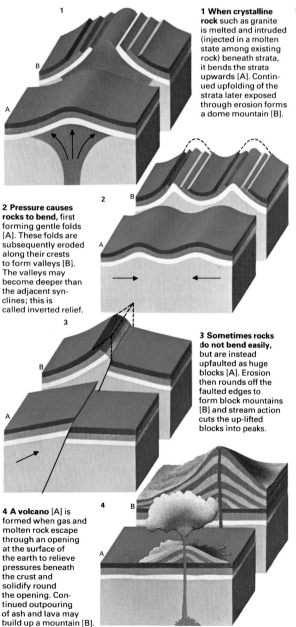

1 When crystalline rock such as granite is melted and intruded (injected in a molten state among existing rock) beneath strata, it bends the strata upwards [A]. Continued upfolding of the strata later exposed through erosion forms a dome mountain [B].

2 Pressure causes rocks to bend, first forming gentle folds [A]. These folds are subsequently eroded along their crests to form valleys [B]. The valleys may become deeper than the adjacent synclines; this is called inverted relief.

3 Sometimes rocks do not bend easily, but are instead upfaulted as huge blocks [A]. Erosion then rounds off the faulted edges to form block mountains [B] and stream action cuts the up-lifted blocks into peaks.

4 A volcano [A] is formed when gas and molten rock escape through an opening at the surface of the earth to relieve pressures beneath the crust and solidify round the opening. Continued outpouring of ash and lava may build up a mountain [B].

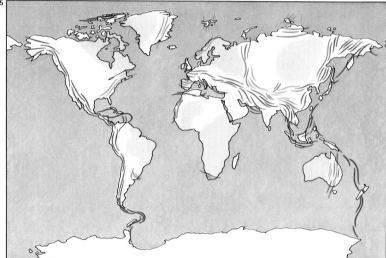

5 Continental areas of the world consist of stable shields surrounded by mobile belts. Between the shields material has been crushed up into mountain chains by collision between plates and continents. The Ural, Alpine and Himalaya mountains were formed by continental collisions, while the American cordillera was formed by collision between a continent and a mobile plate. The East African rift valley indicates that the shield is splitting up to form new oceanic areas.

▤	Cenozoic mobile belts
▤	Mesozoic mobile belts
▤	Upper Palaeozoic mobile belts
▤	Lower Palaeozoic mobile belts
▤	Precambrian shield areas

6 Geosynclines are the birthplaces of mountains. They are large troughs where thick layers of sediments can accumulate [A]. Where geosynclines develop between two colliding crustal plates [B] the sediments can be squeezed up as broad ridges known as geanticlines [C]. Further compression creates mountain ranges [D]. The whole process is usually accompanied by pressure-induced recrystallization or melting, which then forms metamorphic, plutonic and volcanic rocks; examples of each of these are are gneisses, granites and rhyolites.

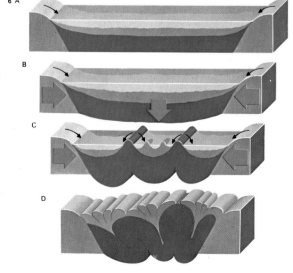

another tends to maintain an upward pressure, rather like a submerged cork seeking to regain the surface: in time the stationary plate is levered upwards and the attached fold mountains move with the plate. The Himalayas were formed when the northern edge of the Indian continental plate collided with and slid under the Asian plate; the Asian plate was then lifted and the world's highest mountain range was created.

Death of mountains
Mountains are sculptured and destroyed by the climatic forces of frost, water (in the form of snow, ice and rain) and wind. Frost may shatter and break up rocks to form screes (masses of debris at a cliff base) and snow and glaciers gouge out rock debris and transport it down the mountainside, leaving the debris as an elongated moraine at the tip of the glacier. Lower down, rivers cut into the mountainside and form zigzag valleys with interlocking spurs. These spurs may in turn be sliced off by glaciers making their way to lower levels down the mountain. In short the erosion of mountains is the continuing story of the breakup of rocks and their gradual descent under the influence of gravity.

In time, weathering and erosion destroy mountains by lowering them so much that they are eventually transformed into broad plains cut by slowly meandering rivers. In arid climates wind erosion may finish the work by sand-blasting the remaining hills into a bare desert, leaving a surface known as a peneplain [Key] – that is, almost a plain. This stage is rarely reached; more commonly, renewed earth movements uplift the area again, so beginning a new geological phase.

The study of mountains
Mountains help geologists to understand plate structures and to learn more about how rocks behave when they are compressed by moving continents. Mountains also mark the positions of ancient plate boundaries in, for example, Mesozoic-Cenozoic times when great ranges such as the Himalayas were being formed. Similarly, the study of ancient mountain ranges also reveals the sites of ancient oceans, enabling scientists to reconstruct the past geography of the planet.

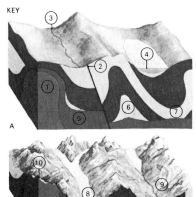
KEY

Three stages in the life of a mountain are shown here. [A] Idealized young complex mountain: [1] granite batholith; [2] major fault offsetting strata; [3] formation of a young stream; [4] sea-level; [5] metamorphic rocks; [6] anticline of upfolded strata; [7] syncline of downfolded strata. [B] Mature complex mountain: [8] glacier scouring U-shaped valley; [9] glacial meltwater forming active stream; [10] erosional "Matterhorn" peak – top of granite batholith exposed by erosion. [C] Peneplain of old complex mountain: [11] peneplain due to total erosion of mountains; [12] rivers reworking sediments of the peneplain; [13] remnant of eroded mountain.

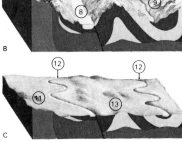

7 The Mont Blanc massif, a lofty mountain range in the French Alps, is typical of the popular conception of mountains. The cold climate due to the high altitude allows frost to split the rocks and, with the aid of glaciers, to carve the mountains into serrated ranges of jagged peaks.

8 This peneplain in the Northern Territory of Australia was formed when a mountain range was eroded to an almost flat surface. Monadnocks rise above the plain showing the former positions of peaks or hard rocks. The surface is scoured by wind and this forms a sandy desert.

9 The Canadian Rockies, in their western part, consist of intensely metamorphosed strata [1]. High pressures here melted granites. These granites, expanding with heat, became lighter than the overlying rocks and rose up through them as intrusions [2]. The uplift sheared the Palaeozoic strata [3] to the east along low-angle faults [4] which also separate these strata from the underlying crystalline basement [5]. The piling-up of these slabs by thrust-faulting finally led to a considerable thickness of sedimentary rocks.

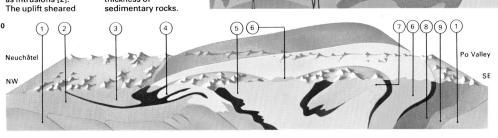

10 The highly folded strata of the western Alps have not only been crumpled and folded but also shoved great distances northwards as nappes. Where pressure and heat were sufficient the sedimentary strata were also transformed into metamorphic rocks such as gneiss and schist [5, 6, 7, 8]. The areas shown here are [1] basement; [2] flysch zone; [3] Pre-Alps; [4] Helvetic nappes; [5] Dent Blanche nappe; [6] Mont Rosa nappe; [7] Ivrée zone; and [9] Dinaric Alps.

Earth's water supply

Water is essential to all life on earth. The study of the earth's water supply, called hydrology, investigates the distribution of water, how it is used by man and how it circulates from the oceans to land areas and back again in the hydrologic or water cycle.

The water cycle – from oceans to land
About 97 per cent of the world's available water is in the oceans [Key]. Oceanic water is salty and unsuitable for drinking or for farming. In some desert regions, where fresh water is in short supply, seawater is desalinated to make fresh water. But most of the world is constantly supplied with fresh water by the natural process of the water cycle [1] which relies on the action of two factors: the sun's heat and gravity.

Over the oceans, which cover about 71 per cent of the earth's surface, the sun's heat causes evaporation. Water vapour, an invisible gas, rises on air currents and winds. Some of this water vapour condenses and returns directly to the oceans as rain. But because of the circulation of the atmosphere, air bearing large amounts of water vapour is carried over

land where it falls as rain or snow (precipitation).

Much of this precipitation is quickly re-evaporated by the sun. Some soaks into the soil where it is absorbed by plants and partly returned to the air through transpiration. Some water flows over the land surface as run-off, which collects into rills and flows into streams and rivers. Some rain and melted snow seeps through the soil into the rocks beneath to form ground water.

In polar and high mountainous regions most precipitation is in the form of snow. There it is compacted into ice, forming ice sheets and glaciers. The force of gravity causes these bodies of ice to move downwards and outwards and they may eventually return to the oceans where chunks of ice break off at the coastline to form icebergs. Thus all the water that does not return directly to the atmosphere gradually returns to the sea to complete the water cycle. This continual movement of water and ice plays a major part in the erosion of land areas.

Of the total water on land, more than 75 per cent is frozen in ice sheets and glaciers as

in Greenland and Antarctica [Key]. Most of the rest (about 22 per cent) is water collected below the earth's surface and is called ground water. Comparatively small quantities are in lakes, rivers and in the soil. Water that is held in the soil and that nourishes plant growth is called capillary water. It is retained in the upper few metres by molecular attraction between the water and soil particles.

Ground water and the water-table
Ground water [4] enters permeable rocks through what is called the zone of intermittent saturation. This layer may retain ground water after continued rain but soon dries up. Beneath this lies a rock zone where the pores or crevices are filled with water. It is called the zone of saturation and usually begins within 30m (98ft) of the surface, extending downwards until it reaches impermeable rock through which water cannot percolate. This impermeable rock layer, lying below the water-holding layer, or aquifer, is called a ground water dam.

The top of the saturated zone is called the water-table. This is not a level surface. It is

CONNECTIONS

See also
112 Rivers and lakes
110 Caves and underground water
168 Water and irrigation

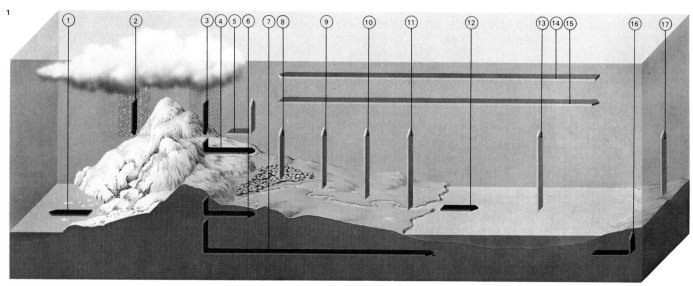

1

1 The hydrological or water cycle is the process whereby water, in some form, circulates from the oceans to land areas and back again. Fresh water is present on the earth as water vapour in the atmosphere, as ice and as liquid water. The elements of the cycle are precipitation as rain [3]; surface run-off [4]; evaporation of rain in falling [5]; ground water flow to rivers and streams [6]; ground water flow to the ocean [7]; transpiration from plants [8]; evaporation from lakes and ponds [9]; evaporation from the soil [10]; evaporation from rivers and streams [11]; evaporation from the oceans [13]; flow of rivers and streams to the oceans [12]; ground water flow from the ocean to arid land [16]; intense evaporation from arid land [17]; movement of moist air from the oceans [14] and to them [15]; precipitation as snow [2]; ice flow into the sea [1].

2 Sandstone (shown here in cross-section) is a highly porous rock through which water percolates easily.

3 Limestone is a permeable but non-porous rock. Water can percolate only through the joints and the fissures.

4 Ground water seeps through the zone of intermittent saturation [1] until it reaches an impermeable layer above which it forms the zone of saturation or aquifer [2, 10]. The upper surface of the aquifer forms the water-table [3, 13], above which is the capillary fringe [6]. Wells [7] must be sunk to the water-table because the capillary fringe is not saturated. Impermeable dikes [8] block the flow of ground water. In uniform material the water follows paths [4] that curve down and up again towards the nearest stream. If an aquifer is part of a series of strata including several impermeable layers [11] a perched water-table [12] may result. If it lies between two impermeable strata it is said to be confined [14]. Its recharge area [15] is where water enters the confined aquifer. A stream below the water-table is called a gaining stream [5] while a stream flowing above it is known as a losing stream [9] because it loses water by seepage.

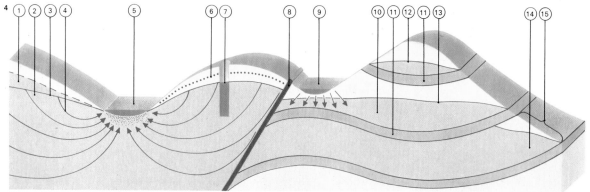

often arched under hills, but beneath plains it generally lies closer to the surface. The water-table also varies in level during the year, depending on the amount of rainfall.

In some places the water-table intersects the surface, forming such features as oases [6] in desert hollows, swamps, lakes and springs. Springs [5] are gushes or seepages of water that may occur along the base of a hillside or in a valley in the hills. They are found where the water-table or an aquifer appears at the surface or where the aquifer is blocked by an impermeable rock such as a volcanic dike. Spring water is usually fresh and clean because it passes through the fine pores of rocks such as sandstones [2] where impurities are filtered out in much the same way as domestic water is purified by sand filtration.

Limestone is not a porous rock but it is permeable – that is, ground water can seep through the maze of fissures, joints and caves in the rock [3]. These apertures are enlarged by rainwater containing dissolved carbon dioxide – a weak carbonic acid that dissolves limestone. In limestone ground water is not filtered in the same way as in porous rocks. In

the late 1800s epidemics of cholera and typhoid often occurred in France in areas where springs emerged from limestone areas. It was finally established that the spring water had been contaminated miles away by rubbish thrown into pot-holes.

Some springs contain so much mineral substances in solution that their water is used for medicinal purposes and spa towns have grown up around them. Occasionally these springs are thermal.

Water from artesian wells

The lowest level of the water-table, reached at the driest time of year, is called the permanent water-table. Wells must be drilled to this level if they are to supply water throughout the year. In artesian wells [7] water is forced to the surface by hydrostatic pressure – this results from the rim of the well being below the level of the water-table in the catchment area. Artesian water is obtained from porous sandstone aquifers that underlie the Great Artesian Basin of Australia. These aquifers are supplied with water from the rain that falls on the Eastern Highlands.

13,000 cu km
230,250 cu km
8,637,000 cu km
29,200,000 cu km
1,322,000,000 cu km

The total water supply of the world is estimated to be about 1,360,000,000 cubic kilometres, and 97.2 per cent of it forms the oceans. Of the remainder, 2.15 per cent is frozen in ice caps and glaciers, and most of the rest is in rivers and lakes (0.0171 per cent) or under land areas as ground water (0.625 per cent). Water vapour represents only 0.001 per cent but this quantity is vital: without it there would be no life on land.

5 Springs appear where the water-table meets the surface. [A] Springs may occur where a fault brings an aquifer into contact with an impermeable layer. [B] Water pressure creates artesian springs at points of weakness. [C] Water seeps through jointed limestone until it emerges above an impermeable layer. [D] Springs form where permeable strata overlay impermeable rock. [E] An impermeable barrier may lead to the formation of a spring line.

6 An oasis is an area in a desert that is made fertile by the presence of water. Some oases are found along rivers crossing the deserts, such as the Nile. Others owe their moisture to underground waters reaching the surface or near-surface. Wadis are intermittent streams that flow only after heavy rainfall but they often have a hidden flow under their beds which can reach the surface to create oases. Aquifers with recharge areas outside the desert can "pipe" water to oases under long stretches of arid desert. These recharge areas are usually mountains suitably sited for catching rain. The natural flow can be increased by pumping the ground water but if the rate of pumping exceeds the water flow into the aquifer in the recharge area the wells will dry up.

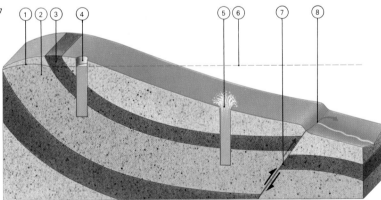

7 Artesian springs and wells are found where ground water is under pressure. The water-table [1] in the confined aquifer [2] lies near the top of the dipping layers. A well [4] drilled through the top impervious layer [3] is not an artesian well because the head of hydrostatic pressure [6] is not sufficient to force water to the surface. In such wells the water must be pumped or drawn to the surface. The top of an artesian well [5] lies below the level of the head of hydrostatic pressure and so water gushes to the surface. Artesian springs [8] may occur along joints or faults [7] where the head of hydrostatic pressure is sufficient to force the water up along the fault. Areas with artesian wells are called artesian basins. In the London and Paris artesian basins the water has been so heavily tapped that the water level has dropped below the level of the well heads.

Caves and underground water

Most of the earth's surface has been mapped, but in many areas vast networks of caves, largely unexplored, lie beneath the ground. There are several kinds of caves, including coastal caves, ice caves and lava caves. The largest cave systems occur in carbonate rocks (limestone and dolomite), most of them forming in massive layers of limestone.

Formation of caves

Limestone is a fairly hard rock formed from calcium carbonate. Although insoluble in pure water, limestone is dissolved by rainwater containing carbon dioxide from the air and from the soil. Rainwater reacts chemically with limestone and converts it to soluble calcium bicarbonate. Limestone is riven by joints (vertical cracks) and by bedding planes, which are usually horizontal. When limestone is exposed on the surface, rainwater widens the joints into "grikes", dividing the limestone into blocks called clints. This broken pavement surface is a feature of karst scenery, named after the limestone Karst district of the Dinaric Alps in western Yugoslavia

Some authorities suggest that limestone caves are formed when rainwater slowly enlarges the joints and bedding planes as it seeps down to the water-table. Eventually, streams flow into the enlarged joints which form sink-holes or swallow-holes. Such streams may flow underground for many kilometres, dissolving vertical chimneys and horizontal galleries.

However, other authorities do not consider that this explanation accounts for cave networks that have underground chambers with high roofs. Such caves, they argue, must have been formed when the land surface was far higher than it is now and when the limestone was completely saturated with ground water [1]. They believe that, under pressure, the ground water seeped through the rock, until it finally emerged at the surface as a spring. Eventually, the forces of erosion planed down the land surface, the water-table dropped and air entered the dissolved caves. Sink-holes might have been formed when the roofs of caves collapsed.

The Mammoth Cave National Park in Kentucky, USA [Key], is the site of the world's most extensive cave network, with a total mapped passageway of 231km (144 miles), linking it with the Flint Ridge cave system. One of the deepest-known caves is the Gouffre de la Pierre St Martin in the western Pyrenees in France, which drops 1,174m (3,850ft). The largest underground chamber is the Big Room in the Carlsbad Caverns, New Mexico. At a depth of 400m (1,320ft), the Big Room is 1,300m (4,270ft) long, 100m (328ft) high and 200m (656ft) wide.

Features of caves

Limestone caves contain many features formed from deposits of calcium carbonate, including icicle-like stalactites [6] and pillar-like stalagmites [5]. Stalactites develop when water that is highly charged with dissolved calcium bicarbonate seeps through holes in the roofs of caves. Drops of water that hang on the roof are partly evaporated and a tiny quantity of calcium carbonate is precipitated and sticks to the roof. Another drop of water deposits a second film of calcium carbonate in the same place and, in this way, stalactites slowly develop.

CONNECTIONS

See also
108 Earth's water supply
102 Sedimentary and metamorphic rocks

In other volumes
26 Man and Society

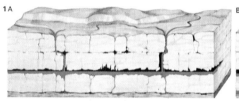

1 A   B   C

1 As rain falls, it dissolves carbon dioxide from the atmosphere and becomes a weak carbonic acid that attacks carbonate rock (limestone and dolomite) by transforming it into the soluble bicarbonate. Carbonate rocks are criss-crossed by vertical cracks and horizontal breaks along bedding planes [A]. Some geologists believe the caves were formed when the rock was saturated by water. Others believe they formed gradually by solution [B] into a major cave network [C].

2 Limestone surfaces are often eroded into blocks called clints [1]. Surface streams flow into dissolved sink-holes [2], that lead to a deep chimney [3]. Pot-holes [7] are dry chimneys. Gours [4] are ridges formed as carbonate is precipitated from turbulent water. Streams flow at the lowest level of the galleries [17]. Abandoned galleries [13] are common. A siphon [12] occurs where the roof is below water-level. Streams reappear at resurgences [20]. Abandoned resurgences [19] may provide entrances to caves. Stalactites [5] include macaroni stalactites [6], curtain stalactites or drapes [11] and eccentric stalactites [16], formed by water being blown sideways. Stalagmites [14] sometimes have a fir-cone shape [15] caused by splashing, or resemble stacked plates [8]. Stalactites and stalagmites may merge to form columns [10]. Signs of ancient man [18] have been found in caves and blind white fish live in the pools [9].

Drops of water that fall on the floor may also be partly evaporated, leaving small deposits of calcium carbonate that grow upwards into stalagmites. Often it is the impact of the drop hitting the floor that forces the carbonate out of solution. The splashing of the water can give rise to stalagmites that resemble stacks of saucers. Stalactites and stalagmites sometimes meet to form a continuous column [6].

The growth of stalactites and stalagmites is usually extremely slow. Some take 4,000 years to increase by only 2.5cm (1in) in length. However, stalactites in Ingleborough Cave, Yorkshire, have been known to increase by 7.6cm (3in) in ten years.

Another deposit, caused by water seeping through a long crack in the roof of a cave, is a wavy band of calcium carbonate which grows across the ceiling like a fringed curtain. Water flowing down a wall or across a floor of a cave may build up a flowstone. Delicate thread- or finger-like formations called helictites sometimes jut out from a stalactite. Their origin is the subject of dispute. On the roofs of some caves are anthodites, which are branching, flower-like formations. In its pure state, calcium carbonate is transparent or white, but these and other cave features are often coloured by impurities.

Life in caves

Caves harbour a variety of animal life specially adapted to the dark environment, including blind, colourless, almost transparent shrimps, worms, mites, insects and sightless newts, often called blind fish. These creatures live permanently in caves. Bats, also common in cave systems, have weak eyes and depend mainly on their sonar systems to guide them through dark tunnels. Every night hundreds of thousands of bats emerge from the Carlsbad Caverns in New Mexico. Within 15 minutes' flight of the caverns is the Pecos valley, where the bats feed on insects, returning to the caves shortly before dawn.

In prehistoric times the most important inhabitant of caves was man. Archaeologists have found many traces of man's occupation – tools, bones, hearths and, usually well inside the caves, rock paintings which may have had ritual or magical significance.

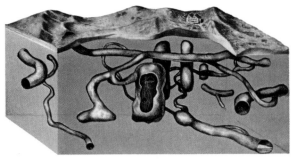

The Mammoth Caves in Kentucky comprise the world's largest underground cave network. The Mammoth Cave National Park consists of a limestone plateau, whose surface is pitted by more than 60,000 sink- or swallow-holes. Surface water drains into the sink-holes that link the caves to the surface. Some sink-holes are connected to the caves by vertical chimneys. The underground caves are interlinked by a maze of passages. The Great Mammoth Cave has more than 48km (30 miles) of continuous passages. In the caves, water seeping through the limestone rock collects into rivers, which finally emerge into the open at the base of the plateau, in the Echo River valley. This system is typical of the arrangement of galleries, caverns and pot-holes found in many limestone areas where acidic water has seeped into the rock and dissolved it.

3 Gours are formed when carbonate-rich waters flow over an irregular surface. The turbulence deposits calcite on the irregularities which grow into a series of ridges perpetuating the process

4 Balcony stalactites are formed by water dripping from the side of a cave wall. They are called stalactites because they grow downwards, but ordinary stalactites hang from the roof of a cave.

5 Stalagmites build upwards from the floor of a cave. They are generally shorter and thicker than stalactites. The tallest known stalagmite is 29m (95ft) high, in the Aven Armand Cave, in Lozère, France.

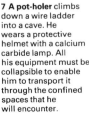

6 Stalactites, stalagmites and columns of calcium carbonate are present in this cave. The columns are formed when stalactites and stalagmites meet. Stalactites are fragile structures and, as they easily break off, do not usually grow to great lengths. The longest known stalactite is supported by a wall in the Cueva de Nerja, near Málaga in Spain. It is 59m (195ft) long and extends from the roof down to the floor.

7 A pot-holer climbs down a wire ladder into a cave. He wears a protective helmet with a calcium carbide lamp. All his equipment must be collapsible to enable him to transport it through the confined spaces that he will encounter.

Rivers and lakes

When rain falls on the ground it is either absorbed in the soil or runs downhill over the surface in small temporary gullies called rills. These unite to form a stream. Other streams start from springs, where water that has sunk into the ground comes to the surface, or from melting glaciers. Streams and rivers usually begin in mountains or hills and the downwards pull of gravity gives them energy to cut away the land and form valleys.

Erosion and transportation
The stones and sand formed by erosion of the rock are transported downstream and are finally deposited at the mouth of the river. Erosion, transportation and deposition are the main work of a river and most rivers can be divided into three sections: an upper course in which erosion dominates; a middle one where transportation occurs; and a lower one where deposition takes place [1].

Stream water erodes in two ways: chemically and physically. Weak acids such as carbonic and humic acids in the water help to decompose limestone and other rocks. The ability of a stream to erode mechanically is closely related to its speed. During normal flow little physical erosion takes place, but during flood the movement of water becomes turbulent and this causes eddies which in turn cause rapid changes in the pressure on the rocks. Sometimes the pressure is so low that a vacuum is formed on a small part of the stream bed; as the eddy changes this vacuum implodes (collapses inwards). Much of the babble of a brook is the sound of implosions. Repeated implosions cause part of the rock to be sucked into the stream and carried away. Erosion mainly takes place at this stage by stones banging into the stream bed and sides and so wearing them away. During the process the stones are broken up into smaller pebbles, so that the boulders in the upper course of the stream provide the sand grains that are present in the lower course.

The faster the stream flows the larger the fragments it can carry. It can also carry more of them. This is why most erosion and transportation occurs during floods. The finest particles are carried in suspension, kept up by the turbulent motion of the water. Eddies bounce the sand from the bottom and it is carried downstream a small distance by the current before it falls to the bottom again. Coarser material is rolled along the bed.

Deposition of sediment
As the river enters more gentle slopes, some of its sediment is dropped. Where there is a sudden change in gradient and therefore water speed, as when a river leaves the mountains and runs out onto a plain, nearly all the sediment will be dropped, forming an alluvial fan [2]. More usually the material is deposited *en route* as the river current slows up. The coarsest sediment is dropped first.

During a flood, however, river water moves at different speeds. In the river channel the current is fast moving, but where the river spreads over its banks on to the surrounding land (the flood plain) the current slows down and mud and very fine sand are deposited as the water leaves its channel. This forms a ridge or levee along each bank.

The long profile of a river [Key] – the plot of the elevation of the river against distance travelled with a suitable vertical exaggeration to show significant features – is theoretically

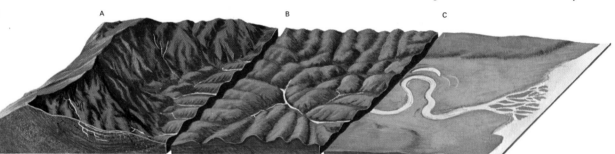

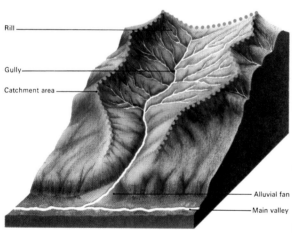

Rill

Gully

Catchment area

Alluvial fan

Main valley

2 Mountain streams are fed by rills meeting in catchment areas to form gullies which carry fast-flowing water to the main valley. Here velocity decreases and sediment is deposited as an alluvial fan.

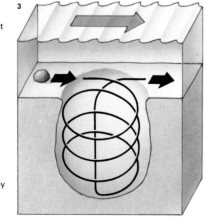

3 Pot-holes occur in the beds of swift rivers or streams. If a small depression is formed in the stream bed a pebble may be caught in it and swirled around by the water, enlarging the depression into a circular pot-hole.

1 A river changes from a small stream in the mountains to a slow meandering river near the sea. The course of the river is divided into three stages. In its upper course [A] the river is fast flowing and able to wear away the rocks. In its middle section [B] it flows slowly and carries sediment to the lower section [C].

4 A valley is formed by two processes. The river cuts downwards, taking out a narrow slice of rock immediately below its bed to form a V-shape [A]. Weathering widens the valley by changing the rocks forming the sides to soil [B]. As the velocity of water decreases, lateral erosion widens the valley floor [C, D]. In its advanced stage [E] the river flows slowly through a flat plain with deposited material forming levees or dikes.

5 San Juan River, Utah 1941

9 September Normal

15 September Surface raised 3m

14 October Surface raised 4·5m

Normal surface

Bed scoured and lowered 1m

26 October Surface lowering

Bed filling

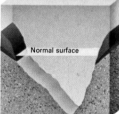

5 Rivers in their lower courses run over deep channels cut in the bedrock. These channels are normally filled with sand and the river runs in a shallow channel on top. During floods the river deepens its channel by moving the sand below it. Only during the largest flood is the river able to scour the rock bottom. In a large flood the river may be ten times deeper and carry 100 times more water.

a part of a hyperbola, being steepest at the source and flattening at the mouth. This is an equilibrium curve towards which the stream tends to adjust its gradient, digging into its bed and removing material from the upper course and depositing it as the speed drops in the lower. However, this is highly idealized and in practice any number of factors can affect it – differences in rock types in the river bed and the addition of water from tributaries, for example, may produce many irregularities in the hyperbola profile.

The course of a river

Stream beds in their upper section are often bare rock patchily covered by pebbles. Here the stream has greatest capacity to erode and transport farther downstream all but the largest stones. The valley in the upper course has steep sides and a V-shaped cross-section and most pools, rapids, waterfalls [6] and pot-holes [3] occur here, caused by the stream wearing away softer rock more quickly than hard rock. This results in rapids such as the cataracts of the Nile, and where a river flows from a hard bed of rock to a soft

one the latter will be eroded away and a waterfall will be formed as a result.

In the middle section most of the irregularities have been worn away, allowing the river to flow freely in a fairly flat channel. The current is just strong enough to carry most of the sediment supplied to it from higher up. It does not erode downwards and most of the time runs on its own sediment.

In the lower section the river has a very low gradient, often less than 10cm per kilometre (2in per mile). It flows across a broad floodplain. Where the river is flowing slowly, it cannot move stones, even in flood, but because it is large it is able to move a huge amount of fine material. The Mississippi carries about 500 million tonnes of fine sand and mud past New Orleans each year. The river there meanders over a thick layer of its own sediment [7].

When the river reaches the sea the sediment it is carrying is deposited. In some areas tidal currents are strong enough to remove it and the river ends in an estuary. Where more sediment is brought down than can be removed by the sea a delta is formed [10].

Natural obstruction renewing the graded profile

A river conforms roughly to a convex upward curve that is nearly flat near the sea and gets steeper and more curved inland. This shape is called a graded profile. Waterfalls, lakes and deltas may vary the shape of the stream bed without fundamentally altering the profile.

6 Pools and waterfalls are both caused by hard bands of rock spanning the river bed. The softer rocks below a pool have been eroded away leaving a hollow and the hard rock stands up like a dam. Lakes are usually caused by landslides blocking the course of the river or by ice (during the Ice Age) scraping deep hollows. The lakes in the English Lake District and the Great Lakes of America are hollows left by the ice. Other large lakes such as those of East Africa were caused by earth movements. Water flowing over a hard bed erodes the softer beds below, causing a waterfall with a plunge pool beneath. Over time, erosion causes the face of the waterfall to retreat, leaving a gorge downstream of it. Niagara Falls is formed from a hard bed of nearly level rock and has a gorge 10.4km (6.5 miles) long below it.

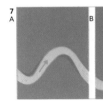

7 Meanders occur where the slope is shallow. In a river bend, the water flows more slowly along the inner bank, depositing sediments and building up the bank, but flows faster along the outer bank, eroding it away. Thus the meander becomes more pronounced [A, B, C] until the arms intersect, allowing the flow to take a shorter route [D]. The abandoned arm silts up [D], forming an oxbow lake [E].

8 In its middle stage a river flows through gently sloping areas. Its eroding and transporting powers are considerably reduced and it runs over a broad flat valley bottom formed by its own deposits of alluvium. Erosion takes place only during floods. The river meanders [2] and the beginnings of floodplains [1] and levees [3] are evident. An oxbow is shown in the process of formation [5] by the river cutting through a meander [4].

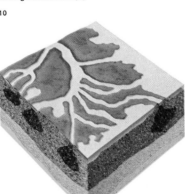

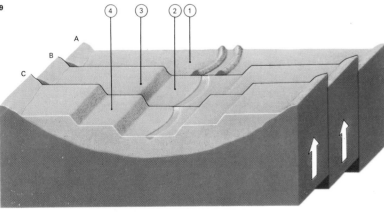

9 A terrace is a flat strip of land along the valley side just above the floodplain. A terrace is formed when the land is uplifted or the sea level drops [B] and the river begins to cut into its floodplain [1] and forms a new one [2] at a lower level. The old floodplain becomes a terrace [3]. Another uplift [C] would cause a new terrace [4]. [A] represents the river valley before uplift.

10 A delta is formed where a river enters the sea or a lake. Here all the sediment is dropped, forming a huge, gently sloping mound on the sea-floor. This builds up, causing the river to flow over it to get to the sea. The river branches into separate streams called distributaries. Deltas are found at the mouths of such rivers as the Nile, Mississippi and Ganges. Some rivers have no deltas as sea currents carry away the rivers' sediments.

113

Land sculptured by rivers

Heavy and prolonged rain may make level ground waterlogged. But once the rain has stopped, the ground will dry out as the water sinks into it. In hot weather standing water will evaporate and plants will absorb water through their roots, transpiring it from their leaves. Sloping ground drains quickly, for the water that cannot sink into the ground flows downhill in rills, then in streams and finally in rivers. That part of the rain that has percolated into the ground will emerge later, at a lower level in the terrain, as a spring and flow away as a stream or river.

Erosion of the land
Water moving downhill will carry with it any particles that it can move. So moving water wears away – erodes – the ground over which it flows. In the course of time rivers have sculptured out their valleys in this way.

In some areas man's activities have greatly increased the erosive effects of rainfall. Too intensive cultivation of southern areas of the United States in the 1700s broke up the protective cover of vegetation that the settlers found there. Heavy rains, falling on

the cleared ground, ripped out rills that quickly widened and deepened into a mosaic of gullies. Strong winds, blowing away the soil, hastened the development of such areas, known as "badlands" [5].

Landforms and drainage pattern
As soon as an area of the earth's crust is uplifted above sea-level, the process of erosion begins. The rain falling on it will develop a river system. The rivers will deepen and widen their valleys until in the course of time the whole area is reduced to a low surface – assuming, that is, that there has been no further uplift to rejuvenate the drainage and start a new episode of vigorous downcutting. The inner gorge or canyon of the Colorado River was cut into a much wider, older valley. The drainage pattern and the landforms produced are determined by the composition and disposition (structure) of the underlying rocks [10].

Rivers will quickly emphasize any differences in the hardness of the rocks over which they flow. In their upper reaches, the more resistant bands of rock form waterfalls and

rapids in the narrower parts of the valley. If the rocks are lying horizontally, the topography developed is characterized by flat-topped hills [7]. But if the beds are tilted, scarpland topography is produced, in which the more resistant layers form cuestas whose steeper sides face up the inclination (dip) of the rocks and vales are worn out on the outcrop (strike) of the softer beds [8]. The trellised drainage pattern may undergo minor changes. A particular river, perhaps because it has more powerful springs at its source, or greater runoff from the valley sides, or a shorter course to the sea, may cut down the level of its valley floor more quickly than its neighbour and eventually capture it [11].

In areas of gently folded rocks, inverted relief may develop, the river valleys being eroded along the line of the upfolds (anticlines), while downfolds (synclines) underlie the higher ground [9]. Snowdon in North Wales is an example of a synclinal mountain.

Where the beds are more tightly folded, or where near-vertical bodies of igneous rock have been intruded into gently dipping strata, hogsback ridges, steep on both sides, will be

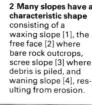

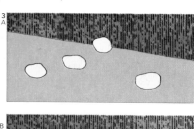

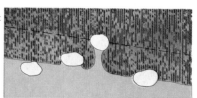

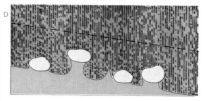

1 The rock formation of a hillside will be gradually broken down into stones and finally soil by water, wind, chemicals and changing temperatures. This loose material will move downhill under the force of gravity. Slopes often show signs of such soil creep, the commonest indication being terracettes [1]. Some slopes have countless little steps that look like sheep tracks across them. Other signs are leaning walls [2], trees with bent trunks [3], a much higher level of soil on the uphill side of a wall [4] and vertical strata curled over where it has been exposed [5].

2 Many slopes have a characteristic shape consisting of a waxing slope [1], the free face [2] where bare rock outcrops, scree slope [3] where debris is piled, and waning slope [4], resulting from erosion.

3 Earth pillars are formed where large rocks occur in the soil [A]. These shelter the underlying soil from erosion and form pillars [B, C]. Once the stone has fallen [D] the soil is easily washed away.

4 The river erosion of an area involves several stages. A newly raised plateau is usually quite flat; the rivers cut into it to form deep gorges. The picture shows the Blue Nile, which is cutting downwards through the African plateau.

5 When rivers have deeply dissected a plateau, their valleys widen so much that the plateau areas between are reduced to isolated peaks. As this picture of the US Bad Lands shows, the peaks tend to be the same height – that of the original surface.

6 The ultimate stage in river erosion is the peneplain – a flat area of land from which most traces of the original plateau have disappeared. In the state of Utah in the USA the desert floor is a good example of a well-eroded peneplain.

produced, while the rivers will erode belts of weaker strata or the line of faults. The Great Glen cut through the Highlands of Scotland from Fort William to Inverness is an example of such a fault-guided valley, but glacial erosion has greatly deepened the valley that had been cut by rivers in pre-glacial times.

Superimposed drainage

Not all river systems are clearly related to the geological structure of the area across which they are now flowing. The drainage system of the English Lake District is clearly radial in plan, but the strike of the lower Palaeozoic rocks (570–395 million years old) runs southwest to northeast. Surrounding the Lake District is a ring of gently outwardly dipping upper Palaeozoic (395–225 million years old) strata. The present drainage system must have originated when these upper Palaeozoic rocks were uplifted to form a dome. Millions of years of erosion have removed all trace of these rocks and the drainage of the Lake District is now superimposed on the lower Palaeozoic rocks of different structure. In the future, the rivers will

gradually change and adjust to this structure.

More extreme superimposition, sometimes called antecedent drainage, is found in India, where the River Brahmaputra has flowed from the Asian plateau to the Indian Ocean since early Tertiary times (about 60 million years ago), before the formation of the Himalayan mountain chains. But their rate of uplift was slow enough for the river to maintain its course across the rising mountains and now it flows through them in stupendous gorges.

In many limestone areas, including chalk downlands, there is a complete valley system, but most of the valleys are now dry with no flowing streams in them. Limestone is a highly permeable rock, so that any rain quickly seeps into it to add to the ground water at depth. That is the position under the present climatic conditions, but in the past rainfall may have been much greater. The level of the ground water would then rise and springs break out higher up the valley sides. During glacial episodes, rainfall or meltwater could not seep into the frozen ground, but must have flowed away, carving the valleys.

Landforms are the result of two conflicting processes. Movement deep within the earth may uplift areas of the crust, while weathering and erosion continually sculpture the surface of the land, wearing it down again. The shapes of individual hills depends on the climate, the structure of the rock and the rate of the lateral and also the longitudinal erosion.

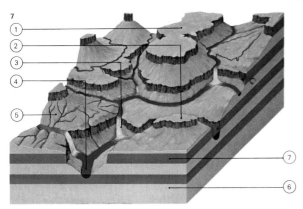

7 In an area underlain by horizontally bedded rocks, rivers follow a dendritic pattern. Their valleys are often steep sided but stepped where erosion has had greater effect on weak strata than on resistant beds. Mesas – isolated tablelands – may form, which may then be eroded to narrower buttes. Landscapes of horizontally bedded rocks are more distinctive in arid regions where rain falls in sharp bursts causing the rivers to swell to raging torrents. Features of such a landscape are: [1] mesa; [2] butte; [3] waterfall; [4] canyon; [5] badlands; [6] weak strata; and [7] more resistant strata.

8 Sedimentary rocks laid down in horizontal layers are often tilted by later earth movements. Main rivers flow down the slope (dip) of the beds and erosion etches out the difference in hardness of the rocks to produce scarpland topography. Tributary rivers flow along the strike vales (running at right-angles to the dip of the rocks) on the outcrop of softer beds [1], while harder rocks [2] are weathered to form features called cuestas [3] with a steep scarp face and a gentle dip slope parallel to the dip of the beds. An intrusion of steeply dipping resistant rocks forms a hogsback [4].

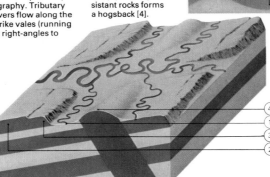

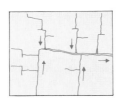

9 In folded-rock areas erosion attacks the raised anticlines [1] more readily than the trough-like synclines [2] because anticlinal flexing of rocks tends to form cracks open to the weather. If this process goes far enough, the result is an inverted relief where the deeper valleys follow the anticlines [3] and the former troughs form the summits [4].

10 The pattern of a river and its tributaries is related to the rocks on which it formed or now flows. On rocks of equal resistance, a dendritic drainage pattern [A] develops. In areas of alternating hard and soft rock, the stream follows the soft rock, forming a trellis [B]. A radial river pattern forms on volcanoes and rock domes [C].

11 River capture is the result of one stream [1] eroding the land at its source. In the process this stream eats into the catchment area of a lesser stream [2] and eventually drains it completely. This leaves a large stream with a sharp elbow of capture and a small "misfit" stream running through a large valley [3].

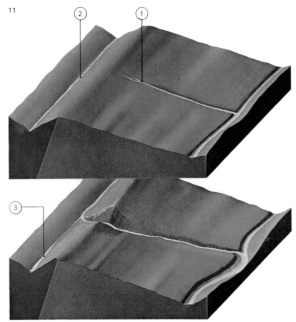

115

Rivers of ice

Ten per cent of the earth's land surface is covered by glaciers, the relentless and irresistible rivers of ice that are the sculptors of dramatic landscape – the peaks and valleys of the high mountains, the fiords and sea lochs of northwestern Europe, of Greenland, Canada, Chile and New Zealand. Many existing landforms were created by the action of ice (which both destroys old features and creates new ones) during the ice age of the Pleistocene, when as much as 30 per cent of the land surface was glaciated.

Glaciers are formed wherever there is perpetual snow; in other words, they are found in polar regions and on high mountains. As the snow accumulates year after year the older layers are compressed into a granular mass called névé, which later becomes firn when all air is expelled from it. Under the force of gravity this mass starts to move down the slope. As it does so it becomes further compacted into clear, compressed glacier ice.

There are three main types of glaciers: the mountain or valley glaciers, which have their sources in the mountains above the snow line; the piedmont glaciers, formed by the joining of valley glaciers as they spread out at the foot of the mountains; and finally the ice caps, which spread over their source area.

The movement of mountain glaciers

In 1788 the Swiss physicist Horace de Saussure (1740–99) lost an iron ladder on an Alpine glacier. It was found, 44 years later, 4,350m (14,250ft) lower down, thus demonstrating glacial movement.

Ice is a crystalline solid, but it can deform and flow when subjected to a sustained pressure. In glaciers this occurs by slippage of the ice crystals, which are lubricated along their boundaries by a thin layer of liquid water melted by the pressure. The downward motion [2] of the glacier can be seen at its very top, where it is separated from the permanent snow by a deep crevasse known as the bergschrund. Lower down, the movement can be observed by taking sights from fixed points on the mountainside along rows of stakes planted across the glacier. These also show the differential movement of the ice, for a glacier moves faster in its centre

than it does along its edges, where it is slowed by friction. Along a vertical section the speed is fastest on the surface slab, which behaves in a rigid fashion (it breaks, forming crevasses) [4], and the speed decreases towards the bottom. Longitudinal crevasses appear in the surface slab owing to the increasing rate of flow towards the axis of the glacier tongue or owing to the widening of the glacier; transverse crevasses are formed where the slope suddenly increases. Where transverse and longitudinal crevasses intersect, a spectacular ice topography of blocks or pinnacles of ice known as séracs is created.

The rate of movement of a glacier varies considerably and is dependent on the slope, thickness, cross-sectional area, roughness of the bottom and the temperature. Rates can vary from a few centimetres a day to several hundred – 66m (216ft) in the case of an Alaskan glacier.

A glacier can be divided into an upper section, where the temperature prevents melting and more ice is formed, and a lower section where the temperature is higher and ice is lost through melting. A steady rate is

CONNECTIONS

See also
118 Ice caps and ice ages
106 Life and death of mountains

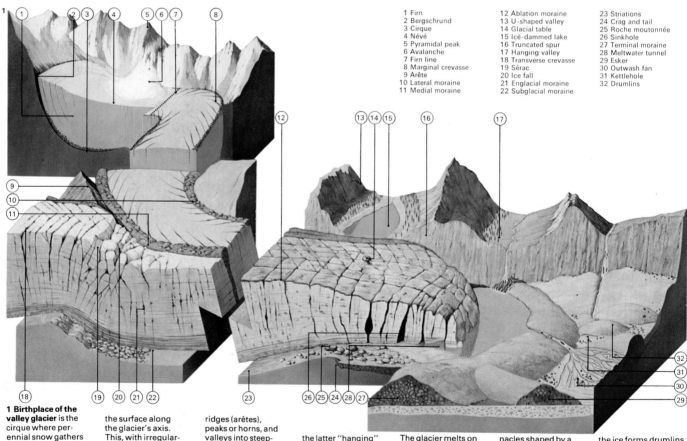

1 Firn	12 Ablation moraine	23 Striations
2 Bergschrund	13 U-shaped valley	24 Crag and tail
3 Cirque	14 Glacial table	25 Roche moutonnée
4 Névé	15 Ice-dammed lake	26 Sinkhole
5 Pyramidal peak	16 Truncated spur	27 Terminal moraine
6 Avalanche	17 Hanging valley	28 Meltwater tunnel
7 Firn line	18 Transverse crevasse	29 Esker
8 Marginal crevasse	19 Sérac	30 Outwash fan
9 Arête	20 Ice fall	31 Kettlehole
10 Lateral moraine	21 Englacial moraine	32 Drumlins
11 Medial moraine	22 Subglacial moraine	

1 Birthplace of the valley glacier is the cirque where perennial snow gathers as névé and is compressed to firn. Pulling away from the valley head, the glacier forms a crevasse, the bergschrund. The ice flows fastest at the surface along the glacier's axis. This, with irregularities of the glacier bed, creates crevasses; where these intersect séracs are formed. Glaciers carve the mountains into ridges (arêtes), peaks or horns, and valleys into steep-sided troughs. U-shaped valleys, where glaciers once flowed, have floors that are deeper than those of tributary valleys, leaving the latter "hanging" and often draining by a waterfall. The load of moraine (rock debris) carried away by the glacier deposited at its snout is called till. The glacier melts on the surface along its lower section. The amount of melting can be judged from the height of the glacier tables – unmelted ice pinnacles shaped by a moraine boulder. Meltwaters form subglacial streams which deposit long, winding piles of rubble called eskers under the snout. Other material under the ice forms drumlins and terminal moraines are deposited by retreating glaciers while stationary.

2 The material in a glacier moves constantly downhill. The position of the glacial snout, however, may move downhill [A], remain stationary [B] or move uphill [C] depending on the rate of melting of the ice [2] compared with the rate at which it accretes [1].

achieved when the accretion of ice in the upper section roughly balances that lost lower down. In these conditions the foot of the glacier is stationary. If the climate becomes cooler, the foot of the glacier advances until a new state of equilibrium is reached; under warmer conditions the foot of the glacier retreats to an equilibrium.

Erosion and transport
A glacier is one of the most powerful agents of erosion. Its ice erodes by abrasion and by plucking away at the bedrock. Blocks embedded in the ice are scraped along the bottom, producing grooved (striated) rocks, and resistant rocks are polished into roches moutonnées. The source area is enlarged into an amphitheatre known as a cirque or corrie, and where two such cirques meet they are separated by a sharp ridge called an arête. Three cirques overlapping produce at their centre a pyramidal peak or horn, of which a classic example is the Matterhorn [Key] on the Swiss-Italian border. Mountain glaciers carve their valleys into deep U-shaped troughs; bigger glaciers have deeper valleys than their smaller affluents, giving rise to "hanging" valleys after the glaciers have disappeared.

Glaciers can carry huge loads of debris or moraine [6]. Rocks falling from above on to the sides of a glacier form lateral moraines and where two glaciers converge the inner lateral moraines merge to form a medial moraine. Some debris falls into crevasses to form an englacial moraine and can work its way down to the rocks plucked off the bed and become part of the subglacial moraine.

Glacial deposits
The glacier's debris is eventually deposited at its foot, forming the terminal moraine which is made of totally ungraded material ranging from clay to huge boulders. If the glacier retreats, the abandoned frontal moraine often makes a dam retaining a lake and other lakes appear in the hard-rock depressions carved by the glacier. Rapidly retreating glaciers dump their loads as they go, leaving large rocks as clues to their former size; they also give characteristic and valuable information about former ice ages.

Europe's most photo-genic mountain, the Matterhorn, was carved by glaciers into its pyramidal shape.

3 One of Europe's biggest glaciers is the Mer de Glace in the Mont Blanc massif. It is 13km (8 miles) long and 2km (1.2 miles) wide, with a thickness of 150m (500ft). It is formed by the joining of the Talèfre, Leschaux and Géant glaciers, which in turn have several smaller tributaries. Crevasses and surface moraines are conspicuous in this picture, plus a small cirque (with its névé) in the background.

5 Ice caves in glaciers may be formed by incompletely closed crevasses, by the margins of the glacier when it skirts around an obstruction or by meltwaters. Ice melting on the surface drains into crevasses. When a crevasse closes through ice movement, the drain hole is maintained by the water and forms a smooth-sided well. Meltwaters running beneath the glacier can form a network of caves and tunnels.

6 Moraine is glacier-borne debris – the rocks, gravel and silt carried by the rivers of ice. It is also used to describe the material once deposited by the melting ice. The terminal moraine at the snout of the glacier illustrated shows the large volume of ungraded matter which ranges in size from huge boulders to fine powders produced by rocks grinding together. This lack of grading distinguishes glacial deposits from those of rivers.

4 The great mass of a glacier flows in a plastic manner, but its surface layers are always rigid and brittle. As a consequence, whenever a glacier flows round a corner or over a hump, or changes its speed, cracks called crevasses appear on the surface at points of tension and wrinkles or pressure ridges are seen at points of compression. The alignment of these depends on the direction in which the pressure acts; they appear in parallel swarms. Two different sets of pressures may produce intersecting swarms, leaving tall pinnacles of ice-séracs between them.

7 Fiords are characteristic of recently glaciated coasts such as those of western Scotland, Norway, southern Chile, British Columbia, southern New Zealand and Greenland. They are long and narrow inlets, steep-sided and often very deep. The deepest in the world is in Chile, with a depth of 1,288m (4,225ft); the Sogne fiord in Norway is the deepest in Europe measuring 1,210m (3,969ft). Shown here is Hardanger fiord, also in Norway. The fiords were the sites of large valley glaciers during the Ice Age. The reason they have been gouged out to such great depths is that glaciers descended to the sea-level which was then very much lower than at present owing to the large amount of water locked into the ice caps.

Ice caps and ice ages

The polar ice caps hold just over two per cent of the earth's water; a small amount compared to the oceans (97.2 per cent) but sufficient to raise the level of the oceans by some 40m (130ft) if they were to melt.

Apart from their size, ice caps differ from mountain glaciers in that they flow outwards in all directions from their centres. The largest ice caps are referred to as ice sheets; only two exist nowadays, in Antarctica [2] and in Greenland.

The polar ice sheets

The Greenland ice sheet occupies 1,740,000 square kilometres (670,000 square miles), 80 per cent of the island's area, and has a volume of 2,800,000 cubic kilometres (672,000 cubic miles). Its average thickness is 1.6km (1 mile) and it reaches as much as 3km (1.9 miles) at the centre.

The Antarctic ice sheet has an area about one and a half times as large as the USA – 13 million square kilometres (5 million square miles), and it holds nine times more ice than the Greenland ice sheet, 25 million cubic kilometres (6 million cubic miles). The ice thickness reaches up to 4km (2.5 miles). This mass flows north and reaches the sea through outlet glaciers and ice shelves that are floating extensions of the ice sheet. The shorelines of the shelves are constantly moving, shedding huge tabular icebergs that are sometimes known as ice islands.

Successive ice ages

The Greenland and Antarctic ice sheets are the last remnants of an ice age that ended, for the mid-latitudes, about 12,000 years ago. Features such as vast amounts of coarse sediments (now referred to as drift), erratic boulders, river terraces and raised beaches had been noted by the early geologists but they were ascribed to the biblical Flood ("Diluvian" deposits). It was not until the mid-nineteenth century that there was widespread belief in the Ice Age.

During the past two million years there have been five major glacial advances and five glacial retreats, the last of these being our present period, the Holocene [1].

Large ice sheets covered the northern continents: most of the British Isles, the North Sea, The Netherlands, northern Germany and Russia were part of an ice sheet centred on Scandinavia and the Baltic, while mountain glaciers descended from the Alps and the Pyrenees. Siberia and Kamchatka were glaciated, as well as mountains to the south, and in North America the sheet reached Montana, Illinois and New Jersey, and the Rockies had extensive mountain glaciers. There was also an ice sheet covering Argentina up to a latitude of 40°S, large glaciers over the Andes and an ice cap over New Zealand.

The chronology of the periods of glacial advance and retreat is established by the study of periglacial lake sediments (forming annual layers known as varves), of fossil pollens of plants (showing the climatic conditions), of fossil soils between two glacial layers, and of ancient beaches and river terraces which reflect former sea-levels. Other dating techniques involve radio-isotopes and tree rings, while micro-fossils in deep-sea sediments and oxygen isotopic ratios from marine fossils provide clues to the then prevailing temperatures.

CONNECTIONS

See also
116 Rivers of ice
108 Earth's water supply
26 Continents adrift

In other volumes
224 The Natural World
28 Man and Society

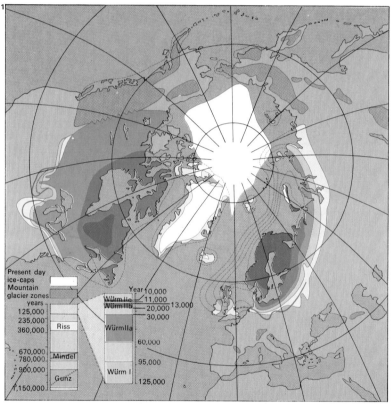

Present day ice-caps
Mountain glacier zones
years
125,000
235,000
360,000
Riss
670,000
780,000
Mindel
900,000
1,150,000
Günz

Year 10,000
Würm IIc — 11,000
Würm IIb — 20,000 13,000
— 30,000
WürmIIa
— 60,000
— 95,000
Würm I
— 125,000

1 The last ice age, the Pleistocene, consisted of several periods of glaciation separated by interglacial periods of mild climates. The earliest traces of the glacial advance have been found in Europe in sediments 2,500,000 years old, and represent the Donau glacial stage. This was followed by the Günz (equivalent to the Nebraskan in the American system), the Mindel (Kansan), the Riss (Illinoian) and the four glacial stages of the Würm (Winconsin).

3 During the ice ages, much of northern Europe and North America looked like Antarctica today – a frozen and totally inhospitable world. Antarctica contains 90% of the world's ice and is the only continent without an indigenous human population. It is protected from land grabs both by its climate and by international treaties. The only human beings are found in scattered scientific stations provisioned and relieved from the outside.

2 Continental ice sheets are now found only in Antarctica and Greenland. In Antarctica the ice [1] covers not only the land [2] but also permanently frozen sea [3]. Beneath the ice the terrain is rugged and variable in height. Because of the weight of the ice, about 40% of the land is depressed below sea-level. If the ice were to melt, there would be a gradual rising of the land, just as the Baltic area is now rising to compensate for the loss of its ice sheet some 12,000 years ago. Because of the melt water, the sea would at first rise faster than the land, drowning much of Antarctica.

Glacial deposits and ice-grooved and polished rock have also been identified in older geological formations, leading to the discovery of former ice ages. Three are known to have occurred during the late Precambrian (940, 770 and 615 million years ago), one during the Devonian (nearly 400 million years ago) and one during the Permo-Carboniferous (295 million years ago).

Origin of ice ages

Some 60 or 70 theories have been forward to account for the origin of ice ages. Those based on purely terrestrial phenomena call for a predisposed position of the land masses such as the present positions of Antarctica and of the landlocked Arctic Sea, which prevent the temperature-evening effects of the sea from reaching the poles; or else they presuppose changes in the atmosphere, such as a decrease in carbon dioxide content (allowing a faster rate of heat loss to outer space) or an increase in atmospheric dust due to paroxysmal volcanic eruptions, so preventing the warming effect of some of the sun's rays from reaching the earth.

Other hypotheses are astronomical. They propose variations in the sun's output and changes in the sun-earth relationship. One such hypothesis relates the glaciations to the passage of the Solar System through the dust clouds of the two spiral arms of the galaxy: this implies an ice age lasting a few million years every 250 million years. Scientific evidence supports this theory except for the time about 250 million years ago when no ice age occurred but there was, nevertheless, a distinct cooling of the climate as evidenced by fossil faunal changes.

According to the latter hypothesis, we should now be moving out of the last ice age although there is currently much talk of an impending renewal of glaciation. This is based on a southward shift, over the last ten years or so, of the Northern Hemisphere's climatic belts, leading to less sunny summers in the temperate latitudes and to droughts in the Sahel and in Ethiopia and Somalia. But the study of the climate over the past 10,000 years shows many such fluctuations and the present state of knowledge about the origin of ice ages makes forecasting difficult.

☐ Area of Pleistocene glaciation ● Devonian tillite ● Precambrian 2 tillite (770 million years)

▲ Permo-Carboniferous tillite ■ Precambrian 3 tillite (615 million years) ▲ Precambrian 1 tillite (940 million years)

Ice ages have several times swept over various parts of the earth and, although of short duration by geological standards, have left a durable impression on its crust. The map shows the limits reached by the ice sheets during the last ice age in the Pleistocene (2 million to 12,000 years ago). Evidence for other ice ages comes from tillites, which are consolidated glacial deposits. Also shown are the main locations of Precambrian, Devonian and Permo-Carboniferous tillites. These were laid down in high altitudes but have since been moved out of place by continental drift.

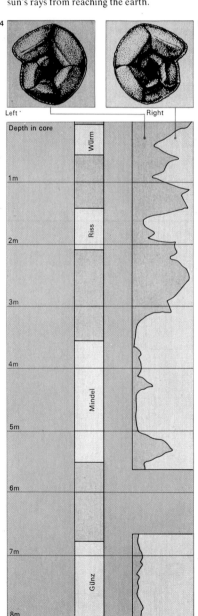

4 The sediments on the ocean floor accumulate very slowly and a few metres can represent millions of years. Microscopic fossils of such animals as Foraminifera can record changes of the climate. The coiling of *Globorotalia truncatulinoides* [top] varies according to the temperature of the water, being predominantly to the left during cold periods and to the right under warm conditions. Analysis of specimens from many cores has provided a good indication of the changes in ocean temperatures. This method is often used with other investigations and particularly with studies on the abundance of other Foraminifera such as *Globorotalia menardii* and *Globigerina pachyderma*, which are also sensitive to changes that affect the sea's temperature.

5 Icebergs form in several ways. When a glacier reaches the sea it floats away from the bed. The movement of waves and tides exerts pressures on this floating ice causing lumps to break away [A]. If the glacier is moving rapidly when it reaches the sea, a projecting shelf of ice forms under the water. The buoyancy of this shelf exerts an upward pressure causing pieces to break off [B]. The snout of the glacier may be above the level of the sea and hence lumps may break off under the force of gravity and fall into the water [C]. The forming of new icebergs is known as "calving". Northern icebergs come from the Greenland ice sheet, but the largest ones originate in Antarctica. The largest iceberg ever seen was 336km (208 miles) long and 97km (60 miles) in width.

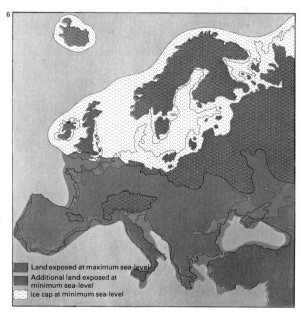

6 At periods of maximum glaciation, sea-levels were 180m (590 ft) lower than at the present time because of the large amount of water frozen in the ice. Many of today's islands would be joined to adjacent continental masses – the British Isles, for example, would have been part of Europe. Land bridges appeared, especially in such areas as the Bering Strait. These bridges helped the spread of mankind round the world.

■ Land exposed at maximum sea-level
■ Additional land exposed at minimum sea-level
▒ Ice cap at minimum sea-level

Winds and deserts

The most obvious characteristic of the desert is its emptiness for it is almost devoid of plants, animals, mankind and water. Hot deserts are places where the heat of the sun [3] is capable of evaporating all the water that falls as rain; most of them have less than 10cm (4in) of rain a year [2] which falls heavily and in rare showers. Many areas have no rain for years – then a sudden cloudburst causes temporary, fast-flowing streams called flash floods. This water usually drains into shallow lakes that have no outlet to the sea, and there it soon evaporates.

The face of the desert
The hot desert is not all sand; in fact only about 20 per cent is covered with sand. Much is bare rock, often cut by deep wadis (intermittent riverways) [5] or carved into fantastic shapes by winds [10]. The landform of rock deserts is very angular. The rounded hill shapes of more humid lands are missing because there is no steady downwash.

In many desert tracts the sand has been blown away, leaving a surface of boulders [6]. Rock or boulder desert may grade into sand

desert [Key, 4]. At first there are dunes with no sand between them; then these pass into areas where the whole surface is sand-covered and is known as a sand sea.

The geographical locations of most deserts lie within the belts of high air pressure centred on the tropics of Cancer and Capricorn in which the air is always very dry [1]. The deserts of Asia and North America lie in the interior of those continents and are cut off from the rain-bearing winds by surrounding mountain ranges. The world's largest deserts stretch across Africa and great parts of Asia. Europe is the only continent with no large deserts. Almost all deserts, such as the Sahara and the Kalahari, are hot, but a few, such as those in the Arctic and on high mountains, owe their lifelessness to extreme cold.

Weathering, the process by which rock is broken down into sand and clay, takes place very slowly in the hot desert compared with more humid lands. On many days the surface temperature of the rocks varies from about 5°C (41°F) in the cool of the night to about 40°C (104°F) at midday. This results in a daily expansion and contraction of the rock

surface, setting up strains that gradually break it up into sand. The small amount of moisture, mostly in the form of dew, may also cause slow chemical weathering.

Water and wind effects
Water from occasional downpours drains rapidly into the wadis, carrying with it loose stones, sand or mud. It rushes down as a flash flood and the stones it carries erode the sides of the wadis. At the end of the wadi is a cone-shaped pile of stones and sand called an alluvial fan [4]. The water sinks into this, leaving the coarse sediment it is carrying on top, and eventually drains out of the bottom taking with it only the finest material, mostly mud and dissolved salts. It then runs into large, flat areas and forms temporary shallow lakes [4]. Within a few days the water has evaporated, leaving a mixture of salt and mud in what are called salinas.

The main effect of the wind in the desert is to move sand and dust. In humid areas vegetation protects the soil from the wind but in the desert there is no vegetation and moreover the sand and dust are completely

CONNECTIONS

See also
68 Winds and weather systems

In other volumes
218 The Natural World
220 The Natural World

1 Deserts are formed in areas where the rate of water loss by evaporation is greater than the rate of water gain by precipitation. Temperature, as well as rainfall, is very important – forests grow in cool latitudes in rainfall that would give only scrub and semi-arid conditions in the tropics. Approximately 25 per cent of the earth's surface is characterized by dry climates, and deserts themselves cover a large proportion of the land between latitudes 10° and 35° north and south. On this map desert areas are red and the regions of semi-arid climates are buff.

2 The rainfall in a desert area such as Cufra, in Libya, is vastly different from that in a wetter area such as Greenwich, in England. Desert rainfall is usually less than 10cm (4in) a year and is very irregular with cloudbursts and long dry spells.

3 The daily temperature range in a desert is very great due to the lack of an insulating cloud layer. At higher latitudes maximum temperature is lower but the temperature range is also small since cloud cover tends to contain the heat.

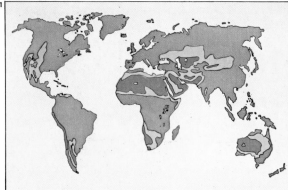

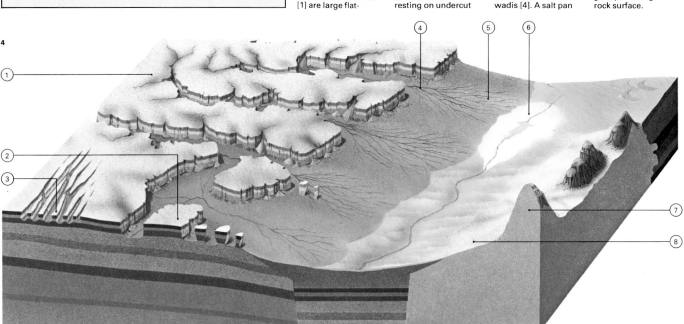

4 The topography of a desert is characterized by the relative absence of chemical weathering. Most erosion takes place mechanically through wind abrasion or the effect of heat. Mesas [1] are large flat-topped areas with steep sides. The butte [2] is a flat isolated hill also with steep sides. Yardangs [3] are wind-eroded features consisting of tabular masses of resistant rock resting on undercut pillars of softer material. They are elongated in the direction of the wind. Alluvial fans [5] are deltaic pebble-mounds deposited by flash floods, usually found at the end of wadis [4]. A salt pan [6] is a temporary lake of brackish water also formed by flash floods. An inselberg [7] is an isolated hill rising abruptly from the plain. A pediment [8] is a gently inclining rock surface.

dry and easily moved. The wind takes the sand and dust from the surface of the alluvial fans, plus any sand produced by weathering, and blows it into dunes.

Sand grains are not carried far in the air, but the strongest wind causes the grains to move in series of bounces [10]. Although they are never lifted more than 1m (39in) above the ground, the wind-blown particles "sand blast" any rock or pebbles in their path and polish the surface of any pebble facing the wind. If the direction from which the wind comes varies, the pebble will acquire several flattened surfaces giving it a pyramidal shape. Such a pebble is called a ventifact or a drei-kanter. The sand also erodes the solid rock over which it is blown, etching out any softer or weaker parts and leaving the harder rock standing up in ridges called yardangs, or in "rock mushrooms" called zeugens [6].

How sand dunes are formed

There are two main types of sand dunes; barchan and seif dunes. Barchan dunes [9] are usually found on the edge of the desert where there is a relatively small amount of sand and

often some scrub vegetation. These dunes are crescent-shaped, with their points facing downwind, and between them there is only gravel or bare rock. The wind blows the sand up the gently sloping windward side of the dune and when the grains reach the top they roll down the steeper leeward side. Therefore the grains at the back of the dune are constantly being brought to the front; in this way the dune slowly advances. Large barchan dunes move extremely slowly, while small barchans may advance 15m (50ft) a year. Where there are many barchan dunes they may line up to form a transverse dune.

Seif or longitudinal dunes [8] cover a much larger area of the desert. They are long ridges of sand separated by strips of rock or stones kept clean of sand by eddies of wind. Barchan and seif dunes merge into areas where all the desert floor is covered with sand and the dunes lose their shape and become part of an irregularly rolling sand surface.

The finest dust may be lifted thousands of metres into the air and carried for hundreds of kilometres and, if blown out of the desert, forms a very fertile soil called loess.

Deserts are extreme-ly dry areas that support only a much-reduced vegetation and a few nomadic tribes scattered in small encampments such as the Saharan one shown here. A main geological agent in deserts is wind. Its effects are empha-sized by the lack of vegetation and of moisture that holds fine-grained particles together. The wind sweeps some areas clean, leaving bare rocks; sand-laden winds sculpture the rocks and the sand is eventually depos-ited, forming dunes.

5

5 Wadis are steep or vertically sided valleys in which water runs only during rare flash floods. They start with random de-pressions in desert mountain areas or arid plateaus. They are deepened and widened by the floods which, because the water is moving so fast, are able to prise away and move large slabs of rock. During a flood the water may cover the whole width of the wadi.

6

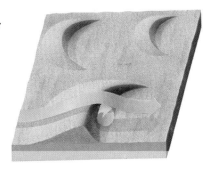

6 Rocky surfaces are far more common in deserts than sand seas. When loose material containing pebbles or larger stones is exposed to wind, all the fine dust and sand particles are blown away leaving a desert pavement or reg. The surfaces of mesas and larger plateaus are scoured clean by the wind, and form rock deserts called hammadas showing wind-eroded features – yardangs and mushroom-shaped rocks called zeugens.

7
A

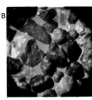
B

7 Grains of desert sand [A] are large-ly spherical and appear "frosted" because they have been rounded by countless collisions with other grains. River sand grains [B] are less polished, having suffered fewer collisions. River sand also con-tains many grains of soft minerals, such as mica, which would have been ground to dust in the desert. Desert sand is more uniform in size than river sand.

8

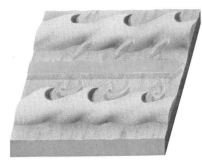

8

8 Seif dunes are long ridges of sand with bare rock be-tween. Each dune may be up to 40m (130ft) high, 600m (1,960ft) wide and 400km (250 miles) long. The pre-vailing wind blows parallel to the ridges.

9

9 Barchans are iso-lated and crescent-shaped dunes that slowly migrate down-wind, horns forward. They occur only in areas such as Turk-estan where the wind always blows from the same direction.

10 A pedestal is a large lump of rock supported only by a thin neck. In a sand-storm the wind is only able to make the sand grains bounce up to about 1m (39in) above the ground. When the sand collides with the rock, it sand-blasts it and wears it away. The dust and finer particles, which are carried higher, are too light to abrade the rock. Therefore the rock is eroded only at its base, which gives it the appearance of a mushroom.

10

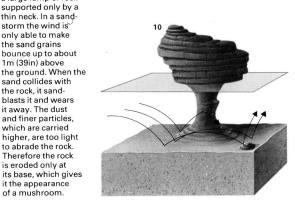

11 Rain falling on the Atlas Mountains drains into porous rocks underlying the Sahara. The water seeps through these rocks which, where-ever they come to the surface, give rise to oases.

11

12 In prehistoric times continuous rivers ran in the Sahara. Their former courses can still be seen from the air. But today it is much drier and habitation is restricted to a few oases.

12

Coastlines

Coasts are constantly changing [Key], sometimes at a dramatic rate. During a North Sea storm in 1953, powerful waves battered eastern England. Near Lowestoft in Suffolk, the sea undercut an 8m-high (26ft) cliff and removed 11m (36ft) of land in about two hours.

The rate of erosion, which is caused by waves [1], currents and tides, depends on the nature of the rock. Tough outcrops of granite are much more resistant than, for example, the glacial boulder clays, gravels and sands in Massachusetts where erosion of the cliffs of Martha's Vineyard island is taking place at a rate of 1.7m (5.5ft) a year and where a lighthouse has had to be resited three times.

The causes of coastal erosion

The force exerted by waves in the Atlantic has been estimated to be about 9,765kg per square metre (2,000lb per square foot) and this force may be three times as great during severe storms, when blocks of concrete weighing more than 1,000 tonnes have been dislodged. The hydraulic action of water is seen when high waves crash against a rock face. Air compressed by the water in cracks and crevices expands as it is released, sometimes with explosive force, enlarging cracks or shattering rock faces.

Another form of marine erosion is corrosion, when waves are armed with sand, pebbles and, during storms, boulders. The waves lift up these materials and hurl them at the shore, bombarding and undercutting the bases of cliffs. Such action may hollow out sea caves within which erosion continues. Sometimes the roof of a cave collapses to form a small opening or blowhole. When waves pound through the cave, jets of spray spurt through the blowhole.

Because rocks differ in hardness, sea erosion may create a series of bays, cut in relatively soft rock, separated by headlands of fairly hard rock. The exposed headlands are battered from both sides. Sea caves forming on each side of the exposed headland may eventually meet in an arch. When the arch collapses, an offshore pillar of rock, called a stack [2], remains behind. In this way, even headlands of hard rock are finally worn away.

Another form of marine erosion, attrition, occurs when sand, pebbles and rocks collide and are rubbed together by the moving water. Loose, jagged material is smoothed and ground down into finer and finer particles. The sea also erodes land by the solvent action of seawater on some rock.

The movement of eroded material

Eroded material is transported along the coast mainly by wave action. Waves usually strike the shore obliquely. As they move forward, they sweep material diagonally up the beach. As the water recedes, the backwash pulls the loose fragments down the steepest slope at right-angles to the shore. This zigzag movement, called longshore drift [3], moves sand and pebbles along the coast. Currents and tides also contribute to the movement of eroded material.

Because of the importance of coastlines to man, attempts are often made to control longshore drift and erosion. Common methods include the building of groynes [5] (low walls usually at right-angles to the shore) and sea walls to protect the coast against storm waves.

CONNECTIONS

See also
114 Land sculptured by rivers
82 The sea-bed

In other volumes
236 The Natural World
82 The Natural World

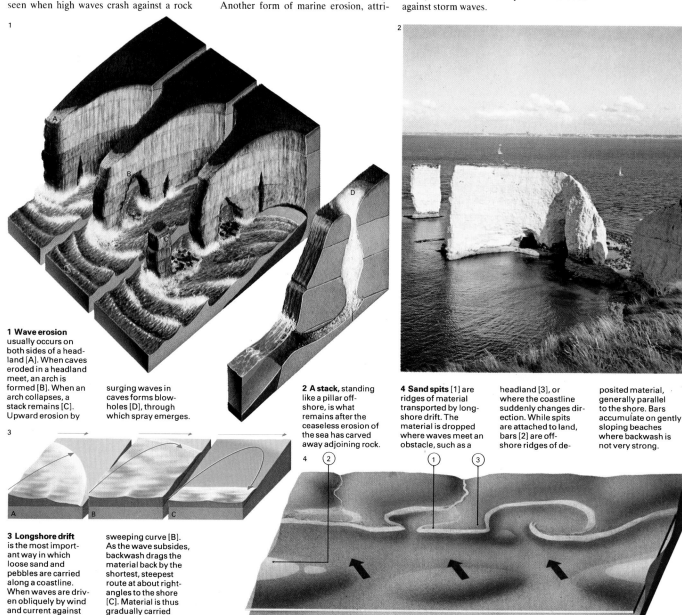

1 Wave erosion usually occurs on both sides of a headland [A]. When caves eroded in a headland meet, an arch is formed [B]. When an arch collapses, a stack remains [C]. Upward erosion by surging waves in caves forms blowholes [D], through which spray emerges.

2 A stack, standing like a pillar offshore, is what remains after the ceaseless erosion of the sea has carved away adjoining rock.

3 Longshore drift is the most important way in which loose sand and pebbles are carried along a coastline. When waves are driven obliquely by wind and current against the shore [A], debris is swept up the beach in a forward sweeping curve [B]. As the wave subsides, backwash drags the material back by the shortest, steepest route at about right-angles to the shore [C]. Material is thus gradually carried along the coast in a zigzag path to be deposited elsewhere.

4 Sand spits [1] are ridges of material transported by longshore drift. The material is dropped where waves meet an obstacle, such as a headland [3], or where the coastline suddenly changes direction. While spits are attached to land, bars [2] are offshore ridges of deposited material, generally parallel to the shore. Bars accumulate on gently sloping beaches where backwash is not very strong.

When material moved by longshore drift meets an obstacle, such as a headland, or where the coast abruptly changes direction, the transport of material may slow down and it may pile up in narrow ridges called spits [4]. Spits often curve part of the way across bays and estuaries. Baymouth bars are spits that seal off a bay completely. Other bars, unlike spits, are not attached to land. They are formed in the sea and run roughly parallel to the coast. Similar features are tombolos – natural bridges joining an island to the mainland or linking one island to another.

Other characteristic coastlines

Since the end of the Pleistocene ice age, melted ice has increased the volume of the oceans and many coastlines have been flooded [7]. These coasts of submergence include flooded river valleys, called rias, and flooded glaciated valleys, called fiords. Other coasts have been raised up by earth movements. Coasts of emergence can be identified by such features as raised beaches and former sea cliffs that are now inland.

Some coastlines have a special character

related to the geological structure of the coast. The two main kinds of coastlines in this category are concordant or longitudinal coastlines and discordant or transverse coastlines. A concordant coastline occurs in Yugoslavia along the Adriatic Sea, where the geological structures parallel the coast. Following submergence, the sea has occupied former valleys while former mountain ranges have become offshore islands. Discordant coastlines occur where the coast cuts across the direction of the geological structures, as in the ria coastline of southwest Ireland.

A special feature of coastlines in tropical seas results from the growth of coral. Coral polyps live in warm water with plenty of sunlight and cannot grow in depths greater than 45m (150ft). Fringing coral reefs develop in shallow water near the shore. Barrier reefs lie some distance away from the shore. They may be built on a non-coral foundation or they may have increased in depth as the depth of the sea increased. The most intriguing coral features are atolls [8], circular or horseshoe-shaped groups of coral islands in mid-ocean.

KEY

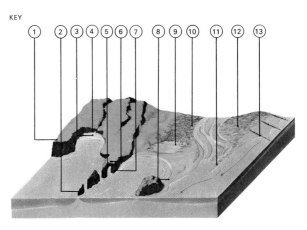

Coastlines are shaped by erosion and deposition which are the result of wind, waves, currents and tides interacting with the rocks and sediments of the land. Among common coastal features

are headlands [1] of relatively hard rock, isolated rock pillars called stacks [2], cliffs [4], natural arches [5], caves [7] and blowholes [6] in the roofs of caves. Features resulting

from deposition are beaches [3], tombolos [8], lagoons [9], salt marshes [10], spits [11] and sand dunes [12]. To slow down the drift of eroded material along a coast, groynes [13] are often built.

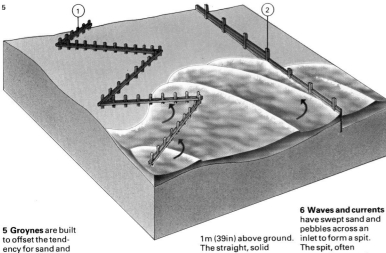

5 Groynes are built to offset the tendency for sand and shingle to be gradually carried sideways along a beach. There are two main kinds of groyne.

The zigzag timber pile type [1] has piles driven 2m (6ft) below the ground and standing

1m (39in) above ground. The straight, solid groyne [2] consists of heavy planks bolted to piles that are also sunk 2m (6.5ft) into the ground

6 Waves and currents have swept sand and pebbles across an inlet to form a spit. The spit, often called a baymouth bar, has cut off the inlet from the sea, leaving behind it an enclosed lagoon.

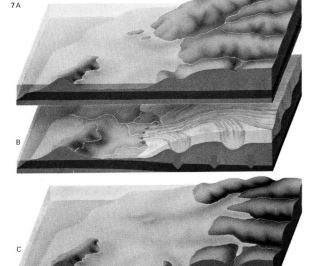

7 Changes in sea-level caused by an increase in the volume of ocean water, or by earth movements, determine the character of some coastlines. When a coastal area unaffected by glaciation [A] is overtaken by an ice age, glaciers and ice caps form on the land [B]. The sea-level drops and the weight of the ice eventually depresses coastal valleys. With the end of the ice age, melted ice returns to the sea. Even though the land begins to rise, recovery is not fast enough to offset a considerable rise in sea-level [C]. Flooded river valleys or fiords (drowned glacial valleys) characterize such a coastline – a coast of submergence.

8A

Volcanic island | Reef

B

Reef lagoon

Low islands | Reef and detritus

C

8 Atolls and coral reefs are found only in tropical seas. The most striking coral feature is the atoll, a ring or horseshoe-shaped group of coral islands. Corals grow in warm, fairly shallow water to depths of about 90m (300ft). But the depth of coral in many atolls is much greater than this. One theory is that the coral began to form as a reef in the shallows of a volcanic island [A]. Then the sea-level began to rise and the island slowly sank [B]. But the coral growth kept pace with these gradual changes, leaving an atoll of low islands [C]. In this way, depths of coral to as much as 1,600m (5,250ft) can be explained.

The record in the rocks

The earth's rocks form a tattered and fragmented manuscript of our planet's past. Although sketchy in places, the historical record can be traced nearly 3,800 million years back in time. This date corresponds to the age of the oldest known rocks. The sedimentary strata in particular are ideal signposts through the past since their features identify the circumstances under which they were deposited and made into rocks.

Clues to the past

The record of geological history, as preserved by sedimentary rocks [1], is known as the "stratigraphical column". It refers to the total succession of rocks, from the oldest to the most recent. The study of its distinguishing features is the science of stratigraphy, which deals with the definition and interpretation of stratified sedimentary rocks.

The science of geology was rescued from the futile perusal of biblical texts for clues to the earth's past by the exposition of the principle of "uniformitarianism" by the Scottish geologist James Hutton (1726–97). This was

propounded in Hutton's work, *Theory of the Earth*, published in 1795, and maintained that the nature of present geological processes was fundamentally similar to those of the past and that rock folds and tilts revealed in strata were caused not by violent upheavals, such as earthquakes, but by gradual pressures within the earth. Uniformitarianism implies that the characteristic features of erosion (wearing down of rocks) and the transportation and deposition of sediments produced by erosion are the same throughout history. Since these features can easily be recognized in sedimentary rock laid down millions of years ago, past events are best interpreted in the light of what is known about the present. In Hutton's words, "the present is the key to the past".

Superposition and correlation

The fundamental concept of stratigraphy is the law of "superposition", which states that in any horizontal, undisturbed sedimentary sequence, the lowest rocks are older than those lying above them. This law applies not only to the relative ages of different strata but

also to the minerals and fossils found within a specific layer. Lower-lying ones were deposited before those above them. Superposition is essential to establishing the comparative ages of the various beds in sedimentary formation and is the single most important prerequisite of geological mapping. In archaeology the same principles apply. Artefacts and fossils found in deeper layers of earth predate those in layers above them.

Igneous rocks (formed from molten magma), unlike sedimentary ones, are not laid down in neat successive layers. They can reach their underlying positions only by filtering up through existing formations while still in a molten state. These intrusive rocks, such as granite, gabbro and diorite, are always younger than the rocks that surround them. However, when they reach the surface to form lavas – such as basalt, obsidian and rhyolite (igneous rocks of very fine texture) – they are, as are sedimentary rocks, always younger than the rocks below them.

Since the complete stratigraphical column has never been discovered in any one site (it would have a thickness of many hun-

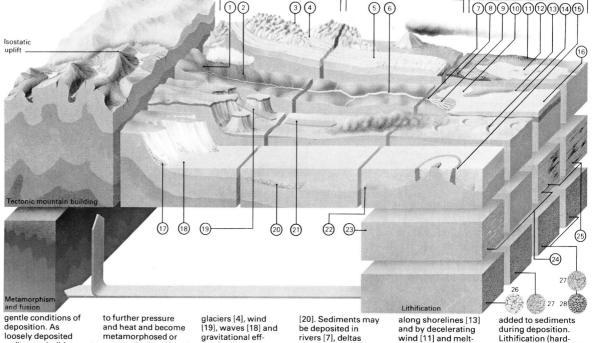

1 The sedimentary cycle begins when sedimentary, metamorphic or igneous rocks are thrust towards the earth's surface, at various times in their history, by mountain building, folding and faulting or the vertical uplift of land freed from the weight of glaciers after an ice age. Rock that becomes exposed to the chemical and mechanical agents of weathering will be rapidly eroded and deposited in new sites by a host of transporting mechanisms. Accumulating sediments possess characteristics that can positively identify the environment and the kind of rock found in their area of origin. They also acquire features that are clues to their environment of deposition, be it desert, swamp, lake bottom or sea-shore. Thus, black shales derived from finely-graded silt and mud can indicate a warm, humid climate and

gentle conditions of deposition. As loosely deposited sediments build up they are transformed into hardened rock. The newly formed rock may be subjected

to further pressure and heat and become metamorphosed or even melted and fused into igneous rock. Agents of erosion include rivers [1], rain [2], frost [3],

glaciers [4], wind [19], waves [18] and gravitational effects [17]. Transportation is performed by glaciers [5], rivers [6], wind [21] and ocean currents

[20]. Sediments may be deposited in rivers [7], deltas [9], lagoons [10], lakes [12], deserts [14], coral reefs [15], shallow and deep seas [16, 22],

along shorelines [13] and by decelerating wind [11] and melting glaciers [8]. Shells [24], plant debris [25] and remnants of other living organisms [23] may be

added to sediments during deposition. Lithification (hardening process) occurs by compaction [28], cementation [27] and recrystallization of fragments [26].

2 The Grand Canyon of Colorado reveals a massive section of the geological history of the earth. Here, the swift-flowing Colorado River, laden with an estimated daily burden of some 500,000 tonnes of debris, has gouged a scar 1.9km (1.2 miles) deep in the surrounding plateau. This erosive activity has been continuous since the Tertiary. The gradual but prolonged uplifting of the area caused the Colorado to cut a deep canyon in order

to maintain its graded profile. In the canyon's plunging walls, hundreds of metres of sedimentary strata are exposed, consisting largely of marine limestones, freshwater shales and wind-deposited sandstones. The lowest Palaeozoic rocks rest unconformably upon a basement of plutonic and metamorphic Precambrian rock. The granites and schists of this rock are the roots of ancient mountains, their tops long ago eroded

away. Radiometric datings have established these rocks as being some 1,600–1,800 million years old. Even a cursory inspection of the canyon is almost a complete course in palaeogeology. Here can be witnessed the successive ages when submergence, uplift, erosion, deposition and folding and faulting have occurred in the plateau. The fossil record ranges from primitive algae and trilobites through dinosaurs to remains of early camels and horses.

dreds of kilometres), assembling its highly fragmented sections in correct order requires geological detective work to correlate widely scattered beds of rocks.

The most obvious tactic is to follow the layers of a specific outcrop as far as possible. This method is usually only possible over a limited area since erosion, deposition and folding or faulting will tend to interrupt or obliterate rock outcrops [3]. Another method of determining the extent of a formation is by searching for various similarities in the rocks. Various features of deposition, weathering and mineralization all identify rocks that belong to the same formation. Correlations can also be established by comparing vertical sections of rock.

Using fossils as clocks
Fossils are an excellent tool for correlation [2]. The most useful are those that are widely distributed but limited in their vertical range, thus indicating that they flourished for relatively brief periods of time. These are known as index or guide fossils.

As living organisms have always under-gone continuous evolution, their fossil remains can be used to identify rocks of comparatively similar geological time. The fossil litter within sedimentary rocks enables palaeontologists to recognize different strata of the same age. And it can be logically deduced from the law of superposition that the remains of primitive life forms occur in rocks lying beneath those containing more advanced forms.

Most fossils are the remains of organisms that lived in the same area and at the same time as the rock in which they are found was being deposited. They are excellent guides to the then existing environment. For instance, fossils of reef-building corals [5] are an indication that clear, warm, shallow sea conditions prevailed.

One of the first attempts to relate fossils to the rocks in which they occur was made by the English surveyor William Smith in the late eighteenth century. He established the lateral or horizontal continuity that exists between scattered outcrops of rocks by identifying strata through their fossil content, as well as by their texture, colour and position.

Towering buttes and mesas in Monument Valley, Colorado, reveal the character of the geological ages of the region. Individual beds within these outcrops can be traced throughout the area, although the rock in between is missing, thus demonstrating the use of lateral continuity. The scale of erosion in the Colorado Plateau is matched by few regions in the world. Several thousand metres of rock have been eroded, leaving only isolated, flat-capped outcrops here and there. Not all the outcrops are sedimentary rock for some are the solidified plugs of ancient volcanoes whose sloping flanks have long ago been worn away.

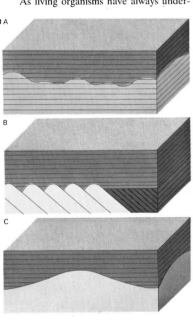

3 "Unconformities" occur when there is a break in sedimentation due to erosion. This creates a gap in the geological record of the earth's history. Unconformities are of three varieties. Disconformities [A] are recognizable because the older strata have not been tilted or in any way deformed before younger rock was deposited above them on the same horizontal plane. Angular unconformities [B] occur where the lower-lying strata have been tilted, deformed and eroded before the deposition of other rocks. Where bedded rock layers overlie non-bedded igneous mass, a nonconformity [C] is said to occur.

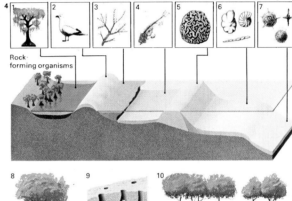

Rock-forming organisms

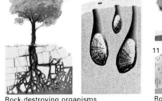

Rock-destroying organisms

Rock-accreting organisms

4 **Living organisms** may both create and destroy rocks. Some rocks are formed when decaying vegetation [1] becomes coal or when the droppings of bird colonies [2] accumulate as phosphate deposits. In marine environments calcareous algae [3] form limestone deposits while fish skeletons [4] may accumulate as beds of phosphate. Corals [5] and tiny globigerina skeletons [6] form limestone sediments while radiolaria [7], which build their hard parts with silica, create siliceous deposits. Tree roots [8] and piddocks [9] hasten rock destruction. Mangrove [10] and dune grass [11] trap loose sediment that may harden.

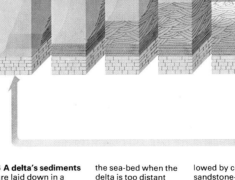

5 **Entire islands** have been built by small organisms, corals and algae. The wreath-shaped reef and sheltered lagoon of a coral atoll is built upon the crown of a subsiding peak. The symbiotic relationship between certain algae and coral polyps is responsible for the formation of coral islands. Coral itself is too fragile to form a reef unless it is reinforced by the carbonate-producing algae *Zooxanthella* spp. Reef corals thrive in water no deeper than 45m (150ft) or colder than 20°C (68°F). Coral formations in a sedimentary sequence are excellent indicators of the prevailing climatic conditions at the time when the rock was laid down.

6 **A delta's sediments** are laid down in a specific order that may be endlessly repeated if the region of deposition is sinking. Limestone deposits [4] cover the sea-bed when the delta is too distant to be influential [A]. As the delta encroaches [B], fine-grained muds that will become shale [3] are deposited, followed by coarser, sandstone-forming sediments [2] as the advance continues [C]. As the water shallows, current bedding [D] indicates that sand is being deposited. Once the delta builds above water level [E] it can support swamp vegetation which forms coal [1]. When the region sinks [F] the cycle restarts.

Clues to the past

The earth's rocks are in a state of constant change. Mountains are worn away by wind, rain and frost and the debris formed is transported by rivers, streams, glaciers, wind and sea currents to be deposited on valley floors and sea-beds. There the sediments are buried and subjected to processes that turn them into sedimentary rocks, later to be uplifted as mountains to start the process all over again. One of the tasks of a geologist is to determine the sequence of these events in particular areas and to do this he uses the tell-tale features preserved in the rocks themselves. This study is known as stratigraphy.

The concept of facies
The term "facies" encompasses all the features of a particular rock or stratum of rock that indicate the conditions in which those rocks were formed. Such features include the mineral content, the shapes and sizes of the particles, the sedimentary structures, the fossils, the relationship to the beds above and below, the colour and even the smell – everything that can be described about the rock.

The mineral content can show whether a sedimentary rock was precipitated out of salts dissolved in seawater or built up from material washed off already existing terrains, and can show the nature of these original terrains. The presence of grains of garnet, for example, show that the original rocks were metamorphic whereas olivine crystals indicate the existence of original rocks that were igneous. The shapes of the constituent particles indicate how far the material has been transported from the source – angular fragments have not travelled far but rounded grains have been carried long distances and have had their corners and edges broken off by the violence of their journey. If a rock consists of particles that are more or less the same size then it can be deduced that the particles have been moved about (or sorted) by currents before coming to rest. A mixture of particle sizes denotes a rapid transportation and dumping of material.

Sedimentary structures give a direct indication of the conditions under which the sediment was accumulated. Ripple marks [7] are formed under shallow water conditions, rain pits [13] and mud cracks show the drying out of shallow pools, small-scale cross-bedding shows the presence of currents and the direction of the currents can be determined by the attitude of the bedding.

Stories told by fossils
Animals are selective about which environments they inhabit and a recognizable fossil in a rock can be the most important clue a geologist can have to the environment under which the rock was formed [3, 9]. The modern bivalve *Scrobicularia*, for example, lives only buried in oxygen-deficient mud. When a fossil *Scrobicularia* is found in shale it can be inferred that the shale was laid down in an oxygen-free environment. Such organisms are usually very sensitive to environment and when they suddenly disappear in a geological succession it can be deduced that conditions have changed drastically.

The condition of the fossils is also important. If the remains are broken up and the fragments are well sorted any deductions made from them must be fairly suspect. It probably means that the dead bodies have been washed about by currents and in some

1 A cliff face gives a cross-section through the layers, or strata, of rocks that comprise part of the earth's crust. If the rocks are sedimentary a geologist can use the cliff as the means of determining the history of that area. Most of the evidence is small-scaled but a number of broad observations can be made by taking the outcrop as a whole. First, as there is no evidence of major disturbance, the law of superposition may be invoked, which states that the oldest beds are at the bottom. Next, since there are no unconformities (breaks in the sequence), it is evident that the beds were laid down continuously with no pauses in the sedimentation. There are three major divisions corresponding to three different environments of deposition that succeeded one another during the history of this area. Starting at the bottom with the oldest rocks (as is customary in dealing with geological successions) there is a massive bed of limestone, the thickness of which is unknown as the base cannot be seen. This indicates a long period of marine deposition and is followed by an alternating sequence of shales, sandstones and coal, suggesting a delta environment. A thick deposit of cross-bedded sandstone is found at the top, indicating a desert environment. Closer inspection of individual beds is needed before a detailed history of the sequence can be worked out.

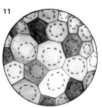

6 Shale, the lowest bed in the next division, is made up of fine mud particles showing that a river was emptying into the sea nearby. The finer debris was carried farther away from the shore.

7 Above the shale is found a bed of sandstone, formed as the river mouth encroached and deposited sand. Samples obtained from the cliff show ripple marks – typical structures

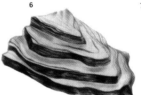

2 The limestone is found to contain a large number of fossil fragments showing that deposition was slow. The broken nature of the fossils shows that constant currents did not allow a dead organism to lie in one place for any length of time.

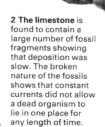

11 The solid mosaic of the sandstone, magnified, reveals the original rounded shapes of each grain within the crystals. This roundness, with the red coloration and constant size, suggests desert sand.

12 The conclusion that a desert existed when the sand was deposited is borne out by the presence of large-scale cross-bedding. This is formed when sand dunes advance over one another, removing their tops and preserving the bedding on their downwind sides. The red tinge is due to the oxidation of iron – a reaction that takes place under very dry conditions, similar to those found in the deserts of today.

of sands deposited in shallow waters, revealing that the sands were built up almost to sea-level. The ripples are aligned at right-angles to the current and so the direction of the river mouth is determined.

3 Most of the fossils are of crinoids (sea lilies). These are related to the starfish and are sessile, being anchored to the sea-bed by a flexible stalk. Their presence indicates a shallow sea environment rich in floating organic food particles.

cases they will have been brought into the area from a completely different environment. On the other hand, if the fossil is obviously in its life position, as for example a burrowing creature in its burrow or a sedentary organism still attached to its substratum, then it is quite certain that this is the environment in which it lived and deduction can be made accordingly.

Occasionally a derived fossil may be found. This is a fossil that has been eroded from an original rock and laid down with other rock debris to form the new deposit. Fortunately this occurrence is rare and is unlikely to confuse a trained geologist.

The evidence is put together
After analysing the numerous features of sedimentary formations, a geologist must piece them together painstakingly to form a coherent whole. Some features are confusing and difficult to interpret; others speak for themselves. Thus a cross-section of rock grading from limestone to shale and sandstone [1] is a typical sequence indicating that the sea was encroached on by river sedi-

ments, eventually building the area up to above sea-level. Similarly, periods of glaciation are typified by distinct striations or gouge marks in the rocks where ice sheets, charged with an enormous load of debris, have ground their way across the land. Another feature of glaciation is the random embedding of irregularly shaped broken rock fragments in finer material. This occurs where the glacier has deposited its load and the resulting rock is known as a tillite.

Palaeogeography is the branch of historical geology that is concerned with the past distribution of land and sea. With the relevant items of data obtained from the study of sedimentary formations, a geologist can construct palaeogeographic maps that give a picture of the ancient world. But such maps tend to be somewhat speculative. Information is often sparse and there are enormous difficulties in interpreting the relative ages of outcrops and thus plotting the ancient boundaries between the land and sea. What these maps do reveal with startling effect is the tremendous transformation that every region has undergone.

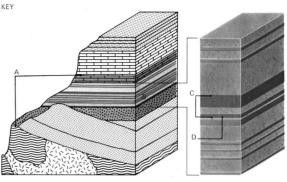

KEY

Rocks can be considered in terms of their stratigraphic units, of which the largest is the group. The Tonto Group of the Grand Canyon [A] is separated by distinct unconformities from the groups directly above and below. It in turn consists of three formations – bodies of rock that share certain generic features. The Bright Angel Formation [B] consists mainly of shales Distinct from the limestone and sandstone formations above and below it. Formations can be subdivided into members such as the dolomite member [C], which are characterized by a distinct lithology. The smallest stratigraphic units of all are beds. The lower dolomite bed [D] is readily identifiable by the division planes between it and its neighbours both above and below. A bed may range in thickness from a few millimetres to several metres and some beds may contain even finer structures.

13 Structures may also show what happened to the sediment immediately after it was laid down. A bed of fine mudstone in the dunebedded sandstones. (itself indicating a temporary flooding in the desert) shows reptile footprints and marks caused by falling rain. Structures such as these provide valuable clues about the climate of the area at the time of the formation of the rocks and give some information about the fauna.

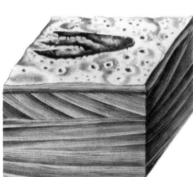

14 A desert region was formed after the area was built up above sea-level by the delta. Rain storms occurred, causing local flooding, and the area was populated by reptiles. By now the volcano would have been eroded to a stump. Conditions such as these were common during the Permian and so the whole cliff face can be said to show a transition from marine to desert conditions in upper Carboniferous and Permian times.

8 Coal, found above some of the beds of sandstone, indicates the presence of thickly forested swamps in which dead vegetable matter accumulated as thick beds of peat. Land had encroached upon the sea.

9 One of the commonest fossils in coal measures is that of a tree stump. The stump and roots keep their shape, being buried in the underlying sand, while the trunk and branches are turned to coal.

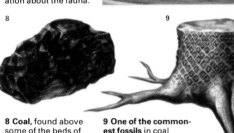

10 The sequence of rocks seen in this part of the cliff – shale, sandstone and coal repeated again and again – is characteristic of sediments laid as a delta advances over a marine area. The fact that the sequence is a repeating one indicates that the area was subsiding and the delta was constantly advancing and retreating, building up the sediments. The absence of volcanic material suggests that the volcano was extinct by this time.

4 A fossil goniatite is also found in this limestone. This was a nautilus-like mollusc with most of its body inside a coiled shell. The shell was divided into chambers by zig-zag partitions that differed from species to species. As each species was a free swimmer and common throughout all the seas of the world at different times, whenever a recognizable species is found in a rock, that rock can be dated. This species shows the limestone to be upper Carboniferous.

5 All the evidence indicates that the limestone at the base of the sequence was laid down during the upper Carboniferous in a shallow, limy sea housing goniatites and beds of crinoids. The crinoid fossils are called "facies fossils", since they show the nature of the environment, and the goniatite is an "index fossil" giving a date to the bed. Microscopic investigation of the rock reveals small quantities of volcanic ash, indicating the presence of a nearby active volcano.

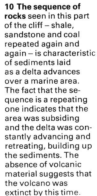

Geology in the field

The structure of rock formations is not always immediately apparent, especially where they are hidden beneath overlying layers of soil or are obscured by vegetation. One of the best ways of determining the relationships between rocks and the processes at work within the earth is by mapping. Geological maps therefore provide the key to understanding the geological history of any particular region.

The basis of geological maps

A geological map [2] shows the boundaries, or contacts, between various rock units, as if the topsoil and vegetation had been stripped off. Maps also reveal the size and extent of any rock formation.

Formations are the basic units of geological maps. They can be recognized by their well-defined contrast to surrounding layers and by the fact that they can be readily traced in the field. The basic criterion of a formation is that it must be a rock layer of sufficient importance and of sufficiently distinct identity for geologists to agree about its characteristics – in other words, it must form a unit which can be mapped.

Where formations have been partly obliterated by erosion, or obscured beneath overlying rock and soil, their shape must be pieced together from isolated and often widely scattered outcrops. A single outcrop is usually insufficient to reveal the complex inter-relations of the various formations in a region. A geologist [6] must make detailed examinations of numerous rock exposures before he can draw a map that assembles his scattered findings into a coherent picture. Such a map reveals the disposition of the rocks and is the basis for understanding a region's structure and history.

The geologist will also draw up cross-sections of the map, showing a vertical slice through the rocks. Canyon walls and coastal cliffs are natural cross-sections of this kind. However, because of their rarity, the geologist must construct his own interpretative section. Cross-sections are derived from interpreting the contours and the attitude of surface outcrops and by making test borings. These sections are essential for determining the commercial importance of ores and in preparing to dig tunnels and mines.

With the knowledge of the fundamental principles that apply to the formation of rocks, a geologist can set off into the field to decipher the structures of a specific region.

Techniques of mapping

There are many techniques for correlating rock formations but the best and most obvious method is to trace a continuous exposure over a distance. In most cases, however, rocks are only sporadically revealed, so the geologist must look for lithological similarities in outcrops. Rocks of the same formation are usually similar in colour, mineral composition and texture although, because most strata change gradually over a distance, other means of identification are also used. Certain characteristics of deposition are especially helpful in identifying separate outcrops as belonging to the same formation. These include ripple marks, formed in sand under very shallow water and later preserved in stone; cross-bedding, which is sand deposited on the underwater slopes of a delta; and graded bedding, beds in

CONNECTIONS

See also
124 The record in the rocks
126 Clues to the past
130 Earth's time chart

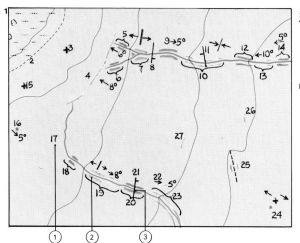

1 The field data map is usually a relief map of the area which the geologist annotates as he works. Such annotation may take the form of numbers [1] referring to an entry in a notebook, colour keying [2] giving a quick visual reference to the rock type, and conventional symbols [3] that describe the folds and faults in the rocks.

2 Geological maps are interpreted from field maps. They show the surface disposition of the rocks as if the topsoil and the vegetation had been removed [B]. The same area is also shown in cross-section [A]. A granitic intrusion [1] adjoins a basal sandstone [2]. Fossils in a bed [3] identify a Carboniferous formation that grades laterally from limestone to shale and then sandstone. Much of this bed is buried by desert sandstone [4] whereas dolomite [5] and the later shale [6] complete the sequences. The two faults are shown on the map by plain lines. A line with converging arrows is a syncline axis and with diverging arrows an anticline.

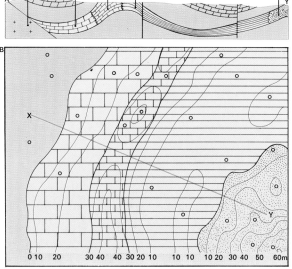

3 Facies maps reveal the variations of the rocks within a single formation, which is mapped as if all the overlying rocks were removed. The example shown is taken from bed 3 of map 2. The cross-section [A] shows the present disposition of this formation or bed, the facies map of which [B], shows that it consists of a deep-sea limestone [2] cut by a granite intrusion [1] and grading into reef limestone [3]. Beyond this, the formation shows shallow-water shales [4] and deltaic sandstones [5]. This facies map also shows the thickness of the formation by means of contour lines of equal thickness known as isopachs. These are determined by drilling or by seismological methods.

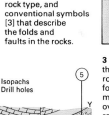

Isopachs
o Drill holes

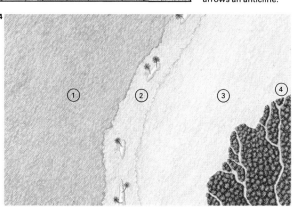

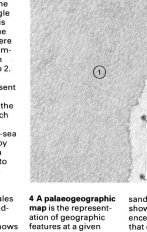

4 A palaeogeographic map is the representation of geographic features at a given geological time. The previous facies map of the formation [3] of map 2, which was deposited during the Carboniferous, can be translated into this map. The shales and sandstone to the SE show the former existence of a land in that direction and of a river flowing from it and building a delta [4] into a shallow sea [3]. Farther out to sea, in waters less than 45m (150ft) deep and more than 20°C (68°F) in temperature, corals and algae built barrier reefs and low-lying islands [2]. In the zone of open water [1] beyond the reef, calcareous plant and animal remains accumulated in deposits of limy mud which, in time, became limestone.

which the coarser material lies at the bottom and the finer material rests on the top.

A highly reliable technique of correlation is that of finding a similarity of sequence. The position of a layer between other readily identifiable layers is an ideal means of correlating scattered outcrops. The fossil contents of rocks are other excellent tools of correlation. Fossils can be characteristic of specific environments and of specific periods in history. They not only identify the formation in which they are found but also help to determine its age.

The structure of an area is important in determining the history of the rocks since their formation. Not all beds are horizontal. Many have been tilted, folded and faulted into a variety of twisted positions. In the field, geologists may notice that the bedding planes of the strata in a particular outcrop slope diagonally into the ground. The acute angle between the plane of this rock and the horizontal surface of the earth is known as the "dip". The angle of dip is measured with a clinometer and is stated in degrees. The "strike" of a rock is the direction in which the

face of its bedding plane lies – this is given as a compass direction. Strike and dip together measure the attitude of a formation. The attitude of a rock provides one of the most reliable indications of the sub-surface structure of a particular region.

Palaeogeographic maps
By interpreting geological maps and examining rocks for clues about the environment in which they were originally deposited, it is possible to piece together clues to the earth's past. This information can be represented in palaeogeographic maps, which portray the surface features of the earth as it existed during any given era in history [4].

Maps can also be constructed so as to show the past distribution of climatic zones. The fossils of organisms that flourished only in specific environmental conditions are an important means of identifying palaeoclimates, but more direct indicators of climate can also be found. These include such features as ice-grooves in erosion surfaces and rain pits in sandstone, each telling their own story of the geological history of the earth.

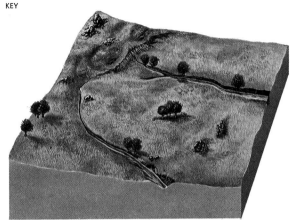

A geologist coming into an area for the first time is rarely faced with a completely exposed sequence of rocks as in the Grand Canyon. More often the rocks in his area of study are concealed beneath

soil and vegetation and exposed only occasionally where the natural covering has been removed by the action of water or the weather. The geologist studies each outcrop and from his notes, and

the samples he has collected for further study, he can reconstruct the area's geological history. The different steps in the study of the area shown here are illustrated in maps 1 to 5.

5 The geological history of the area can be unravelled by the study of all the formations shown in map 2A, which are referred to here by the same numbers. The lowest formation [2] is a sandstone deposited in a desert [A]. The sea then advanced [B] producing marine conditions [3]. When the sea withdrew a new desert [C] and sandstone formation [4] occurred. A return of the sea [D] brought calcareous sediments [5] and muds [6]. Later faulting and folding occurred [E] and granite [1] was intruded into the sedimentary layers. Since then erosion produced today's landscape [F]. We can now look at the Key illustration with greater understanding.

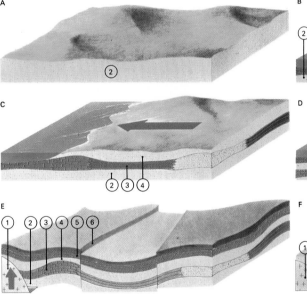

6 The geologist who sets out to do fieldwork must carry with him all the tools and measuring devices that he will need to make his observations. Typically, he will have a compass and clinometer to take the dips and bearings of bedding planes, faults and other features. A hand lens is used to examine details in rocks and a camera is useful to record the attitude and structure of rock outcrops. A hammer is essential for breaking open rocks to examine their mineral composition and to chip off samples that are collected for the specimen bag. Most important is the pen and notebook in which all his observations, rough drawings and maps are recorded.

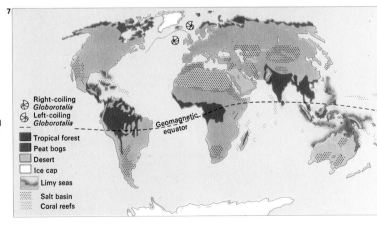

7 Palaeoclimatology is the study of ancient climates through traces left in contemporary rocks. The present-day formations of such preservable climate-related features is shown. The foram-

inifernan *Globorotalia* is an indicator of sea temperature. It coils right in warm waters and left in cold waters. Coral reefs and major carbonate deposits are both typical of warm, shallow seas.

Common to desert environments are evaporite deposits (salt basins) and reddish-hued sandstones. The lush plant life of tropical forests and swamps is the raw material from which coal is formed. Ice

sheets groove and scratch the face of rocks and leave characteristic deposits of glacial till, and peat bogs are typical of the tundra environment along the fringes of the ice caps.

Right-coiling *Globorotalia*
Left-coiling *Globorotalia*
Tropical forest
Peat bogs
Desert
Ice cap
Limy seas
Salt basin
Coral reefs

Earth's time scale

In the mid-seventeenth century, Archbishop James Ussher (1581–1656) of Ireland reached the conclusion that the earth was created at precisely 9am on 23 October 4004 BC. His findings were arrived at after diligent study of religious texts. Not until well into the nineteenth century did efforts to establish both absolute and relative techniques of geological dating meet with any semblance of success. In 1897, the renowned British physicist William Thomas Kelvin (1824–1907) attempted to deduce the earth's age from the temperature difference between the young molten planet and its present state, assuming that the rate of heat loss was constant. His estimate of 20-40 million years was more than a hundred times lower than current estimates. Radioactivity was then unknown, so Kelvin failed to take into account the fierce heat generated by the decay of radioactive elements within the earth.

The law of superposition
Although early efforts to find an absolute dating system repeatedly failed, a relative time scale proved far easier to develop. Such a system merely seeks to establish the order in which rocks were laid down [1]. It does not make any reference to fixed units of time. The entire sequence of rocks deposited since the beginning of time is known as the geological column. Once the law of superposition (which states that in strata which have remained undisturbed since deposition older rocks lie beneath younger ones) had been elucidated by William Smith (1769–1839) late in the eighteenth century, piecing together the geological column was simply a long, arduous task of identifying and correlating rocks and slotting them into their appropriate order in the stratigraphic sequence. The entire column was subdivided into units based on events that were taken to be natural breaks between one geological era and the next. The major divisions, therefore, are of widely differing lengths of time [5].

The correlation of rock strata was made easier by observing the fossils they contained. Organisms of any particular time in history possess quite distinct characteristics that can be used to identify the widely scattered rocks in which these fossils occur [3]. Strata containing similar fossil assemblages can be assumed to have been deposited during the same period of geological history.

The search for absolute dating
The breakthrough to an absolute time scale was finally achieved with the discovery that radioactive decay proceeds at a constant pace. In 1907, a chemist at Yale University, Bertram Boltwood (1870–1927), found that the decay of radioactive minerals could be thought of as a convenient yardstick of time. He recognized the regular relationship that existed between decay products and their parent elements and that progressively older specimens possessed increasing amounts of stable end products [8].

The most useful concept in radiological chronology is the notion of a half-life, the time it takes for half a given amount of material to decompose, or decay, into a radiogenic product. The half-life of uranium-238, for example, is 4,510 million years. After a lapse of this time, only half an original given quantity of uranium remains, the rest having been transformed into a series of radioactive

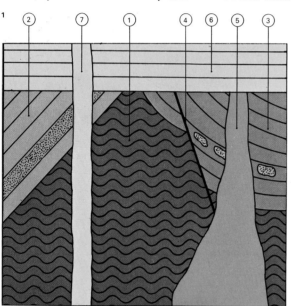

1 The relative ages of rock structures can be understood from clues in the rocks. In this cross-section the oldest formation is of metamorphic basement rock [1]. It was tilted and heavily eroded before being buried by a sedimentary sequence [2] that in turn weathered and was buried by a later sequence [3], shown by fragments of 2 found in 3. Tectonic activity caused a fault [4] to displace earlier rock [3 and 1] followed by an intrusion of magma [5]. Erosion, followed by a sea inundation, deposited a new layer [6]. The most recent structure is an igneous intrusion [7] to the land's surface.

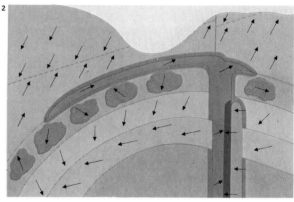

2 The earth's magnetic field can provide a useful tool for the dating of rocks. When a rock is formed, magnetic particles in it align themselves in the direction of the earth's lines of magnetic force acting at the time. If the changes in direction and position of the lines of force are established the ages of the rocks of an undisturbed sequence can be determined by investigation of their magnetic alignment. If the rocks have been disturbed the variations in their alignments indicate the nature of the movements involved. The complex history of the example shown can be interpreted by a study of the alignments of the magnetic particles.

3 Index fossils are useful for determining the ages of rocks. The age ranges of the various families of trilobites are known. If proetid and agnostid trilobites are found in the same rock, it is Ordovician in age.

1 Redlichiida
2 Asphidea
3 Ilanidae
4 Proetidae
5 Trinucleidae
6 Agnostida
7 Odontopleurida
8 Lichida

4 Orogeny is the process of mountain building by folding and thrusting. In the geological past, there were several major orogenic climaxes, which make ideal reference points as they form breaks in the stratigraphic column through an increase of erosion and changes in sedimentation patterns.

isotopes, eventually decaying to the lead isotope Pb-206. Thorium-232 has an even longer half-life, of some 14,100 million years, while that of carbon-14 is of only 5,570 years.

The age of a rock specimen is arrived at by comparing the ratio of decay elements to the remaining amount of parent material. The known half-life of the element in question is then used to calculate the sample's age. This technique has only become reliable since the 1950s when mass spectrometers, instruments that can analyse and measure elements in quantities of only a few millionths of a gramme, were developed.

The process of decay is extremely complex. The disintegration of unstable atoms is spontaneous and has never been shown to be affected by surrounding physical or chemical conditions. This is one of the only processes on earth which can be assumed to have had a constant rate throughout time, making it an ideal standard for the measurement of absolute age.

Absolute dating establishes the age of a specimen from the time it crysallized into a mineral, not the age of the element itself. Once a crystal has crystallized, chemical composition is fixed. The decay products within it are the result of the disintegration of a radioactive parent element.

Finding the age of the earth

Uranium, and its close cousin thorium, are not the only elements that are suitable for absolute dating. Potassium-40, with a half-life of 1,300 million years, occurs throughout the earth's crust in measurable quantities. It decays into argon-40, an inert gas found drifting in the atmosphere. A comparison of the ratio of these two elements in the crust and air yield a figure of some 4,600 million years for the age of the earth. The oldest rocks of the great Precambrian shields of North America, Greenland, Africa and Australia yield dates of up to only 3,500 million years. The discrepancy between the two figures is perfectly plausible since a long cooling period must be allowed for before any major rock systems could have formed a crust. Despite all these up-to-date techniques, many rocks cannot be dated absolutely.

KEY

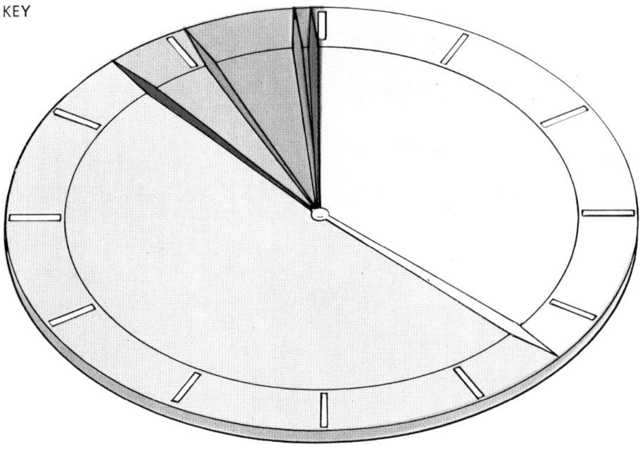

The colossal time-span over which geological processes operate is emphasized if one compresses the 4,600 million years of earth history into 12 hours on a clock face. The first 2 hours 52 minutes are obscure. The earliest rocks occur at about 02.52 hours but the planet remains a lifeless desert until 04.20, when bacterial and algal organisms appear. Aeons of time drag by until just after 10.30 when there is an explosion of invertebrate life in the seas. Dinosaurs wander the land at 11.25 only to be replaced by birds and mammals 25 minutes later. Hominids arrive about half a minute before noon. The last tenth of a second covers the history of civilization.

5

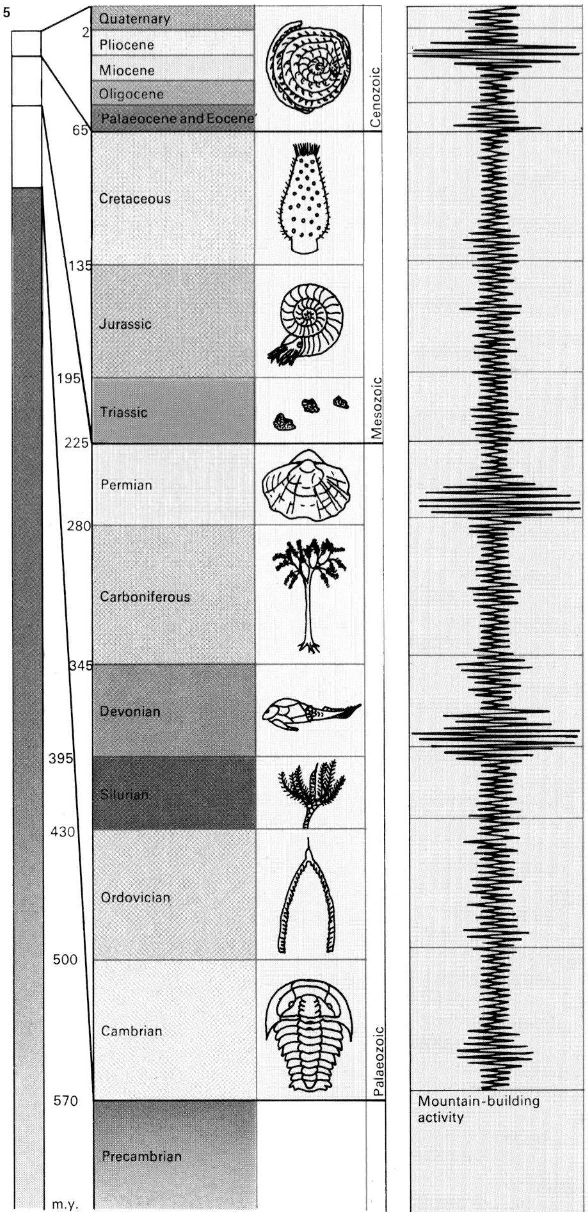

Quaternary
Pliocene
Miocene
Oligocene
Palaeocene and Eocene
Cenozoic
Cretaceous
Jurassic
Triassic
Mesozoic
Permian
Carboniferous
Devonian
Silurian
Ordovician
Cambrian
Palaeozoic
Precambrian
2
65
135
195
225
280
345
395
430
500
570
m.y.
Mountain-building activity

6

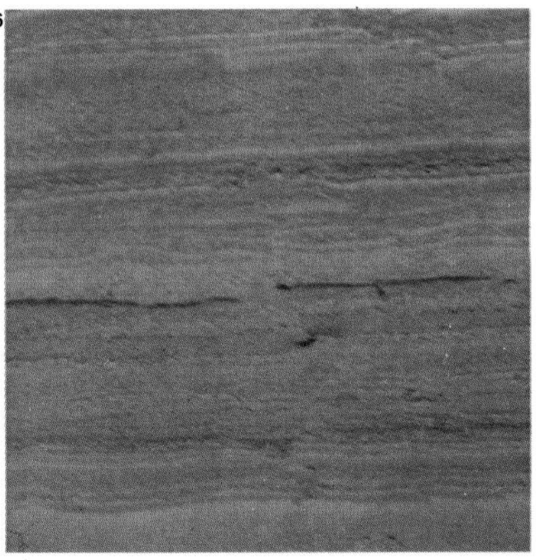

7

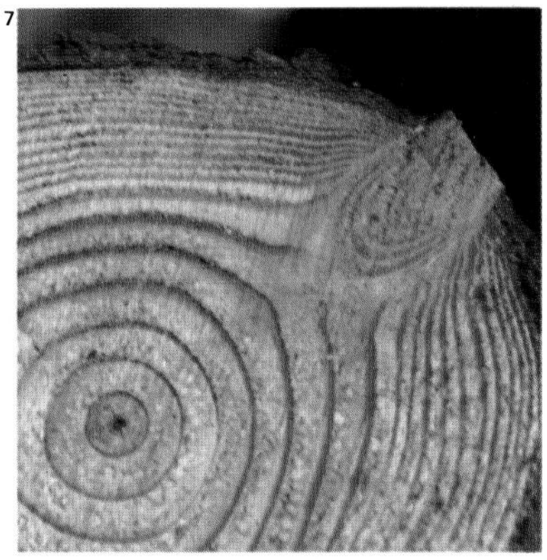

5 The age of the earth is some 4,600 million years, although only the last 570 million years show an abundance of plant and animal life. The most widely found fossil remains from any specific period are called index fossils and are used to correlate various rock formations of the same age. Mountain building took place mostly during specific periods in the geological time scale.

6 Varves are thin bands of sand, clay and silt deposited as easily recognized annual layers in glacier-fed lakes. Varves can be read to determine the dates of the retreat of the last ice age.

7 Dendochronology is the use of annual growth rings of trees to measure time. The innermost rings of recent trees can be matched to the outer rings of older trees and a chronology can be set up.

8

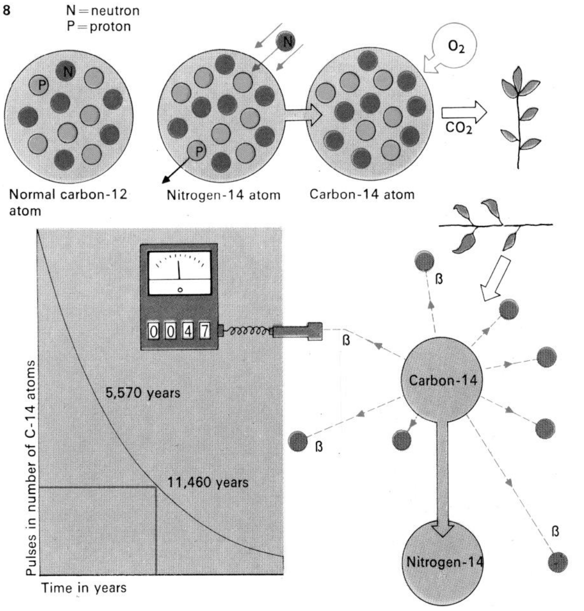

N = neutron
P = proton
O_2
CO_2
Normal carbon-12 atom
Nitrogen-14 atom
Carbon-14 atom
0047
Pulses in number of C-14 atoms
5,570 years
11,460 years
Time in years
Carbon-14
Nitrogen-14
ß

8 Radiocarbon dating is typical of the techniques that use radioactive decay to estimate age. Carbon (C) posesses two isotopes, C-14 which is radioactive, formed by cosmic ray bombardment of nitrogen-14 (N-14), and C-12 which is not. The C-14 combines with oxygen to form carbon dioxide which is absorbed by living organisms. A constant ratio of C-14 to C-12 is established in each organism during its life. After death no more carbon dioxide is absorbed and the C-14 decays steadily, by emitting beta particles, to N-14, falling to half its original quantity every 5,570 years. The age of organic remains can be estimated by comparing the ratio of C-14 to C-12 in them.

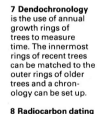

Mineral resources of the land

Minerals are the building blocks of rocks. Some rocks are made up of only one mineral while others contain many of them. Only rarely do useful minerals occur in sufficient concentrations to make commercial exploitation worthwhile [Key]. Increasingly sophisticated technology means, however, that deposits that were uneconomic to work a few years ago can now be profitably exploited. New techniques have also made it possible to re-work the waste heaps of some mines. Scarcity due to an increased demand or to depletion of richer reserves can also make the extraction of low-grade ores profitable without necessarily involving a change of the basic mining techniques being employed.

The formation of ores

Concentrations of minerals that contain economic quantities of such metals as copper, tin, tungsten, lead and others are called ores. These are formed in many different ways. The various ores of the same metal can have dissimilar origins.

Magma [2] or molten rock is the origin of many mineral deposits. Some deposits are formed within the cooled and consolidated igneous mass itself. The minerals become concentrated by magmatic segregation. Examples of deposits formed by this process are the chromite [5] deposits of South Africa, the famous iron ore deposits of Kiruna, Sweden, and the nickel sulphide of Sudbury, Ontario. Deposits formed at the same time as the surrounding rock are called syngenetic. When a mineral deposit is formed later than the surrounding rock it is known as epigenetic.

During cooling of the magma, hot gases and liquids may be forced, under great pressure, into surrounding rocks. These mineral-rich solutions cool and are forced to deposit minerals as pressure is reduced. Sometimes they find their way into small cracks to give veins, or a collection of veins (sometimes called a lode), containing both economically important minerals and worthless ones (gangue). Veins have no great thickness but may run for considerable distances and penetrate to great depths.

Deposits formed by gases are called pneumatolytic. The apatite (phosphatic mineral) deposits of Norway were formed in this way as were some of the tin deposits once mined in Cornwall. Those that originate from hot aqueous solutions are known as hydrothermal deposits.

The heat of the igneous intrusion changes surrounding rocks, particularly those in direct contact with it. Where permeation of these by hot gaseous and mineral-rich solutions replaces some of the original rock the deposit is called pyrometasomatic and is often a source of copper, zinc and lead. Such interactions are most evident at points of contact between granite and limestone.

Heat zones and weathering

Mineral-rich solutions sometimes replace only certain elements in the original rock; at other times the whole mass may be affected, forming large deposits such as the pyrite deposit of Rio Tinto in Spain and those of the Copper Belt of Zambia. The type of mineral deposit is determined by the temperature of the solution and associated gases. Specific minerals are associated with hotter and cooler areas. Tin, copper, lead, zinc and iron

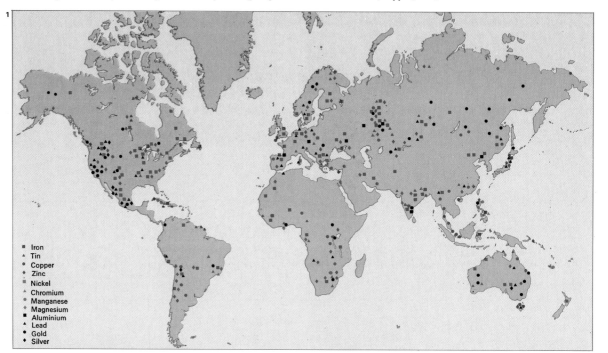

1 **Reserves of important metals** are widely distributed throughout the world. The map provides a guide to major ore-producing areas but many other locations could be plotted if areas were studied in greater detail. Information about reserves in communist countries is often withheld and it is likely, therefore, that reserves are greater than is indicated. The USSR leads world production of iron, chromium and manganese and is a leading producer of most other metals. Except for Antarctica, major discoveries of minerals have been made in every continent since World War II. Some areas such as Australia and the Sahara have seen considerable mining development since the mid-1960s.

- ■ Iron
- ▲ Tin
- ● Copper
- ◆ Zinc
- ■ Nickel
- ▲ Chromium
- ◆ Manganese
- ◆ Magnesium
- ■ Aluminium
- ▲ Lead
- ● Gold
- ◆ Silver

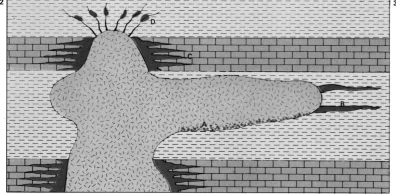

2 **Magmatic ore bodies** may result from several different causes: settling of denser minerals or elements to the lower part of a magma chamber during cooling [A]; injection of late-crystallizing magma components along fissures [B]; contact metamorphism in which minerals of the wall rocks are replaced by other minerals derived from the magma [C]; and hydrothermal deposits filling the fissures with minerals from the magma and transported by hot watery solutions [D], as in the copper deposits of Butte, Montana, USA.

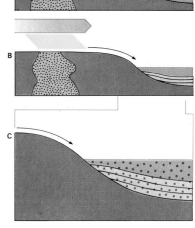

3 **Sedimentary ore bodies** are originally derived from metal-rich igneous rocks such as an iron-rich pluton [A]. On exposure to weather the igneous body is eroded and iron is dissolved away in the form of bicarbonate [B]. The solution of bicarbonate reaches a basin of deposition (a lake or the sea) where it becomes a hydrous oxide and is precipitated into the sediments [C]. Iron oxide rolled about by currents before settling may form small rounded aggregates. During consolidation of the sediment the water is squeezed out of the hydrous oxide, giving haematite.

are characteristically found in that order, working outwards from the zone of most heat. Hydrothermal solutions may surface as hot springs to produce deposits of the mercury ore, cinnabar, while sulphur deposits [7] are often found near volcanoes.

Rocks at the surface of the earth are subjected to weathering. Downward percolation of ground water may leach some useful elements from upper layers and bring them to the water-table level. Many valuable deposits of copper, for example, have been created from low-grade ores in this way. Below the layer of secondary enrichment the ore remains low grade. In tropical areas weathering may affect the top 75m (240ft) of the surface. Under such conditions aluminium silicate rocks are broken down to give deposits of bauxite, the chief ore of aluminium. Some iron ores and some manganese deposits originated from weathering.

Sedimentary rock deposits

When the rocks are broken down the particles are carried away by streams and rivers [3]. Minerals that are not easily changed and that are heavy may collect in workable quantities as placer deposits. Such deposits occur in streams and on beaches. Gold is one metal mined from placer deposits. Most of the world's tin, which comes from Malaysia and Indonesia, is dredged from placer deposits.

Many other sedimentary rocks are important sources of minerals. Evaporation of sea-water and inland lakes can give precipitation deposits of gypsum [Key, 8], anhydrite, halite (common salt) and potash [6]. Many of these are mined conventionally but some are exploited by solution mining – hot water is pumped into the deposit and the resulting brine forced to the surface. Important phosphate deposits are the result of vast accumulations of bird droppings called guano. Clays are sedimentary rock deposits important in paper and brick manufacture and in pottery and ceramics. China clay, or kaolin, is a clay formed as a result of action of hydrothermal solutions on the felspars in granite. Marls are mixtures of clay and calcium carbonate and are quarried for the cement industry; limestone is exploited for building stones or for lime preparation.

Open-pit mining of gypsum, near Paris, produces the raw material known as plaster of paris. Where an exploitable mineral occurs near the surface it can be extracted in this way by removing the overburden of soil and rock.

4 Garnierite, also known as Noumeite, from Nouméa, New Caledonia, where it is found in veins, is a source of nickel. The New Caledonia deposits were once the most important in the world.

5 Chromite is found as a mineral in ultra-basic igneous rocks. In some areas, however, ore bodies have formed by segregation, as in New Caledonia, where this specimen was found, and in Rhodesia.

6 Potash salts are the last precipitates of an evaporating sea. Potash is used mainly as a fertilizer but also in the explosives industry and in various metallurgical processes.

7 Sulphur is often found in the native (pure) state around the vents of volcanoes; it is deposited by sulphurous gases. A major source of sulphur today results from the refining of crude oil.

8 Gypsum, like the potash salts, is an evaporite. It is found in Britain, Germany and the USA and is used in plaster and cement and also in the paper and paint trades.

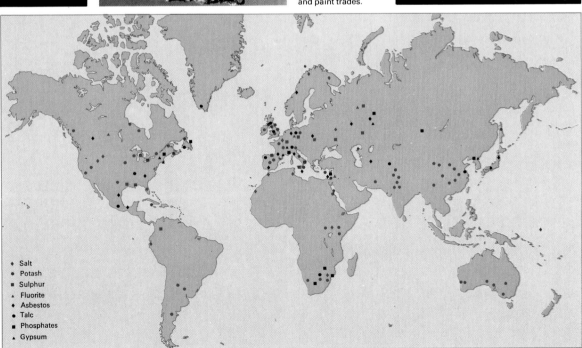

9 Reserves of non-metallic minerals of economic importance are distributed irregularly. This is partly due to lack of information about finds in certain areas and partly to the fact that some areas are inaccessible for exploration. It is likely that vast reserves will be discovered as new methods of exploration are developed. No country will ever have all the mineral resources it needs within its own boundaries. The "energy crisis" may be followed by a "mineral crisis", particularly if less developed nations advance rapidly along the road to full industrialization. They will then require their own resources rather than being able to export them for use by the big industrial nations.

- ◆ Salt
- ● Potash
- ■ Sulphur
- ▲ Fluorite
- ◆ Asbestos
- ● Talc
- ■ Phosphates
- ▲ Gypsum

Mineral resources of the sea

The oceans are a vast storehouse of minerals [2]. Seawater itself contains almost all the elements, but many of them are present in such extremely low concentrations that the cost of extraction is high compared with the cost of mining the same elements on land.

Chemicals from seawater

A few substances can be extracted economically from seawater – namely, common salt, magnesium and bromine. Salt has been obtained from the sea since ancient times [3]. The traditional method is to flood coastal pans with seawater. As the water evaporates in the sun, some impurities are precipitated out. The concentrated brine is then passed to another pan where the salt is precipitated. Today about 33 per cent of the world's total salt output comes from ocean water. Sea salt is especially important in countries that have no other source of salt, such as Japan.

Magnesium is an important metal in the lightweight alloys that are used to manufacture aircraft, missiles and precision instruments. In World War II, it was used in incendiary bombs. In the 1930s, Germany pro-duced nearly 66.6 per cent of the world's output so American and British scientists began experiments to extract it from seawater. British efforts proved successful in 1939 with a process that removed the magnesium as a hydroxide after seawater had been mixed with lime. Today, magnesium from seawater accounts for 66.6 per cent of the world's annual production.

Bromine, an important element in the photographic and pharmaceutical industries and in the production of high-octane petrol, is also largely obtained from seawater. However, the prospects for extracting other elements from seawater are influenced by two main factors. First, there must be a shortage of the element on land to justify the high level of investment that would be required and, second, successful exploitation relies on technological developments.

Water itself is a valuable resource in arid coastal areas [7]. Fresh water can be produced from seawater by evaporation and condensation – this is the source of rain and fresh waters. But the artificial speeding-up of the process in desalination plants uses large amounts of energy, usually oil or nuclear, and it is justified only in those parts of the world where energy is cheap, as in Arabia.

Continental shelf resources

Apart from resources in seawater itself, exploitation of ocean floors is already yielding substantial supplies of many minerals. Most of these minerals are obtained around coastlines or in the shallow waters of continental shelves.

Sand, gravel and limestone, which are used in building, are taken from beaches and coastal waters. Some beach deposits contain metals in quantities worth extracting. For example, the black sands along the Atlantic coast of the United States from New Jersey to Florida contain ilmenite and rutile, ores from which titanium is obtained, and monazite, a rare earth mineral. The black sand of New Zealand's west coast beaches is rich in iron. However, over-exploitation of some deposits may result in serious coastal erosion.

Mining materials on the sea-bed in shallow water has also become important. Around Japan, underwater iron-bearing

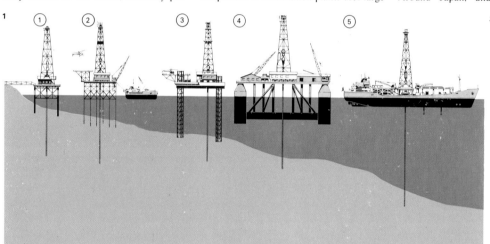

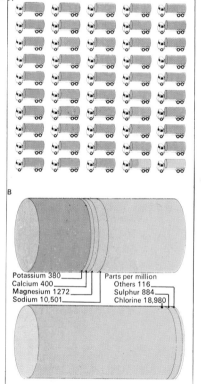

1 Early marine oil rigs [1] were essentially land rigs on wooden piles. Later designs [2] have multiple legs sunk into the sea-floor. Jack-up rigs [3] have legs which are extended to the bottom. Large semi-submersible rigs [4] are supported by their buoyancy and tethers secure them to the bottom. Ships [5] are used in deep water. The drill is lowered through a hole in the hull.

2 Seventy-three of the 93 natural elements are known to exist in measurable amounts in seawater. The 50 tankers [A] represent 1 million litres (220,000 gals) of seawater (each tanker holds 20,000 litres). The last two tankers [B] contain the elements present in this volume and the proportions of each are indicated.

Potassium 380
Calcium 400
Magnesium 1272
Sodium 10,501

Parts per million
Others 116
Sulphur 884
Chlorine 18,980

3 Salt has been extracted from seawater since the days of the Minoan civilization. Seawater is first trapped in coastal pans. As the water evaporates in the sun, the salt is precipitated from the brine.

4 The element iodine is present in seawater in very low concentrations (0.006%). It is concentrated by seaweeds which, botanically, are green, red and brown algae. Iodine was first discovered in 1811 when it was extracted from the ashes of seaweed. Kelp, a ribbon-like brown alga, was one of the earliest sources. Much iodine is now obtained elsewhere, for example from Chile saltpetre and brine from oil wells, although seaweed remains a valuable source of iodine.

sands are piped ashore and, off the coast of southwest Africa, gravels containing diamonds have been mined since 1962.

Other minerals are extracted from rocks under the sea-bed. Prospectors use similar methods to those used on land, aided by such techniques as underwater photography. Today mining is effectively carried out in water up to 183m (600ft) deep. For example, about 20 per cent of Japan's coal comes from submarine mines. Having located a coal bed, engineers build an artificial island and then drill down to the coal layers. In the Gulf of Mexico, sulphur is flushed out from beneath the sea-bed with superheated water.

The sulphur deposit in the Gulf of Mexico was discovered when prospectors were searching for oil. Currently, oil and natural gas are the most important minerals being extracted from under the sea [Key]. Oil rigs are located round the world [1], in such places as the Bass Strait between Tasmania and the Australian mainland, the Black Sea, the Gulf of Mexico, the Arctic Sea, off the coast of California, the North Sea and the Sulu Sea in the Philippines. The setting-up of

rigs in the sea has, however, met with many problems, involving risks for the operators. The rigs may have to withstand severe storms and even the drift pressure of pack ice. But the technology of drilling in the sea-bed has developed quickly. Today about 20 per cent of the world's oil production comes from offshore operations and this proportion is bound to increase as the land oilfields dry up.

Deep-sea deposits

Also present on the deep ocean floor are large numbers of so-called manganese nodules [5]. The presence of these potato-shaped nodules, which also contain cobalt, copper, iron, nickel and other metals, has been known for 100 years [6]. In parts of the Pacific, the nodules are concentrated at more than 54kg per square metre (100lb per square yard), and they also occur in the Atlantic and Indian Oceans. The origin of the nodules is unknown. Scientists believe that they must be formed by some biological process that is increasing the number of nodules. Manganese, important in steelmaking, is unevenly distributed on land.

Oil and natural gas are two of the most important products now being extracted from beneath the sea-bed. The technology of drilling into the sea-bed has developed rapidly since 1945 and today about 20% of the world's total oil production originates from off-shore oil wells.

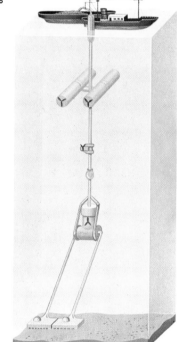

5 Ways of mining the rich deposits of manganese nodules in the deep waters of the ocean have been discussed for some time. The diagram shows one of the methods proposed. This device employs the principle of the vacuum cleaner. It picks up the nodules and pumps them to the surface. On reaching the surface, the nodules are transferred to a separate barge. The device is driven by a series of propellers located between the surface and the bottom. Other proposed devices include self-propelled rigs with air-lift systems, and a bottom-crawling miner attached to a surface vessel by articulated arms. Such systems may incorporate closed-circuit television to aid in locating the nodules. Other devices return to the surface loaded.

7 One important oceanic resource is water. Desalination of seawater to obtain fresh water is achieved by electrical, chemical and change of phase processes (by condensing steam from the water or by freezing it to remove the salt). However, all these processes require power and complicated machinery which makes them expensive. Today desalination is carried out on a large scale only in areas where fresh water is desperately short, as in Israel. The profits from petroleum have made it possible for some Arab nations to do this in arid areas.

6 Manganese nodules [A] are potato-like lumps of minerals, possibly of plant or animal origin, found in great concentrations on parts of the ocean floor. The composition of the nodules varies to some extent. However, in addition to manganese (Mn), a valuable metal used in steelmaking, the nodules contain other valuable metals [B], including cobalt (Co), copper (Cu), iron (Fe) and nickel (Ni).

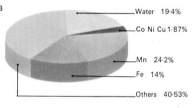

B

Water 19·4%
Co Ni Cu 1·87%
Mn 24·2%
Fe 14%
Others 40·53%

Oil and gas exploration
Sedimentary basins favourable for petroleum
▲ Oil
● Gas

8 The maps show the areas where mineral resources can currently be extracted from the sea-bed. Most of the economic mining is in the shallow waters of continental shelves.

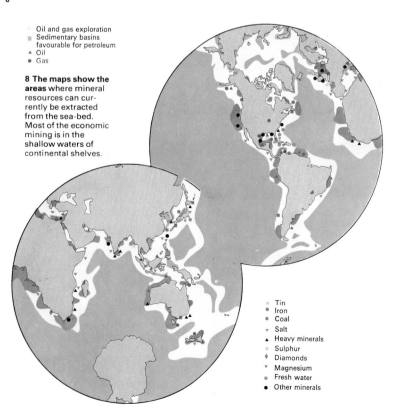

Tin
Iron
Coal
Salt
▲ Heavy minerals
Sulphur
◆ Diamonds
▼ Magnesium
● Fresh water
● Other minerals

Energy resources: coal

Coal was the first fossil fuel to be exploited on a large scale. It made possible the Industrial Revolution which in turn benefited the coal industry by providing it with a superior technology, thus enabling coal to be mined at even greater depths. Coal consumption has been on the decline for more than 20 years. In 1960 coal provided about half the world's energy; by 1970 its share had fallen to a third, oil and gas, the other two main fossil fuels, providing an ever-growing share of the world's needs. However in the near future the situation is likely to be reversed as oil and gas reserves are depleted. There is estimated to be in excess of eight million million metric tonnes of coal in world reserves, which is plenty to last into the twenty-first century.

What is coal?

Like oil and gas, coal is organic material that has been slowly broken down by biological and geological processes but, whereas oil and gas are mainly composed of animal matter such as plankton, coal consists of plant remains. The fossil fuels are a part of the carbon cycle [1] that is basic to life on earth.

There is probably about 50 times as much carbon locked up in fossil fuels as there is in all the living matter on earth.

Coal is a sedimentary rock consisting of carbon, water and volatile gases, with small amounts of mineral impurities that produce ash when the coal is burned. It is variation in the proportions of the constituents present that produces the different types of coal and is responsible for their different calorific values. Lignite (brown coal) contains a high proportion of water (43 per cent) and is therefore of lower calorific value than bituminous coal (ordinary household coal) which contains only three per cent water. Anthracite on the other hand is almost all carbon (96 per cent) and contains hardly any water or volatiles. It is of higher calorific value but less readily inflammable than bituminous coal which has a high proportion of volatiles (32 per cent). The mineral impurities present are mainly clays, chlorides and sulphides. The latter in particular are troublesome on combustion, oxidizing to sulphur dioxide, an important cause of atmospheric pollution.

The creation of coal is dependent upon abundant plant growth. This condition was fulfilled during the Carboniferous, about 345 million years ago, when large areas of swamp forest covered low-lying regions [Key]. Most commercial deposits of coal date from this period, although there are deposits dating from most ages in the geological time scale.

How coal is formed

Coal is formed when decaying plant material is building up faster than it can be destroyed by bacterial activity. The swamp environment is ideal for coal formation. The water tends to be stagnant and therefore deficient in oxygen which inhibits bacterial activity – preventing the vegetable matter from being completely broken down. Once a certain stage is reached acids produced in the process prevent the bacteria from further activity. The initial product so formed is peat. If it is then buried under other sedimentary material the peat will be compressed, water and gases being squeezed from it forming coal. If the sedimentary material is 1km (0.6 mile) deep a 20m (66ft) layer of peat yields a 4m

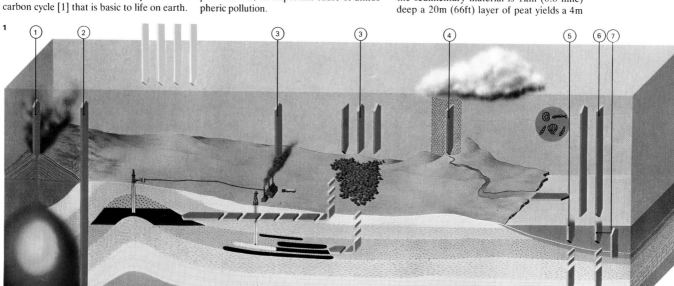

Sunrays　　Photosynthesis　　Carbon dioxide transfer　　Carbonate transfer　　Burial and rock formation

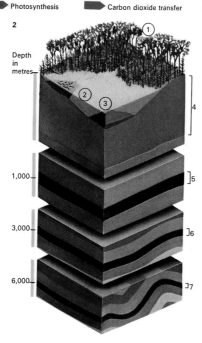

1 Coal and other fossil fuels are part of the complex carbon cycle that maintains life on earth. Carbon dioxide (CO_2) is produced in the earth's interior [1] and in rocks [2], when animals breathe and when organic material – including fossil fuels – burns [3]. It is converted into living matter by the process of photosynthesis carried on by plants. Carbon dioxide exists in the atmosphere and in solution in seawater. It is exchanged between these two elements when rain washes it out of the air [4]; it can also be dissolved by seawater [6]. Some of this dissolved gas is then deposited chemically [7] and organically [5] as carbonates.

2 The process of making coal starts with plant debris [1]. Dead vegetation lies in a swampy environment and forms peat [4], the first stage of coal formation. Underwater, bacteria remove some oxygen, nitrogen and hydrogen from the organic material. Debris carried elsewhere and deposited by water forms a product called cannel coal [2]. Algal material collected underwater forms boghead coal [3]. If the dead organic material is buried by sediment, the weight on top of the peat, and higher temperature, will turn the peat into lignite [5]. With more heat and pressure at increasing depths, lignite becomes bituminous coal [6] and then anthracite [7].

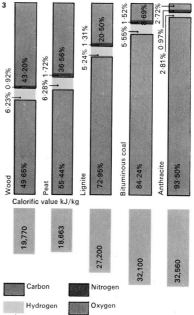

3 The economic value of a fuel is largely dependent on its energy content – that is, the amount of energy that can be recovered from burning a standard quantity of coal. However, this must be offset against cost of extraction and combustion. Shown here is a comparison of the energy content of the most commonly used kinds of coal. In fuel value alone anthracite appears to be the most attractive coal, but its disadvantage is that it lies deep underground and in thin seams. Lignite can be more attractive because it occurs in thick seams fairly close to the surface, and therefore can be retrieved by a "strip mining" process.

(13ft) layer of lignite. If the material rests under 3km (1.9 miles) of earth, the product is a 2m (6.6ft) layer of bituminous coal. If the depth is greater – say, 6km (3.7 miles) – and the temperature higher, the 20m peat layer becomes a seam of anthracite 1.5m (4.9ft) thick. Anthracite is so hard that it is classed as a metamorphic rock.

The geology of coal is simpler than that of oil and gas insofar as, unlike these, coal cannot travel from place to place underground, or seep to the surface. This means that estimates of coal reserves are more reliable and easier to calculate than those of oil and gas. Coal is found in seams of any thickness from only a few centimetres up to 30m (100ft) thick. Characteristically the seams are found interbedded with shales, sandstones, clays and limestones of both marine and non-marine origin in a regular sequence. This phenomenon is known as cyclic sedimentation and indicates to the geologist that the coal was likely to have been laid down in a deltaic area. Many of the main coal measures have been formed in this delta-type environment such as the Pennsylvanian coal measures of the United States and the Carboniferous coal measures found in various parts of the United Kingdom.

The prospects for coal

Although world reserves of coal are more than enough for our needs in the near future, only about half of the coal is exploitable. The rest has either to be left where it is to prevent adjacent coal mines from collapsing, or it is so inaccessible that mining would be a very difficult and expensive operation and uneconomic at present coal prices. The Soviet Union is estimated to have the largest coal reserves amounting to 68 per cent of the remaining world resources; the United States has about 14 per cent of world reserves. Of the Third World countries, India has the largest resources – one per cent of world reserves – but there are large areas yet to be properly surveyed both there and in the African continent. In Antarctica and Greenland there are known to be large deposits of coal that await improvements in mining technology and a rise in the economic value of coal to make their exploitation worthwhile.

KEY

The Carboniferous period, 345 to 280 million years ago, was a time of extensive freshwater and coastal swamps covered in forests of ferns, horsetails and other large tree-like plants. The vegetation of these swamps produced most of today's coal. The decaying plant matter gradually sank and became compressed to form seams of coal.

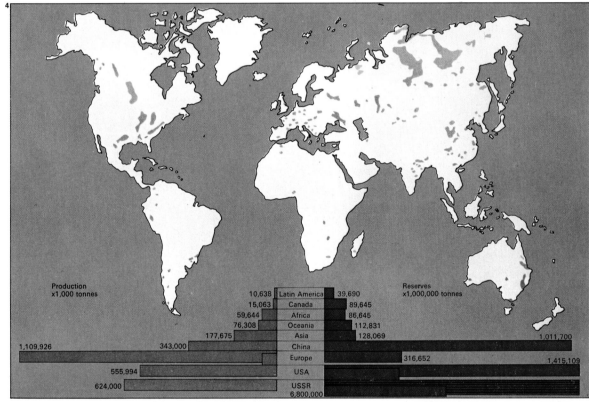

Production x1,000 tonnes

Reserves x1,000,000 tonnes

Production x1,000 tonnes	Region	Reserves x1,000,000 tonnes
10,638	Latin America	39,690
15,063	Canada	89,645
59,644	Africa	86,645
76,308	Oceania	112,831
177,675	Asia	128,069
1,109,926 / 343,000	China	
	Europe	316,652 / 1,415,109
555,994	USA	1,011,700
624,000	USSR	6,800,000

4 Coal was the major fossil fuel for two centuries until the rapid increase in oil production pushed it into second place. This has led to great changes in the coal industry during the last few decades. Europe and Japan have cut back their production because of competition from cheaper oil, while other countries such as the USA, China and the USSR are still increasing their output. The figures shown indicate the known reserves of the world but the actual reserves will be much greater since exploration is not yet complete. There is enough coal to last for at least another 200 years at present rates of consumption. Total reserves must not be confused with the recoverable reserves, which are often calculated as half the total. But advances in techniques and fluctuations in the economic value of coal mean that the recoverable fraction will become greater.

5 Underground coal seams may be thin and distorted and this makes mining a hazardous and difficult business. Today's sophisticated machinery is reducing the numbers of miners who have to work underground.

6 Open-cast mining of coal and peat is relatively inexpensive. In some countries notably Ireland, Finland and the Soviet Union, peat is a useful source of fuel. Peat has to be dried before it can be burned.

Energy resources: oil and gas

Oil and gas provided 64 per cent of all the energy used in the world in 1974: oil accounted for about 45 per cent and natural gas for 19 per cent of the total energy consumption. In 1962, by comparison, they accounted for just less than half the world's energy consumption.

Rise of oil power

In 1859 (the conventional date for the birth of the modern oil industry), Edwin L. Drake (1819–80) struck oil at Titusville, Pennsylvania, USA, when drilling a well with a derrick. The contribution of oil and gas to world energy was, at that time, negligible. The speed and magnitude of the rise of oil as an energy source is staggering, and its consequences are even more spectacular. Modern civilization and technology and present-day population levels would be inconceivable without the cheap and plentiful energy provided by oil and gas. Petroleum is also the raw material for the petrochemical industries that produce plastics, synthetic fibres and hundreds of organic compounds used as drugs, pesticides and detergents.

Drake's venture was not, however, the first use of petroleum. Noah is said to have used pitch (a form of asphalt, one of the petroleum products) to seal the seams of the Ark, a practice followed by boat builders up to the present day. As early as 1000 BC the Chinese were using natural gas for fuel and the streets of Babylon were surfaced with asphalt in 600 BC.

The world's proven recoverable oil reserves are 666 million barrels (1 barrel=159 litres=35 Imperial gallons) – only 36 years' consumption at the 1972 production levels [4]. Even accounting for undiscovered fields and improved extraction, oil is a finite resource. It is therefore not surprising that oil, with its economic importance and with its declining reserves, has become a key issue in world politics.

How oil and gas are formed

Oil and gas are considered by geologists as "minerals" because like coal they are part of the make-up of the earth's crust. Collectively they are termed hydrocarbons because they are made of molecules consisting entirely or

essentially of hydrogen and carbon. Natural petroleum or crude oil is a liquid ranging in colour from yellow to black, including reds, browns and dark greens. It is a mixture of many hydrocarbon compounds and its viscosity ranges from very fluid to highly viscous (as in pitch). Gas contains smaller, lighter hydrocarbon molecules and is colourless.

Hydrocarbons are stored solar energy. Organic matter is synthesized by living plants using the sun's energy converted by chlorophyll. Swarms of tiny plants and animals feeding on these plants lived in the seas and their dead bodies fell to the bottom. Under normal conditions, ordinary decay by bacteria breathing oxygen would "burn up" the organic matter, producing carbon dioxide and water. But because oxygen was absent the process of decay performed by bacteria was not complete and hydrocarbons and other organic compounds were produced [1].

Clays and silts were also deposited with the organic matter and in time this sediment, rich in both carbon and hydrogen, was buried under newer layers and was compacted into rocks such as mudstone and shales. Pressure

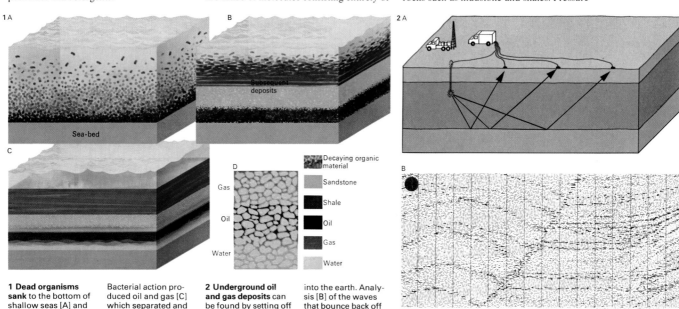

1 Dead organisms sank to the bottom of shallow seas [A] and were covered by sandstone and shale [B].

Bacterial action produced oil and gas [C] which separated and rose through the sandstone [D].

Decaying organic material
Sandstone
Shale
Oil
Gas
Water

2 Underground oil and gas deposits can be found by setting off an explosion [A] which sends sound waves

into the earth. Analysis [B] of the waves that bounce back off underground strata can locate deposits.

3 Oil and gas are found in reservoir rocks but these are seldom the rocks in which they were formed. Reservoirs are bound by impervious rocks forming oil or gas traps. Anticlines possessing an impervious layer form traps [A], as do faults that bring an impervious rock in contact with a reservoir rock [B]. Salt domes [C] are similar to anticlines. Other traps are formed by impervious layers resting unconformably on tilted strata [D] and by reservoir rocks "pinching out" [E]. Where no water is present the oil may form pools in the hollows of a basement rock [F]. Fossil coral reefs [G] and fossil river-beds [H] can form long and winding reservoirs.

and heating caused by burial is thought to have completed the bacterial process that finally led to the formation of hydrocarbons.

Compaction and the eventual crustal movements often expelled oil and gas from their source rocks and an upward movement of the hydrocarbons occurred. If they reached the surface as a seep they were lost. Many seeps exist today, including the natural underwater oil seeps near Santa Barbara in California. Sometimes, however, the upward movement was stopped by an impermeable layer that formed a trap [3]. The hydrocarbons accumulated in the permeable rock under the trap, which is known as a reservoir rock. Reservoir rocks are most usually sands, sandstones and limestones.

Scarcity of oil and gas
Hydrocarbons were formed throughout the period of organic evolution and small quantities are still being formed today, but not sufficient to replenish current oil reserves. The conversion efficiency of solar energy into recoverable oil is exceedingly low. It took several millions of years to deposit the

organic matter at any one place because most of it was lost through decay, and several tens of millions of years to transform it into petroleum, most of which seeped away through the ground. Only about 30 per cent of oil held in exploited reservoirs can actually be extracted by normal means.

Much thought is now being given to secondary recovery procedures to increase the yield of existing oilfields, to perhaps 70 per cent. Secondary recovery procedures include pressurizing the reservoir rock (by pumping the gas back into it or by pumping water), fracturing it with explosives, attacking it with acid (in the case of limestones), heating it with pumped-in hot water or pumping in chemicals that are effectively able to lower the viscosity of the crude oil.

Prospecting for oil consists first in determining those areas that have sedimentary rocks – the only rocks likely to contain oil [2]. Then the structures and successions of strata are studied by field geologists and geophysicists to locate possible traps and reservoir rocks. It is not possible to say whether a trap structure contains oil without drilling.

Offshore drilling is becoming a necessity if new oil and gas fields are to be found to replace the rapidly dwindling reserves from the land. It started in shallow and protected waters but the drillest ever. The first oilfield was found off the southwest coast of Norway in November 1970. Soon after this British Petroleum found the first oil in the British sector of the North Sea — the ing industry is now moving to deep and stormy waters and the cost of offshore oil will be high.

5 The North Sea between Britain and Norway has been one of the most successful oil areas to be discovered in the recent past. Natural gas was found first in 1965 and the gas fields were among the larghuge Forties field. These discoveries have been followed by a steady flow of new finds of oil and gas in the North Sea. Production from these oilfields is now well under way in the United Kingdom.

4 "Proven" oil and gas reserves together with recent levels of oil consumption are shown here. We can be reasonably sure that this much oil exists and more should be discovered. The true extent of the world's oil "crisis" is revealed by dividing the reserves by the annual consumption. This shows how long the known reserves will last at present consumption levels. One recent estimate suggested that the world's oil could last less than 40 years, and that does not allow for any growth in consumption. The USA has enough oil to last 13 years and gas for 12 years while the Middle East and Asia have enough oil for 50 years and gas for 80 years at the levels of 1976.

Gas
⬤ % of world reserves
⬤ Annual consumption
Oil
⬤ % of world reserves
⬤ Annual consumption

3% | 21% | 4% | 14% | 22% | 36%

Western Europe | North America | South America | Africa | Middle East | USSR, Asia and Australasia

4% | 7% | 6% | 9% | 36% | 38%

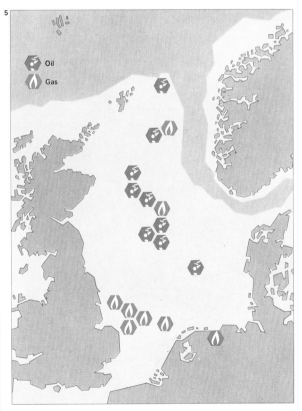

Oil
Gas

623·8 904·6
NORTH AMERICA

WESTERN EUROPE
22·6 699·2

USSR and ASIA
548·6 500·4

1077·5 69·7
MIDDLE EAST

AFRICA
272·5 47·0

CARIBBEAN and SOUTH AMERICA
221·9 147·8

SOUTH-EAST ASIA and AUSTRALASIA
84·0 374·2

Figures in million tonnes

6 The oil production and consumption in different regions are shown on this map. It reveals that oil production [yellow] and consumption [orange] are uneven. The Middle East produces a great deal of oil but consumes relatively little while in Europe consumption is high and production is small but rising, due to the discovery of oil in the North Sea. In 1974 the world's oil industry produced nearly 58 million barrels of oil a day which added up to a total production of 2,850 million tonnes of oil in the year (7.3 barrels of crude oil weigh a tonne). The imbalance of production and consumption around the world means that there is a large and vulnerable international trade which can lead to energy crises.

139

Energy supplies

The energy resources that man is using up today took many millions of years to create [Key]. Coal, oil and natural gas – the fossil fuels – contain solar energy that reached the earth perhaps 500 million years ago. They are produced from plant material that has been processed for millions of years.

There is not much doubt that oil, gas and coal are bound to run out one day and there is a great deal of debate about how much fossil energy might still be available for man's use.

Energy from coal

Coal was the first of the fossil fuels to be exploited in quantity. The gradual industrialization of Western society was associated with the growth of coal mining. New technology made it possible to expand coal mining to greater depths underground and the availability of coal spurred the development and widespread introduction of steam engines and other labour-saving inventions in the 18th century.

It is almost impossible to know exactly how much coal there is in the ground; but a reasonable estimate is that there are probably about 600,000 million tonnes of coal that could be dug up using the mining techniques employed today. Some idea of how long that amount of coal might last can be gleaned from the fact that in the early 1970s the world's coal mining industry was digging up over 2,000 million tonnes of hard coal a year. About 25 per cent of this was mined in the United States, while coal production in Europe was just below the American level. From these figures it is evident that the coal industry could survive, and even thrive, until well into the next century and maybe for many years longer than that.

The outlook for oil

Nearly half the world's energy is now supplied by oil, and a fifth of the energy comes from natural gas. The two together provide about 67 per cent of our energy.

The long-term future of the oil and gas industries looks less certain than the future of the coal industry. In relation to demand oil reserves are much smaller than coal. A recent estimate put the world's oil reserves at about 90,000 million tonnes. In 1973 production was running at 2,765 million tonnes. Clearly, if man continues to use up oil as fast as he has in the recent past, and if there are no new major discoveries of oilfields, the oil industry will be a thing of the past by the second quarter of the twenty-first century.

The situation is even more dramatic in the USA, which has something approaching 5,500 million tonnes of known oil reserves (a figure that is the subject of heated argument) and which produced about 500 million tonnes in 1973. At this rate the USA could exhaust its oil reserves well before the end of the twentieth century.

Nobody expects the oil industry to fail completely in its search for new oilfields. For example, there is confidence in the USA that, as with Britain, the offshore continental shelf will turn out to have some large underwater oilfields that could be exploited.

Other sources of energy

Hydroelectricity is an attractive energy source [5]. It is regularly renewed by rain and snow falls, thus hydroelectric power stations do not depend on an exhaustible fuel supply.

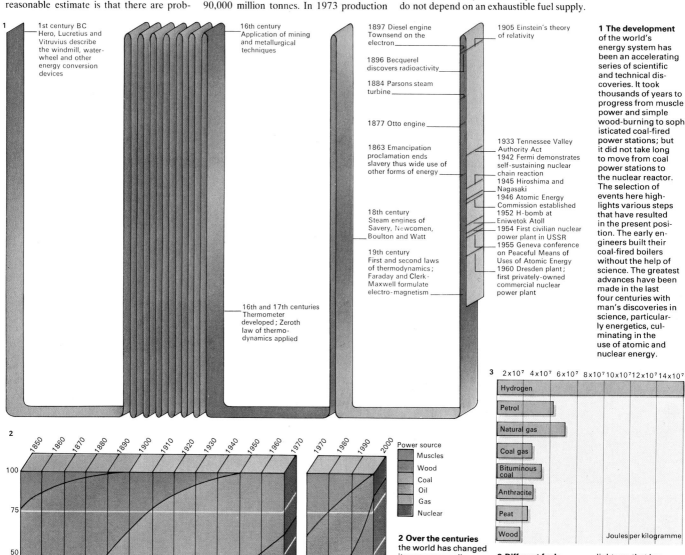

1 1st century BC Hero, Lucretius and Vitruvius describe the windmill, water-wheel and other energy conversion devices

16th century Application of mining and metallurgical techniques

16th and 17th centuries Thermometer developed; Zeroth law of thermodynamics applied

1897 Diesel engine Townsend on the electron

1896 Becquerel discovers radioactivity

1884 Parsons steam turbine

1877 Otto engine

1863 Emancipation proclamation ends slavery thus wide use of other forms of energy

18th century Steam engines of Savery, Newcomen, Boulton and Watt

19th century First and second laws of thermodynamics; Faraday and Clerk-Maxwell formulate electro-magnetism

1905 Einstein's theory of relativity

1933 Tennessee Valley Authority Act
1942 Fermi demonstrates self-sustaining nuclear chain reaction
1945 Hiroshima and Nagasaki
1946 Atomic Energy Commission established
1952 H-bomb at Eniwetok Atoll
1954 First civilian nuclear power plant in USSR
1955 Geneva conference on Peaceful Means of Uses of Atomic Energy
1960 Dresden plant; first privately-owned commercial nuclear power plant

1 The development of the world's energy system has been an accelerating series of scientific and technical discoveries. It took thousands of years to progress from muscle power and simple wood-burning to sophisticated coal-fired power stations; but it did not take long to move from coal power stations to the nuclear reactor. The selection of events here highlights various steps that have resulted in the present position. The early engineers built their coal-fired boilers without the help of science. The greatest advances have been made in the last four centuries with man's discoveries in science, particularly energetics, culminating in the use of atomic and nuclear energy.

3 2x10⁷ 4x10⁷ 6x10⁷ 8x10⁷ 10x10⁷ 12x10⁷ 14x10⁷

Hydrogen
Petrol
Natural gas
Coal gas
Bituminous coal
Anthracite
Peat
Wood

Joules per kilogramme

Power source
Muscles
Wood
Coal
Oil
Gas
Nuclear

1850 1860 1870 1880 1890 1900 1910 1920 1930 1940 1950 1960 1970 1970 1980 1990 2000

100 75 50 25 0

2 Over the centuries the world has changed its energy supplies. Muscles and wood were the most important until coal took over. The figures show the estimated pattern of energy use based on the consumption of various fuels from 1850 onwards.

3 Different fuels contain different amounts of energy. The graph shows how much energy is produced when a standard weight of fuel is burned. Hydrogen looks like the best fuel but it is a light gas that has to be pumped through pipes. This means that to carry a quantity of energy through a pipe some energy must be used to power the pump, reducing the net energy transported.

Unfortunately most of the promising hydro-electric sites that have yet to be developed are in remote parts of the world where the demand for power is low.

Around the world there are more than 60 hydroelectric projects with a power generating capacity of 1,000 MW or more, either in operation or planned for the near future. (One of the largest is the Grand Coulee Dam in the USA, with a capacity of nearly 10,000 MW.) These projects will ease the pressure on world resources of fossil fuels, for the electricity generated by hydroelectric stations saves the equivalent of about 360 million tonnes of oil a year.

Today hydroelectric power stations produce more electricity than the world's nuclear power stations – which generated power equivalent to burning 60 million tonnes of oil in 1974. But nuclear power output is increasing each year and it may not be long before it overtakes hydroelectric power.

The amount of electricity that can be generated from the world's reserves of uranium [8] and other nuclear fuels depends very much upon the types of nuclear reactors

that are built. One type of reactor – the fast breeder reactor – can produce about 60 times the energy that the same fuel would generate in a thermal reactor of the kind now being built. Without breeder reactors, which have yet to be developed to a commercial stage, the world could find itself short of uranium within 40 years on present sources.

Geothermal energy [7] is derived from the heat locked up in the earth's core. Most of it was, and still is, produced by the slow decay of radioactive elements that occur naturally in all rocks. The combined output of geothermal stations is about the same as the amount of power produced by one large nuclear reactor. Italy, Japan, New Zealand, the United States and Mexico have the largest installed capacity for geothermal electricity generation. Geothermal energy can also be used to provide hot water. In New Zealand geothermal hot water is used in the paper-making industry, and the Icelandic capital of Reykjavik is almost wholly heated by a district heating scheme fed from geothermal wells. The water reaches the city's buildings through pipes by flow of gravity.

KEY

Oil 45%
Gas 19%
Coal 29%
Nuclear power 1%
Hydro-electricity 6%

The relative proportions of the diverse sources of energy used in the world in 1974 reflect their respective convenience, availability and cost. By far the most important primary source of energy is oil, which in 1974 accounted for nearly half the world's energy consumption.

Coal, the source that fuelled the Industrial Revolution, was in 1974 only a poor second although its long-term prospects are brighter than those for oil, as the reserves of coal are expected to last a lot longer than those of oil. Natural gas is a clean and popular source of energy but it might

well be the first of the fossil fuels to be exhausted. Hydro-electric power is most suited to mountainous areas and is a renewable source of energy. The most promising primary source is the 1% supplied by nuclear reactors. A similar graph for 1964 would not have shown this source at all.

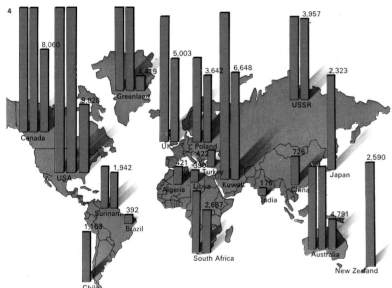

Energy consumption kilogrammes per capita

4 Energy consumption is most unevenly spread amongst the peoples of the world. The USA has the highest energy consumption per head of population, but it also has some of the world's richest energy deposits. The British also consume massive amounts of energy, but they do have to import a great deal. Europe's oil comes from the Middle East and other oil-producing regions.

5 Hydroelectric power is a renewable source of energy. It is fed by the waters caught by a dam. In some countries, hydroelectric power provides a large share of the energy, such as 77% of Sweden's in 1973.

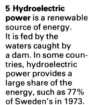

7 Geothermal energy is derived from the earth's internal heat which powers geysers and steam springs. This steam and hot water is used to generate electricity in New Zealand, Italy and Iceland.

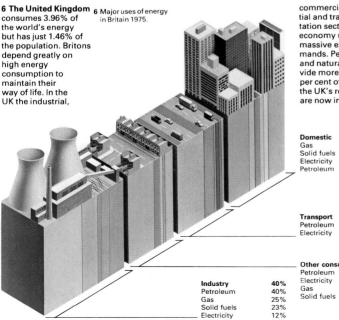

6 The United Kingdom consumes 3.96% of the world's energy but has just 1.46% of the population. Britons depend greatly on high energy consumption to maintain their way of life. In the UK the industrial,

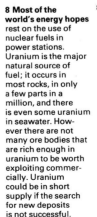 6 Major uses of energy in Britain 1975.

commercial, residential and transportation sectors of the economy make massive energy demands. Petroleum and natural gas provide more than 44 per cent of this and the UK's resources are now increasing.

Domestic	26%
Gas	40%
Solid fuels	29%
Electricity	21%
Petroleum	10%

Transport	22%
Petroleum	99%
Electricity	1%

Other consumers	12%
Petroleum	48%
Electricity	22%
Gas	20%
Solid fuels	10%

Industry	40%
Petroleum	40%
Gas	25%
Solid fuels	23%
Electricity	12%

8 Most of the world's energy hopes rest on the use of nuclear fuels in power stations. Uranium is the major natural source of fuel; it occurs in most rocks, in only a few parts in a million, and there is even some uranium in seawater. However there are not many ore bodies that are rich enough in uranium to be worth exploiting commercially. Uranium could be in short supply if the search for new deposits is not successful.

141

Energy for the future

Is there an energy crisis? In principle, no. The world's known reserves of energy are more than large enough to meet all foreseeable needs of mankind for ever, if ways can be found of making use of all of them [1]. The oil shales of Colorado alone, for example, contain more oil than all the Middle East reserves added together. And the heat contained in a cube of rock with a 5km (3 mile) edge lying beneath the Jemez Plateau in New Mexico is as much heat as the whole world uses in an entire year.

Types of energy potentials
Potential sources of energy – oil shales, tar sands, solar and geothermal sources, waves, tides, winds and nuclear fusion – could in principle provide far more energy than the world is ever likely to need. But many difficult problems have to be solved before any of these neglected sources of energy can be tapped successfully. The problems are not simply of a technical nature, but are of many different kinds.

The most important is the fact that most untapped sources of energy are diffusely spread across the surface of the earth, and not concentrated into neat packages as are coal, oil or gas. The sunlight falling on any country in the world – even northerly Britain – would be more than enough to meet all power demands if it were not so thinly spread [6]. Short of covering very large areas of land with solar collectors, there is at present no way of using more than a minute fraction of the sun's energy. Similarly, the energy of the waves [3] or the wind [4] is enormous, but so diffuse that huge structures or thousands of small ones would be needed to capture it. So far only one tidal power station exists, utilizing the great tidal range in the Rance estuary in northern France.

Diffusion and cost of alternative energy
Since most alternative energy sources are diffuse, it would be extremely costly to provide structures capable of concentrating them for use. The energy consumed in building these structures may well exceed, over their potential lifetime, the amounts of energy they would provide. Careful "energy analysis" is therefore needed in each case to make sure that the investment will make a profit both in financial and in energy terms. Energy analysis of the rich oil shales of Colorado, for example, shows that the return in oil produced by mining and processing the shale would not exceed by a very wide margin the amounts of oil that would be consumed by mechanical mining equipment, transportation of shale, process heat to extract the oil and refinement to turn it into a useful form. Unless a less expensive technology can be found, therefore, oil shales are unlikely to provide a large source of oil.

A third difficulty is what economists call a "rate and magnitude" problem. What matters, if the world is to avoid running out of energy, is not the potential amounts available but the rate at which they can be brought into use. Past energy sources – coal, oil and natural gas – have been brought into use successively, each more quickly than the last. However, all the present alternative sources of energy are harder to bring into use than oil or natural gas – they need bigger investments and produce smaller returns. This means that it would in theory be possible for a world rich

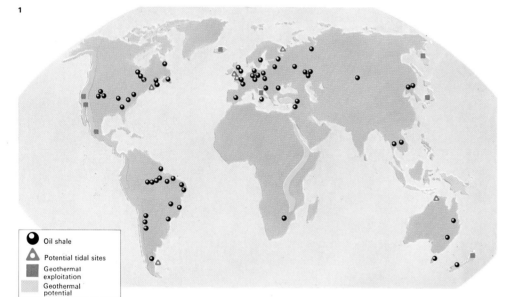

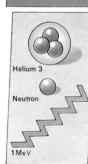

2 Fusion is a nuclear reaction that can take place between certain elements with low atomic numbers. It involves a re-arrangement of the nuclear particles and a release of energy. The attraction of this as a potential power source lies in the lack of radioactive waste products (the drawback in other nuclear reactions), the great quantities of energy released (it is the reaction that takes place in the sun and stars) and availability of fuel (deuterium occurs naturally in a concentration of 1 atom for every 5,000 atoms of hydrogen). In the example shown two deuterium nuclei give a helium 3 nucleus, a neutron and 1 million electron volts (MeV) of energy.

Oil shale
△ Potential tidal sites
■ Geothermal exploitation
Geothermal potential

1 Alternative energy sources are unevenly distributed through the world. The North American continent is best equipped, with large oil shale deposits in Colorado, tar sands in Alberta, good solar radiation and extensive geothermal potential in the geologically active west of the USA. Britain has one of the best tidal sites in the world in the Severn estuary, where the tide rises and falls more than 6m (20ft) every day. The UK also has good wave-power prospects, with the 4,800km (3,000 miles) "fetch" of the Atlantic driving waves on to the western coast. Alternative energy sources are likely to be exploited first where the source coincides with a large market (eg the USA, Japan and parts of Europe). Many developing countries have good solar and geothermal prospects and their needs are growing.

3 Waves have enormous power, sufficient to destroy jetties and quays weighing thousands of tonnes. They are strongest along the edge of great oceans, because their size depends both on the strength of the wind and on the distance it blows across the sea (the "fetch"). Waves grow and decline more slowly than wind, producing a less "peaky" power output. Thus they smooth out the rapid variations of strength and direction of the winds that drive them.

1 Wind generator
2 Electrolysis cell
3 Oxygen storage
4 Hydrogen storage
5 Fuel cell
6 Inverter
7 Recycled water
8 Pump back control
9 Rectifier

4 The wind is a capricious and undependable source of power. To use it requires a means of storing energy produced in windy periods that can then be used on calm days. This diagram shows one possibility; the wind turbine is used to generate electricity, which produces hydrogen gas and oxygen gas in the electrolysis cell by the electrical decomposition of water. The two gases are then combined in a fuel cell to produce electricity in the form of direct current. An inverter converts it into alternating current and feeds it to the grid. Power produced elsewhere could also be used to run the electrolysis cell when the wind is still. This would produce electricity on a continuous basis.

in potential energy sources to be suffering from an energy crisis, simply because new resources of energy could not be introduced quickly enough to replace existing ones.

Gravitational, solar and nuclear sources

The large sources of energy so far unused fall into three categories: those derived from gravitation, those derived from the sun and those derived from nuclear processes.

Only one potential source of energy, the tides, derives from gravitation. The pull of the moon and the sun on the world's oceans shifts the water to and fro, creating a potential for hydroelectric plants in places where the tidal range is greatest.

Solar energy sources include the familiar wood, coal, oil and natural gas, all products of plant or animal life which could not exist without the sun. They also include solar energy itself and, less obviously, wind [4], hydroelectric, wave and ocean thermal gradient [5] sources.

Nuclear processes provide three sources of energy: nuclear fission, which is already in use, nuclear fusion and geothermal energy.

Nuclear fusion would make use of the energy produced when two light elements fuse together to produce a heavier one [2], so far employed only by the sun and other stars and by the hydrogen bomb. Two approaches are being tried in the effort to tame fusion. The first is to bring light elements together in the form of a "plasma" at a very high temperature and to prevent them from escaping by magnetic fields. The second is to force two light elements together by using an intense laser beam as a sledgehammer, pouring in energy so fast that the minute fuel pellet "implodes", producing a fusion reaction and giving off more energy than is originally used to drive the laser.

Geothermal energy, finally, uses the heat generated by nuclear processes deep in the earth. It is in limited use already and has considerable future potential. Power plants worked from geothermal heat are usually driven by steam in hot-spring areas such as in Iceland and New Zealand. Other schemes are being devised that would use hot water from the earth to evaporate a low-boiling-point fluid and pass it through a turbine.

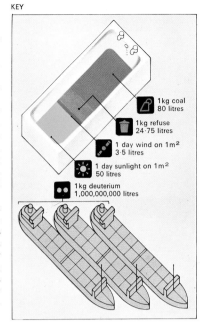

1kg coal
80 litres

1kg refuse
24·75 litres

1 day wind on 1m²
3·5 litres

1 day sunlight on 1m²
50 litres

1kg deuterium
1,000,000,000 litres

The light element deuterium, which would be used in a fusion power station, is the most concentrated of the alternative energy sources. Fifty microgrammes of deuterium are equivalent in energy content to a whole day's solar radiation at average intensity on a square metre of the earth's surface. The diagram shows the volume of water boiled by comparable units of the energy sources dealt with below. In general, fuels of medium-energy concentration, such as coal, are the easiest to exploit. Diffuse sources of energy, such as wind and solar power, are difficult to exploit in a traditional central power station. It may be better to use small collectors suited to the needs of a local area.

5

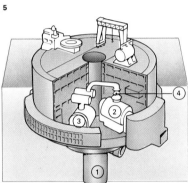

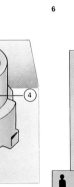

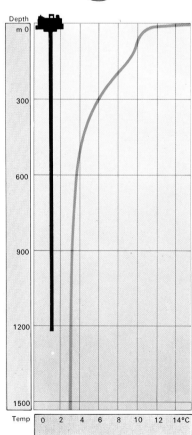

6

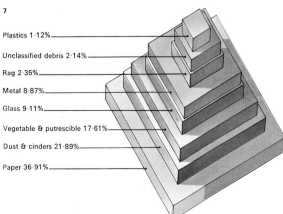

Greenland 600m²
United Kingdom 460m²
Algeria 20m²
Japan 212m²
Brazil 17m²
India 8m²
Australia 250m²

5 Sea thermal power plants may use the difference in temperature between the water on the surface of the ocean and the much colder water in the deeps. In the design shown, the power station floats on the surface with a 1,200m (4,000ft) pipe [1] descending into deep water. Cold water from the depths is pumped up and used to condense ammonia in a heat exchanger [2]. This liquid ammonia flows to a second heat exchanger [3] where it is evaporated by the heat of the surface water and passed back to the beginning of the cycle. As it flows round this closed system the ammonia drives a turbine [4] and generates electricity. This system can work from fairly small temperature differences.

6 If the world were dependent on solar energy alone, the greatest sufferers would be those with high standards of living in northern latitudes. Each British citizen would need, for example,

a solar collector 100% efficient and more than 460sq m (4,950 sq ft) in surface area to capture as much energy from the sun as he uses today. A Greenlander would need an even bigger collector, almost

600sq m (6,460sq ft). An American would need a large area because although solar radiation is high in the USA, consumption of energy is enormous. The developing countries such as

India could, on the other hand, meet their present energy demands with comparatively small solar collectors. No solar collector, however, is efficient enough to collect 100% energy.

7

Plastics 1·12%
Unclassified debris 2·14%
Rag 2·35%
Metal 8·87%
Glass 9·11%
Vegetable & putrescible 17·61%
Dust & cinders 21·89%
Paper 36·91%

7 Every household in a developed country throws away energy every day as rubbish. The amount of paper thrown away by British households has almost doubled since 1935 to about 5kg (11lb) a week, while the amount of ash and cinders in the rubbish is less than a third of what it was then. Domestic rubbish can be burned to provide a district heating scheme or generate electricity. Such schemes serve to recover wasted energy.

143

Resources at risk

The developed countries of the world are using up valuable resources at a rate unprecedented in human history. Fossil fuels, minerals, metals, water, timber and even soils are treated as if supplies were infinite and the larder easily replenished. But most of the earth's resources are not an income to be spent, but a stock of capital built up over thousands of millions of years before man evolved. If present rates of use continue, that capital will be exhausted in a period of no more than a few centuries.

How long will resources last?

Opinions differ about the rate at which the earth's resources are being exhausted. Calculations of the "lifetime" of any given fuel or mineral resource [1] can vary widely depending on the assumptions used. But most experts, including bodies such as the US Geological Survey, now concede that present rates of fuel and mineral use can be sustained for only another generation or two. The timetable [1] depends both on the particular resource under consideration and on such imponderables as the future of the world

economy, the possibility of finding new supplies, the geographical distributions of those supplies and the extent to which the lifetime of any resources can be extended by more careful use [4, 5] or by recycling.

The consumption of most mineral resources was, until recently, increasing at an exponential rate. This rising curve of consumption meant that demand for crude oil, for example, doubled every decade. Other primary commodities showed a slower growth, with doubling times of between ten and 20 years. The oil crisis and subsequent economic depression, which began late in 1973, slowed down demand for fuels and metals and stopped the steady increase in consumption for the first time in two decades. Most economists believe, however, that this is merely a pause in the growth of the world economy rather than a turning point towards zero growth.

Whether or not growth resumes is of crucial importance in assessing how long resources are going to last, for the effect of such growth on the demand for fuels and minerals is remarkable. In the case of crude

oil, consumption for the decade from 1960 to 1970 was as much as in the entire period between the drilling of the first oil well in 1859 and 1960. The consequence of such a rapid growth rate is that discovery of new reserves must continue and keep ahead of demand.

Not all resources are in equal danger. Coal, for example, is more plentiful than oil and natural gas, and some metals such as aluminium, iron and magnesium are relatively abundant. Others, including copper, vanadium, lead and zinc, are less plentiful but not in imminent danger of running out. But a few, including mercury, silver, tin [3], tantalum and platinum, are scarce.

Alarms about declining resources are not new; but so far the pessimists have been wrong. Improving technology, better methods of prospecting, a bigger market and a better price have been enough to keep production ahead of demand.

Reasons for pessimism

There are a number of reasons for believing that the pessimists cannot be wrong for ever. The first is the fact that minerals are distri-

CONNECTIONS

See also
132 Mineral resources of the land
134 Mineral resources of the sea
140 Energy supplies
142 Energy for the future

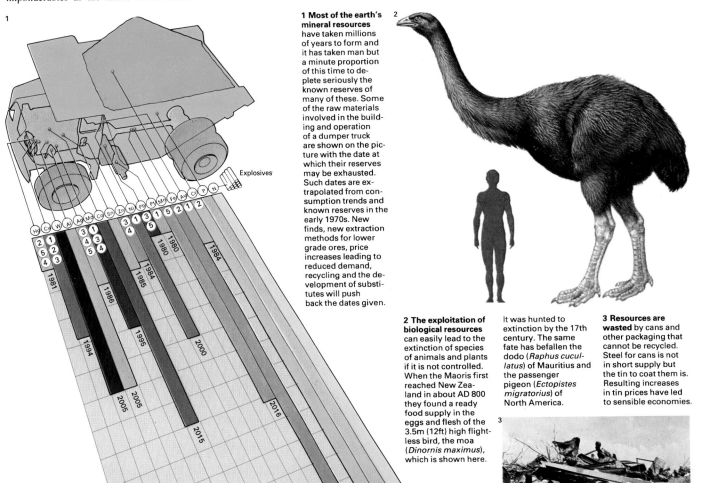

1 Most of the earth's mineral resources have taken millions of years to form and it has taken man but a minute proportion of this time to deplete seriously the known reserves of many of these. Some of the raw materials involved in the building and operation of a dumper truck are shown on the picture with the date at which their reserves may be exhausted. Such dates are extrapolated from consumption trends and known reserves in the early 1970s. New finds, new extraction methods for lower grade ores, price increases leading to reduced demand, recycling and the development of substitutes will push back the dates given.

Explosives

Hg	Mercury
Cu	Copper
W	Tungsten
Al	Aluminium
Ag	Silver
Mo	Molybdenum
Co	Cobalt
Sn	Tin
Zn	Zinc
Ni	Nickel
Pb	Lead
Pt	Platinum
Mn	Manganese
Fe	Iron
Au	Gold
Cr	Chromium
P	Phosphates
N	Nitrates

Environmental problems associated with specific recovery:
1 Waste disposal
2 Land reclamation
3 Aerial pollution
4 Water pollution
5 Poison control

2 The exploitation of biological resources can easily lead to the extinction of species of animals and plants if it is not controlled. When the Maoris first reached New Zealand in about AD 800 they found a ready food supply in the eggs and flesh of the 3.5m (12ft) high flightless bird, the moa (*Dinornis maximus*), which is shown here.

It was hunted to extinction by the 17th century. The same fate has befallen the dodo (*Raphus cucullatus*) of Mauritius and the passenger pigeon (*Ectopistes migratorius*) of North America.

3 Resources are wasted by cans and other packaging that cannot be recycled. Steel for cans is not in short supply but the tin to coat them is. Resulting increases in tin prices have led to sensible economies.

buted very unevenly over the earth's surface. Other things being equal, a rich ore is more profitable to exploit than a poor one, so that most of the best ones have already been mined out. Progressively thinner or less accessible ore deposits now have to be exploited, thus increasing costs and reducing returns. In the past geologists have consoled themselves by arguing that as the quality of an ore declines, its quantity increases; that there is ten to 100 times as much of a low-grade ore to be extracted as there is of a high-grade ore. For some metals, including copper, this is true; but for others, it seems, there is either rich ore or unproductive rock and nothing in between. For many metals the huge tonnages of low-grade ore once counted on to take over from the rich ores simply do not exist. Once present grades of ore have been exhausted, there may be little left.

The second reason for pessimism is that mining is consuming ever increasing quantities of energy, to extract the same amounts of metal from ores of declining quality. Since energy is likely to be in short supply, it is unlikely this process can go on for ever. The same applies to extraction of metals and other minerals from seawater. It is true that the oceans contain huge resources of minerals, but most occur in such small concentrations that the cost of extraction would be prohibitive. The energy needed to extract uranium, for example, from the sea, even using the most up-to-date technology, would exceed the energy the uranium would produce in a nuclear reactor.

Geographical and political factors
A final difficulty is related to the geography of resources. When ownership of a crucial resource is held in few hands its price can be manipulated to a level far above the cost of production. This has already happened to oil and could in future happen to a wide range of other resources, including perhaps tungsten, of which China controls 75 per cent of the known world deposits; chromium, of which South Africa controls 75 per cent; or mercury, of which Spain controls 33 per cent. Phosphate rock, now mainly in the hands of a few North African countries, and essential for fertilizers, is an even better example.

KEY

The motor car is a prime energy waster. Too heavy, too powerful and usually under- occupied, it converts little fuel energy into motion. At full power the theoretical effi- ciency of an average combustion engine is only about 25% and in practice is 10%.

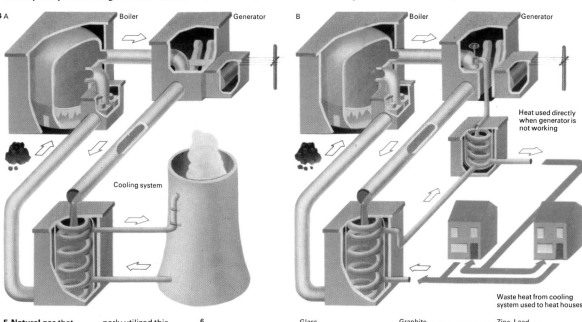

4 A · Boiler · Generator · Cooling system

B · Boiler · Generator · Heat used directly when generator is not working · Waste heat from cooling system used to heat houses

4 District heating schemes can make much better use of diminishing fuel resources. In a conventional system [A] electricity is generated in a power station at an efficiency of 35 to 38%, distributed to houses with a further loss and used for lighting and for heating at 55% efficiency. In a district heating scheme [B] electricity and hot water or steam are generated at the power plant and distributed for domestic consumption, the electricity in the usual way and the hot water or steam in insulated pipes. The overall efficiency of this system can be as much as 80 to 85% if demands for heat and electricity coincide.

5 Natural gas that occurs in association with oil is often "burned off" as a waste product since the price it fetches is insufficient to justify its collection and distribution to consumers. But pro- perly utilized this gas could be an important source of heat for domestic and small-scale commercial purposes, for it has a higher heat-producing or calorific value than manufactured coal gas.

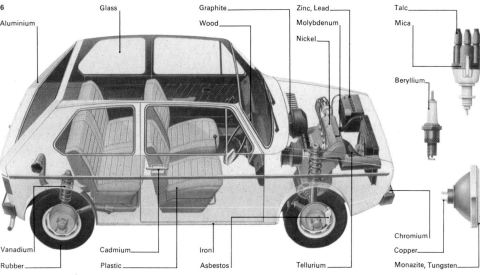

6 · Glass · Graphite · Zinc, Lead · Talc · Aluminium · Wood · Molybdenum · Mica · Nickel · Beryllium · Vanadium · Cadmium · Iron · Tellurium · Chromium · Rubber · Plastic · Asbestos · Copper · Monazite, Tungsten

6 An everyday object like a motor car depends, for its manufacture, on a very large number of different minerals and raw materials. Some, such as iron and glass, have obvious structural applications while others, such as tellurium and talc, are used as fillers in rubber and paint. Such an intimate mixture of different ma- terials makes their recycling difficult when the car's useful life is finished.

Pollution of the air

Clean, fresh air is a mixture of gases with tiny particles suspended in it. By volume it is made up of roughly 78 per cent nitrogen and 21 per cent oxygen, with much smaller amounts of carbon dioxide and argon and traces of several other gases. Water vapour can be present in varying amounts according to the temperature and history of the air. The particles, which often help to make air visible and add colour to the clouds and sky, include fine water droplets (forming mist, fog and low cloud) and ice crystals (forming high cloud). Air that has blown over the sea may be loaded with tiny crystals of salt, and wind off the land often contains dust that may include fine sand, pollen, plant spores and a host of other substances both organic and inorganic. These gases and particles are not generally harmful to plants or animals, and human beings are able to tolerate a wide range of atmospheric conditions.

How is the air polluted?

Air pollution is usually a result of man's activities [1]. Polluted air, of the kind found in cities and industrial areas, usually contains higher-than-normal concentrations of some of the "trace" gases and other gases that are not generally present in the atmosphere.
It may also contain more particles, which thicken the atmosphere, darken it and reduce visibility. Polluted air is often unpleasant to breathe and can be extremely harmful to living creatures of all kinds. Man suffers from it directly, through deterioration of the environment and his health, and indirectly through damage to crops and property.

The chief cause of air pollution is the large-scale burning of fossil fuels. Coal burned in household grates, industrial furnaces and railway engines blackened the cities of Europe and North America during the eighteenth, nineteenth and early twentieth centuries and fumes of oil products – especially petrol and diesel fuel – contribute heavily to the foul air of cities today.

Burning fuels, especially in grates or engines that are not properly maintained, produces a wide range of pollutants that affect the air in different ways. Prominent among pollutants is sulphur dioxide. This is an acrid gas that dissolves readily in water to form acid solutions that kill plants and damage buildings. In the high-temperature conditions of a furnace or engine cylinder, oxides of nitrogen may also be formed, again producing choking fumes that yield acid solutions. Nitrogen oxides and sunlight also react to form photochemical smog, which is particularly prevalent in Los Angeles, USA [4]. Unburned hydrocarbons and other particles enter the atmosphere in large quantities, forming soot and fly-ash. These make greasy, black deposits on buildings and fabrics and also form nuclei on which water droplets condense, causing smoke haze and fog. "Smog" [4], a thick, persistant fog that forms when polluted air cannot rise because of air layers of different temperatures that hang over a city, combines these features and is a well-known hazard to human life.

The role of the motor car

The petrol used in most countries today contains lead compounds, which enter the atmosphere in enormous quantities from motor car exhausts [2]. Together with carbon monoxide, a poisonous gas produced by

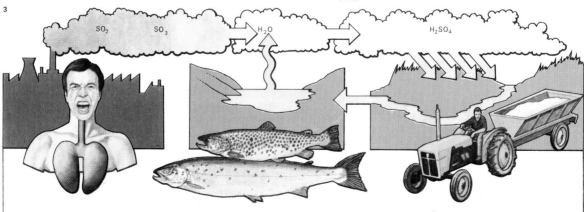

1 Some of the main sources and proportions of air pollution in an industrial region are shown in these diagrams. Gases and other products of transport account for 51% of all atmospheric pollution. This is followed in decreasing proportions by pollution from domestic fires, forest and other open fires such as bush fires, industrial smoke and the incineration of domestic waste.

Transport — 51%
Domestic heating — 16%
Miscellaneous — 15%
Industrial pollution — 14%
Solid waste disposal — 4%

Carbon monoxide
Hydrocarbons
Solid particles
Sulphur dioxide
Nitrogen oxides

2 The motor car is a major source of atmospheric pollution producing oxides of nitrogen, unburned hydrocarbons, carbon monoxide, lead compounds and other poisons, in the proportions shown.

Proportion of pollutants produced by various components

Petrol tank
Exhaust
Carburettor
Crank case

Carbon monoxide
Nitrogen oxides
Lead
Hydrocarbons
Solid particles

Pollution produced by a car in 20,000km
Lead — 0·775kg (12,000 miles)
Nitrogen oxides — 40·75kg
Hydrocarbons — 234kg
Carbon monoxide — 765kg

Idling | Acceleration | Cruising | Deceleration
Nitrogen oxides
Hydrocarbons
Carbon monoxide

3 Acid rain, believed to originate from industrial pollution in western European countries, is damaging buildings, forests and streams and affecting the health of man in Scandinavia. Sulphur dioxide (SO_2), produced during the burning of coal and oil, is oxidized in the atmosphere to sulphur trioxide (SO_3). These gases dissolve in water in the clouds, forming sulphuric acid (H_2SO_4), which falls in rain many miles downwind of the industrial areas from which the gases originated.

SO_2 SO_3 H_2O H_2SO_4

burning hydrocarbon fuels, these compounds may reach intolerable concentrations in city streets and present a special health hazard.

The motor car produces other air pollutants, notably asbestos fibres from brake linings; these may enter the lungs and set up chronic irritation that may lead to cancer and other respiratory diseases. Asbestos fibres also enter the atmosphere from building materials such as fire-proof wadding. Many industrial chemical processes add to the atmosphere pollutant gases and particles.

Recent sources of atmospheric pollution
A small but growing source of atmospheric pollution is the release of radioactive particles, both from nuclear weapon testing and from peaceful uses of atomic energy [6]. Crop-spraying adds local concentrations of organic poisons to the air, sometimes to the detriment of plants and animals on neighbouring land.

Many new organic chemicals released into the atmosphere come under suspicion of causing long-term environmental damage. For example, the freons, used as refrigerants

and aerosol propellants, may interfere with the protective ozone of the upper atmosphere and allow dangerous concentrations of ultra-violet light to reach the earth's surface; these may harm the skin and body tissues. Increased quantities of carbon dioxide, released into the atmosphere from burning fossil fuels, may be upsetting the heat balance of the earth.

Most forms of air pollution have arisen from the activities of industrial man and many are readily curable. Legislation in the wealthier industrial countries has been effective in reducing smoke and smog pollution, keeping airborne effluent from factory chimneys under control and generally improving the quality of the atmosphere. New factories are better equipped to reduce pollution and ways are being sought to keep down the pollutants from motor car and aircraft engines. These remedies are costly; industrial man finds that he has to pay more to maintain and improve his standard of living. The developing countries, now on the brink of industrialization, still have the problem of air pollution in front of them.

Pollution is a global problem affecting the land, rivers, the sea and the atmosphere. It is especially acute in industrially developed countries, where for generations man has been dumping his waste without thought for the environment. The level of pollution now causes concern.

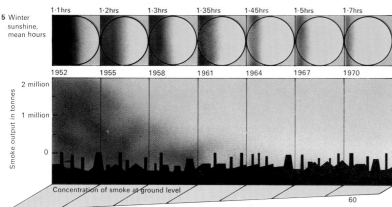

4 Smog caused 4,000 deaths in London in 1952 and is still a major hazard in Japan and the USA, particularly in the area of Los Angeles as this picture illustrates. In warm climates smog is caused by oxides of nitrogen and hydrocarbons from car exhausts undergoing chemical reactions influenced by sunlight. Photochemical smog is a poisonous mixture lethal to those with lung complaints.

5 Winter sunshine, mean hours

1·1hrs 1·2hrs 1·3hrs 1·35hrs 1·45hrs 1·5hrs 1·7hrs
1952 1955 1958 1961 1964 1967 1970

Smoke output in tonnes
2 million
1 million
0

Concentration of smoke at ground level
180 microgrammes/m³
60

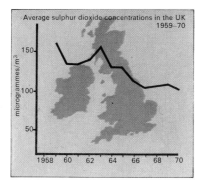

Smoke emissions in the UK   1953
Domestic
Railways
Industrial

Railways : nil   1970
Domestic
Industrial

Average sulphur dioxide concentrations in the UK 1959–70
microgrammes/m³
150
100
50
1958 60 62 64 66 68 70

5 Britain's air was relatively pure until the Industrial Revolution. In the 18th and 19th centuries the coal that powered industry and provided domestic heat progressively blackened buildings and harmed vegetation and health. The lethal winter smogs of 1952 spurred government action and in 1956 the Clean Air Act reached the statute book. This reduced the smoke output. In certain areas only smokeless fuels could be burned. Within a few years there was a distinct improvement in the quality of the air. Since the mid-1960s British industry has spent £400 million on reducing pollution without economic damage to production. Overall the big clean-up attempts have been successful. London's last smog occurred in 1962 and, as the diagram shows, there has been much more sunshine in winter in Britain since 1950.

6 Nuclear explosions in the atmosphere have exposed man to damaging amounts of radiation. During the 1950s it was found that strontium-90 dispersing in the atmosphere was being absorbed by vegetation. Grazing animals eat contaminated plants and thus human beings drink milk and eat meat containing strontium-90. This partly replaces calcium in bones and, by emitting beta particles which interfere with blood cell production, can eventually cause leukaemia.

147

Misuse of the land

Mark Twain, the American writer, when asked for investment advice, replied: "Buy land. They've stopped making it." A century later it is clear that, far from being "made", land is being used in a manner that is often increasingly harmful to the interests of the earth's population.

The dangers of industrial expansion
Land fulfils many roles, some of them mutually incompatible. Increasing disregard for the true value of land has been shown in recent assessments of agricultural and industrial requirements. The implications are ominous.

Failure to recognize the priorities in the use of land has often made it more valuable as a support for concrete than as a support for growing foods. Loss of agricultural land occurs annually in virtually every industrialized country. The cities of the world are spreading across the landscape, often encroaching upon all the open land in their path, by the construction of houses, factories and roads. Industrial expansion seeks "green field" sites for new developments rather than reclaiming old sites, because the land is less expensive. The old sites may be left to waste.

When a "green field" site is taken over for industry the effects may be felt over an area much larger than that eventually occupied by the plant. The fallout from stack gases may blight the surrounding countryside for many kilometres. Motorways and their access routes cut through the rural scene; the traffic they carry emits exhaust fumes and lead over a still wider area. They also carve wildlife habitats into smaller units that make life impossible for many of the larger species.

The effects of excavation and mining
To produce the materials for construction work requires yet more land. The fill for foundations under buildings, roads and dams is excavated from surrounding areas and hills are levelled to produce aggregates. Even larger scars are left by open-pit mines for extracting valuable metals from the rock and strip mines for extracting coal. An open-pit copper mine – such as that at Broken Hill in Australia, or that at Bougainville in New Guinea – may gouge into the ground for more than a kilometre (0.6 mile) in width and almost as deep, involving earth-moving on a geological scale. Such mining practices lead to the leaching of metals by rainwater which, as run-off, spreads its poisonous load in the surrounding region; and metal-poisoned soil is difficult to reclaim.

Strip-mining for coal does not go so deep but it can cover a vast area. With care, strip-mining [1] – or, as it is called in Britain, open-cast mining – can be carried out without causing permanent damage. The topsoil can be set aside and replaced on levelled terrain after the coal has been extracted from underneath, but the techniques are costly. Strip-mining that is not followed by refill and replanting produces an acidic water run-off (through leaching by rain) which blights the downslope vegetation. Mine operators concerned primarily with achieving high rates of production have, since the mid-1950s, created widespread devastation with strip-mining in the Appalachian region of the eastern United States. If the coal reserves of the western United States are brought into production by strip-mining, as presently

1 Mechanical interference with soil takes place in many ways. Road building [1] not only uses up land for the graded surface but also for the embankments where the soil is mechanically destroyed by the heavy road-building machinery; furthermore the drainage pattern is often upset, with the run-off draining directly into waterways instead of replenishing the soil moisture and the water-tables. Agricultural machinery [2]

compacts the soil and affects its biological properties. Open-pit and strip-mining [3, 4] can create widespread land erosion and unearthed minerals can be leached away by rainwater, poisoning the land. The topsoil from pits and strips should be replaced over the filled-in excavations after the end of the mining operations.

2 Chemical degradation of the land can occur both by direct and indirect means. Substances applied directly to the soil in agricultural processes in order to produce some beneficial result may have lethal side-effects. Fertilizers and pesticides sprayed on crops [1] will pass into the soil and be distributed by ground water. This can alter the biochemical balance of the soil and kill off earthworms and other organisms that play a key role in maintaining its porosity and aeration. Sewage sludge is often spread as a valuable organic fertilizer [2] but it is impossible to remove from it metals such as copper and zinc which come from industrial effluent and these build up, altering the chemistry of the soil. More indirectly the soil can be polluted by industrial processes when gases such as sulphur dioxide are passed into the atmosphere by chimneys [3]. These are washed down in the rain as acid, changing the acidity of the soil to the detriment of the plants and animals that feed on it. Unburned hydrocarbons, lead and other noxious substances from motor vehicles [4] may build up in fields near roadways. Excessive use of heavy agricultural machinery can crush a well-drained soil structure [5] into a non-porous waterlogged mass [6].

proposed, it will be necessary to enforce erosion-control measures and to prevent water pollution in order to rehabilitate the land for agricultural purposes. Otherwise the result will be a barren and eroded wasteland.

Land management and deforestation
Even when land is valued for its fertility, modern techniques of agriculture may have harmful effects. Excessive use of chemical pesticides [2] may kill off the tiny organisms in the soil that help to maintain the soil's vitality. Chemical fertilizers make it possible to grow the same crop year after year, but if there is no organic matter in the soil its fine structure may break down. This breakdown may be accelerated by the repeated passage of heavy agricultural machinery [1], which compacts the soil.

In extreme cases, under tropical conditions, delicate soil structures may be replaced by a solid, impenetrable mass called laterite or hardpan. In recent years some countries have proposed the opening up of tropical and subtropical regions to cultivation by clearing away rain forests, such as those of Amazonia

in Brazil. The obliteration of the luxuriant but fragile vegetation of the rain forest, which plays a major part in maintaining the oxygen balance of the planet, might lead to an irreversible global catastrophe.

Ruthless deforestation of the plains of North America, even to the extent of wiping out almost all trees from horizon to horizon, resulted in the dust bowl of the 1930s, where many millions of tons of precious topsoil were blown away by the winds. A misjudged programme of irrigation in Pakistan raised the salt level in hundreds of thousands of hectares, making them almost useless for crops. The Sahara Desert [4] is advancing southwards at some 100m (328ft) a year – partly because of careless, primitive agricultural practices on its fringe.

On a local scale, townspeople have come to value land so little that gaping ugly rubbish tips [Key] are left suppurating in the neighbourhood of countless settlements. Although the amount of available useful land should be increasing, mankind, through sheer numbers and population pressure, is destroying it on a geological scale.

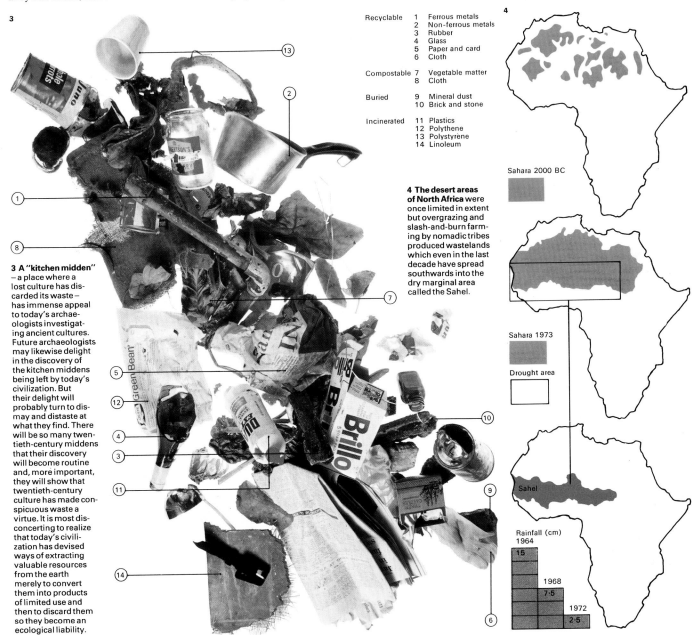

KEY

The rubbish tip is symbolic of man's use of natural resources. First it shows the amount of raw materials used up and discarded and second it shows how valuable land can be lost in the creation of an unsightly dump. This land can be reclaimed but the problems involved are great.

Recyclable	1	Ferrous metals
	2	Non-ferrous metals
	3	Rubber
	4	Glass
	5	Paper and card
	6	Cloth
Compostable	7	Vegetable matter
	8	Cloth
Buried	9	Mineral dust
	10	Brick and stone
Incinerated	11	Plastics
	12	Polythene
	13	Polystyrene
	14	Linoleum

3

3 A "kitchen midden" – a place where a lost culture has discarded its waste – has immense appeal to today's archaeologists investigating ancient cultures. Future archaeologists may likewise delight in the discovery of the kitchen middens being left by today's civilization. But their delight will probably turn to dismay and distaste at what they find. There will be so many twentieth-century middens that their discovery will become routine and, more important, they will show that twentieth-century culture has made conspicuous waste a virtue. It is most disconcerting to realize that today's civilization has devised ways of extracting valuable resources from the earth merely to convert them into products of limited use and then to discard them so they become an ecological liability.

4 The desert areas of North Africa were once limited in extent but overgrazing and slash-and-burn farming by nomadic tribes produced wastelands which even in the last decade have spread southwards into the dry marginal area called the Sahel.

Sahara 2000 BC

Sahara 1973

Drought area

Sahel

Rainfall (cm)
1964
15
1968
7·5
1972
2·5

Pollution of rivers and lakes

Water moves through a continuous cycle of evaporation, cloud-forming, rainfall, collection in waterways and then more evaporation. In doing so, the water is able to purify itself naturally of the impurities it acquires during the cycle: these include decaying organic matter, dissolved gases and minerals, and suspended solids.

Where people and animals are concentrated in large numbers the self-purification capacity of fresh water can easily be overloaded, especially if water is used to collect and transport sewage wastes from settlements of any size. If sewage is deposited on soil in small quantities, soil organisms break it down, re-use the nutrient components and allow almost pure water to filter through into nearby waterways.

But if the sewage is flushed directly into a waterway, its breakdown must take place in the water. This requires a supply of dissolved oxygen in order to oxidize the waste. The biochemical oxygen demand (BOD) thus imposed may severely lower the level of oxygen available to other organisms living in the water, especially fish and plants [1].

In extreme cases, the lack of oxygen may suffocate all the living organisms in the water. The water is then biologically dead, except for the so-called "anaerobic" bacteria which thrive in the absence of oxygen, and produce noxious gases, such as hydrogen sulphide which smells like rotten eggs. The result may be a waterway both dead and foul smelling that is of no use to man or beast.

Life-giving oxygen

The same result can occur in water supplied with an excess of nutrient material, such as nitrate or phosphate, perhaps from the run-off of agricultural fertilizer or waste water containing detergent. The nutrients encourage the growth of organisms such as algae; but this growth also requires oxygen, and may exceed the capacity of the water to supply it. If this occurs, the algae die, and the decaying remains impose a further BOD, once again bringing the life of the waterway to an unpleasant end.

A lake may have a life-span of some 20,000 years before it gradually silts up and disappears. But the effects of excess nutrients can accelerate the ageing process, called eutrophication, and reduce the life-span of a lake, making it a much less welcome feature of the landscape [2].

The life-giving oxygen dissolved in waterways is less soluble at higher water temperatures. Some industries, notably the electricity generating industry, use enormous volumes of water for cooling purposes. The heated water is discharged back into waterways, further upsetting the biological balance of the aquatic system. The lower level of oxygen handicaps some species in favour of others; so does the higher mean temperatures. Species that become accustomed to thriving in the warmer water, however, may suffer disastrous consequences if a plant is shut down and its heated water flow interrupted.

Alien substances

Organic wastes, nutrients and heat become major problems to freshwater ecological systems only when too abundant for the systems to handle. But in recent years the systems have had to deal with vast quantities of substances that are totally alien, and

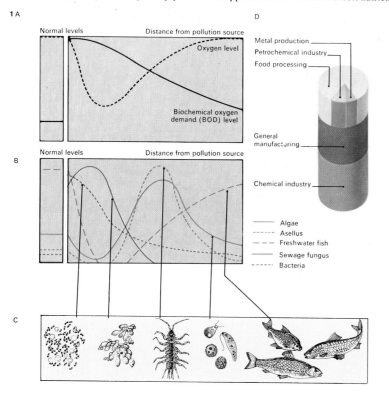

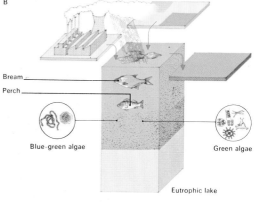

2 An oligotrophic lake is a young lake [A] with clear water and a low density of algae and plankton, in the form of diatoms and desmids. Typical fish are trout and char. As the lake ages, it becomes more fertile, or eutrophic. Nutrients in the run-off water from the surrounding land [B] stimulate the algal growth, especially of green and blue-green algae. The water becomes muddy and the fish are mainly perch, bream and roach.

1 Freshwater plants and animals almost all depend for their survival on oxygen dissolved in the water. Sewage entering the water also needs dissolved oxygen for its decay, thus competing with the aquatic life. The pollutants cause the oxygen level in the water to fall [A] by creating a biochemical oxygen demand (BOD), which is the amount of oxygen needed during the time it takes for the waste matter to oxidize. Living organisms are also affected by the wastes [B]. Some thrive on them, for example sewage fungus and algae, while others, such as freshwater fauna, may be suffocated. Some measure of the extent of pollution in fresh water can be gained by examining the peaks in populations of the plant and animal life [C] and the distribution of the organisms in the polluted waters. Industrial wastes discharged into waterways [D] are mostly responsible for causing pollution.

3 Waste hot water fed into rivers from the cooling systems of factories can kill native fish, such as salmon [1] and trout [2] by raising the water temperature above toleration levels. They may be replaced by unwanted species, such as the guppy (*Lebistes reticulatus*) [3] and the cichlids (*Tilapia zillii*) [4]. Upstream water [A] with a temperature of 5°C (41°F) holds indigenous fish. The water enters a factory and emerges hot, the temperature rising to 21°C (70°F) [B]; the native fish die and undesirable fish thrive. [C] Downstream the water cools and the original species reappear.

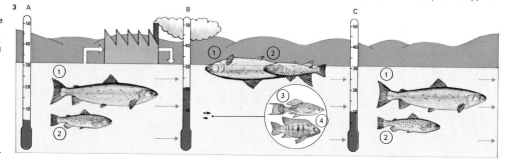

to which they may be acutely vulnerable. Pesticides from agricultural run-off, and metals and process chemicals from industrial effluent, have managed to enter aquatic food chains, with unpredictable consequences. Species high up in the food chain may build up dangerous concentrations of such substances, making them yet more open to other environmental pressures [4].

Polluted water can be cleansed. Even the natural water cycle will accomplish this, given the chance. But polluted basins – riverbeds, lake bottoms and such – may take much longer to recover their biological vitality once it is undermined. To allow natural systems a chance to restore themselves it is necessary to stop adding further burdens to waterways. Sewage works can allow suspended solids to settle out to form sludge, reducing by half the BOD of waste water, and can even reduce the level of nutrients. But industrial effluents may poison sewage works; and agricultural run-off does not even enter them. Furthermore the works are costly to build, and their value in providing cleaner water is still not adequately recognized. Since urban settle-

ments and industries find it convenient to dispose of their wastes into nearby waterways, they are reluctant to stop doing so, even when the wastes render the water useless or dangerous. Another problem arises in waterways that serve as regional boundaries, especially those between nations. In North America, the deterioration of the Great Lakes has been a diplomatic problem for many years [5].

Who is responsible?

Canada and the United States each find it difficult to justify effluent controls on their respective shores while the other still permits copious discharges of pollutants to enter the lakes from the opposite shores. Even more desperate problems confront those attempting to prevent the biological death of the Rhine. Power stations, potash mines and chemical industries line the river banks.

All over the world, freshwater habitats – ponds, streams, rivers as large as the St Lawrence and the Volga, and lakes as large as the Great Lakes and the Caspian Sea – are succumbing to poisonous pollution.

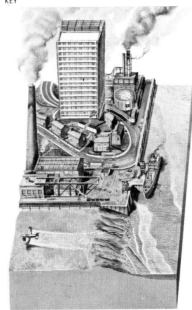

Water moves in an endless cycle, carrying with it a variety of dissolved or suspended substances, and purifying itself repeatedly by evaporation. Many of the substances in fresh water arise in nature, and reach waterways by natural means such as rainfall or ground water run-off. Some man-made pollutants take the same routes. Smoke, ash and industrial gases can be brought down by rain; and chemicals and wastes spread on soil can be carried by ground water into waterways. Other wastes follow man-made routes, through drains and sewerage outfalls. These wastes are usually more noxious, but easier to control than those following natural pathways.

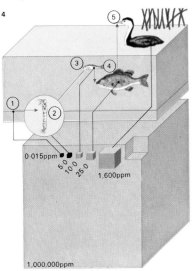

4 Pesticides applied to the land, even in small doses, are poisonous to many animals. The concentration of poison increases along the food chain, finally becoming lethal to animals at the end of the chain. A pesticide [1], such as DDT, is applied to water at 0.015 parts per million (ppm) to control midge larvae, but the plankton [2] accumulates 5ppm. The fish population [3, 4] builds up still higher concentrations, and finally a grebe [5] which feeds on the fish accumulates as much as 1,600ppm of the pesticide in its body fat, which is quite enough to kill the bird.

0·015ppm
5·0·0
10·0
25·0
1,600ppm
1,000,000ppm

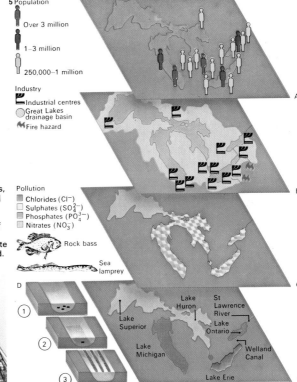

5 Population

Over 3 million

1–3 million

250,000–1 million

Industry

Industrial centres

Great Lakes drainage basin

Fire hazard

Pollution

Chlorides (Cl^-)
Sulphates (SO_4^{2-})
Phosphates (PO_4^{3-})
Nitrates (NO_3^-)

Rock bass

Sea lamprey

Lake Huron
St Lawrence River
Lake Ontario
Lake Superior
Lake Michigan
Welland Canal
Lake Erie

5 The Great Lakes are the largest freshwater lakes in the world, but their physical condition has deteriorated in the present century. Population increases [A] and industry [B] have been the main factors. Pollution from domestic and industrial wastes [C] has wrought changes in the chemical composition and fauna of the lakes. Local fish life has been devastated by the parasitic sea lamprey, which has invaded the lakes via the Welland Canal, opened in 1932; and new species of fish, such as rock bass, are becoming dominant. Eutrophication [D, E], the natural ageing process of all freshwater lakes, occurs in three stages, oligotrophic [1], mesotrophic [2] and eutrophic [3]. In the Great Lakes, the process has been artificially accelerated by human activity, and Lake Erie has already entered its final, eutrophic phase.

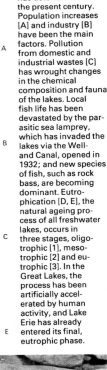

6 Natural waterways have long been regarded as obvious dumping grounds for man's wastes. In the modern technological society the wastes from industry have been more than the waterways can handle and, instead of the water being able to break down the polluting substances, these substances tend to accumulate and form poisonous suspensions or unsightly scum [A] in the water and kill off the animal and plant life that originally existed there [B].

Pollution of the sea

The world's oceans make up a vast expanse of water – the Pacific alone is greater in area than all the continents together. Man has accordingly looked upon them with awe – and then casually dumped into them every imaginable form of waste – solid, liquid and gas. Ships and barges carry solid refuse and liquid effluent out of sight of land and unload them into the sea. Galley rubbish is tossed over the sides of ships, water closets are flushed directly into the sea. The world's rivers carry their burden of effluents, nutrients and suspended solids into the coastal seas. The atmosphere wafts pesticides, lead compounds and many other pollutants far out to sea where they settle or are rained out to add to the pollution of the oceans.

Accidental oil spills

Oil and water do not mix, but a staggering total of oil – millions of tonnes a year – is deposited in the ocean, by accident and by intent. Accidental oil spills are the most notorious. The wreck of the *Torrey Canyon* [4] in 1967 released nearly 100,000 tonnes of crude oil into the waters off Land's End,

coating many kilometres of English and French coastlines with thick, black sludge, and killing many thousands of sea birds, especially those, such as the auks, which feed from the surface.

Every year adds to the toll of tanker wrecks. Sometimes these wrecks – such as those of the *Pacific Glory* and the *Allegro* in the English Channel in 1970 – occur in crowded waterways. Others, such as that of the *Metula* in the Strait of Magellan in 1974, occur in remote waters. But all add to the patches of oil slowly spreading across the surface of the oceans. There are other kinds of accidental spillage. One of the worst oil spills so far came from a Union Oil drilling platform not far off the California coast near Santa Barbara in 1969.

As tanker sizes continue to increase, accidents release greater deluges of oil into the oceans. But accidents do not release as much oil as do the irresponsible actions of a small minority of the world's tanker fleet operators. Tankers clean their tanks by flushing them down with seawater. Reputable fleets have adopted a procedure called

load on top (LOT) in which tanker washings are stored on board to be unloaded at special shore facilities rather than poured straight overboard. Overwhelmingly, however, the largest source of oil pollution at sea comes from the land – from industries, and motor vehicles. Their oily residues are often discharged indiscriminately into the sea or carried there by rivers [2].

Death in the sea

The quantity of oil pollution is troublesome but at least oil is organic in origin and can in time be broken down by marine organisms. Heavy metals such as lead, cadmium and mercury remain toxic indefinitely. Indeed, marine organisms may make them even more toxic. Mercury dumped into the coastal waters of Japan was thought to be of low toxicity. But there it was converted into methyl mercury, a powerful poison to the central nervous system. As is often the case, the poison was concentrated in fish and shellfish, producing an outbreak of what came to be known as Minimata disease [6], whose origins remained obscure for nearly a decade.

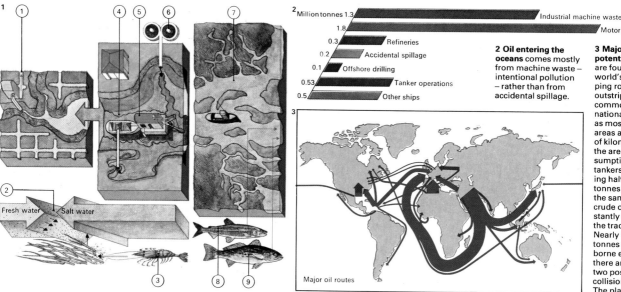

2 Oil entering the oceans comes mostly from machine waste – intentional pollution – rather than from accidental spillage.

3 Major areas of potential oil pollution are found along the world's main shipping routes. Oil now outstrips all other commodities in international traffic and, as most exporting areas are thousands of kilometres from the areas of consumption, fleets of tankers, some weighing half a million tonnes and carrying the same weight of crude oil, are constantly moving along the trade routes. Nearly three million tonnes of oil are seaborne every day and there are, on average, two possibly serious collisions every week. The places which run the greatest risk of oil pollution are the major shipping bottlenecks such as the English Channel, the Cape of Good Hope and the Malay islands, but now that oil is being exploited in Arctic regions tanker fleets will soon be operating in areas where the hazards of navigation are magnified by ice and poor visibility for the greater part of the year and serious incidents can be expected there. The escape of oil close to populated areas prompts most outcry but accidents elsewhere are just as dangerous to the local ecology. Bodies such as the Intergovernmental Maritime Consultative Organization have taken steps to reduce pollution by redesigning tankers and restricting dumping to certain specified areas.

1 Man's influence on the sea may begin upstream. He uses fresh water from an estuary to irrigate the land [1] and so alters the salinity gradient – the rate at which fresh water becomes salt at a river mouth [2] – killing organisms such as shrimps [3]. Reclamation of saltings or mud flats [4] destroys one of the most productive zones in the biosphere. Dredging [5] stirs up mud, reduces surface sunlight and oxygen levels. Raw sewage [6] overloads biological systems. A clean estuary [7] is important to the fishing industry as species like menhaden [8] and spotted sea trout [9] return to fresh water to spawn. Sewage can hinder the ability of fish to "smell" their way to the breeding grounds or return to the sea.

4 Torrey Canyon, a supertanker carrying Kuwait crude oil, ran aground off Land's End, Cornwall, in 1967. Her back broken, she disgorged nearly 100,000 tonnes of oil. Emergency measures to combat the huge slick, including fire bombing, were unsuccessful. Sea birds were coated in oil and died in their thousands. The slick went ashore in quantities in southwest England and drifted to the south and east to coat the beaches of the Channel Islands and France. Spraying of detergent in an attempt to clean up proved even more harmful to marine life than the oil. Species affected included flounder [1], shore and edible crabs [2, 3], blenny [4], prawn [5], lobster [6], and oyster [7].

Fishing will never be safe in Minimata Bay and some other Japanese bays where mercury still remains in the water.

The effects of marine dumping are still poorly understood. It took prolonged detective work to establish the cause of a notorious episode in the Irish Sea which involved the deaths of thousands of sea birds. The key factor was found eventually to be polychlorinated biphenyls (PCBs), organic chemicals in industrial effluent dumped in the estuary of the River Clyde. Some seaborne organic chemicals travel much farther from their origins; birds and fish can carry the powerful insecticide DDT to remote parts of the globe. Photosynthesis in marine plant life, which makes an important contribution to the earth's oxygen balance, can be inhibited by even quite low levels of DDT.

Closed systems

Some seas, such as the Mediterranean and the Baltic, are nearly closed systems having only limited interchange of water with the rest of the world's oceans. Such seas are already showing signs of serious biological breakdown [7]. But their plight, while worrying, is only a foretaste of what happens in the entire marine environment.

In a very real sense the whole world of the ocean is a closed system but the nations of the world have been unable to unite behind any effective national measures to administer the oceans for the benefit of all. In the early 1970s the United Nations Inter-Governmental Maritime Consultative Organization (IMCO) met in conference and drew up a convention for the control of maritime pollution by shipping. However, as always, a convention is only as strong as the steps taken to enforce it. The IMCO convention remains in essence a pious hope. The UN Conference on the Law of the Sea has dragged on since 1958 in session after session, with very little real progress.

National interests are generally more jealously guarded than "the common heritage of all mankind" and until there is more international co-operation the oceans will continue to be dumping grounds for sewage, pulp waste, radioactive wastes and all the other refuse of civilization.

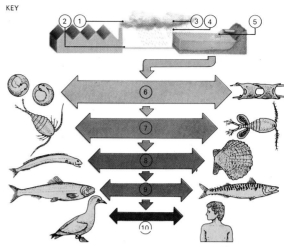

KEY

Waste returns to its originator. Smoke [1] and effluent [2], diluted or apparently dispersed in cloud [3], may, however, return in concentrated form. Pollutants reach the sea in rain [4] and run-off from the land [5] and they are absorbed by marine organisms – first simple [6, 7] then larger species [8]. Fish [9] may be eaten by birds or by man [10] whose place at the end of the food chain makes him especially vulnerable to the dangers of the waste that he creates.

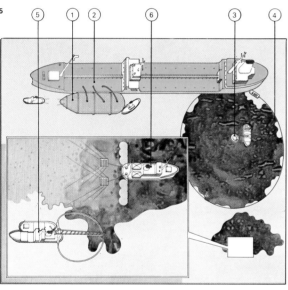

5 Tanker disasters of *Torrey Canyon* magnitude are spurring new techniques to minimize subsequent damage. A collapsible bladder [1] is towed alongside a stranded tanker [2] and used to store oil pumped from the ship. If any oil has been spilled a "skimmer" [3] collects the drifting oil which has been confined by a floating boom [4]. Oil can be mopped up by a moving absorbent belt [5] also known as an "oil scrubber", or sprayed with a dispersant [6] to speed its breakdown by microorganisms. These techniques are not yet widely available.

6 The effect of toxic pollution hit back at man on the shores of Minimata Bay, Japan, in the 1950s and 1960s. Inorganic mercury discharged into the bay as waste by industry [A] was converted by marine organisms into methyl mercury. This highly toxic substance accumulated in fish and shellfish [B] which were taken from the bay by local fishermen [C]. Fish is a staple diet [D] of the region. Townsfolk became ill [E], died or showed signs of brain disorder; children were born malformed. The plant has closed and fishing has not resumed in the bay – it has 600 tonnes of mercury on its bed.

7 Life in some coastal waters of the Mediterranean is dying. This sea has three "lung basins" where cold air blown down from the mountains [1], forms a cold, oxygenated layer of water which sinks [2], releasing oxygen. The basins are fed by effluent-poisoned rivers [3], the pollutants collecting in deep waters [4], killing all life. Tanker waste [5] films the sea with oil, denying sunlight to phytoplankton, the primary sea food. Man's work has killed sardine fishing off the Nile delta [6] by making the area unfavourable to the sardines. The Aswan Dam [7] reduces the fresh water input and salinity is increased by a flow of water from the Red Sea via the Suez Canal [8].

Areas of lung action

Deep water areas

5km band of dead sea

Mountainous regions

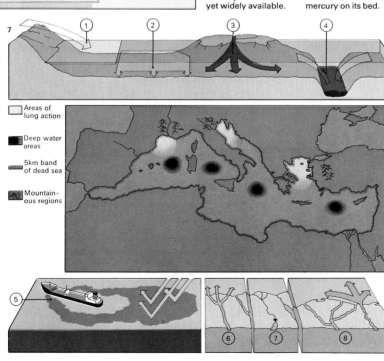

Land uses

Man is beginning, at last, to realize that if he is to continue to live on this planet he must conserve its resources. For thousands of years his impact on the land was negligible. But since the middle of the eighteenth century, intensive agriculture, the search for minerals and energy sources and the growth of population and transport systems have led man to interfere increasingly with the balance of nature and to compete for available land in ways that are often destructive.

The eras of change

Until about 10,000 years ago man left little trace of his presence on the landscape. He hunted, fished, trapped, sheltered in caves and searched for suitable materials for his tools. Early mines, such as those at Grime's Graves in Norfolk, were mere pits dug with picks made from the antlers of deer. But with the agricultural revolution of Neolithic times nomadic peoples settled down to cultivate fields, cut trees to build houses and established for the first time a pattern of land use that altered the natural ecological balance. Despite this, communities were small,

averaging perhaps one person per square kilometre even in settled areas. By the year AD 1000 world population had reached only about 350 million.

The growth of larger towns in medieval times, the spread of roads and the quarrying of materials such as clay, stone and iron gradually extended man's impact on the landscape. Large areas of forest were cleared for grazing or cultivation of crops. The Industrial Revolution of the eighteenth century was a major turning point in land use. The development of the steam engine based on power from coal greatly contributed to the establishment of a whole range of new industries and to the rapid growth of towns to house an expanding population of factory workers. As the transport of manufactured goods became more important, canals, railways and roads cut across the land. The search for building materials accelerated as world population grew from about 900 million in 1800 to 1,650 million in 1900.

At the same time an agricultural revolution began, bringing great changes in farming methods. Crops and livestock were improved

and food and raw material production developed. Agriculture spread to newly discovered lands and farming techniques became steadily more mechanized. Throughout the years of the agricultural and industrial revolutions of the nineteenth century disturbance of existing vegetation and wildlife, and pollution of the water and atmosphere, occurred with little awareness of the damage being done.

Man's adverse impact

Despite examples of the sensible conservation of land – such as reclamation in The Netherlands [3] and terrace cultivation in Asia – agricultural areas elsewhere were often mismanaged on a vast scale. Overuse of land in the western prairies of the USA broke up a stable ecosystem so that during a period of drought and high winds in the 1930s the topsoil was removed and a dust bowl created. In other areas thoughtless removal of vegetation allowed heavy erosion of hilly areas – rainwater carved deep gullies and carried good topsoil into rivers, causing further problems through silting.

CONNECTIONS

See also
148 Misuse of the land

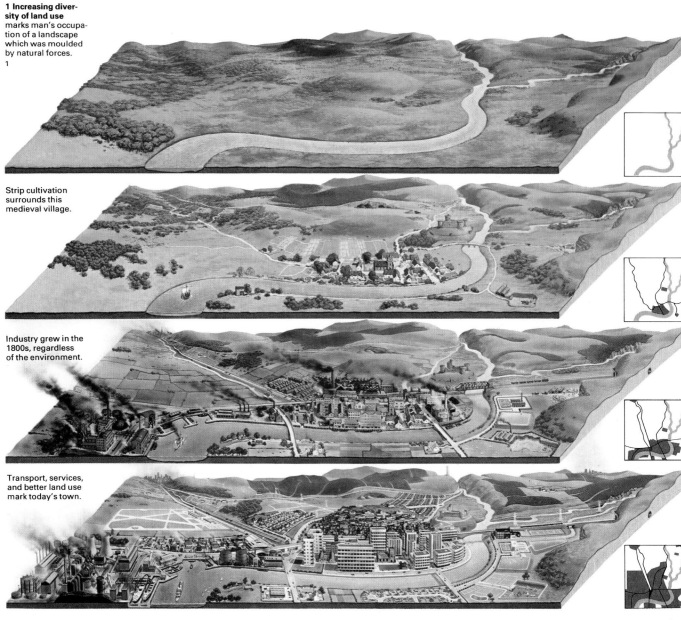

1 Increasing diversity of land use marks man's occupation of a landscape which was moulded by natural forces.

1

Strip cultivation surrounds this medieval village.

Industry grew in the 1800s, regardless of the environment.

Transport, services, and better land use mark today's town.

Relatively large areas of country on which man has had little or no impact remain but the landscape in developed countries has become more and more formalized. With world population soaring to some 4,000 million in 1975, cities continue to grow at the expense of the countryside. New roads sweep through previously impassable territory [Key], while in Europe and America extensive motorway systems devour the land. Airports take agricultural land and ubiquitous electricity pylons stride over hills and valleys.

Man's impact on rivers and lakes is already immense in terms of both pollution and engineering works and his ambitions indicate still more spectacular interference in future. The Soviet Union has begun to implement a policy to divert its northward-flowing rivers to the south in order to make desert areas fertile. A recent survey of the North American continent has suggested the expenditure of $86,000 million on 370 projects to effect the diversion of rivers and prevent "loss" of water to the Arctic Ocean.

Such schemes involve the building of dams, an activity almost as old as terracing in

agriculture. Although dams can control water flow, help irrigation and generate electricity, they also have their drawbacks. They may hold back valuable nutrients, drown vegetation in scenic areas, alter local climates and even affect seismic activity.

New techniques of monitoring land resources are now being developed, the latest being satellite surveys by means of microwave, ultra-violet and infra-red radiation [2]. Information emitted and reflected from the earth's surface gives a more detailed picture than aerial photography [4] alone.

Conservation of land

As well as the recent emergence of strong political lobbies in favour of land conservation, more attention is now being paid to the establishment of a balanced urban environment through replanting of trees and reclamation of previously waste areas. Old mines and tip heaps have been levelled and landscaped, overgrown canals cleaned and used for recreation and national parks created to maintain natural landscapes and limit further encroachment.

Transcontinental roads such as the Pan-American Highway, which links North and South America and covers a total distance of 47,516km (29,525 miles), can in many ways deeply influence land use. Apart from their immediate physical impact such roads open up new areas for settlement.

2 Accurate monitoring of the world's remaining resources of land is essential if the best possible use is to be made of them. Remote sensing by infra-red photography provides pictures showing ground heat as shades of red and blue; it also "sees" through haze. Healthy vegetation appears red; harvested fields are purple and ripe wheat fields are blue. Satellite surveys are now widely used. A National Aeronautics and Space Administration (NASA) project in the USA orbited an Earth Resource Technology Satellite (ERTS) in 1972 to provide data of benefit to man the world over.

3 In areas of dense population man has attempted to increase land area by reclaiming it from the sea. The Dutch are the most famous makers of land, having perfected their methods over hundreds of years. Embankments or dikes are first built across inlets of the sea. The area so enclosed is then drained and desalinated. Agricultural land is created and some freshwater lakes are left to provide fisheries and areas for general recreation.

4 Land-use surveys are now being carried out on a national scale in many countries of the world. A map based on aerial photography of the crowded coastline in the Kobe region of Japan shows clearly the confinement of the population to a narrow strip by the steep hills of the hinterland [buff]. Other obvious features are the residential area [red], heavy industries [pale blue], docks [pale pink], land fills reclaimed from the sea [yellow], large city-centre offices [purple] and wooded areas [green]. The first modern land-use survey was made by L. D. Stamp in Britain during the 1930s. Now the whole world is being surveyed in a series of maps by the UN Food and Agriculture Organization.

155

World food resources

The average adult male needs between 2,300 and 2,700 calories and 37–62 grammes (1.3–2.2 ounces) of protein each day to stay in good health. In addition, he needs adequate quantities of the appropriate micronutrients such as minerals and vitamins. Whether or not the world population receives this level of nutrition depends not only on the quantity of food produced but also on how that food is distributed.

Early food resources
It is only in recent times that man has had control over his food resources. Originally a hunter of wild animals and a gatherer of nuts and berries, his life was occupied with the constant quest for food. Because he had no control over his sources of food, starvation posed a constant threat.

Then about 10,000 years ago, man began to till the land for the first time. In the fertile floodplain of the Tigris and Euphrates he planted crops and grew his own food. He also learned to domesticate and raise animals for food to supplement the fish he caught from the rivers, lakes and seas.

In relying less on the bow and arrow than on the furrow and seed, man gained a greater control over his food resources. By planting a certain area of crops and raising a certain number of livestock, he could be reasonably sure of having enough to eat the following year. And yet, he still found himself at the mercy of elements over which he had no control. One year drought might wither the crops; another year locusts destroy the plants; and yet another year, disease decimate the livestock. In years when natural disasters struck, starvation could be a very real threat to man's survival.

Man also found that certain constraints affected not only how much food he could grow, but also where he could grow it. Crops needed fairly flat and fertile land if they were to flourish. They also needed adequate rainfall and appropriately long growing seasons. Since the moisture requirements and growing seasons of various crops differed quite widely, certain crops grew better in particular regions of the world than in others. For example, rice grew better in the tropical regions where temperatures and rainfall were

high than in temperate regions where both temperature and rainfall were considerably lower. Nevertheless, limitations imposed by topography, soil fertility, rainfall and temperature meant that at least 70 per cent of the world's dry surface area was unsuited for the purposes of arable farming.

Despite these constraints, man succeeded in greatly increasing his food production over the centuries. He was able to do this with the help of several major innovations which included (in roughly chronological order): the development of irrigation; the exchange of crops, such as wheat, barley, maize and cassava [1], between the New World and the Old; the development of artificial fertilizers [9] and pesticides; the discovery of the laws controlling plant and animal genetics; and the invention of the internal combustion engine to assist agricultural practice.

Feeding the population
While these innovations, along with many more minor ones, enabled man to grow greater quantities of food, it seemed that the number of mouths that needed feeding grew

CONNECTIONS

See also
252 The future of food

In other volumes
314 Man and Society

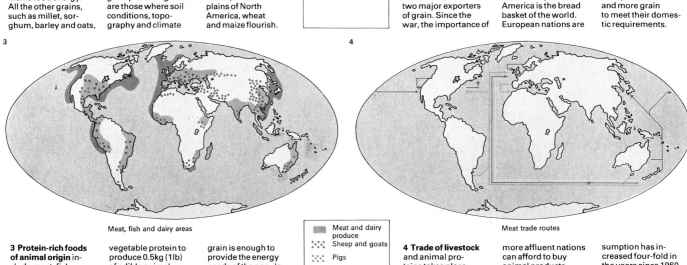

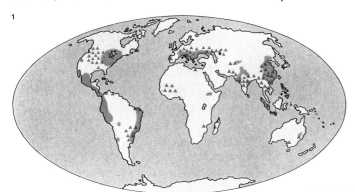

Main wheat, rice and maize areas

Grain trade routes

Maize	
▲▲▲ Wheat	
▲▲▲ Rice	
→ Maize	
→ Wheat	
→ Rice	

1 Wheat, rice and maize are of vital importance to the world because together they supply nearly a half of man's food energy. All the other grains, such as millet, sorghum, barley and oats, supply only 10%. In addition, maize and, to a lesser extent wheat, are also important animal foodstuffs. The major grain producing areas are those where soil conditions, topography and climate favour the growing of a particular crop. Thus the river deltas in Asia favour the growing of rice, while on the cooler, drier plains of North America, wheat and maize flourish.

2 World grain trade patterns have changed radically since the 1930s. Prior to World War II, Latin America and North America were the two major exporters of grain. Since the war, the importance of Latin America as a grain exporter has declined as her population growth began to catch up with her ability to produce food. Today North America is the bread basket of the world. European nations are the main importers of grain. However in recent years, developing nations, previously self-sufficient in grain, have begun to import more and more grain to meet their domestic requirements.

Meat, fish and dairy areas

Meat trade routes

Meat and dairy produce	
Sheep and goats	
Pigs	
Fish	
→ Meat	
→ Dairy products	
→ Sheep	
→ Pigs	

3 Protein-rich foods of animal origin include meat, fish, eggs and dairy products. Animal proteins cannot be entirely replaced by vegetable proteins in the diet; moreover their production is inefficient. Cattle require 9kg (20lb) of crude vegetable protein to produce 0.5kg (1lb) of edible animal protein while the broiler chicken requires 2kg (5lb). About 33% of the world's grain production (400 million tonnes) is currently being fed to livestock. This grain is enough to provide the energy needs of the populations of India and China. The major livestock producing regions are those where grain is abundant (the American Midwest) and where grassland is plentiful (Argentina).

4 Trade of livestock and animal proteins takes place primarily between countries of the developed world. In 1971, they accounted for over 90% of the world market economy's trade in meat. This reflects the fact that only the more affluent nations can afford to buy animal products, which are usually more expensive than other foods. Over the past century, there has been a growth in the eating of meat in the developed countries. In Japan, meat consumption has increased four-fold in the years since 1960. As Japan does not possess the resources to grow sufficient food to satisfy her needs, she has had to increase her imports of grain as well as her imports of meat.

almost as rapidly as (and sometimes more rapidly than) his agricultural skills. It is estimated that when man first began farming, the world's population numbered no more than ten million. By the time of Christ, this figure had grown to 250 million; by the mid nineteenth century, 1,000 million; by 1900, 1,650 million; and by 1975 the world population was nearly 4,000 million [7].

With population growing almost as rapidly as man's ability to produce crops, famine and starvation has continued to threaten major segments of the world's population, especially in parts of Asia and Africa.

Malnutrition past and present
The threat of famine tended to be especially severe in years when natural disasters, such as floods or drought, caused sudden reductions in world food production.

In the past, major famines included one in Italy in 436 BC and another in India in AD 1291 when large numbers of people died. In more recent years, the Irish famine of 1846, which was due to the failure of the Irish potato crop, forced 800,000 Irishmen to migrate to the United States.

In recent years, famine has been reduced because world food production has tended to outstrip the growth in population and because certain nations have been able to build up reserves which they have distributed to other nations in times of disaster. Nevertheless, malnutrition remains widespread. The UN estimates that 460 million people do not receive an adequate amount of the right kind of foods. The diet of these people is frequently lacking in calories, protein and the essential micronutrients. Protein deficiency alone severely affects the physical and mental development of children.

The widespread incidence of malnutrition is not primarily due to inadequate world food production. Rather it is due to the inequitable distribution between (and for that matter, within) nations of the food that the world produces. Today, the affluent one-third of the world's population eats well over half the food that is produced. It is not surprising then that amongst the remaining two-thirds malnutrition is rife and that a solution to this imbalance is urgently needed.

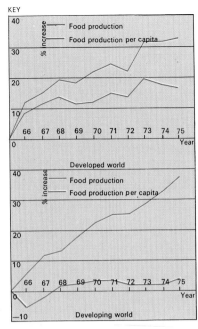

KEY
Food production
Food production per capita
Developed world

Food production
Food production per capita
Developing world

The production of food has risen almost equally in the developed and developing world over the past 20 years. In the developed world, population growth has tended to lag behind the growth in food production with the result that *per capita* food production has risen by 21%. But in the developing world, population has grown almost as rapidly as food production so that *per capita* food production has remained virtually unchanged. Because of population growth, attempts to improve diet and the level of nutrition have not been very successful. It is possible that *per capita* food production could even begin to fall in developing countries.

5A
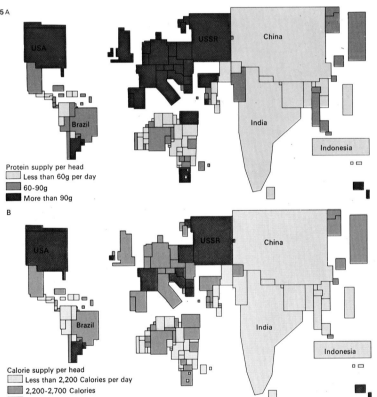

Protein supply per head
Less than 60g per day
60-90g
More than 90g

B

Calorie supply per head
Less than 2,200 Calories per day
2,200-2,700 Calories
More than 2,700 Calories

5 These maps of food supply and population show the area of each country proportional to its population. They measure the food supply in terms of available protien in weight [A] and number of calories [B].

6 Arable land accounts for a small slice of the earth's total dry land surface area. Over 70% of the land area is unfit for agricultural purposes which limits increases in food production.

Forested land

Meadow and pasture

Arable land

Others

1971
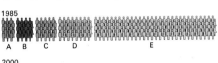
A B C D E
= 20,000,000

1985
A B C D E

2000
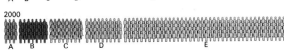
A B C D E

7 The world's population will number approximately 6.5 thousand million by the year 2000. UN figures predict it is unlikely to stabilize before 2125 when there may be 12.3 thousand million.

A North America
B Latin America
C Africa
D Europe and USSR
E Asia and Oceania

8 Intensifying production could increase overall world food output. Unfortunately many of the more modern methods of food production, which produce more food from a given area, require large inputs of fuel energy for power and fertilizer production. It is estimated that if the whole world were to adopt the food production methods used in the USA, as well as its diet, the world's known reserves of petroleum would be exhausted within 29 years.

8

9 Millions of tonnes of phosphorus
8
7
6
5
4
3
2
1
1905-6 1915-6 1925-6 1935-6 1938-9 1945-6 1955-6 1965-6 1969-70

9 Chemical fertilizers are being used in increasing quantities, as shown in this graph for phosphates, expressed in millions of tonnes of phosphorus (the drop after 1938-9 was due to World War II). Unfortunately the reserves for two of the three major plant nutrients, phosphorus and potassium, are limited and the processing of the third, nitrogen, requires large amounts of either natural gas or oil for energy. Thus fertilizer shortages will worsen.

History of agriculture

When *Homo sapiens* first evolved, he lived in the same way as other animals, by gathering vegetable food – grain, fruits, roots and nuts – and by killing animals for meat. It was only about 10,000 years ago that man domesticated animals and discovered how to cultivate nutritious plants and harvest them.

Primitive farming methods

About 7000–9000 BC, the invention of the digging stick and a light plough (the ard) [3], made it possible to prepare the soil to grow crops. The basin of the rivers Tigris and Euphrates and the valley of the Nile became the centres of great civilizations, founded because there was enough food to supply the non-farming population. Similarly, farming evolved in northwest India, in north China, in the southwestern parts of the United States and in Central and South America. Various plants were cultivated, such as rice in the East and maize and gourds in the Americas. Such primitive farming continues to this day in many parts of the world.

The people of ancient Greece and Rome imported grain from Egypt and Africa, but both also grew some food – grapes, figs and olives – of their own. The Greeks and the Romans ploughed with ards, and possibly used animal-drawn harrows. Sickles were used for harvesting the grain they cultivated (wheat and barley, not oats or rye). Domesticated animals included horses, cattle, sheep, pigs, goats, poultry and bees. The system of farming was crop and fallow; that is, a crop was grown one year and the land left fallow during the next in order to allow it to recover its fertility.

In northern Europe the forests were cleared from the land chosen for cultivation. First the larger trees were felled, the brush cut and the whole area burned. High yields (for the times) of rye and oats were sometimes harvested. After a year or two the land was allowed to revert to scrub. This system, known as "slash and burn", is still practised in parts of Central Africa.

Following the fall of the Roman Empire the Anglo-Saxons conquered England and brought with them a new type of farming that was spreading all over western Europe. This was the open field system [4]. The arable land was divided into three great fields cropped in a rotation of wheat or rye; barley, oats, peas and beans; and fallow, although ploughed three times a year. Sheep grazed meadows after the hay harvest and possibly stubble-fields. The new heavy plough, hauled by two to eight oxen [5], turned over the long furrows in difficult soil.

The Moors had conquered most of Spain and brought with them new crops – sugar cane, rice and subtropical fruits – as well as Merino sheep, ancestors of the great flocks now bred in Australia. The silkworm was introduced into Italy and fed on mulberry trees, while rice was grown extensively in the south and also in Piedmont.

New crops, methods and machines

From America and the Indies new crops and new foodstuffs, the most important of which was the potato, were brought to Europe. Maize was rapidly adopted in Spain, southern France and Italy, as was tobacco. Tomatoes and gourds were important novelties. It was in Europe that the most significant progress was made in farming techniques.

1 Primitive agriculture involves the whole family in the task of providing enough food for subsistence. The field work is mainly done by women, after the men have completed the heavier work of clearing the land with oxen. While men tend the animals, women plant, harvest and wage a constant battle against weeds. Work patterns of this sort persist in many peasant communities. The surplus of crop production over the subsistence level is often small; a year of crop failure from drought will mean hunger and possibly even starvation in this inefficient system of agriculture.

2 The mattock or heavy hoe was the basic and sometimes only tool used in early cultivation. It was used to break new ground or shift stones and it served as a hoe while the crop was growing.

3 The ard or scratch plough, still in use among peasant cultivators in some parts of the world, does not invert the soil but breaks the surface for sowing. Originally dragged or pushed by men, it was later adapted for pulling by oxen, which doubled the area that could be ploughed. Various methods of attachment were devised, usually with a rigid shoulder yoke pegged or strapped in place.

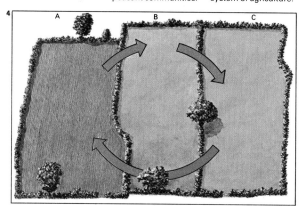

4 Crop rotation farming, introduced in Europe in the Middle Ages, was based on three fields. One [A] would be left fallow, the second [B] would grow wheat or rye and the third [C] would be in barley, oats and peas. Medieval fields were vast, unfenced areas arranged round the village. Each field was divided into a large number of strips and each villager was allocated strips according to his status. As well as working his own land the villager would also work that owned by the feudal lord of the manor. Despite the principle of division of land, this system had many disadvantages, including the loss of land because of pathways between the strips and the possibility of a badly farmed strip affecting other strips.

5 Ox teams provided the basic power units of the farms of the Middle Ages in northern Europe. The sturdy ploughs used had extended farming to the heavy soil, which had defeated earlier cultivators. Cattle were still bred as work animals but their milk might be made into butter and cheese. The plough ox was fattened for slaughter when seven to nine years old. The working day of the oxen plough team was limited by the need for the animals to graze. The horses that replaced them could work longer hours and also pull the plough faster.

By the fourteenth century farmers in the Low Countries had begun more intensive cultivation of crops on the fallow land, mainly grains, root crops and fodder plants – clover, lucerne and rye-grass. The dye crops, madder, woad and saffron, were also cultivated. As mechanical skill developed, a method of drilling seed by machine instead of broadcasting by hand was invented and this was to be one of the foundations of modern mechanical farming.

The intensive cultivation of fodder crops made it necessary to abandon the open field system and to make fenced enclosures, both to protect the growing crops and to contain the grazing animals. The four-course rotation [6] replaced the three-course rotation of open fields. This allowed higher productivity and the feeding of more domestic animals, which were also being slowly improved by selective breeding.

It was the opening up of the American plains in the nineteenth century that provided the real impetus for mechanization. The plains could grow millions of tonnes of wheat to feed the booming population of industrializing Europe, but local labour was both scarce and expensive. By 1850 a reaping machine was already in use, just before a steam engine was produced to apply power to ploughing, harvesting and threshing.

Scientific farming and increased yields

The work once done by animals is today accomplished by the tractor, while combine harvesters and other complex machines gather the crops. A hectare of wheat in AD 1200 might yield 420–530 litres (12–15 bushels); today 1 hectare (2.5 acres) of wheat in Europe, America, Argentina, Australia or Canada might yield 3,500–7,000 litres (100–200 bushels). In 1200 the yield of one hectare was enough for four or five people for a year; today, in the advanced countries of the world, a hectare can supply bread sufficient for between 20 and 50 adults for a year as well as seed for the next sowing.

Stock farming has also made unprecedented progress, particularly in the last hundred years. Farm animals have been improved, producing more offspring, meat, eggs or milk in less time than before.

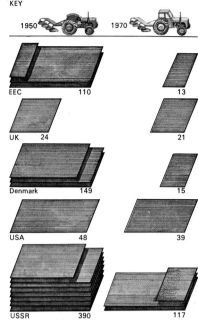

KEY

1950   1970

EEC 110	13
UK 24	21
Denmark 149	15
USA 48	39
USSR 390	117

A rapid replacement of man and animal muscle power by machines has taken place since World War II. This can easily be illustrated by dividing the area of farmed land by the number of tractors in developed countries to give the number of hectares worked by each tractor. In 1950 the USA and the UK were already mechanized, while parts of Europe were using horse-drawn machines and the USSR was recovering from war losses. By 1970 there were proportionately more tractors on the tiny farms of Western Europe, and those on larger holdings in the USA and the USSR had become more powerful and versatile machines.

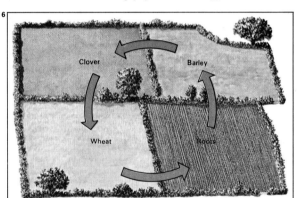

6 By introducing new crops the Norfolk four-course rotation eliminated the fallow year. Clover and turnips fed more stock and the manure from the stock helped in the growing of improved crops.

7 Stock breeders of the 18th and 19th centuries considerably improved beef cattle. The new breeds produced the extra meat that was needed to feed the growing population of northern Europe.

8 Modern farming is growing increasingly specialized. The flat, fertile silts of South Lincolnshire, for example, seen here from the air, carry no livestock. This is an area of "cash crops": potatoes, sugar beet, field vegetables and bulbs, with wheat as a break in the rotation. Some farms are large and make much use of machines. The business side is also well developed with large, private and co-operative investments in crop processing and marketing.

9 Large numbers of beef cattle are kept throughout the year in open yards such as this US feed lot. Several thousand cattle may be assembled for fattening on a balanced diet of maize, silage, lucerne, grain and a mineral supplement. With mechanized feeding, labour requirements are small and each animal's growth is rapid. The capital required is large and the owners are more likely to be corporate companies than traditional private farmers.

The small farm

Most of the world's farming is carried on in units that are small by the standards of North America or Australia. Land area is not the best measure of productivity in farming, however. On very fertile soil, with plenty of labour available, output can be high and special buildings and equipment can enable a livestock farmer or market gardener to earn a good income single handed.

Traditional peasant farming

Much traditional peasant agriculture was carried on to keep the farmer's family alive rather than to earn a cash profit. The farmer often paid his rent in kind as a form of share cropping. From the 7 hectares (17.3 acres) of a typical old-style Tuscan farm [1], 47 per cent of the produce went to the landlord and the remainder fed the eight adults and seven children working and living there, with only a small surplus for sale. This holding grew cereals, potatoes, beans and peas, other vegetables and salads for market, vines, olives, apples, pears, peaches, apricots, cherries, plums and walnuts. It had four milking cows, two draught oxen, two pigs for meat

and a horse. Wine from the grapes and oil from olives were used on the farm and the main cash income was derived from the daily sale of vegetables in the local market. Such farms were still found in Italy until recently.

In northern Europe small farms of a different type developed during the last century as a result of increasing job opportunities outside farming and of a growing demand in the cities for meat and milk products. Areas of 10 to 50 hectares (25 to 123.5 acres) with dairy cows, pigs and hens were worked by a farmer and his son, or one hired man, the farmer's wife assisting with the poultry and the calves. In Scandinavia, a network of local co-operatives grew up to undertake the manufacture of butter, cheese and other products from milk and of bacon from pigs. A somewhat similar pattern developed in parts of Ireland but was based mainly on grassland farming. Also using grass, but rather differently, small farms in The Netherlands and New Zealand could compete effectively in world markets for butter and cheese.

The majority of small farmers in Scandinavia [2], Ireland and New Zealand were,

or became, owners of their land. Government policies often encouraged the break-up of large holdings and provided loans on easy terms for small farmers to buy the land they worked. In countries where cereal growing was not protected by import duties, livestock farmers had the advantage of cheap feed from abroad, which enabled them to keep more animals than their land would carry.

In some of these countries, and in many areas where agriculture was once run on traditional peasant lines, small farms have now been absorbed into larger units. Since 1958 the number of people working in Europe's agriculture has been halved from about 18 million to 9 million. Farm wages have risen continuously and the very small farmer finds it difficult to earn sufficient to pay a hired worker.

Help from the government

Voluntary amalgamations of small farms have been accelerated in some European countries (including France, The Netherlands, West Germany and the United Kingdom) by official schemes to provide

CONNECTIONS

See also
162 The farming corporation
164 Farm machinery and buildings

1

1 The buildings and yard of an Italian farm of 7 hectares (17.3 acres) near Florence had to serve a variety of needs and enterprises. The farm was divided into 20 small enclosures, some of them terraced on a steep slope. Much of the cropping was in ground under permanent fruit trees. The crops provided a meagre living over the greater part of the year for the 15 people living on the farm. Tools and equipment were simple – the tenant's only capital was in the form of animals and growing crops. This situation was not unusual in the south of Europe, in areas where there was only small prospect of alternative employment. Attempts to rationalize farming in these areas have created political problems.

2 Scandinavian farms have been designed to be worked by at most two men, with family help at peak periods of the year. Severe winters call for substantial buildings to house the livestock, as on this horse farm. Animals are the main source of income and most arable land is devoted to grain and fodder to feed them. Milk from the cows goes to the creamery for butter making and the skim is returned; mixed with barley it feeds the best bacon pigs. This system is giving way to greater geographical specialization in Denmark. Other Scandinavian farmers often work in forestry during the harsh winter months.

retirement pensions for older farmers and retrain younger men to work outside agriculture. The aim, not always successful in practice, is to bring the minimum reward of the full-time farmer up to that of an employed man. In assessing the viability of small farms and their qualification for government assistance United Kingdom economists use a unit that takes into account other things beside mere land area. Each enterprise on a holding is classified in terms of "standard man-days" [Key], based on average performance over the whole country; ten days are allotted per year for each milking cow, for instance, and five for each hectare of wheat or barley. If these add up to less than 275 the holding is not considered to be a full-time farm. Standard man-days measure the farmer's capital investment in equipment, stock and land.

Part-time small farms have been an important element in rural life over the centuries. They are still a common, though diminishing feature of parts of Europe and the United States. Depending on geographic location, farming may be combined with fishing (in Scottish crofting), with forestry (in

Scandinavia), with small-scale mining or country crafts (in south Germany) or with part-time and seasonal work in nearby factories. Production from this form of agriculture may not be vital to the national economy, but for many people part-time farming satisfies a social need for an outdoor way of life and a certain independence [5].

Farming and the environment

Preservation of viable small farms, even if they are worked only part-time, is increasingly seen as important to tourism and to the wider enjoyment of the countryside. The traditional picture that many people have of farming and their knowledge of the countryside often come from small mixed farms on which they have stayed as summer guests. In the more difficult, if picturesque, environment of the mountains and uplands the disappearance of small farms could leave a depopulated wilderness. Year-round life is hard on such farms, however attractive they may seem in summer. Some countries have therefore introduced special schemes of financial assistance.

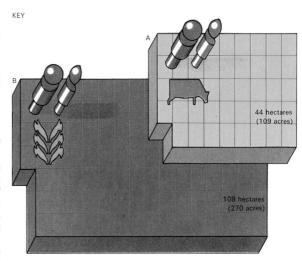

Two contrasting farm patterns both need the labour equivalent of a man and a half. The livestock farm [A] will have help from the farmer's wife and family at times, and possibly from a pensioner or student. The arable farmer's wife will not work on the farm [B] but contractors will be used for some work and casual labour will be hired at busy periods.

1 Sugar beet
2 Fodder maize
3 Summer fallow
4 Milk tanker
5 Cattle in pasture
6 Barley
7 Oats
8 Glasshouse crops
9 Pigs
10 Silo
11 Wheat
12 Fowl
13 Ducks
14 Clover
15 Orchard
16 Farmhouse
17 Vegetables

3 A wide range of enterprises could be carried out on a small farm if they were all gathered together. In practice the number on a small mixed farm is not likely to total more than about half a dozen. In areas of high rainfall the emphasis will be on grazing livestock; where summers are dry crops will be of most importance.

4 Farming in Switzerland is restricted to a series of small holdings. The high altitude of most of the agricultural land and the high rainfall prevent farming from being a major part of the Swiss economy; in recent years it has employed only about 20% of the population.

5 New England is one part of the United States where small mixed farms of an earlier traditional pattern survive. Many are now owned by city dwellers as places for relaxation and a change of scene, with caretaker workers in charge. This pattern is also spreading in Europe.

The farming corporation

Organized agriculture planned to provide the balanced food requirements of a community is probably as old as settled man and his farms and first towns in the Fertile Crescent. High production farming or "agribusiness", involving capital investment as well as scientific management and business techniques, is a response to the demands of the huge industrial populations of the Western world and hopefully its benefits may be extended ultimately to countries that are still greatly undernourished in the late 1900s.

Surplus, not subsistence
The experience of English farming, notably in the southern and eastern counties in the eighteenth century, pointed the way by producing crops surplus to the needs of a simple subsistence economy (this still remains the occupation of farmers in the greater part of the world) in order to supply the burgeoning population of London with sufficient food and for export to Europe.

The Industrial Revolution in western Europe provided an historic impetus towards the modern practice. In countries where the

rapidly increasing urban population had led to a demand for food that could not be met by traditional farming methods in Europe, a market emerged for large-scale farming production and for imported food from such developing production areas as South America and Australia. Yet meat and dairy produce from countries in the Southern Hemisphere could not be shipped in good condition through the tropics to European markets.

The answer to the problem was the introduction of mechanically refrigerated ships. From that first important beginning, modern techniques of refrigeration have developed to play a fundamental role in food marketing and in the organization of "agribusiness" with its huge output of deep-frozen, polythene-wrapped food for storage in domestic freezers.

Modern farming techniques
The modern farm may be one of a group of several operating under the control of a central organization, each perhaps averaging several hundred hectares; not large farms,

but each rather larger than the national average. Economic studies have shown that farms of medium size tend to be more efficient than large ones. This is particularly true of dairying in which there has to be a close relationship between the milker and the cows he or she handles and of farms with mixed cropping in which timing calls for close supervision. Further limitations may be set on the size of some single livestock units to prevent the risk of disease and to aid the disposal of waste products.

Difficulties of supervision have been most successfully overcome with poultry, both for meat and egg production. In this area "factory farming" [1] has developed to its greatest extent. Birds bred specifically for a controlled environment and for rapid maturity spend their whole lives in buildings and are kept in very large numbers. The buildings are insulated against changes in outside temperature, with forced ventilation. For egg production, lighting simulates an even day length, to eliminate the natural spring flush and autumn decline in laying that occurs out of doors. In this totally artificial

CONNECTIONS

See also
160 The small farm
164 Farm machinery and buildings

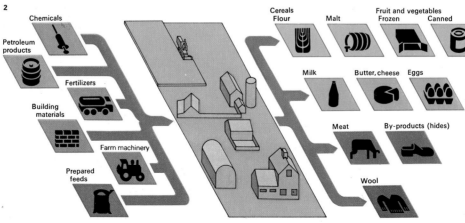

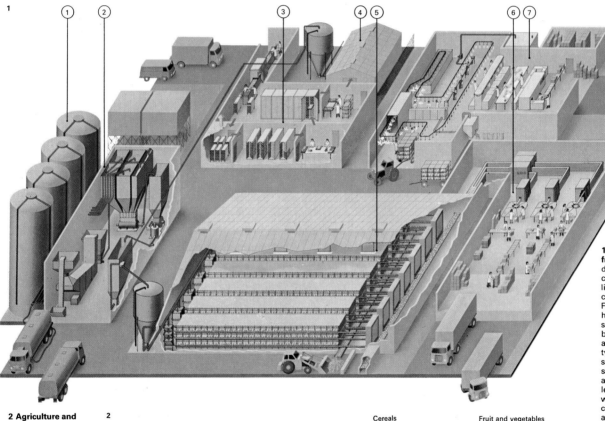

2 Agriculture and industry have become highly interdependent. The farm, once self-sufficient, is now an important market for suppliers of petroleum products, chemicals – crop dressings, synthetic fertilizers – machinery and building materials. The farm in turn supplies food – cereals, fruit and various vegetables (fresh, frozen and canned) – as well as dairy products, meat and many animal by-products.

Chemicals

Petroleum products

Fertilizers

Building materials

Farm machinery

Prepared feeds

Cereals Flour   Malt   Fruit and vegetables
                          Frozen   Canned

Milk   Butter, cheese   Eggs

Meat   By-products (hides)

Wool

1 Far from the free-range life, hundreds of thousands of chickens spend their lives in an artificial environment. Factory farming has made its greatest strides in poultry breeding – chickens, and turkeys. In a typical plant, food stored in grain silos [1] is mixed and milled into pellets [2]. The eggs, which have been checked for fertility, are incubated [3] and the chicks reared [4]. From here the young birds pass to the battery house [5] with its tiers of cages, automatic feeding and droppings disposal. The eggs are graded and dispatched [6], while birds destined for the table are transported to the broiler house [7], slaughtered and prepared before being deep-frozen and packed in polythene for dispatch to the chain stores and their customers' deep freezers.

environment birds are hatched, grown to maturity and slaughtered in their hundreds of thousands before passing through a deep-freezing process to keep the plant's output steady throughout the year. This applies even to turkeys, where the market is seasonal, as at Christmas and Thanksgiving.

Production and processing

Pigs are also raised for slaughter in large numbers but the breeding units are not on the scale of those used for poultry and are often in the hands of specialists other than farmers. Slaughtering and processing are performed by outside commercial companies or by co-operatives, which may be wholly or only partly owned by farmers.

The integration of beef production and processing along similar lines is mainly confined to the United States, where the trend towards larger units for many types of livestock production was originally fostered by feed supply companies to increase their own sales. In Australia and New Zealand [3] much farming has always been on a large scale and geared to the requirements of

export markets. This has led to highly developed methods of organization, often on a co-operative basis, with powerful producer boards and finance companies as well as firms specializing in the processing, packing, exporting and marketing of food products.

Not all farming enterprises lend themselves to organization as large units – a fact discovered in places where agriculture has been nationalized or collectivized. In the USSR, much of the national supply of fresh meat and vegetables still comes from private plots worked by individual members of the collective. Elsewhere the commodities most successfully produced by large units have been cereals, sugar, tea, tropical industrial crops (tea, cotton and rubber), vegetables (where there is a ready access to large markets) and salads (in The Netherlands, where land is at a premium, multi-storey greenhouses are in use) and orchard fruit.

The techniques of modern profitable farming call for larger amounts of capital than previously, which enable more raw material to be bought or products to be processed for the market [2].

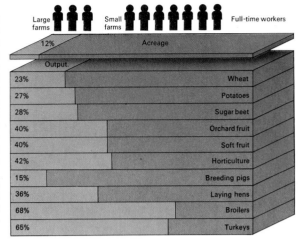

KEY
Large farms occupy 12% of agricultural land in England and Wales, representing 1.2% of the holdings, and employing 30% of the full-time farm.

workers (1973). The production statistics indicate that the large farm is much more efficient than the small farm especially regarding

fowls that are factory farmed. The figures do not reveal the full picture as they do not compare the capital input of the two types of farms.

	Output	
23%		Wheat
27%		Potatoes
28%		Sugar beet
40%		Orchard fruit
40%		Soft fruit
42%		Horticulture
15%		Breeding pigs
36%		Laying hens
68%		Broilers
65%		Turkeys

3 New Zealand farming was for many years geared almost entirely to the export of dairy products, lamb and wool to Great Britain. The climate of the islands favours grass growth for much of the year and large-scale farming is therefore almost all pastoral. Sheep farming, based partly on the Scottish experience of early settlers, utilizes three different environments. Mountain country in the South Island [A] is devoted almost completely to wool production from Merino [1] and Corriedale [2] sheep. The fertility of the high country has been greatly increased by using aircraft to spread lime and fertilizers, thus encouraging the growth of grasses and clover and improving their quality. The lower hill country [B], where conditions are less severe, is mainly stocked with Cheviot ewes [4] which, mated with Romney Marsh rams [3], produce crossbred ewes [6] that are hardy, productive and good milkers. The other main product of this zone is wool. When the crossbred ewes are moved down to the plains [C], where the stocking rates are higher, they are mated with a Southdown [5] or other Down breed ram and the offspring [7] are fattened for slaughter, as are the male crossbreds brought from the hills. This well considered system makes possible the production of uniform carcasses, which are carefully graded for weight and "finish", and has given the product a high repu-

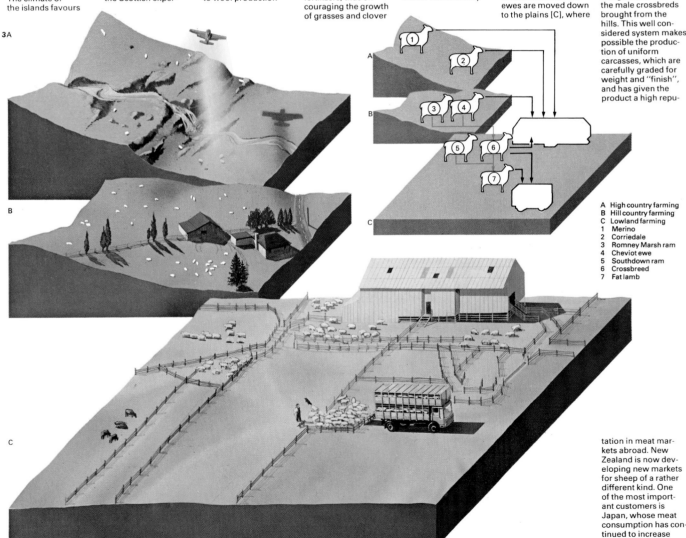

A High country farming
B Hill country farming
C Lowland farming
1 Merino
2 Corriedale
3 Romney Marsh ram
4 Cheviot ewe
5 Southdown ram
6 Crossbreed
7 Fat lamb

tation in meat markets abroad. New Zealand is now developing new markets for sheep of a rather different kind. One of the most important customers is Japan, whose meat consumption has continued to increase in recent years.

Farm machinery and buildings

Mobile farm machinery first came into general use during the nineteenth century. Before this time one farmer with manual farming methods could grow little more than enough food for himself and his family. Some attempts had been made to mechanize such jobs as threshing, using a horse driven in a circle as the source of power. Steam-powered ploughing, although widely adopted on difficult soils, involved winching the implement from one side of the field to the other.

The most successful horse-drawn farm machine was the reaper invented by Cyrus McCormick in 1831 which cut grain crops and tied them into sheaves. It made large-scale growing possible where labour was limited, as on the North American prairies.

The rise of modern machinery
The development of the internal combustion engine replaced the horse teams by tractors [3]. At first these merely pulled trailed implements or replaced steam engines for stationary work when fitted with a belt-drive from the side. Hydraulic lifts, powered by a small rotary pump, enabled other implements to be attached at the front or rear; these implements could be raised for transport and lowered for work.

Grain harvesting with a reaper-binder still involved manual work when assembling the sheaves for drying, carting, stacking and handling for threshing. The combine harvester [2], controlled by one man, delivers the threshed grain in a single operation. As successive models have increased the work capacity, they have come to be fitted with their own power units, as have other large field machines. Grain is conveyed mechanically from reception pit to dryer and then in and out of bulk stores.

Other types of farm machines include tillage machines, which break up the soil and prepare it for the seed. The plough is the best known of these and they do the heavy work. Rollers are used to flatten the larger clods of earth and harrows make the soil surface fine and soft so that the seed can be easily buried. Sowing machines place the seed in the soil. Broadcasting machines sow seeds such as grass and clover at random, while drills that place seed in neat rows and at a given depth are used for crops such as wheat and barley. For root vegetables precision drills space the plants correctly and protect the seed from birds. Cultivators do the hoeing; they kill the weeds and keep fallow land from being overrun with weeds. The last main category comprises livestock machines. Milking machines, worked by suction-vacuum pumps, have replaced the hand milker and the milk is delivered by machine direct to the cooler. Animal feed is mixed and carried by air flow to the feeding point and manure is also handled mechanically in the buildings and in the fields.

Development of farm buildings
Buildings have undergone radical changes in design and layout over the last 30 years. Their traditional pattern consisted of a group of structures around a yard or a number of yards, with solid outside walls for protection against bad weather and for the security of the stock. Typically, such an assembly would include a barn for storage of grain and fodder, a shed for tied cows where the milking took place and a dairy. There would also be

CONNECTIONS

See also
160 The small farm
162 The farming
corporation

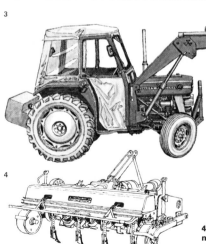

1 **Horse-drawn reaper-binders** opened up the grain lands of the New World to large-scale export production and had a far-reaching effect on world trade. Seasonal workers were still needed to assemble the sheaves into stooks for drying and to load them on to carts for transport back to the farm for threshing.

2 **The combine harvester** does its own threshing in the field, separating grain from straw, sieving out small weed seeds and blowing away the chaff. Grain is collected in a bulk tank, then discharged into a servicing truck. The main parts are the cutter-bar assembly [1]; the threshing area [2] with beaters, sieves and a fan to drive out the chaff; the straw walkers [3], where more grain is shaken out and the straw discharged; and the grain tank and discharge [4].

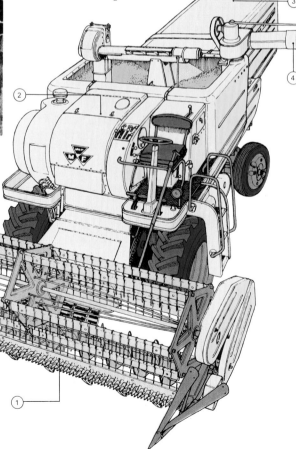

3 **A modern high-power tractor** can carry and work implements both in front and at the rear. The high-level scoop is carried on the hydraulic system and the operator, using a single lever, can manipulate it from his seat. The cab provides protection against overturning.

4 **Most cultivating machinery** is mounted on the 3-point linkage at the tractor's rear and connected to the power-take-off shaft. It can be raised by the hydraulic system for transport and adjusted by it to control the depth or height of working. Implements such as this rotary cultivator can carry out several tasks at one pass of the field, reducing time and soil consolidation from the passage of the tractor wheels.

separate stable boxes for horses, sties for pigs and open-fronted shelter for the beef cattle which trod straw into manure in the yards during the winter months.

Modern farm buildings are of two main types. The first and most usual was made possible by the development of wide-span metal, concrete or laminated wood beams which carried lightweight roofing material without the need for an intermediate support. This meant that tractors and other machines could be driven anywhere in the building without obstruction. The basic "Dutch barn" [6] can be subdivided in any way that suits the farming system, given exterior cladding (for insulation) against the weather, and it can also be enlarged at the sides by adding lean-to extensions where height is not essential.

Parts of the covered area can be used for different purposes in the course of the year; for example for housing stock at one time and maintaining machinery at another. For dairy cows, the cowshed has been replaced by an open area with individual cubicles for lying down, a loafing area and a bulk feeding area. With the aid of modern machinery, cows are milked in a parlour adjoining the dairy to which the milk is subsequently piped, and concentrated rations may be automatically allocated to each cow by computer.

Controlled environment house
The second main type of building is the controlled environment house for intensive production [7] of pigs, broiler chickens and eggs. This is normally a fairly low structure, with the hens housed in wire cages. The eggs roll onto conveyor belts and are removed to a special egg-packing room. The interior is completely isolated from the outside world, and lighting is operated to make "days" and "nights" so that the hens lay their eggs more often. There is also forced ventilation which controls the temperature; except for piglets and young chicks, housed animals generate their own heating.

The closed yard has given way to a grid of all-weather concrete roads to carry the in-farm animals and materials, and the delivery and dispatch of feed, fertilizer and produce [Key]. It is the traffic flow that is the determining factor in a sound modern farm layout.

Traffic round the buildings of an arable farm with livestock is considerable at any time of the year, particularly at harvest. A modern layout will have to take account of this. A hard surface round the buildings [1] not only allows free access for vehicles and machinery but also keeps animals clean (important for milking cows). Traffic concentration points include reception and loading areas [2] for grain and potat-oes, the area where silage crops are unloaded from the field into towers or clamps [3] and points where livestock are handled [4]. Few farmers are able to start such a layout from scratch and so must compromise.

5 Traditional farm buildings and layouts are often very old and embody the needs and practices of different epochs. This one belongs to the farming days of the 19th century with modern additions such as the Dutch barn [1]. There are storage buildings [2], an open-fronted shelter at the rear of the yard for bullocks to be fattened on roots and cattle cake in the winter [3], a cowshed [4], a small range of pig sties [5] and stable boxes for horses [6]. These substantial and expensively constructed buildings are often difficult to adapt to modern mechanized agriculture.

6 Modern general purpose buildings consist of a light roof carried on wide beams high enough to admit fully laden vehicles and to allow for storage. The lean-to is an economical addition, needing only one set of uprights. The rest of the weight is taken by the original building. Cladding may be used to keep weather out and as temporary insulation, for bulk storage of potatoes, provided by a layer of straw bales on the inside. Cattle may be given access to concreted yards for bulk feeding.

7 Intensive animal production calls for special housing to save on labour and the cost of feed. Ventilation and heating are both needed to keep the animal at an optimum temperature; otherwise the feeding is affected. With a high concentration of numbers in a small space there is an added risk of infectious diseases and therefore a high standard of hygiene is essential. With hens, egg laying is related to day length; to maintain regularity, hours of artificial daylight are kept fairly constant. Buildings of this kind are normally low and have special arrangements for dealing with the large quantities of manure produced.

The living soil

All forms of life on land depend directly or indirectly on soil [Key]. Soil is a result of all the processes of physical and chemical weathering on the barren, underlying rock mass of the earth that it covers, and varies in depth from a few centimetres to several metres. The depth of soil is measured either by the distance to which plants send down their roots or by the depth of soil directly influencing their systems. In some places only a very thin layer is necessary to support life.

Pedology [2], the study of soil and its unique biological, chemical and physical properties, is a science that first came into its own at the close of the nineteenth century when the Russian geologist Vasilii Dokuchaev (1846–1903) identified the basic determining factors in the morphology – structure or form – of soils.

The organic content of soil

If soil is the outcome of time and weather at work upon rock it remains an unconsolidated mass of inorganic particles until it acquires a minimum organic content and plants take root and deposit their litter within. As the

organic matter accumulates fine humus builds up in the upper soil horizons [1], enriching them chemically and providing an environment for a wide variety of life forms. In the course of time plants, fungi, bacteria, worms, insects and burrowing animals such as rodents and moles reproduce in the soil and thrive in the complex ecosystem of a mature soil [3].

The formation of soil is the result of the interaction of five major elements – the parent rock, land relief, time, climate and decay. The parent rock is the source of the vast bulk of all soil material. During weathering this rock is physically reduced to a mass of gravel, sand, silt and clay. Soil is not always identical to its parent material because of the numerous chemical transformations and physical disturbances it may undergo during its formation.

Land relief is another factor influencing the creation of soils. On steep slopes only thin, dry layers accumulate because of the rapid water run-off. In level, high country, soil layers dense with clay tend to accumulate. Where organic decay is slow, in poorly

drained regions, thick layers of dark organic soils build up. A hillside receiving direct sun will acquire a different soil from that hidden from direct exposure, because of the differences in moisture content. Time is another passive agent in the formation of soils. Young soils have almost no distinct horizon markings, whereas mature ones acquire a well-marked profile that undergoes only minimal changes with the passage of time.

Climate and soil composition

The single most important factor in the development of soil is climate [4, 5]. Water is essential to all chemical and biological change in soil; as it percolates through, it leaches the surface layers (this is "eluviation") and deposits material in the subsoil ("illuviation"). In areas of heavy rainfall the soil undergoes extreme leaching and is rendered relatively sterile. In contrast, excessive evaporation in arid climates results in the building up of salt deposits in the soil.

Temperature directly affects the rate of chemical and biological activity in the soil. In tropical climates where such activity is high,

CONNECTIONS

See also
170 Improving crops

In other volumes
88 The Natural World
98 The Natural World
102 The Natural World

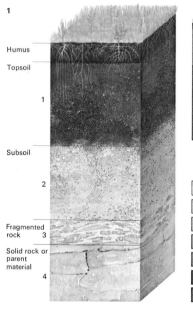

1 The profile of the soil reaches down from the most recently deposited topsoil to the parent bedrock revealing the various recognizable layers or horizons. Beneath a horizon of grey-black topsoil [1] and its upper layer of humus there lies the subsoil layer [2] which, while poorer in organic

materials, is richer in accumulated minerals than both the topsoil and humus above it. Horizon 3 is made up of the partially weathered particles of the lifeless parent bedrock [4] at the bottom. Profile A is of acid brown earth found in temperate climates – this one on sandy rock – and B is a culti-

vated brown earth of the same climatic region. Grey leached podzol [C] is typical of wet, cool climates – for example, the taiga in Russia – while oxisol [D], a thick red soil containing iron compounds, is found in humid, tropical lands where chemical and biological activity are both high.

Humus
Topsoil

Subsoil

Fragmented rock

Solid rock or parent material

A B C D

Leached acid horizon
Organo-mineral horizon
Ploughed or cultivated
Fresh litter and humus
Oxidized iron enrichment
Mineral humus enrichment
Weathered parent material

2 The composition and colour of a soil identifies it to a pedologist. This tundra soil [A] has a dark, peaty surface. Light-coloured desert soil [B] is coarse and poor in organic matter. Chestnut-brown soil [C] and chernozem [D] – the Russian for "black earth" – are humus-rich grassland soils typical of the steppes and prairies of North America. The reddish, leached latosol [E] of tropical savannas has a very thin but rich humus layer. Podzolic soils are typical of northern climates where rainfall is heavy but evaporation is slow. They include the organically rich brown forest podzol [F], the grey-brown podzol [H] and the grey-stony podzol [I] that supports mixed growths of conifers and hardwoods. All are relatively acid. The red-yellow podzol [G] of pine forests is quite highly leached.

3 The soil is a complex ecosystem. A square metre of fertile soil teems with more than 1,000 million

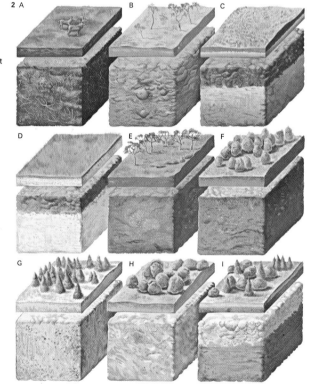

individual forms of life, from microscopic organisms to insects, worms and large animals such as burrowing rodents.

In the steppes, for example, these include marmots, susliks, hamsters and mole rats. All play an important

part in helping to aerate the soil and to accelerate the processes of decay and of humus formation.

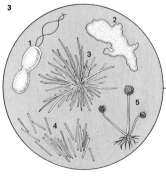

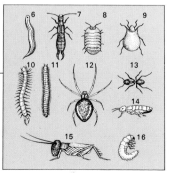

Slug
Snail
Mole
Earthworm

1 Bacterium
2 Protozoan
3 Alga
4 Virus
5 Fungus
6 Eelworm
7 Earwig
8 Woodlouse
9 Mite
10 Centipede
11 Millipede
12 Spider
13 Ant
14 Springtail
15 Cricket
16 Cockchafer larva

decay is rapid and the soils are poor in humus. In tundra regions, where the topsoil is frozen for more than half the year, and the subsoil is permanently below freezing point, the reverse holds true: organic matter accumulates in thick layers.

Lateritic soils are an excellent illustration of the effects of climate. In hot, wet environments such soils are highly leached, and contain little other than deposits of iron and aluminium oxides. If directly exposed to the fierce tropical sun, these soils become a baked, brick-like mass known as laterite to which vegetation can never return.

The importance of decay
A variety of biological factors influence soil formation. Plants stabilize the earth by reducing erosion and surface run-off. They also maintain soil fertility by concentrating organic material and nutrients back at the surface after they have been washed down. As they decay the plants also provide the fine organic humus litter vital to soil life.

The role of soil bacteria is crucial for they not only fix nitrogen from the air in a form

that plants can use but also promote the processes of decay. Animals whose homes are in the soil have an important if largely mechanical function in shifting and aerating the soil – it has been estimated that earthworms alone can turn over between one and ten tonnes of soil per hectare per year. As they eat and excrete the soil they also change its texture and composition.

Under normal conditions soils naturally replenish themselves. Yet where ruinous agricultural practices prevail, soils can easily deteriorate in fertility. A dramatic illustration of this is the dust bowl, a man-made desert from which the valuable topsoil has been removed by the wind in conditions of continuing drought. This is brought about by certain farming practices in regions with unsuitable climates.

Soil conservation is the effort to avoid this destruction of the soil and to maintain it at the most productive level possible. This requires a combination of all the techniques of soil science in preparing the land, irrigating it, fertilizing it and planting the right crops to stabilize the soil and prevent erosion.

KEY

The whole structure of life on earth, with the enormous diversity of plant and animal types – herbivores, carnivores, man himself – is utterly dependent upon a thin, nutrient- and moisture-rich mantle of soil.

4 Soil groups can be correlated with the various climatic and vegetational zones of the earth. Podzolic soils are found in moist, cool climates. Latosols and black or dark grey soils are common to the equatorial regions between the tropics of Cancer and Capricorn. In sub-humid and semi-arid temperate zones of the world, where the land is at its most fruitful, the prevalent soils are chernozem, chestnut-brown and reddish prairie. Grey or red soils are typical soils of the near-barren desert regions of the earth, while sub-polar climates are characterized by the scanty tundra soils bearing only sparse vegetation.

4

Ice

Podzolized soils
Podzol soils (incl. brown)
Grey-brown
Red-yellow
Terra rossa and brown forest soils

Lateritic soils
Latosolic
Black and dark grey

Grassland soils
Prairies and chernozem soils (incl. denegrated chernozems)
Chestnut and brown soils
Reddish prairie, reddish-chestnut and reddish-brown soils

Soils of arid regions
Grey desert soils and red desert soils
Tundra (incl. Arctic brown forest soils)
Undifferentiated highlands

5

Trees
Cereals
Cattle
Sheep
Sparse vegetation
Reindeer
Camels
Goats

Podzol
Grey-brown, red-yellow and terra rossa
Chestnut, reddish and brown
Latosolic, black and dark grey

Tundra
Prairie and chernozem
Grey desert and red desert
Undifferentiated highlands

6
Recycling
Condensation
Input
Output 8
Output
Alcohol burnt for energy

1 Carbon dioxide
2 Oxygen
3 Water
4 Nutrients
5 Energy
6 Heat
7 Light
8 Harvest
9 Waste
10 Potash from burnt waste
11 Alcohol from waste
12 Recycled nutrients
13 Chemically inert support
14 Plastic mesh support
15 Air space
16 Water culture solution

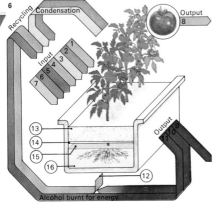

5 Climate exerts the strongest influence on the nature of soil on which plant and animal life depends, and this, in turn, determines which species will flourish best in its particular area. Tundra and desert soils can support only a slender indigenous plant and animal population, but ephemeral life – in the tundra, during its brief summer, or the desert regions after exceptional rainfall – will attract a considerable visiting population of animals. Other soils such as latosol support a wide range of plants and animals: in the African savannas, for example, with their huge populations of herbivores and attendant predators.

6 Hydroponics is the science of making substitutes for soil and has been successful in growing plants in totally soil-less conditions. In hydroponic agriculture all a plant's needs, including oxygen, light, water, mineral salts and other nutrients, are artificially provided within the protected and temperature-controlled environments of a greenhouse. These plants are freed from competition with weeds, and damage by insect pests and viruses is reduced.

167

Water and irrigation

Irrigation is the artificial watering of land. It may be necessary for several reasons: where rainfall is too sparse to raise any crops at all; where it is seasonal or erratic from year to year; or simply where the natural supply of water needs to be supplemented in order to increase crop yields.

Systematic watering has been responsible for the transformation of vast expanses of arid, infertile land to highly-productive soil [4]. Some 162 million hectares (400 million acres) of land are irrigated in the world today.

Ancient irrigation methods
Irrigation was crucial to the development of settled agriculture and the rise of the first great civilizations. More than 5,000 years ago, farmers along the banks of the Nile used its waters to irrigate their fields.

Ancient irrigation systems could be immense, even by modern standards. In Iraq, remnants of a canal 120m (395ft) wide and some 10m (33ft) deep have been traced for miles across the countryside. In Egypt, a channel some 19km (12 miles) long connected the Nile to Lake Moeris, which was

used as a reservoir for the river's regular and life-enriching floodwaters.

Ancient forms of irrigation ranged from simple networks of ditches that were filled from wells and rivers to elaborate systems of dams, reservoirs and canals [3].

Traditional methods
Pre-industrial civilizations possessed and still use a wide variety of devices for lifting water to their fields. The Arabs, for example, use a simple contraption called the *shadoof*. It consists of a long, pivoted pole with a bucket at one end and a weight at the other that helps to counterbalance the weight of the water scooped up in the bucket.

The Archimedean screw, while still only a primitive form of water pump, is, even so, a sophisticated irrigational device compared with the *shadoof*. It consists of a helical shaft housed in a long cylinder. One end of the assembly is dipped below the surface of the water and, when the shaft is turned, water is lifted and disgorged from the other end of the tube on to the land.

Another ancient water-lifting device is

the Persian wheel, consisting of an upright waterwheel linked by gears to a horizontal drive-wheel. A draught animal harnessed to the latter circles endlessly to turn it.

Modern systems
Today, three main forms of irrigation are employed to provide the root systems of plants with a uniform supply of water.

The first is underground irrigation, a method that brings water to plants directly through the soil. This method is practical only where the terrain is level and where the soil is highly permeable and situated over an impermeable layer which traps ground water and permits it to seep upwards to the plants by capillary action. Underground irrigation minimizes water loss due to evaporation. Unfortunately, it tends to deposit unwanted mineral salts at the surface. Such accumulations have to be cleaned from the soil, so heavy rains or deliberate flooding is essential.

The second and most common form of irrigation is from the surface [5]. Land on which this method is used is normally laid out in border strips, barns or furrows.

CONNECTIONS

See also
108 Earth's water supply
148 Misuse of the land

In other volumes
60 History and Culture 1
82 History and Culture 1

1 **Dry farming** means growing crops without irrigation where the annual rainfall is below 50cm (20in). Here olive trees are placed at the centre of shallow earth basins which funnel rainwater inwards.

2 **Drought-resistant sorghum** is farmed in arid regions of Israel. Wide-spaced crops, thorough weeding, and well-tilled and organically rich soil all ensure the efficient use of natural water.

3 **Mesopotamian irrigation systems** like this were vital to the civilizations that grew up in that arid area. Without irrigation, cultivation would not have been possible; without a stable society, the complex of canals, ditches, dams and reservoirs would not have been built or maintained. Hammurabi, king of Babylon (1750 BC), actually drew up an elaborate code of rules for use of water in irrigation.

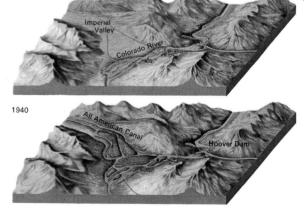

4 **Imperial Valley** in the United States was transformed by large-scale irrigation engineering. It used to be arid [A], like its surrounding area in the Great Basin region, with commercial crops impossible to grow. Now it is an intensely farmed area. In 1905 and 1906 flooding of the Colorado River wrecked first attempts at irrigation. But with the construction of the Hoover Dam (1935), to control the Colorado, and the 130km (80 mile) All American Canal (1940) the valley changed from parched desert to an agricultural garden. A 4,800km (3,000 mile) irrigation network [B] is the area's mainstay; it waters half a million acres and Imperial Valley's productivity is immense [C]. One problem is that water cannot drain away, as the valley is below sea-level. It evaporates eventually, a residue of salt is left and this is a threat to the soil's fertility in future.

The border strip technique entails dividing the land into long, rectangular sections fed by a common supply channel. Water ducted into this ditch flows as a broad sheet over the entire strip, the surplus draining off at the lower end. Border strips are most often used to grow small grains and fodder crops.

Basin irrigation means that water is trapped by low retaining walls around the edge of the field until the earth becomes thoroughly soaked. Rice is the major crop to be cultivated in such a way.

Furrow irrigation is most suitable for row-grown crops like corn or cotton. Furrows up to 500m (1,640ft) in length are ploughed between the rows. They slope very gently away from the water source, so that water running along them does not cause excessive erosion but can soak slowly into the soil surrounding the crops.

The only drawback with surface irrigation is the difficulty of giving all parts of a field an equal amount of water. In the process of ensuring that all parts receive enough, many receive too much and, as a result, valuable water is wasted.

The final watering method is overhead irrigation. Spray lines or sprinklers are used in an effort to simulate natural rainfall. Sprinklers can deliver water in a variety of concentrations – from fine mist to a heavy downpour. They are generally set up in rows and connected by pipes to a central pumping unit. The main advantages of sprinkler irrigation are that the land requires no special preparation and the application of water can be efficiently controlled by the farmer.

A recent innovation has been the use of trickle systems. Water is brought directly to the base of a plant by narrow plastic tubing and steadily dripped from a nozzle. This method is highly efficient in its use of water and returns higher crop yields than any other approach. The high cost of installation is its main drawback.

Just one example is sufficient to indicate how enormously the amount of water needed by a field varies with crop, climate and the nature of the soil. Grain crops that are able to flourish on 46cm (18in) of water in one region may require nearly double that amount elsewhere.

The waterwheel is one of the earliest known devices to have been used in irrigation. Like all the most basic irrigation implements, this was designed to lift water from wells, rivers, pools and reservoirs and transfer it to ditches, which carried it to the fields containing crops.

5 An essential part of any irrigation system [A] must be a high-lying source of water [1]. Lakes and rivers of mountainous regions are obvious choices, with the man-made reservoir [2] as an alternative. A main canal [3] transports water from the primary source to the head of a low-lying, cultivated area [4]. B shows how water brought in by the main canal [3] is ducted into a number of secondary channels [5]. These follow the contours of the land, so permitting a continuous downhill water flow [6], and irrigation without need of any pumping. Usually, irrigated areas are the flat, alluvial floors of river valleys [7] containing settlements [8]. Irrigation of the fields themselves is done in a variety of ways, depending on the lie of the land, the crop and available water. If water is plentiful, fields can be subdivided into basins [C] which can then be flooded as needed. An alternative method is furrowing [D] where controlled amounts of water are allowed in via sluices and siphons. (This is called infiltration.) Natural flow irrigation [E] is best suited to gently-sloping types of field; the water percolates slowly down the slope from an upper channel. If the land slopes very sharply, then terraces can be cut into the hillside as has been done for centuries in China.

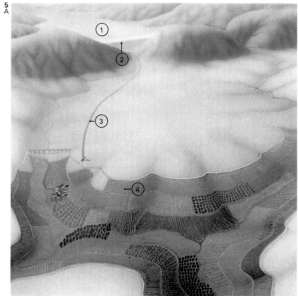

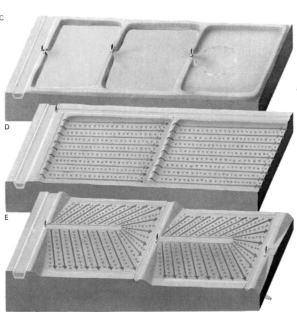

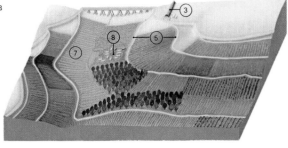

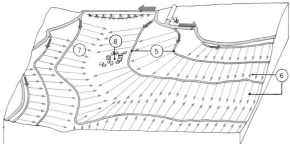

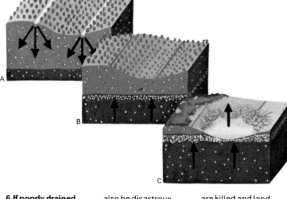

6 If poorly drained land [A] in a dry climate is irrigated, salts in the water used can seriously threaten crops and soil; water containing more than 700 parts per million of salt is directly harmful to plants, but lower levels can also be disastrous. Poor drainage means that the water-table will eventually rise and bring dissolved salts back to the surface [B]. As the water evaporates [C], salt is deposited in the soil in ever-increasing amounts. Finally, the crops are killed and land once fertile becomes sterile desert. Even so, the installation of a drainage system underground and new ditches will lower the water-table once more. The salt can then be washed out of the soil by means of fresh water.

Improving crops

One of the most effective technical measures for increasing the productivity of crops and plants is by the application of fertilizers [Key]. These are substances supplied to plants as nutrients. There are millions of living cells in a plant and to construct and maintain them it has to obtain nutrients from the air and the soil.

Where plant nutrients come from

The plant obtains carbon, hydrogen and oxygen from the air and from water. It then needs nitrogen (N) [1], phosphorus (P) [2] and potassium (K) [3] as primary nutrients, with calcium (Ca), magnesium (Mg) and sulphur (S) as secondary nutrients and a further group of trace elements that are called micronutrients because they exist in minute quantities.

Some of the plant's needs are met by the soil itself. But where the soil is naturally poor, or where the nutrients have been leached by heavy rain or removed by crops, fertilizers can be used to supply the necessary nourishment. Different crops have different needs and the type and amount of fertilizer has to be related to the type of soil and the requirements of the crop.

Some legume crops, such as clover, can obtain nitrogen from the air by the process known as "nitrogen fixation". But most plants need additional inorganic nitrogen and this element is the one most widely used in fertilizers: in some form or other it accounts for almost half of the world consumption of fertilizers. Phosphorus makes up another 30 per cent and potassium, the remaining primary nutrient, accounts for the balance. In recent years world consumption of fertilizers [6] has totalled about 80 million tonnes, but it is very unevenly distributed. Only 15 per cent of this is used in agricultural areas such as China and India, which have to feed 70 per cent of the world's population.

Natural and synthetic fertilizers

In the past farmyard manure was the basis of good farming and it is still an important fertilizer today. Farmyard manure is the solid and liquid excreta of farm animals, usually left for a while in heaps to allow some rotting. Its value lies in both its organic content and in the nitrogen, phosphorus and potassium (with small amounts of magnesium and trace elements) it contains. The organic matter is beneficial to soil structure.

Synthetic fertilizer is most commonly supplied as a powder or in granules but liquids or liquefied gases can also be used. Often it is supplied as a compound fertilizer (known as NPK) containing nitrogen, phosphorus and potassium, in stated amounts, and many countries demand by law that composition should be guaranteed. When the artificial fertilizer contains only one nutrient it is called a straight or simple fertilizer.

Plants that are deficient in nutrients not only have smaller yields but also develop certain characteristic symptoms. Deficiency in nitrogen results in poor plant health; small plants; small, often yellowish-green leaves; and "scorched" lower leaves. Lack of phosphorus may reveal itself in stunted growth, leaves purplish or bronzed at the edges, and in slow-ripening, misshapen fruits. Similarly, leaves with whitish or brown spots, scorched or dead leaves, weak stems and small fruits with splits or injured spots indicate an inade-

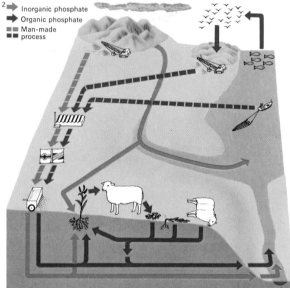

2 Phosphates are derived from granite rocks and can be removed by weather, or they can be mined and converted into fertilizers. Organic sources of phosphate include guano (bird droppings) and fish waste. Phosphates in the soil are taken up by plants, which may be eaten by animals. They return to the soil in dead plants, dead animals and animal excretions. Much phosphate is supplied in compound fertilizer.

■ Nitrogen fixation
■ Nitrate utilization
■ Ammonification
■ Ammonia nitrification
■ Ammonia denitrification
⊙ Micro-organisms

1 Nitrogen, the most important fertilizer, undergoes a natural cycle. Together with its compounds it is involved in five basic processes: fixation of nitrogen from the air by micro-organisms and by lightning; use by plants of nitrates in the soil to make proteins; ammonium compound production in decaying plant and animal matter; nitrification of these to nitrites and then to nitrates; and de-nitrification of ammonium compounds back to nitrogen gas. Nitrogen is removed from the soil whenever man consumes food, but he replaces it by the addition of nitrogenous fertilizers to the soil.

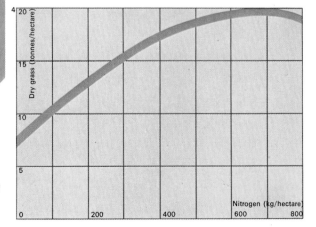

→ Inorganic phosphate
→ Organic phosphate
■■ Man-made process

Much removed in root and leaf crops
Little removed in seed crops
Much supplied as fertilizer
Derived from seawater
Little removed in livestock
Waste
Humification
Held fairly strongly in soil
Potassium
Small amount derived from parent rock (silicates)
Little leaching, except in sand

3 Potassium is an essential plant nutrient. The quantity in circulation is more or less constant, losses being made up by input from rock erosion. But when the plants are physically removed as crops their potassium is lost to the land and needs to be replaced by artificial means; and a soil that is naturally deficient in potassium can be enriched by the application of fertilizer. The major sources of potassium fertilizer are deposits of potash salt (KCl).

4 When nitrogen fertilizer is applied to grassland the dry grass production increases as shown by this curve. Similar increases also occur with other fertilizers for all types of crops. The graph shows that a given quantity of fertilizer will produce a larger output increase on unfertilized or poorly fertilized land than on land already highly fertilized; this illustrates the law of diminishing returns. It follows that giving fertilizer to underdeveloped nations is often better than giving them grain grown on highly fertilized fields, since the fertilizer increases their crop yields more than the donor's.

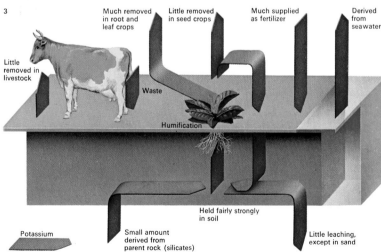

Dry grass (tonnes/hectare): 20, 15, 10, 5
Nitrogen (kg/hectare): 0, 200, 400, 600, 800

quate supply of potassium. These are only rough guides, for many plants have complex metabolic systems in which there may be interactions between an excess of one nutrient and perhaps a deficiency of another. Physical conditions also play an important part in the growth of a plant and for correct assessments soil analysis is vital.

The United Nations Food and Agriculture Organization (FAO) publishes figures relating crop yield to the use of fertilizer. These are based on farm tests, as well as on research carried out in scientific stations, and include studies of the use of different sources of the same nutrient. In India, for example, the FAO has assessed the relative efficiencies of urea, ammonium sulphate, ammonium nitrate and calcium ammonium nitrate as sources of nitrogen. Similarly, different phosphatic materials have been compared as have other important factors such as plant varieties, plant populations, sowing time and use of irrigation, which interact with fertilizer treatments and affect the response of crops. Overall, the FAO, after many years' research, reports an average increase of 58 per cent for

all crops using fertilizers and suggests that many crops could show even higher rises. Fertilizers also improve the yield per unit of water supplied – an important consideration in vast areas of the world where water is in short supply.

In the humid tropics the soil tends to be acidic. Where this is excessive, lime is added to reduce the acidity rather than correct a deficiency. Ground limestone is one of the most effective and cheapest materials that can be applied for this purpose.

Plant hormones and plant growth

The growth of plants can also be greatly influenced by applying materials known as plant hormones [7] or plant growth regulators. Used in very small amounts plant hormones promote, inhibit or modify the physiological processes within the plant and include auxins, kinins, gibberellins and abscisins. Compared with fertilizers they are relatively little used, although they are the subject of considerable research. They can all be produced artificially and their main areas of application are to fruit, vegetables and cereals.

KEY

Treated Untreated

The effect of fertilizer is evident in the growth rates of these kale plants. The one on the left has been treated with a nitrogen fertilizer. The other plant has the stunted growth and discoloured leaves common in plants lacking nitrogen. For healthy growth, plants need ten basic kinds of food. Carbon, hydrogen and oxygen are provided by air and water. Iron, magnesium and sulphur are usually in the soil. The other four important elements, nitrogen, phosphorus, potassium and calcium, may be in the soil – or may need to be added as fertilizer in the form of minerals, manure or man-made chemicals.

5 An aircraft flies low to distribute fertilizer over New Zealand farmland. The use of aircraft in agriculture is especially advanced in New Zealand where hilly pastures need fertilizer in order to produce more grass for livestock. Aerial top-dressing is a quick and efficient method of spreading fertilizer over hilly or extensive areas where distribution in more conventional ways would be difficult and costly. The fertilizer is loaded mechanically into the plane's hoppers. The plane is then flown low over the distribution area and the pilot releases the fertilizer, which spreads evenly in controlled amounts.

6 The deposits of fertilizer minerals – phosphates [yellow], potash [blue] and sulphur [red] – are unevenly distributed. The major phosphate producers are the USA, the USSR and Morocco. The major potash producers are again the USA and the USSR closely followed by West and East Germany and France. Natural nitrates have been superseded since 1914 by chemically produced nitrates that use atmospheric nitrogen. The process requires sulphuric acid and the latter's raw material is sulphur. The world's major sulphur producers are Canada, France, the USA, the USSR and Japan.

6

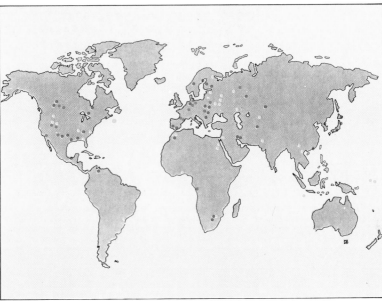

7 Plant hormones are chemicals that occur naturally in very small amounts in plants, where they control growth. They are used extensively in crop production. Their effect on plant cells and the uses to which they are put are shown here. Auxins [A] aid cell division and elongation and are used to stimulate fruiting. Ethylene [B] is a gas that acts as a hormone ripening fruit by partial breakdown of the cell wall. Gibberellins [C] make cells longer and are used to improve fruit quantity. Cytokinins [D] maintain nutrient levels in cells and help to preserve cut blooms. Chlorocholine chloride [E] inhibits cell elongation and causes dwarfism.

5

7 A

B

C

D

E

Protecting crops

Pests and diseases are estimated to destroy up to one-third of the world's agricultural harvests each year. Crops can be ravaged at any stage during growth, reaping and storage. The problem is most severe in the developing countries. For example, the Australian yield of rice is 6.44 tonnes per hectare, while in India the yield averages only about 1.62 tonnes per hectare. (In some African countries the corresponding figure falls as low as 0.5 tonnes.) Although there are many reasons for these substantial differences in yield, the infestation of crops by insects, other pests and diseases and the competition for vital nutrients by weeds and useless grasses are major contributing factors.

Increasing food production

In agriculture a pest is any organism that damages or destroys plants useful to man and the term is taken to apply to harmful insects, mammals and birds. Crops also suffer from micro-organisms that cause disease and competing wild plants or weeds.

The important groups of pesticides are insecticides, fungicides and nematicides – for the eelworms which are microscopic nematodes feeding on plant roots. Herbicides are chemicals used for weed control. In tropical and subtropical areas of the world it is the insecticides that tend to be more important. But in the USA (which uses as much agricultural chemicals as the rest of the world together) herbicides are more widely applied than pesticides.

The amounts of pesticide needed over a unit area are small and they are usually applied diluted from commercial formulations by spraying and dusting. Alongside improvements in fertilizer use, farming techniques and development of better crop varieties, pesticides and herbicides have helped to increase food production and benefited food storage, forestry and gardening.

Although chemical insecticides had been used earlier, it was the discovery of the insecticidal properties of DDT in 1939 that marked the dramatic growth of the use of synthetic organic pesticides. One of the most remarkable examples of enormous gains in yield brought about by an intense and sustained campaign of pest control is provided by Japan. At the end of World War II, Japanese rice output was a little over 1.6 tonnes per hectare. Since then rice cultivation has been transformed into high yield, high quality production giving an average output of almost 6 tonnes per hectare.

Harmful side-effects

Chemical pesticides, despite their great benefits, have their drawbacks. In some cases insecticides have become less potent because the insects developed resistance to their effects [2]. In other cases the insecticide has affected the animal enemies of the harmful insects with the result that the insects have actually multiplied in number. A further possible harmful effect occurs when a persistent pesticide is eaten by a predator as it consumes its prey and then is eaten in turn by the animal preying on that predator. In this way the pesticide passes up the food chain [3] and may accumulate dangerously in the higher animals – including man. Emphasis is now shifting from developing more effective methods of purely chemical control to so-called integrated approaches which attempt

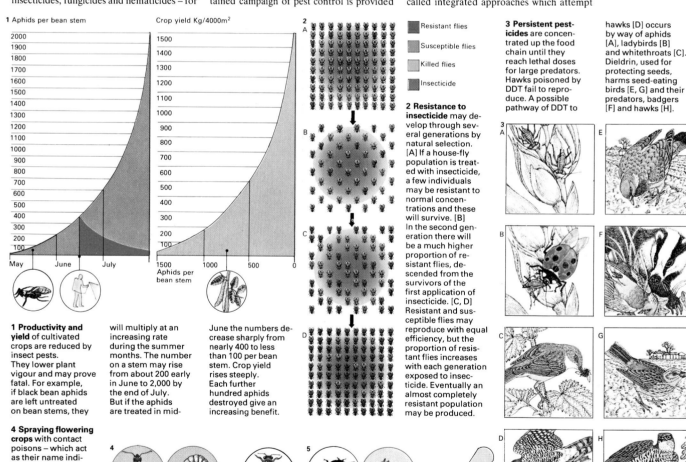

1 Productivity and yield of cultivated crops are reduced by insect pests. They lower plant vigour and may prove fatal. For example, if black bean aphids are left untreated on bean stems, they will multiply at an increasing rate during the summer months. The number on a stem may rise from about 200 early in June to 2,000 by the end of July. But if the aphids are treated in mid-June the numbers decrease sharply from nearly 400 to less than 100 per bean stem. Crop yield rises steeply. Each further hundred aphids destroyed give an increasing benefit.

Resistant flies
Susceptible flies
Killed flies
Insecticide

2 Resistance to insecticide may develop through several generations by natural selection. [A] If a house-fly population is treated with insecticide, a few individuals may be resistant to normal concentrations and these will survive. [B] In the second generation there will be a much higher proportion of resistant flies, descended from the survivors of the first application of insecticide. [C, D] Resistant and susceptible flies may reproduce with equal efficiency, but the proportion of resistant flies increases with each generation exposed to insecticide. Eventually an almost completely resistant population may be produced.

3 Persistent pesticides are concentrated up the food chain until they reach lethal doses for large predators. Hawks poisoned by DDT fail to reproduce. A possible pathway of DDT to hawks [D] occurs by way of aphids [A], ladybirds [B] and whitethroats [C]. Dieldrin, used for protecting seeds, harms seed-eating birds [E, G] and their predators, badgers [F] and hawks [H].

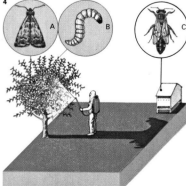

4 Spraying flowering crops with contact poisons – which act as their name indicates by contact with susceptible insects – during the day kills all insects, including bees which pollinate the flowers. If the same poison is sprayed in the evening, when the bees [C] have returned to their hives, only the pests such as the codling moths [A] and caterpillars [B] are killed. The poison is made of a chemical that decomposes in a few hours.

5 Systemic insecticides are chemicals that have the remarkable properties of being harmless to plants that absorb them, lethal to the herbivorous insects that feed on the plants and harmless to their predators. These insecticides, sprayed on leaves or soil, are absorbed by the plants, then act as stomach poisons to pests such as the aphids [B] which are sap suckers but leave ladybirds [A], which prey on the aphids, unharmed.

to combine profitable production with minimum disturbance of the environment. This environment is not, of course, the original ecological system but the landscape as altered by man for producing agricultural and forest crops (the "agro-eco-system").

Government laboratories in many countries have been reappraising methods of pest control. The UK Ministry of Agriculture, Fisheries and Food issues lists of products approved for use in agriculture and food storage, and of those for use in gardens. Jointly with the manufacturers, it has compiled a code of conduct to guide farmers, agricultural advisers and conservationists.

"Biological" controls

In the fight against insect pests there are now several methods which, unlike large-scale chemical spraying, lead to minimal environmental damage. These include biological control, sterilization, the use of pheromones, traps and "antifeedants".

Biological control makes use of the pest's natural enemies to regulate its population density. Sterilization of males is done by radiation or by exposure to "chemosterilants"; when the captive males are released and mate with the females the resulting eggs are infertile. Pheromones are "sex attractants" produced by females to enable the males to locate them. When certain specific pheromones are artificially released, in small quantities, the males can be disoriented – thus preventing the meeting of the sexes. Pheromones are also used to bait traps [6]. Some traps use light or other baits to lure the insects. Others prevent their escape or kill them by poisoning or irradiation. Antifeedants are synthetic chemicals that cause insects to remain on treated foliage without feeding. They eventually die from exposure, predation or starvation. Vertebrate pests – rodents, foxes, rabbits and birds – pose some of the most difficult problems. Poisons have been used extensively, but they can be dangerous, expensive and harmful to the environment. The introduction of predators brings its own problems because some pests keep down other pests; in North America, for example, coyotes, although pests, help to keep down the mouse population.

KEY

Field tests are carried out to establish the value of proposed treatments in protecting crops against various kinds of pests. A section of a field had been treated with a selective weed-killer of the DNBP type to control yellow charlock in peas. Different selective weed-killers are used at various stages of the growth of the plant for best effects. For example, the triazine group of weed-killers are applied before the peas are through the surface of the soil, while DNBP – which kills broad-leaved weeds – is used when the peas are 5cm (2 in) to 8cm (3in) tall.

6 Insect traps play an important role in pest control. One successful type uses a synthetic form of a sex attractant or pheromone which can be dispersed in a water-filled container over-spread by a film of oil. Attracted by the scent, which is the same as that produced by a female, the male homes in on the trap and is drowned. These traps have given excellent results in African rice fields.

8 Crop diseases are produced by viruses, fungi and bacteria. Virus infections are the most difficult to treat. They are often spread by insects and prevention consists in destroying both the infected plants and the insect carriers. Tobacco mosaic virus (TMV) infects tobacco and tomatoes [A]. It also attacks the leaves and fruit of many other plants and causes a drop in production. TMV is extremely contagious and is carried on the fingers and clothes of growers. Bacterial and fungal diseases will respond to chemicals. Bacterial infections cause galls, wilts, spots and rot. The halo blight of beans and peas [C] is caused by a fungal infection; it causes the pods to spot and rot. Fungi are responsible for the greatest number of plant diseases. Downy mildew (*Bremia lactucae*) [B] is a common fungus disease of lettuces. It is controlled by regular applications of fungicide.

7 Traps are widely used for destroying mammal pests, which range from elephants to rodents. Traps are either bait tripped or tripped as they are walked on or through. The tong type [A] is a spring-loaded mole trap with jaws which are placed across the mole run. The Fenn trap [B] is a walk-on trap for grey squirrels, stoats, weasels, rats, mice and other small rodents. It is intended to be a humane trap, which will kill the victim instantly. The snare [C] is an ancient trap design used to catch animals from antelopes down to birds. It is used not only for pest control but also by hunters and poachers.

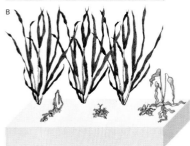

9 Physiological processes within plants can be affected by organic compounds other than nutrients. Some of these are herbicides akin to auxins, the so-called plant "hormones". If these are sprayed on they produce uncontrolled growth and, eventually, death. Their toxicity depends on how readily they are absorbed and also on the plant's stage of development. Broad-leaved plants absorb more readily than narrow-leaved cereals. [A] MCPA spray is applied to broad-leaved weeds [1] and to spring oats between the 3-leaf stage [2] and the beginning of jointing [3]. Results [B] show that the spring oats are unaffected but all the weeds have died.

173

Organic farming

On an organic farm the farmer raises crops naturally, without using the inorganic fertilizers and chemical sprays that have revolutionized farming throughout the developed countries of the world. An organic farmer feeds the soil naturally and so provides the energy to feed the plant life [5]. He uses, therefore, only manures that he is able to produce, while conventional farmers depend on finite resources of minerals and fuels for inorganic fertilizers and pesticides.

Conserving nature's balance

Organic farmers seek to avoid disturbing the balance of nature and they also fear the effects of pesticide residues in human food. Spraying against pests may be successful in the short term but the practice has been known to upset the balance of insect life.

Farmers cannot be divided into two rigidly distinct camps – organic and conventional. While most organic farmers follow the same basic principles of ley farming (crop growing and pastures in alternate periods), sub-soiling and shallow cultivations, composting and a high level of self-sufficiency,

some of them have adopted varying levels of modern chemical technology. Many use chemical weed-killers to replace the hoe because farm labour is scarce. Manpower is one thing the organic farmer cannot do without; his manuring demands more time and effort than for spreading fertilizer.

The organic farm functions best if the farmer has a mixed system [1] with livestock grazing grass and other "break" crops (those that allow the soil a "break" or rest from cereal cropping) supplying manure [4] and slurry in place of bagged fertilizer. Stockless organic farming is possible but the crop rotations [2] need skilful planning, making use of legume crops such as peas and beans.

Looking after the soil

A mixed farmer is more self-sufficient than a specialist. For example, unlike many pig and poultry producers, he has a use for manure; he spreads it back on the land.

Care of the soil is the cornerstone of organic farming. Soil animals and micro-organisms improve soil structure, release plant nutrients and are claimed by some to

combat soil-borne disease. To keep these soil workers active, the farmer supplies organic manures and cultivates carefully to maintain them in the top 10–20cm (4–8in) of soil where most fertility-building work is done. Deep-rooting herbal leys bring up a broad spectrum of nutrients from the lower soil levels. They give nutritious food for stock and improve soil structure. When the leys are ploughed up they provide large quantities of organic matter. A five-year deep-rooting ley may generate enough organic material to grow about three cereal crops without adding chemical fertilizers.

Lower crop yields are obtained on organic farms compared to those where chemical fertilizers are used, and some organic farmers purchase manure from outside to boost their productivity [3]. Arable crops (those that entail ploughing and planting) encourages weeds, and the refusal to treat these with chemical weed-killers prevents the adoption of a crop rotation with a predominance of arable crops – another reason for the comparatively low production.

Some crops are risky on organic farms.

CONNECTIONS

See also
172 Protecting crops

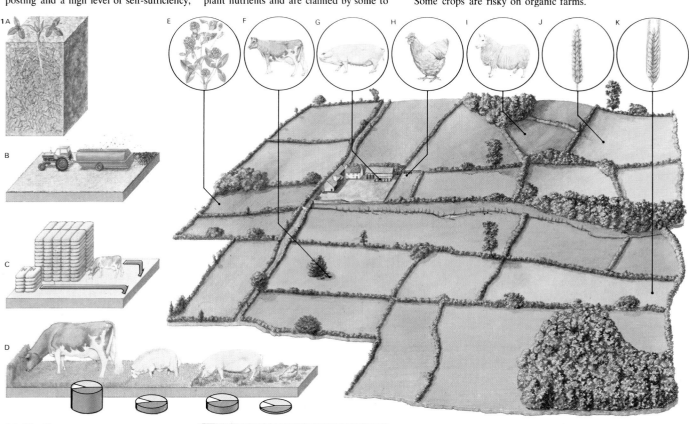

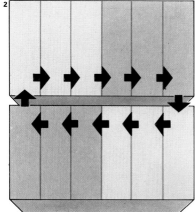

1 Soil fertility on an organic farm is maintained by rotation of crops. Legumes [A] fix soil nitrogen. Manure is spread on the stubble [B] after harvest. Micro-organisms reduce this compost and turn it into topsoil. A tonne of barley [C] fed to livestock can yield as much as 20kg (44lb) of nitrogen. Excreta of livestock returns nutrients [D] extracted by growing crops. Excreta contains a proportion of 120 parts of nitrogen [blue] to 45 parts of phosphorus [white]

and 150 parts of potassium [red], as shown above Organic farmers claim that when these are added inorganically the conventional farmer is oversimplifying nature. Clover planted in pastures [E] increases nitrogen in the soil. Cattle [F] graze on and manure pastures. Pigs [G], hens [H], sheep [I] all contribute to the farm's self-sufficiency. Wheat [J] and barley [K] feed livestock in winter and provide grain for milling and for malting.

2 An organic farmer rotates his crops according to whether or not his farm is mixed, and he adapts practices to the needs of the soil and to the weather conditions. Organic farming is best suited to a mixed farm system. Livestock manure is spread on the arable fields as natural organic fertilizer. In practice, to obtain reasonable yields the organic farmer usually has to buy in extra manure. A rotation system is normally

based on a three- or four-year deep-rooting ley, grazed by at least two kinds of livestock whose various grazing habits help build fertility. Ley mixtures on an organic farm include a wide variety of grasses, clovers and deep-rooting herbs such as chicory and yarrow. Often a ley is followed by two or three years of cereal crops. Variations on this rotation may be a wheat, oats and barley sequence with a break crop of a

legume such as field beans. In the diagram the rotation is for a six-year cycle that is then repeated. It may also use green manure crops such as rape and mustard and green crops grown between cereal crops. Before manufactured fertilizers were provided, farmers had to follow a rotation. In the Middle Ages, many followed a three-year rotation of autumn-sown cereal, spring-sown cereal or legume, and a year's fallow to allow the land time to recover.

Unsprayed potatoes may succumb to blight, sugar beet to virus yellows and beans to aphids. But with cereals and fodder crops a healthy spontaneous ecological balance may be obtained through "unforced" growth and a healthy soil micro-organism population.

Down on the organic farm

Organic farmers plan their holdings in various ways [2]. A farmer with a mixed farm of 80 hectares (200 acres), for example, may employ a team of five workers (three of them part-time). The lower yields are compensated by the savings on fertilizers and pesticides, although this means more work.

The farmer's herd of 30 Jersey cows gives top-class milk and grazes rich pastures containing clovers that "fix" atmospheric nitrogen. Following the cattle, 20 ewes and their lambs graze the pastures and replenish them with their droppings. The livestock includes 200 free-range hens and some pigs.

On the farm there are nine hectares (20 acres) of milling wheat, grown after beans. The beans, being legumes, enrich the soil

with nitrogen. They yield 3,200 to 3,800kg per hectare (25 to 30cwt an acre) and are fed to the cattle. The wheat yields 4,800 to 5,000kg per hectare (38 to 40cwt an acre). Wheat is followed in rotation by oats which are fed to the stock, and the next crop is barley for malting. The barley yields 4,500kg per hectare (35cwt an acre).

The farm has 3.2 hectares (8 acres) of woodland, plus ponds, thickets and hedges which harbour pest predators and pests, thus helping to provide a wider natural habitat for insects and wild plants. The farm also grows mangels and hay and all stock feed is home grown; only salt blocks are bought in. The farm almost meets the organic farmer's ideal of self-sufficiency and therefore outside economic pressures are kept to a minimum.

Farming organically can relieve the demand on the dwindling stocks of chemical fertilizer raw materials (including energy) and reduce the environmental problems created by many fertilizers and pesticides. However, more research is needed before food can be produced organically on the same scale as modern conventional farming.

Muck spreading is a vital operation on most organic farms. Manure returns to the soil some of the nutrients extracted by crops. Soils must have sufficient nitrogen, phosphorus and potassium, and have smaller amounts of calcium, magnesium, sulphur and sodium, and traces of copper, zinc and boron.

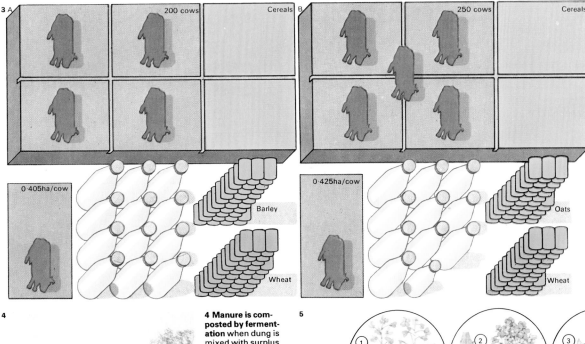

3 A 200 cows Cereals B 250 cows Cereals
0·405ha/cow
0·425ha/cow
Barley
Oats
Wheat
Wheat

3 The lower productivity of an organic farm may be offset by the savings on chemical products. A typical conventional farm [A] consists of 240 hectares (600 acres) of which one-third is in cereals, producing 5,000kg/hectare (40cwt/acre) wheat and 4,500kg/hectare (35cwt/acre) barley. The farm has 400 cows, at a density of 0.4 hectares (one acre) per cow producing 5,500 litres (1,200 gal) of milk daily. An organic farm of the same size [B] can support 500 cows, but at the lower density of 0.42 hectare (1.05 acres) per cow, producing 4,800 litres (1,050 gal) of milk. 4,500kg/ha (35cwt/acre) wheat and 4,700kg/ha (38cwt/acre) oats are produced.

4 Manure is composted by fermentation when dung is mixed with surplus straw and vegetable matter. The compost heap is well watered and is turned several times to ensure thorough mixing.

5 Organic farmers feed the soil which then feeds its plant life. The farmer may leave his land under grass for many years [A], allowing earthworms and bacteria to improve the soil structure. He may plough in green manure crops [B] such as rape [1], mustard [2] and lupins [3]. The soil animals break it down into humus, enriching the topsoil. Animal manure, mixed with chopped straw saved from cereal crops [C], can also be spread on the surface.

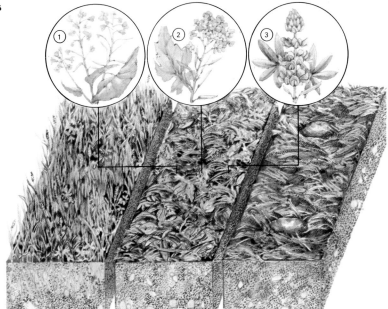

Plant genetic resources

Since the earliest days of agriculture man has been trying to breed better food plants for himself and his domestic animals. For thousands of years he did this simply by saving seed from the best plants of the harvest. Only in the twentieth century has this mass selection been supplemented with planned breeding programmes.

Genes and their actions

The success of scientific plant breeding has depended largely on a knowledge of the inheritance of characters. The unit of inheritance is now accepted as the gene; every plant (and animal) cell possesses vast numbers of genes grouped together on chromosomes and made of the chemical deoxyribonucleic acid (DNA). It is on the plant's genetic make-up (genotype) that its mature appearance depends and the environment modifies the maximum potential of the genotype.

There are so many genes in each plant that the potential variability in the characteristics conveyed from parents to offspring is enormous. The science of tapping this potential is still in its infancy, however, as is success in using plant genetic resources to produce a "green revolution". Simple selection is often not possible and new varieties depend as much on agricultural practice as on the geneticist [2, 4].

The aims of modern plant-breeding schemes almost invariably include an increase in yield. The plant breeder may also wish to improve such genetically determined characteristics as nutritional value, the time it takes the plant to mature, resistance to disease and tolerance of unfavourable climates [Key]. Or he may aim for crops that grow to the same height and mature at the same time so as to be harvested mechanically.

The plant-breeding programme

In nature plants usually change through alterations in their genetic material, called mutations, or through the crossing or hybridization of different species or varieties, as happened with wheat [1]. Modern plant breeding is largely based on planned hybridization combined with rigorous selection in an attempt to speed up evolution.

Experiments to evaluate the respective influences of genes and environment are usually the starting-point of the plant-breeding programme. The characteristics easiest to assess are those that are governed by one or a few major genes, such as the dwarf stature of beans. Characteristics determined by the cumulative actions of many genes are harder to study and control experimentally.

Individual plants selected for the programme are used as parents in controlled mating schemes. The schemes depend on the type of fertilization or pollination mechanism that is employed. Wheat, rice, tomatoes, beans, peas and many other important crop plants are self-pollinated; that is, the pollen that fertilizes the ovule to give rise to seed comes from a flower on the same plant. In self-pollinating species the pollen and ovule share many common genes and their offspring are said to breed true. True-breeding plants are easy to control; individuals in a crop are uniform in size and quality.

Cross-pollinated or outbreeding crop plants, in which pollen and ovule are from different parents, include maize, clover and cabbage. Although these plants rarely breed

CONNECTIONS

See also
224 Farm stock breeding and management

In other volumes
30 The Natural World

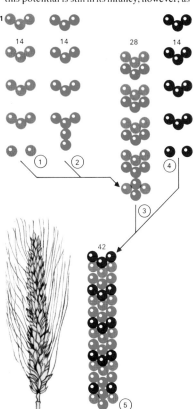

1 Man's daily bread is based on the accidental hybridization of wild grasses in the Near East more than 10,000 years ago. The wild wheat (*Triticum monococcum*) [1], and goat grass (*Aegilops speltoides*) [2], both with 14 chromosomes, had natural mechanisms that enabled their seeds to be scattered by the wind before harvest. A fertile hybrid between them was Emmer (*T. dicoccoides*) [3], which found its way into cultivation, although the seeds were quite easily airborne. A second genetic accident involved Emmer and the goat grass *A. squarrosa* [4], resulting in a full-eared bread wheat (*T. aestivum*) [5], with 42 chromosomes. The natural seed-dispersal mechanism of wheat was bred out of the new hybrids (from which modern wheats derive) to stop the grain from dropping off the ears before harvesting. The hybrids' survival and spread were entirely man's work.

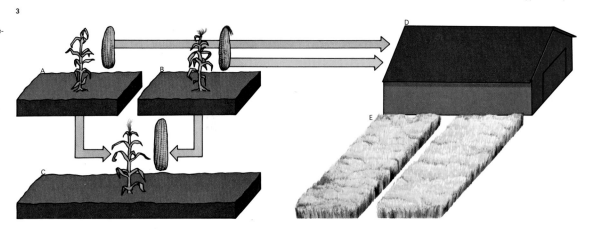

2 Average farm yields still lag a long way behind the genetic potential of existing plant varieties, even in the most advanced agricultural areas. It has been calculated that in northwest Europe wheat could, under ideal conditions, yield up to 15 tonnes per hectare, barley nearly 14 tonnes and potatoes 95 tonnes. The average yields for EEC countries in 1974 were: wheat 4.02 tonnes, barley 3.97 tonnes and potatoes 28.3. Under experimental plot conditions, potato yields near to the potential have been obtained, but the highest cereal yields so far on a field scale are little more than two-thirds of the potential. Achievement of this ideal will depend on planting of the best varieties, adequate and balanced provision of fertilizer, absence of disease and of competition from weeds and the right balance of sun and moisture to suit the type of crop.

3 Gene banks are being established in many countries to preserve basic material for breeders. For a simple cereal crossing, inbred strains [A and B] are tested for hybridization [C] and stored in controlled conditions [D]. They are studied periodically and if germination falls below 95 per cent they are re-crossed to renew the stock. Parent material for crossing is maintained by growing to maturity inbred strains from cuttings [E].

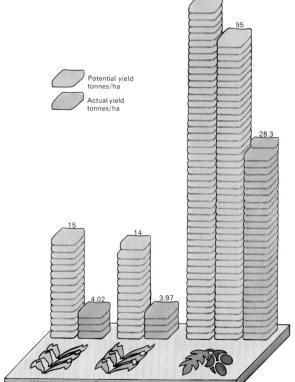

true, their first generation of offspring, known as F_1 hybrids, show particular vigour and uniformity compared with their parents. Increasing commercial use is being made of this so-called hybrid vigour, especially with maize, in which hybridization is relatively easy to control because the male and female flowers are on different parts of the plant. In cross-pollinated species the production of hybrid varieties and mass selection are the two most important methods of plant breeding. Cross-pollination may also be used to hybridize different varieties whose genetic make-ups go well together, thus producing "synthetic" varieties.

In self-pollinated species mass selection has largely been replaced by pure-line selection and hybridization. In pure-line selection, which is chiefly used in the development of new crop species, plants with desirable characteristics are selected to produce offspring and these progeny allowed to self-pollinate for several generations.

Hybridization in self-pollinating plants initially involves the selection of parent stock and their cross-pollination. Each of the chosen parents usually possesses desirable characteristics lacked by the other. The progeny of the cross form the F_1 generation and the best F_1 plants are similarly paired off (out crossed) to produce an F_2 generation. The plants of the F_2 and succeeding generations are generally allowed to self-pollinate and it may take until F_{15} generation for really excellent plants to arise.

Throughout a plant-breeding programme meticulous records must be kept and plants tested in trials against other varieties in a wide range of conditions. From many thousands of crossings only one new line is likely to survive to the F_{15} stage.

Hopes for the future

Current research and breeding programmes have many aims but one of the most important is the increased and improved breeding of protein-rich soya, a tropical species, in temperate areas. Other research is directed towards the development of a greater range of plants able to fix nitrogen from the air and incorporate it into proteins. The only plants able to do this naturally are legumes.

Semi-dwarf wheat such as the Cambridge-bred Maris Fundin is the result of crossing high-yield dwarf wheats bred in Japan and Mexico with European varieties. The short stalks of these varieties help them to withstand storms of wind and heavy rain just before the crop is harvested.

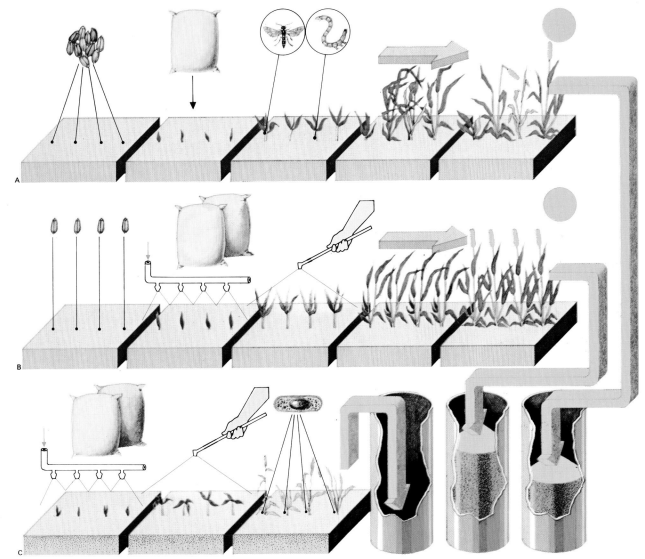

4 "Green revolutions" depend for success on more than simply introducing improved seed and farming techniques, as in this illustration.

By traditional methods [A] diverse crop strains were sown and fertilizer used modestly; some crops were affected by disease and tall plants could be blown over before harvest. Final yields were just adequate. At B a new selected strain has taken over. It has been heavily fertilized and sprayed for pests and diseases. Low-growing plants stand well to a full harvest. Yields are doubled, but the situation does not last. Eventually the continuation of this treatment [C] allows new strains of disease-producing organisms to flourish and kill the crop. Soil may also deteriorate as spray and fertilizer residues build up. Bad-weather yields drop below the original average [A]. Local trials are essential before new varieties become dominant. Even in advanced farming areas no one variety should take more than one-third of the area planted in that crop.

Plant propagation

Flowering plants reproduce themselves by means of seeds and many, in addition, by vegetative means that bypass the normal sexual process. Given the appropriate stimulus, plant tissues, apparently specialized by virtue of their position and function, can change their character. Stem tissue or even leaves, for instance, may develop roots and this adaptability has long been exploited by man. Some wild species spread as much by vegetative means as by seed. Examples include such troublesome weeds as couch grass (*Agropyron repens*) and ground elder (*Aegopodium podagraria*), which form new plants at the joints of underground rhizomes. Such monocotyledons as tiger lily (*Lilium tigrinum*) form bulbils in the axils of their leaves, or – like some onions (*Allium* sp) – in place of a normal flower head.

Growing from clones

A group of plants can be grown vegetatively from a single specimen. Such a group is genetically homogeneous and is called a clone. Every named potato, apple or rose is a continuation from the original plant, which began as a hybrid seed. A hybrid may not be able to produce fertile seed itself; but even if it can, a clone (developed by grafting or budding) is less variable and more predictable than its parents. Grafted shrubs and trees can be brought to maturity more quickly than seedlings.

A potato tuber consists of a large mass of food storage cells, with cells adapted to growth just below the skin at the "eyes". Any small piece containing an "eye" is capable of developing into a new plant [5]. In fact botanists have produced new plants from microscopic cell material taken from the extreme point of a growing shoot. These cells will be free from any virus that may be present in the rest of the plant and can be grown on to develop a virus-free clone. Many crops have been improved by this method, notably the King Edward variety of potato.

Stem cuttings are a convenient way of multiplying many ornamental species, timber trees (such as willows and poplars) and shrubs [17]. The process of rooting is accelerated by the use of synthetic plant hormones, which encourage a change in the function of the cells. Leaf cuttings are used to multiply species of gloxinia and of those begonias [21] grown for their decorative foliage. If the gardener removes the midrib of the leaf and fixes the leaf to a suitably moist base, new plants develop at the inner ends of the cross-ribs.

The family Rosaceae, to which much of our fruit belongs, lends itself well to vegetative multiplication. Strawberries [8] throw out runners from which new plants develop. In the genus *Rubus*, raspberries (*R. idaeus*) spread by way of underground rhizomes. Blackberries (*R. fruticosus* and sub-species) and their hybrid the loganberry, which are climbing plants, will root wherever a trailing stem touches the ground; the deliberate encouragement of this form of growth is known as layering [9].

Growing from grafts

Tree fruits of the rose order (apples, pears, plums, peaches, apricots and cherries) develop quite freely from seed but the seedlings vary in quality. Practically all those in cultivation are clones – often of a chance seedling, such as Cox's Orange Pippin. A

CONNECTIONS

See also
176 Plant genetic resources

In other volumes
70 The Natural World

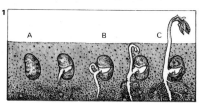

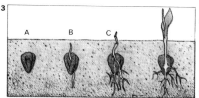

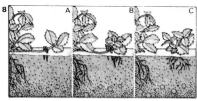

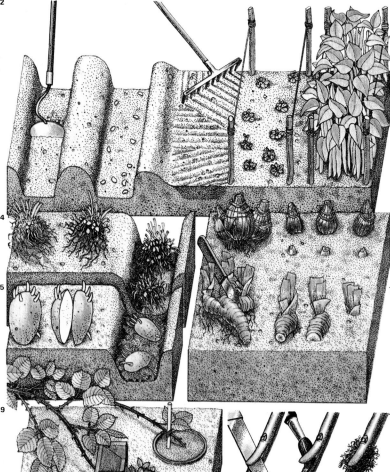

1 Dicotyledons, such as the bean [A], have two seed leaves as food stores for the embryo that lies at their junction. In the ground these cells divide, splitting the seed coat [B] and begin to differentiate into shoot and root. The shoot grows to the surface, carrying seed leaves with it, while the root grows down. At C the root has developed rootlets for anchorage and nourishment.

3 Monocotyledons such as maize [A] have only one internal food body from which the single cotyledon (seed leaf) feeds. A single shoot develops [B, C] but the cotyledon remains below ground.

4 Fibrous rooted perennials, such as phlox, multiply when the roots are cut in two and planted out.

5 Seed potatoes are planted out cut into pieces that each contain at least one eye.

8 Strawberries and some related species spread and propagate mainly by runners (specialized stems each carrying a miniature plant at its end). Nourished at first by a parent plant, runners later form roots of their own for independent growth.

9 Blackberries will root naturally where trailing or damaged shoots touch the soil surface. Gardeners assist the process by weighting down the shoot or pegging its tip into a pot of prepared soil. Loganberries multiply the same way.

2 French beans growing in a garden are cultivated in stages. Drills are cut in the soil with a hoe and in these are sown bean seeds at regular intervals with a few reserves at the end of the row. When the plants emerge they are given support at the side, for their growing period, as on the right. Some beans, like peas, are self-fertile; others are cross-pollinated naturally with the assistance of insects.

6 The bulbs of the daffodil (*Narcissus* sp) develop round those that have flowered. They may be removed and planted separately. For commercial multiplication they are lifted after two years in the field.

7 German irises form a mass of rhizomes near ground level. To grow them best these should be broken up after flowering and replanted singly. Irises include both rhizomatous and bulbous species.

10 Strawberries may be multiplied simply by pegging their runners into peat pots sunk in soil. Well-grown young plants may be lifted when rooted and replanted without removal from the pots.

11 Air-layering is used for some climbing plants whose stems produce roots for support. The stem is nicked [A] and the slit [B] kept open. When roots [C] form they are wrapped in a peat-based medium [D-F] until they grow stronger.

nurseryman grafts a cutting on to a young tree of the same or closely allied species. The type of stock he uses will often determine the rate of growth and the size of the mature tree or bush. In making a graft it is essential to align the cambium layer (a thin layer of stem tissue-forming cells) under the bark of both scion (the new plant material) and stock (the host plant), since it is the active cambial tissue that fuses and that is subsequently responsible for the "in step" production of water- and nutrient-conducting tissues [16]. The scion is normally a short length of twig of the desired type, trimmed to fit exactly into the prepared stock. The practice of grafting trees and shrubs has a long history, dating back to classical times, when it was used particularly for propagating vines.

Propagating roses
Budding, as practised with roses, is a variant of the same technique [12]. A bud of the desired clone, with a small spur of stem sheath, is inserted on to the exposed soft stem of a young briar (wild rose) stock. The species most commonly used is the very variable dog

rose (*Rosa canina*). A standard rose is budded on young lateral shoots at the desired height, close to the main stem of the briar. For a bush rose the budding is done at ground level on two-year-old briar seedlings. Once the bud comes into growth, the gardener trims away the rest of the briar to concentrate the plant's whole activity into the developing scion. He may also graft stem cuttings to the briar; this is in fact normal for ramblers and some other climbing types.

Much research has been carried out in recent years into the selection and breeding of stocks for grafting and budding and the range is now large. Horticulturists have paid particular attention to the production of dwarfing stocks for modern orchards. The dwarf trees permit closer planting with less labour when the time comes for pruning and picking the fruit.

Root cuttings may be used for the propagation of some herbaceous plants [19]. Included among them are species of *Acanthus*, *Anchusa*, *Romneya* and, in the kitchen garden, horseradish (*Armoracia rusticana*) and seakale (*Crambe maritima*).

Pollination is vital to the production of normal fruits. The process is usually left to insects, but they are not always to be relied upon in glasshouses. The family Cucurbitaceae – which includes marrows and melons – offers difficulties because male and female elements are carried by separate flowers whose number may not balance. To ensure that the greatest number of female flowers are fertilized, gardeners generally resort to hand pollination [C]. Pollen from the male melon flower [B] is transferred by hand to the stigma of the female [A]. Hand pollination is also used with marrows, but not with cucumbers, which are best eaten without seeds.

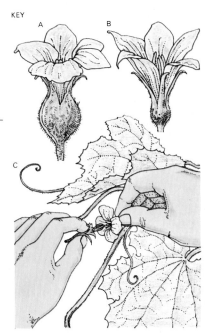

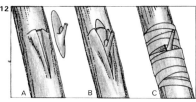

12 Budding, as used for roses, consists in taking a sliver from a leaf axil with a bud and inserting it into a T-shaped cut in the young skin of the stock [A]. When in place [B] it is secured with bast [C]. Once established, the shoot into which it was put is severed above the join. This is a convenient way to propagate, because a healthy scion produces many buds.

14 Whip and tongue grafts are usually carried out in the spring. Scion and stock [A] must match [B] and be at the same stage of annual development. The graft [C] is bound until it is united.

15 Standard roses are budded on to briars grown to the right height in the previous year. Bush roses are budded on younger briars at ground level. Many roses will also grow from cuttings.

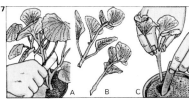

17 Stem cuttings [A] are taken of a variety of plants. They are trimmed just below a bud [B] and the leaves removed to adjust for depth of planting in an open-textured compost [C]. Until roots form, the cuttings must be kept damp.

18 Camellias are normally propagated from leaf-bud cuttings. The leaf, with the bud in its axil, is cut off, together with a section of its bark. This is set with the exposed base of the bud on top of a compost-filled, protected pot.

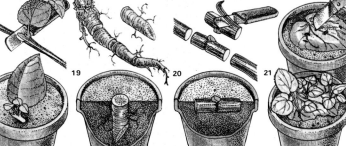

13 Trained fruit trees occupy less space than standards and take advantage of favourable positions, such as wall borders. Here, a peach is trained as a horizontal espalier. All but a few young branches are pruned away and two each year are tied to the wall. Apples are often restricted to one main stem, set diagonally to the wires, while peaches, apricots and cherries are trained in a fan.

16 Grafting into old trees is commonly performed in orchard trees to change the variety without grubbing out the trees. A bark graft [A] is sealed in place with clay or grafting wax. The crown graft [B] is the most commonly adopted form. The cleft graft [C] requires the gardener to split the stock with a wedge so that he can insert the pointed twig of the scion. In all cases grafts must be tightly bound and sealed.

19 The hollyhock (*Althaea*) can be multiplied by root cuttings. This is useful for hybrids that do not seed readily.

20 Vines and vine stocks may be multiplied by bud cuttings. The bud, with its sliver of bark, is placed in a suitable rooting medium.

21 *Begonia rex*, as a foliage plant, forms new individuals at cuts in the main veins of its leaves when placed on moist sandy compost.

Staple foods: grains

The history of civilization has always been closely paralleled by the development of cereals – mainly wheat and barley in the world's temperate zones and maize and rice in the tropical regions

Wheat and barley
Wheat was perhaps the earliest cereal to be brought into cultivation, the oldest authenticated remains of wheat seeds being found in the excavated site of a seventh millennium BC village in the Tigris-Euphrates valley (now eastern Iraq). Grains there have been matched against wild wheats that still grow in the Near East (*Triticum boeoticum* and *Triticum dicoccoides*).

Yields of these early wheats were small compared with those achieved by modern farmers who use more sophisticated varieties and intensive cultivation techniques to harvest up to seven tonnes a hectare (about three tons an acre). The principal wheat-growing areas are the grassland zones, primarily the steppe regions of Russia and China and the great almost treeless prairies of Canada and the United States [3].

Two major types of wheat are cultivated today – winter wheat and spring wheat. The former is sown in the autumn in regions where winters are comparatively mild. It grows strongly in the spring for harvest about ten months after sowing. Spring wheat is used in areas where severe winters would kill off the plants of winter varieties. It is sown as soon as the frost leaves the ground and is harvested in late summer – the same time as winter varieties. Winter wheats are higher yielding than spring wheats but produce a lower quality of flour.

Barley [10] originated in Abyssinia and has been cultivated in Egypt for at least 6,000 years. It grows mainly in the same areas as wheat, although plant-breeding developments have extended barley growing into colder areas. It used to be grown mainly for human consumption as a parched grain or as a meal but today it is used to feed animals and for making malt – which in turn is used for distilling or brewing into alcoholic drinks.

Two other cereals are important in temperate zones – oats and rye [10]. The common oat has been cultivated in western Asia and

the Mediterranean countries for about 2,000 years. The main uses are for cattle fodder or for human consumption as a porridge and breakfast cereal. Most rye is grown in eastern Europe and the Soviet Union, as a forage crop for dairy cattle; it is also pounded into flour and used for rye bread.

Maize and rice
Most important of the tropical cereals is maize, now second to wheat in international importance as a food grain [4, 5]. It was known in America before the Spaniards found the New World, but it probably originated in southern Asia. Today developments in plant breeding are extending the maize-growing areas farther into the temperate zones [6] by producing hardier strains.

In general, however, maize requires a hot, humid climate and will not tolerate any frost during the growing season. Its main use is as a fodder plant and for manufacturing purposes. The grains are used extensively in animal foodstuffs but also for making starch, and for human consumption as maize meal and cornflakes.

Wheat
Triticum aestivum

Wheat
Triticum durum

1 One of the common bread wheats grown extensively throughout the temperate lands is *Triticum aestivum*. It is a soft, mealy wheat used for flour and biscuit manufacture. *Triticum* *durum*, however, is one of the hard wheats. This particular species of wheat is used primarily for grinding into a flour that is manufactured into pasta products such as macaroni.

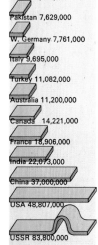

2 Wheat production in tonnes 1974 figures
Yugoslavia 6,283,000
Poland 6,410,000
Pakistan 7,629,000
W. Germany 7,761,000
Italy 9,695,000
Turkey 11,082,000
Australia 11,200,000
Canada 14,221,000
France 18,906,000
India 22,073,000
China 37,000,000
USA 48,807,000
USSR 83,800,000

2 Total world production of wheat is now more than 315 million tonnes. The wheat-producing countries are those such as India and most of Europe, which are self-supporting or need to import to make up

3 Wheat

local deficiencies, and the exporters such as the USA, Canada and Australia. The pattern of trade is for wheat to be exported from these countries to deficient areas such as the tropics.

3 Europe, excluding the USSR, grows about a quarter of the total world wheat crop and the yields are the highest in the world. The Soviet wheat belt produces about one-fifth of the world total.

Key areas are the prairies of America and Canada; India; China; the northern tip of Africa; the temperate zones of Australia; and the South American grain producers – Uruguay, Chile and Argentina.

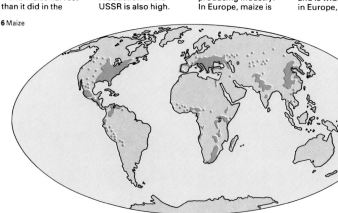

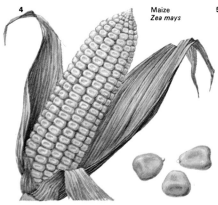

4 Maize (*Zea mays*) is the key cereal in the subtropical zones of the world. It can yield up to five tonnes a hectare (two tons an acre). Modern techniques of harvesting and storage, and the fact that it is relatively free from most major cereal diseases (except blight and fusarium stem rot), have made it a favourite with farmers in developing countries. Hybridization techniques give new varieties that have increased yield potential and disease resistance.

Maize
Zea mays

5 Maize production in tonnes 1974 figures
Italy 5,180,000
India 5,300,000
Hungary 6,664,000
Romania 7,159,000
Mexico 7,784,000
Yugoslavia 8,030,000
France 8,884,000
Argentina 9,900,000
S. Africa 11,035,000
USSR 12,142,000
Brazil 16,065,000
China 31,085,000
USA 118,144,000

5 The world production of maize has almost trebled in the past 50 years, to 300 million tonnes. The USA is the major exporting country, although it now produces a lower proportion of the total world maize harvest than it did in the

early part of the 20th century. Small quantities are produced for export in Brazil, Argentina and South Africa, most of it for animal feedstuffs used in northern and northwestern Europe. Production in the USSR is also high.

6 Maize

6 The Corn Belt of the USA produces nearly half of the world's maize, although less and less of this is exported. Most of it is used on farms and supports a vast cattle and pig-producing industry. In Europe, maize is

similar importance in Africa is sorghum (*Sorghum vulgare*). It is used extensively in the manufacture of beer. Barley (*Hordeum vulgare*) is one of the principal cereals of temperate agriculture and is widely grown in Europe, America

The maize plant may grow to a height of 4.5m (15ft) and with its big, spreading leaves differs greatly from the grass-type cereals such as wheat and barley. In cooler, temperate countries it is now becoming an important forage crop for beef and dairy cattle. The whole maize plant is cut, wilted and stored in a silo for winter feeding.

Although of less importance in world trading terms, rice is perhaps the most vital of cereals [7, 8]. It is said to provide food for more people than any other of the world's cereals. The plant can grow to a height of 1.5m (5ft) and is probably a native of India. It is an annual swamp plant [9]. Unlike the other main cereals, it is not sown directly into the field and left until harvest. Instead, the seeds are sown and when the young plants are big enough they are transplanted into flooded "paddy" fields. These fields are drained at harvest and the rice is then treated like the other small grains and, in developed farming areas, is harvested mechanically with a combine harvester.

The grains of the rice plant are rich in starch. Rice may be ground into a flour but mainly it is husked boiled and eaten, in its original state.

Rice is grown principally in southern and eastern Asia, the great river deltas and alluvial plains forming the main production areas. However, some rice is still produced in hilly areas where the slopes are terraced to allow controlled flooding and draining of the rice field.

The USA has also become an important rice-growing area, and research there has shown that the laborious transplanting of the rice plant is unnecessary and the crop can be drilled directly into the field like other cereals. These techniques are cutting the cost of rice production and encouraging its spread and expansion in other suitable areas.

Millet and sorghum
Another long-cultivated cereal, again native to China, is millet [10]. It grows to a height of about 1m (39in). Millet tolerates prolonged drought and is grown mainly in the rain-deficient tropical areas of Africa and Asia. Sorghum [10] is also vital to tropical Africa, where it is the main bread plant.

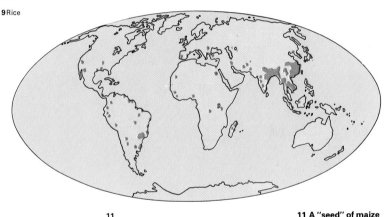

Huge wheat silos on the prairie skyline of midwest North America testify to this region's importance as a wheat-producer. Known as "Kansas cathedrals", they could become bulwarks against global famine from population growth.

7 Rice (Oryza sativa) is the basic food grain in densely populated monsoon Asia. The husk of the grain has a high silicon content and is dangerous to animals. The outer skin and kernel contains the protein and vitamins. The rice plant is commonly known as paddy, hence paddy fields, and only when it has been harvested and the husk removed does it technically become the rice that is sold.

8 Rice production in tonnes 1974 figures	
USA	5,175,000
Port Timor	5,594,000
S. Korea	5,908,000
Brazil	6,817,000
Burma	8,446,000
Vietnam	11,400,000
Thailand	13,175,000
Japan	15,902,000
Bangladesh	17,222,000
Indonesia	22,800,000
India	61,500,000
China	115,275,000

8 In the early 1970s world production of rice totalled over 310 million tonnes, including an estimate for production in China. There is, however, little international trade. There was world over-production of the grain in the early 1970s and developed countries were urged to curb production to avoid damaging the market for countries that relied on rice as a major source of income.

9 Rice

9 The river deltas and plains of China and southwest Asia are the main rice-producing areas of the world and up to three or four crops a year are taken off many fields, although this practice soon depletes the soil. China produces an estimated 100 million tonnes while India, Pakistan and Bangladesh produce 60 million tonnes. Japan, Thailand and Vietnam are the other major producing areas.

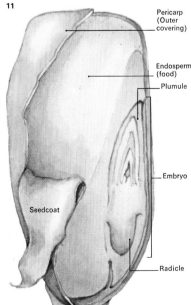

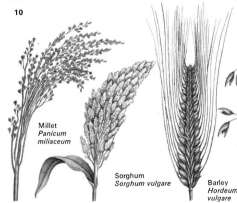

Millet *Panicum miliaceum*

Sorghum *Sorghum vulgare*

Barley *Hordeum vulgare*

Oats *Avena sativa*

Rye *Secale cereale*

10 One of the oldest cultivated small grains in the world is millet (*Panicum miliaceum*), which is grown in tropical Africa. It tolerates prolonged drought and, when exported, is used largely in bird seed preparations. A grain of similar importance in Africa is sorghum (*Sorghum vulgare*). It is used extensively in the manufacture of beer. Barley ((*Hordeum vulgare*) is one of the principal cereals of temperate agriculture and is widely grown in Europe, America and Australia. Most varieties are spring sown, as they will not tolerate frost during the growing season. Oats (*Avena sativa*) and rye (*Secale cereale*) are both thought to have been first noticed growing as weeds in early wheat cultivation. Both cereals are now extensively grown in many parts of Europe, rye proving particularly popular in the glacial soils of central Europe and Russia. Rye grows well in poorer soils than those required for most cereal grains.

11

Pericarp (Outer covering)

Endosperm (food)

Plumule

Embryo

Seedcoat

Radicle

11 A "seed" of maize (*Zea mays*) contains both the embryo of the future plant and also the food to nourish it while it germinates. The main food supply, as in all grain crops, lies in a single seed leaf since the plant is a monocotyledon (as distinct from a dicotyledon such as the bean, whose food is contained in two leaves within the seed) and consists predominantly of carbohydrates. It is this food supply that makes the plant economically important. The embryo, made up of the young shoot, the plumule, and the root – the radicle – contains much of the protein of the seed and this was once lost through primitive winnowing techniques. Nowadays the embryo and its protein are saved.

181

Bread and pasta

Western man is capable of creating ever more complex and exotic gourmet dishes to tempt his palate. These invariably contain large amounts of protein, carbohydrate and fat but most of the people of the earth are concerned from day to day with the way in which they are to obtain staple, flour-based food – bread and pasta – that is the basic staff of life.

The first loaf of bread probably originated from the grass seeds that an ingenious prehistoric man, some 8,000 years ago, gathered, ground into flour, mixed with water, covered with hot ashes and baked on a hot stone. The ancient Hebrews, Chinese and the Egyptians all ate flat cakes made from flour and water but the ancient Egyptians were the first to discover that allowing the dough to ferment produced gases that made the resulting bread both lighter and softer.

Ancient recipes preserved

Despite the many variations that have developed [1], ranging from white bread and black bread to Alaskan sourdough bread and Irish soda bread, and despite modern production methods, in essence bread remains the same as when its Neolithic originator began making it. Bread is simply dough made from flour, moistened with water or milk, usually leavened with yeast or a similar fermenting agent, and then baked, or even steamed.

Bread flour is made from wheat, rye, barley or oats. What makes a soft, spongy loaf is the nature of the wheat's protein (gluten). The different glutens in barley and rye make for heavy bread; maize and rice, having no gluten, cannot be made into bread. Special gluten-free flour is produced, however, for people with certain digestive abnormalities. Today in Britain and America strong flour milled from hard wheat is preferred for its high gluten content, which helps the bread to keep fresh for several days, while in France soft flour is used, necessitating two daily bakings.

Despite the simplicity of using modern yeast [Key] many flat or unleavened breads, usually made from wheat, flour, salt and water, remain popular for reasons of taste, tradition or climate. The Scandinavians favour a wide selection of crisp flatbrods and pliable lefsers. Among India's many flat-breads is the poppadum – a crisp, thin wafer that is sometimes crumbled over food. The Mexican tortilla is made with maize flour instead of wheat, as are many of the African unleavened "breads".

Nutritionally bread is a relatively inexpensive source of calories, protein and vitamins. Modern wheat bread is made from flour fortified with small amounts of thiamin, nicotinic acid, iron and calcium.

Pasta – pipes, shells, rings and bows

Although legend tells that Marco Polo returned from China to Venice more than 700 years ago with the pasta that became Italy's most distinctive national food, equipment for making leganum (pasta similar to tagliatelli) used by the Romans was found in the ruins of Pompeii. Whatever its origins, pasta has remained an Italian speciality although it is popular in other countries also. Noodles still feature widely in Chinese and South-East Asian cooking and with greater or lesser prominence in the cuisines of Japan, France and the United States. In central

CONNECTIONS

See also
180 Staple foods: grains

1 Bread has always had a special meaning. Christians use it at Holy Communion to symbolize the body of Christ; the Arabic words for bread and life have the same origin. Festive breads such as Mexican fiesta bread, Russian kulich and Jewish matzos have long been important in seasonal and religious celebrations.

1 Indian poppadum
2 Jewish matzos
3 Rye crispbread
4 Scandinavian crispbread
5 Pitta with sesame seeds
6 Pitta
7 Indian naan
8 Indian chapati
9 Dark rye
10 German rye
11 Pumpernickel
12 Russian black
13 Wholemeal loaf
14 Rye with caraway seeds
15 Danish rye
16 American sourdough
17 Irish soda
18 French petite baguette
19 American cornbread
20 French petit Parisien
21 French baguette
22 Scottish bap roll
23 English cottage
24 French pain Espagnol
25 Jewish challah
26 French épi de Charente
27 Croissant

Europe they are often served with meat instead of potatoes or rice. But the formidable assortment of Italian pasta and the many ways it is served are incomparable.

There are more than 60 varieties of pasta [2] included in the three main categories – rod or tubular forms, like spaghetti and macaroni; flat sheets or strips like lasagne or fettuccine (which may be flavoured with spinach); and the small, grain-like soup pasta (pastina). Pasta is cooked by boiling or is boiled then baked and is always served with a sauce of some kind.

Food for more than half the world

Rice provides a staple diet for more than half the population of the world, for whom it provides 80 per cent or more of their total diet. It is thought that altogether there are more than 7,000 cultivated varieties of rice of which over 1,000 are grown in India alone [3]. Of the long-grained variety, Patna and Basmati are among the most widely known. Although their names do not necessarily denote their origins, other varieties include Java or Spanish rice, with more oval-shaped grains;

Piedmont rice, which is short, round and very white; and Carolina, which is another long-grained kind of rice but one that is suitable for puddings.

The rice plant grows best in paddy fields [4], although there is a variety that thrives on dry land [5]. Early records refer to Chinese irrigation systems for rice-growing in 770 BC, and in India and the Philippines terraces built 2,000 years ago still support rice cultivation. Today, as centuries ago, rice plants are grown in water until they flower then the field is slowly dried out as the grain ripens.

The grains are then hulled and undergo a polishing process that removes the germ and bran layers. In this process the rice is robbed of much of the proteins and especially the vitamins it contains. In most lands where rice is the main food source only the husk is removed: the "brown rice" that remains retains its valuable nourishment.

While rice cannot be satisfactorily processed into bread it is eaten extensively in many other ways – on its own, in soups, puddings, and cakes or accompanied by all kinds of different vegetables and meat.

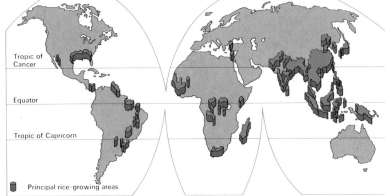

Baking yeast *(Saccharomyces cerevisiae)*, is a microscopic form of life with remarkable powers. Combined with sugar and warmed, yeast ferments; an enzyme in it produces alcohol and the gas carbon dioxide from the sugar. It is the carbon dioxide that makes the yeast-leavened dough rise. Commercial yeast is available in two forms, compressed and granular. Billions of individual cells are contained in compressed yeast, each containing the enzyme responsible for fermentation.

2

3

Principal rice-growing areas

2 All pasta is made from the same ingredients following a basic method. A flour mixture is prepared from dried durum wheat. This is mixed with warm water and kneaded into a stiff dough. Dried or fresh eggs or spinach can be added, although most pasta is a simple flour and water combination. The dough is forced through perforated cylinders to be made into various shapes. Small pasta like macaroni is cut to size as it emerges from the presses while the longer noodles or spaghetti are produced to different lengths by machines. The cut pasta is then dried in ovens or on racks and packaged.

3 The rice plant *(Oryza sativa)* is indigenous to Asia – 90% of the total output of the world is grown in China, India, Pakistan and South-East Asia.

The only Asian countries to produce enough for export are Burma and Thailand. Cultivated rice was first referred to in China over 5,000 years ago.

It was introduced to the Nile delta and thence to Spain and northern Italy in the fifteenth century from where it reached Brazil and North America in 1685.

4

4 The staple diet of most people in the East, rice can be grown where there is sufficient water to flood the fields or paddies in which it is planted. Young plants are set out by hand in fertile mud, which is permanently flooded during the growing period and gradually drained as the rice grains ripen. The plants' nodding golden tassels are a familiar sight in hot countries throughout the world.

5

5 The rice of the paddy fields is much more common than any other variety. But about 10% of all rice is "upland rice". This grows on dry land and is cultivated in much the same way as any other cereal grain.

Staple foods: pulses

All commercially grown species of peas and beans (pulses) are legumes, members of the family Leguminosae. Generally, peas and beans can be eaten raw, but there is one major exception, the soya bean [1].

Soya bean production

Soya beans were first cultivated in China more than 4,000 years ago and today they are still one of the main sources of vegetable protein for people living in the Far East; elsewhere they are used mainly as food for animals. More than half of the world's annual crop of 50 million tonnes is now grown in the United States. But large areas of soya beans are still cultivated in China, Japan and parts of Russia as well as in South America [4].

The importance of soya beans has now spread from the Eastern countries to the West. In the United States the main production area is in the eastern and middle eastern states where the crop is intensively cultivated by grain farmers, who can use the same harvesting and storage equipment for the crop as they use for wheat and barley. The crop is gradually being developed in more northerly latitudes as plant breeders introduce varieties that can resist cold climates. Parts of Europe, especially southern France, Italy and Spain, now produce soya beans, and crops have been tried experimentally in Britain and Sweden. It is expected that varieties suitable for cultivation in northern Europe will be available by the 1980s.

Major livestock-producing nations use vast quantities of soya beans either ground or, more usually, crushed. Crushed soya beans yield large amounts of oil, which is used in the manufacture of margarine and cooking oils and, in a refined form, to make varnishes and other industrial products [3].

Recently soya beans have been used to make texturized vegetable proteins, also known as meat substitutes. Other legumes, such as field beans and field peas [6], can be used in a similar way. The protein content [7] is separated from a flour produced by milling beans or peas. This flour is then extruded into a thin silk-like thread which is woven mechanically to produce a solid form. The technology has now become so advanced that imitations of meat, fish and other natural foods can be manufactured. As animal proteins become more expensive to produce, such texturized vegetable substitutes will become relatively cheaper and may become part of a nutritious everyday diet for much of the world's population.

Peas and french beans

The most important pulse crop grown in cool temperate zones is the pea [5]. Large areas are planted in Europe, the United States and Canada and in parts of the Southern Hemisphere with similar climates, such as New Zealand. The pea came originally from Asia, from the region lying between the Mediterranean and the Himalayas. It reached China from Persia about AD 400 and was introduced into the United States during the earlier years of colonization.

Peas have several uses. In most industrialized countries they are eaten either fresh or in some processed form – usually either canned or frozen. New techniques involving quick freezing of the crop within hours of its being harvested have widened its potential market and peas are now eaten all the year

CONNECTIONS

See also
216 Fibre and oil crops
156 World food resources
252 The future of food

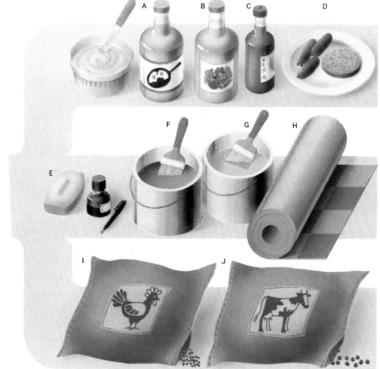

1 **Soya beans**, now widely grown throughout the world, are natives of China where they were first cultivated some 4,000 years ago. They were introduced into North America in 1880. The plants can grow to about 60cm (24in) high, but most varieties attain about 45cm (18in), carrying their pods quite near the ground.

Glycine max

2 **The flowers of the soya bean** can vary from pure white to light purple. The beans themselves are yellow, brown or black, depending on the variety. The crop is now grown on a vast scale in the USA, but it is not high-yielding, producing less than two tonnes of dried beans to the hectare. The soya bean is today the largest source of vegetable oil and protein meal in the USA.

3 **The versatility of the soya bean** as a vegetable means it can be put to almost any use, ranging from direct consumption in less-favoured areas to sophisticated industrial uses in developed nations. Products made from it include bean curd [A]; vegetable oils [B], which are used in cooking and are refined for the manufacture of margarine; soya sauce [C], a fermented flavouring that is used in Oriental cooking; and texturized vegetable protein [D]. The oil is also used in soap manufacture [E] and for inks, cosmetics, dyes, paints [F], varnishes [G] and glues. It is also processed into lacquers used in colouring floor coverings [H]. When the oil is crushed from the bean a high-protein meal remains. Once pressed, this can be fed directly to animals [I] or incorporated with cereals into livestock rations [J].

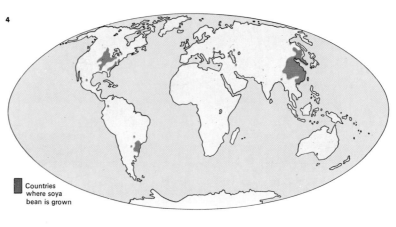

Countries where soya bean is grown

4 **Cultivation of soya beans** has spread in response to the increasing world demand for protein. The main areas are the USA, with more than half of world production, and the Far East, notably China, Japan and Korea. The crop is spreading into parts of South America, especially Brazil, and into developing African states. As new varieties resistant to colder climates become available, more soya will be grown in Europe and northeastern Asia.

round. Some varieties are not harvested in the fresh, immature state, but are left on the field to ripen and dry. They are then harvested with combine harvesters in much the same fashion as grain crops. They are dried further during storage and either sold dry, or soaked and canned as "processed" peas. Britain has traditionally imported Alaskan peas from North America for canning in this way but now new varieties have been developed and the American trade has diminished.

Dwarf french beans are another important legume for the crop farmer and they are grown in the same areas as peas. The crop is also either eaten fresh or processed by canning or freezing.

Pea and bean crops are especially important to crop farmers because most do not need to have extra nitrates added to the soil to aid their growth. As leguminous plants, they have the ability to "fix" nitrogen from the air in small nitrate nodules that form on the roots of the plants (with the help of symbiotic bacteria). In this way they increase the fertility of the soil for the following crop in the arable rotation. The dwarf french bean is

an exception to this general pattern in that it needs the addition of inorganic nitrogen to produce an economic yield.

Other edible pods and beans
Many other members of the bean family are common as garden plants grown for home consumption [6]. Scarlet runner beans, which are picked when immature and can be eaten fresh, frozen or canned, broad beans and various types of haricot beans belong to this category. Black-eye peas and mung beans, however, form staple diets in the areas of the world in which they proliferate. One of the earliest cultivated legumes is the lentil, a native plant of Asia and a traditional source of vegetable protein throughout Middle Eastern countries.

One other bean which is widely grown in the United States and very popular in Britain is the navy bean, a variety of the same species as the dwarf french bean. It is imported into Britain in the dry harvested state and is then processed and canned into the familiar baked bean, usually in a tomato-flavoured sauce.

The pea-viner harvests the crop ready for processing. The peas are separated by the viner from the vines and pods and are delivered to a collecting chamber from where they are transferred to boxes for transport to the factory.

Canning

Freezing

Dehydrating

5 The garden pea (*Pisum sativum*) is eaten all year round. The crop is grown on a field scale with planting dates staggered to allow steady harvesting for processing. For canning, freezing and dehydrating the peas are taken into the factory within two or three hours of harvesting. For canning they are washed, sorted and put into tins before cooking. For freezing, they are sorted, graded and blast-frozen. The whole process takes only about 20 minutes. The dehydration or freeze-drying technique is even quicker. All the water is extracted from the peas, which then have to be re-hydrated by soaking before they are used as food.

6 Peas and beans are vital parts of the staple diet in many parts of the world, whether for human consumption or as a prime feed protein for cattle. This international selection of peas and beans includes green gram or mung beans [A]; butter beans [B]; pea beans [C]; black gram beans [D]; pigeon beans [E]; lablab beans [F]; the traditional American black-eyed peas [G]; purple-coated kidney beans [H]; soya beans [I]; the brown and the white haricot bean [J, K], both varieties of the species *Phaseolus vulgaris*; scarlet runner beans [L]; and chick peas [M], known as garbanzos in the Americas.

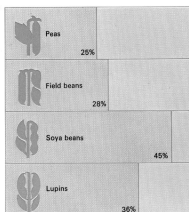

7 Most pulses contain protein in relatively large amounts. The 25% protein content of peas is exploited through processing of the crop in Canada, where a high-protein flour is made for sale to countries short of protein. The field bean is used mainly for animal feed. Soya bean is a dual-purpose vegetable, either for human or animal feed. Lupins are normally poisonous but can be treated to make animal feed.

Peas 25%

Field beans 28%

Soya beans 45%

Lupins 36%

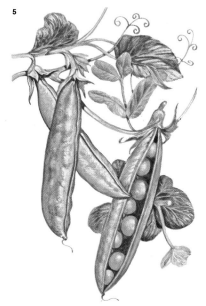

2%

1 Soya beans 59%
2 Dry beans 11%
3 Dry peas 10%
4 Chick peas 6%
5 Dry broad beans 5%
6 Pigeon peas 2%
7 Lentils 1%
8 Other 6%

8 World production of pulses should increase with the continued rise in the demand for both proteins and oils, in which most species are rich. Soya, however, is likely to remain the most widely cultivated pulse crop. Several types of pea crop could increase, especially if present North American breeding programmes succeed in producing types higher in protein than those now available. The extraction of protein and starch separately from peas could have an impact on the areas cultivated. Field beans, or faba beans, will also increase in popularity. Work in Europe on new varieties with improved resistance to disease is also well advanced in an attempt to find an alternative type of bean to soya.

Staple foods: tubers

Tubers are underground stems specially modified for the storage of plant foods. Root tubers have the same function but are modified roots. Rhizomes are another type of storage organ but are modified stems that grow underground. Tubers form part of the staple diet of almost all modern societies from urbanized Europe and North America to the tropics of Africa and Asia.

The most important tuber

To Western peoples, the most important tuber is the potato (*Solanum tuberosum*). It was taken to Europe by the Spaniards in the sixteenth century from South America, where it had been a vital part of the Peruvian diet from about AD 200.

Europe accepted the potato gradually and it was not until the late 1580s that it appeared in the British Isles. Crop development was fastest in peasant economies – notably in Ireland – but after the mid-nineteenth century the potato's importance as a food source was recognized and the vegetable was cultivated more widely. Ironically, it was the major failure of the potato crop in Ireland in 1845–6

and the devastating famine that ensued that alerted society to the potato's food value.

The potato grows best in well-drained, fertile soils and, because of this, its cultivation spread rapidly to the better crop-growing areas of Europe such as the east coasts of England and Scotland, the northern plains of France and the fertile polder areas of The Netherlands. Similarly, its development in the United States has been greatest in the north – Idaho, Washington and Maine.

Modern agricultural techniques, the use of chemical fertilizers and the development of mechanized planting and harvesting have led to increased yields. Farmers on the best soils now harvest up to 40 tonnes of potatoes a hectare (16 tonnes an acre).

The potato crop

The potato crop is planted when the last spring frosts are gone. Tubers are placed about 1m (39in) apart in ridges, well covered with soil. The plant shoots emerge within two or three weeks of planting and the tubers of early varieties are harvested – usually by hand [Key] – about three months after

planting [1]. The main part of the crop is lifted in autumn and a high percentage is stored to ensure year-round supplies.

The principal use of the potato [2] has traditionally been as a fresh vegetable, but now it is increasingly used as the raw material for the production of processed products. Frozen "french fries" (chips), crisps [3], dehydrated potato for human or animal feed and a series of canned potato products are extending its use as a food. In the USA, the trend towards processed and "convenience" foods has been most marked and up to 45 per cent of the US potato crop is now eaten in this form. In the UK the proportion is about 15–16 per cent, but is increasing steadily. In Europe, notably in The Netherlands and Germany, processing into starch and alcohol has become an important industry.

The development of new uses for the potato and the need to adapt the plant to grow in many climates and to mature earlier or later for specific markets, has led to the development of major plant-breeding programmes. Some varieties now available have the characteristics of early maturity, or high

CONNECTIONS

See also
178 Plant propagation
156 World food resources

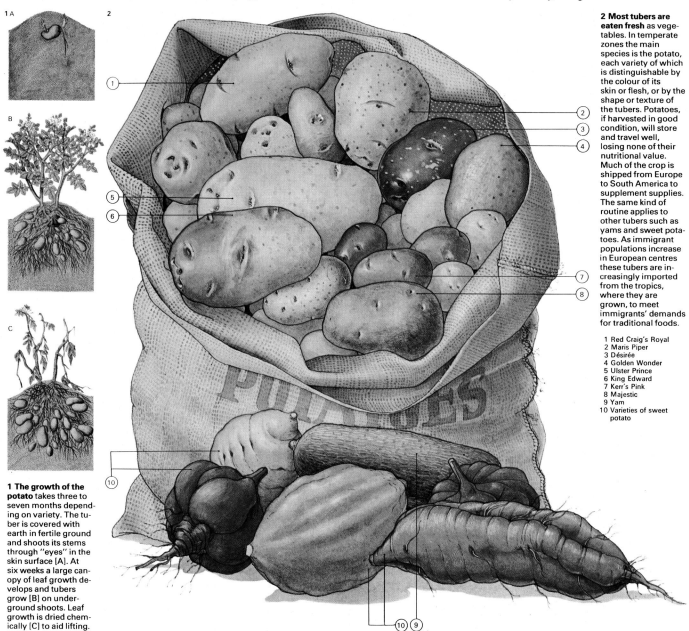

2 Most tubers are eaten fresh as vegetables. In temperate zones the main species is the potato, each variety of which is distinguishable by the colour of its skin or flesh, or by the shape or texture of the tubers. Potatoes, if harvested in good condition, will store and travel well, losing none of their nutritional value. Much of the crop is shipped from Europe to South America to supplement supplies. The same kind of routine applies to other tubers such as yams and sweet potatoes. As immigrant populations increase in European centres these tubers are increasingly imported from the tropics, where they are grown, to meet immigrants' demands for traditional foods.

1 Red Craig's Royal
2 Maris Piper
3 Désirée
4 Golden Wonder
5 Ulster Prince
6 King Edward
7 Kerr's Pink
8 Majestic
9 Yam
10 Varieties of sweet potato

1 The growth of the potato takes three to seven months depending on variety. The tuber is covered with earth in fertile ground and shoots its stems through "eyes" in the skin surface [A]. At six weeks a large canopy of leaf growth develops and tubers grow [B] on underground shoots. Leaf growth is dried chemically [C] to aid lifting.

starch content or a smaller proportion of water than usual in the live weight.

Early maturing varieties of potato have become vital to the economy of many less-developed countries, especially around the Mediterranean. The varieties are ready for harvest in early spring and are exported to cities in northern Europe.

Less-productive tubers

Other tubers – mostly those in tropical areas – have not yet had the benefit of sophisticated plant-breeding programmes and are therefore comparatively less productive. This situation is likely to change, particularly with cassava [6], a tall herb that originated in South and Central America. There are two useful species – *Manihot dulcis*, the sweet cassava, and *Manihot utilissima* (or *M. esculenta*), the manioc or tapioca plant.

The sweet cassava is a root tuber eaten fresh as a vegetable or used as a source of starch. By far the most important species is the bitter cassava. When harvested the tuber is bitter and poisonous, but when the poisonous juice is washed away the tuber is

processed into tapioca for human consumption and into cassava meal (also called manive) for animal feed. The meal's importance as a cereal replacement in animal feeding was highlighted during the world cereal shortage of 1973 and 1974. This shortage, and the need to make tropical countries more self-sufficient in food, is leading to far-reaching plant-breeding programmes for cassava which will increase its importance.

Yam (*Dioscorea* sp) and taro (*Colocasia esculenta*) are tubers that are of importance in their respective geographic areas. Yams are indigenous to tropical areas of both the Southern and Northern Hemispheres and taro, which originated in southeastern Asia, has become an important staple food in the Pacific Islands, particularly Polynesia where it forms the basis of a thick paste called poi.

The sweet potato (*Ipomoea batatas*) bears no relation to its more common namesake. It produces heavy crops of large, irregular tubers and is now widely cultivated as a vegetable in the southern USA, Central America, the warm Pacific Islands, Japan and the Soviet Union.

Potatoes can be harvested mechanically or by hand. Hand picking of early crops continues when mechanical lifting is stopped by bad weather, so allowing the crop to reach the market when the price for early new potatoes is high.

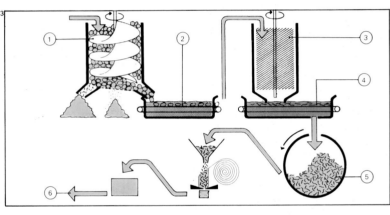

3 Potatoes are manufactured into crisps, one of the chief forms of processed potato eaten in Britain. The tuber must be high in dry matter and low in sugar content to be suitable for crisping. The potato is peeled [1] and washed [2] mechanically, sliced [3] and blanched in boiling water before frying [4]. The crisps are then dried and to some batches flavouring is added [5] before packaging [6] ready for distribution. All this is done automatically.

4 Tannia
Xanthosoma sagittifolium

5 Arrowroot
Maranta arundinacea

6 Cassava
Manihot utilissima

6 The cassava is becoming one of the world's most important tubers because a processed form of meal, produced from its root, is used as a cereal substitute.

5 Arrowroot mostly grows on the West Indian island of St Vincent. It has rhizomes that produce a starchy powder used for thickening sauces.

4 The rhizomes of the tropical tannia plant are collected and ground into meal. The plant grows to 3m (10ft) in height.

Vegetables

Vegetables are important to a healthy human diet even if the energy requirements of the body are met by starches from grains and roots and by protein of animal or vegetable origin. Fresh vegetables provide vitamins that may be lacking in the rest of the diet. They also give the diet the bulk that helps to keep the digestive tract in good working order. Furthermore, they add to the enjoyment of food by providing variety. The seeds, pods, leaves, stems and roots are used.

Pods and seeds

The young seeds or fleshy pods of immature pulses are widely eaten [1]. Peas (*Pisum* spp) and broad beans (*Vicia* sp) originated in the Old World. French, haricot and dwarf beans, all *Phaseolus* sp, originated in America. Most of these are eaten as pods, although some are allowed to mature or may be shelled out like green peas (flageolets).

Most versatile of all the vegetable groups are the brassicas, which include both leaf and root types, and others where the immature flower head is eaten. Many are sub-species of *Brassica oleracea*, whose area of origin is not known; they include cabbages, brussels sprouts, cauliflower and broccoli [3]. Other leafy brassicas are Chinese mustard (*B. juncea*), kale (*B. oleracea*) and Chinese cabbage (*B. pekinensis*). The root species, in which the "root" is actually a modified stem, include kohlrabi (*B. caulorapa*), swede (*B. napobrassica*) and turnip (*B. rapa*).

Other leaf vegetables that are widely grown include spinach (*Spinacia oleracea*) and the Swiss chard (*Beta cicla*), lettuce (*Lactuca sativa*), endive (*Cichorium endivia*) and chicory (*C. intybus*) [4].

Less hardy because of their sensitivity to frost, but cultivated for many years in temperate as well as tropical climates, are the cucumbers (*Cucumis* sp) and several species of the genus *Cucurbita*, which includes the marrow, squash and pumpkin [8]. In all of these it is the immature fruit that is eaten.

Fruiting species

A number of fruiting species botanically related to the potato came with it from America. Of these the most important today is the tomato (*Lycopersicum esculentum*) [6], although it was originally grown in Europe as a purely ornamental plant. Others in this group are aubergine or egg plant (*Solanum melongena*) [6] and sweet pepper (*Capsicum annuum*) [7]. These plants figure increasingly in international trade, and are largely grown in glasshouses.

Root vegetables, which cover a wide botanical range, have always been a most important winter food, because many are biennials, whose roots carry the energy stock for the following year's flowering. Examples are the tap-rooted carrot (*Daucus carrota*) [5] and parsnip (*Pastanaca sativa*), whose wild forms are common all over Europe. The beetroot (*Beta sativa*) is one variety of a versatile species that includes the sugar beet and the mangold. Radishes (*Raphanus* sp) [5] have also been developed from common wild plants in Europe and Asia, where they are still weeds in arable crops on some soils.

The genus *Allium* adds flavour to diets in many parts of the world. Vegetables of this genus include the onion (*A. cepa*) [5] and its varieties, shallot (*A. ascalonicum*), leek (*A. porrum*) from the Mediterranean region

CONNECTIONS

See also
184 Staple foods: pulses
186 Staple foods: tubers
190 Green salads

In other volumes
70 Man and Society
42 The Natural World

1 Legumes and grains can both be eaten as vegetables. Pulses, some having edible pods, include peas [A], broad beans [B] and runner beans [C]. Sweet corn [D] is a variety of maize bred to be cut before the grains ripen. Frost-resistant varieties have extended the range of sweet corn.

2 Young shoots of asparagus [A] are grown as a luxury vegetable in beds that may remain in production for many years. Rhubarb [B], a member of the dock family, has poisonous leaves but the edible young stems are eaten in spring. Growth may be forced in darkened sheds.

3 Cabbage [A], cauliflower [B] and brussels sprouts [D] all belong to the same *Brassica* species although they differ greatly in appearance. Like curly kale [C], they are all hardly and some varieties can stand quite cold winters.

4 Leafy plants for salads and cooking include chicory [A], whose young shoots are forced in winter, lettuce [B] and spinach [C], of which there are two species to cover winter and summer cropping.

and garlic (*A. sativum*) from central Asia. Without one or other of these, regional cookery would be much less interesting.

Plants such as celery (*Apium graveolens*) and sea-kale (*Crambe maritima*) are cultivated for their blanched stems. The young stems of *Asparagus officinalis* [2] are considered a particular delicacy. Rhubarb (*Rheum rhaponicum*) [2] is another stem vegetable of commercial importance in temperate countries. The globe artichoke (*Cynara scolymus*) is cultivated for its edible flower buds – as well as its looks.

The tropics support a wide variety of both root and leaf vegetables, most of which are mainly eaten locally. Two that have become more widely familiar are the green bananas or plantains, which are prominent in so much Creole cookery, and okra or gumbo (*Hibiscus esculentus*), which is a close relation of cotton.

Cultivation and preservation

Vegetables lend themselves to small-scale cultivation in most parts of the world, by amateurs as well as professional cultivators; much of their production is still on small intensive holdings, largely run by family labour. However, with urbanization a larger proportion of the main species is grown on a field scale by general farmers or on big commercial holdings in areas such as California, where the weather is predictable, and irrigation combined with plentiful sun makes for quick growth and high yields for the cultivators.

Sophisticated handling and packing, with refrigerated transport, make it possible to move even highly perishable vegetables into distant markets in a condition as good as that of produce grown on the very outskirts of any conurbation.

Large-scale vegetable production has also been furthered by the technique of deep freezing. A crop with a comparatively short season, such as the green pea, can be made available to consumers all over the world throughout the year. It can also be produced to high standards of quality in the field and in the processing plant, while inferior grades are diverted to other forms of preservation, such as canning and drying.

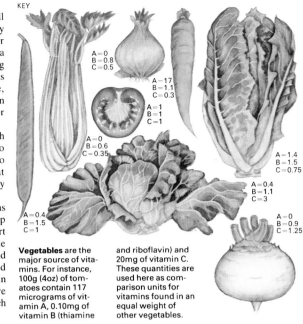

KEY

A=0
B=0.8
C=0.5

A=17
B=1.1
C=0.3

A=1
B=1
C=1

A=0
B=0.6
C=0.35

A=0.4
B=1.5
C=1

A=1.4
B=1.5
C=0.75

A=0.4
B=1.1
C=3

A=0
B=0.9
C=1.25

Vegetables are the major source of vitamins. For instance, 100g (4oz) of tomatoes contain 117 micrograms of vitamin A, 0.10mg of vitamin B (thiamine and riboflavin) and 20mg of vitamin C. These quantities are used here as comparison units for vitamins found in an equal weight of other vegetables.

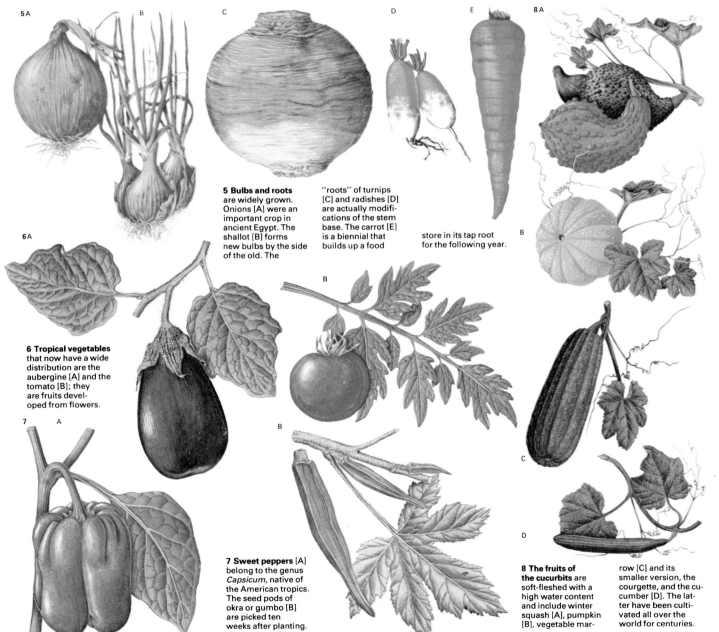

5 Bulbs and roots are widely grown. Onions [A] were an important crop in ancient Egypt. The shallot [B] forms new bulbs by the side of the old. The "roots" of turnips [C] and radishes [D] are actually modifications of the stem base. The carrot [E] is a biennial that builds up a food store in its tap root for the following year.

6 Tropical vegetables that now have a wide distribution are the aubergine [A] and the tomato [B]; they are fruits developed from flowers.

7 Sweet peppers [A] belong to the genus *Capsicum*, native of the American tropics. The seed pods of okra or gumbo [B] are picked ten weeks after planting.

8 The fruits of the cucurbits are soft-fleshed with a high water content and include winter squash [A], pumpkin [B], vegetable marrow [C] and its smaller version, the courgette, and the cucumber [D]. The latter have been cultivated all over the world for centuries.

Green salads

The crispness and flavour of uncooked vegetables, together with their medicinal qualities and nutritional value, have been recognized for centuries. Lettuce was prepared for the kings of Persia in the sixth century BC and the Romans were well aware of its curative powers. Early Greek doctors regarded both lettuce and rocket as beneficial for the stomach.

Lettuce and other leaves

The leaves of salad vegetables were credited with other properties. The spicy flavoured rocket (*Eruca sativa*), some varieties of which are grown for salads, was thought to be an aphrodisiac. Because of the milky juice characteristic of the lettuce (*Lactuca sativa*) [1A], the vegetable was said to be good for nursing mothers. This belief belongs to the doctrine of plant signatures, which associated the physical form of the plant with the condition it was to aid: the word *lactuca* is derived from the Latin word for milk.

A prickly lettuce (*Lactuca serriola*), native to Asia, is thought to have been the ancestor of *L. sativa*, but many varieties

flourished in the East by the first century AD. During the Middle Ages the lettuce spread to Europe and Christopher Columbus (1451–1506) may have taken it to the New World. Modern writers have noted that eating flowering lettuce in quantity induces a coma. In Beatrix Potter's animal tales the Flopsy Bunnies ate this type of lettuce and fell into so sound a sleep that they were oblivious to danger. Even so, lettuce has always been the basic salad ingredient. The Romans treated lettuce in brine, then in brine and vinegar. A spicy vinaigrette (oil and vinegar mixture) was used to dress fresh salads.

Embellishing a salad

The addition of other raw vegetables to the basic green salad lends it flavour and a crunchy quality. Blanched leaves of chicory (*Cichorium inty bus*) (1E) are highly regarded by the French. The closely packed yellowish-white leaves of the chicory were developed in 1850 by the head gardener of the Brussels Botanical Garden and introduced in the United States in the late nineteenth century. The related curly endive (*Cichorium endivia*)

may have originated in Egypt although it is said to have arrived in Europe from the East. There are many varieties, from the very curly – the most frequently grown as a tasty salad ingredient – to the slightly curled broad-leaved Batavian type which is cooked. Another crisp addition is celery (*Apium graveolens*) [1L]. Or a lively embellishment is the distinct aniseed flavour of Florence fennel (*Foeniculum vulgare*) [1K].

Watercress (*Nasturtium officinale*) [1I], a wild aquatic plant belonging to the mustard family, is an ancient salad plant and it has a distinctive pungency. Purslane (*Portulaca oleracea*), which grows in Iran, Africa and Asia, and the leaves of the sorrel (*Rumex acetosa*) [3G], also add a notable flavour. A wide selection of herbs, including basil [3I], tarragon and parsley, is another addition. A dressing of oil and wine vinegar lends the salad a final piquancy.

An ancient Chinese art

The Chinese culinary art is one of skill and distinction. With infinite patience cooks have evolved a style that preserves the natural tex-

CONNECTIONS

See also
188 Vegetables

1 The salad bowl provides an endless variety of fresh vegetables. The cabbage lettuce [A], with its leaves folded into a neat crown, has a mild taste and pleasing texture. Forming a tall, loose head, leaves of the cos or Romaine lettuce [B] are stronger in flavour, while butterhead lettuce leaves [C] are thick and oily. Sea kale (not used as a salad vegetable) [D], chicory [E] and curly endive [F] are all rather bitter vegetables. When chicory is blanched the leaves add a suggestion of sharpness to a green salad. The white and seedy centre of the cucumber (*Cucumis sativus*) [G] is another popular salad ingredient. Mustard greens [H] and watercress [I] are often used in salads as a garnish. The radish [J] adds a colourful finishing touch and Florence fennel [K] and celery [L] a complementary crunchiness.

tures and flavours of foods. The Chinese were among the first to understand and respect the value of partially cooked vegetables. They were using the cultivated lettuce as early as the fifth century BC, but *pe-tsai* or Chinese cabbage (*Brassica pekinensis*) is more commonly grown. Chinese cabbage has a long, sculptured heart, crisp as celery, and makes an excellent salad. The leaves are stir-fried, a classic method of Chinese cooking in which the thinly sliced food is turned quickly over and over in hot oil. *Pak choi* (*Brassica chinensis*), which is also Chinese cabbage, is looser and more floppy. In China cabbage is frequently used in soups. A hot leaf mustard, or mustard greens (*Brassica nigra*), is pickled in salt and vinegar. This species is also cultivated for its seeds.

The radish (*Raphanus sativus*) [1J], often used in Europe merely as a colourful garnish to the green salad, is an important vegetable in the East. An ingredient in many Japanese dishes is the *daikon*, a long, white radish of mild flavour. The radish is probably of Oriental origin and another large, firm winter root much grown in the East is known as

Chinese radish. The delightful pale green *wasabi* is also used for seasoning. The flavour of this root, stronger than horseradish, can be preserved powdered in tins.

Chinese skill in the use of familiar vegetables is admirable, but their imaginative use of other plants is unique. Some of these, if seen in the West at all, serve only a decorative purpose. The tiger lily, for instance, is a plant prized in Europe for its beauty. In China its buds are dried into pale orange strips known as golden needles and several Chinese dishes are enriched by their aromatic properties. The lotus [2D], sacred to Buddha and Kwan Yin, the Goddess of Mercy, is also eaten raw as a vegetable. The seeds of the lotus are preserved in syrup and its leaves are wrapped around spicy food for steaming. All over the Orient young shoots of the bamboo (*Bambusa* sp) [2C] are eaten raw. The water-chestnut [2B] and the Chinese water-chestnut [2F] are other crisp vegetables. Among the most beautiful water vegetables are the many kinds of seaweed. The Japanese have harvested the branched fronds of these nutritious plants for centuries.

Three common, wild plants make pleasant-tasting additions to a green salad. Young leaves of the dandelion (*Taraxacum* sp) [A] have a bitter taste. This persistent plant is especially common in grassland but can be found almost anywhere. Brooklime (*Veronica beccabunga*) [B] is found in slow-flowing streams and ditches. The leaves have a similar flavour to those of watercress. Cultivated fields and dunes are where corn salad (*Valerianella* sp) [C] grows. The leaves have a pungent taste. Other easily found wild salad plants are shown in illustration 3. Leaves are not the only edible part of wild plants. The pig-nut (*Conopodium majus*) has edible tubers.

2 Unusual plants have been in use in China for centuries and the lively regard of the Chinese for texture and flavour is reflected in this array of vegetables commonly used in their cuisine. The

pe-tsai [A] is a loose-leafed cabbage that can be eaten raw. An important vegetable rich in starch, the tasty and crisp waterchestnut [B] grows along the banks of the Yangtze River. Frequently used are

delicately flavoured bamboo shoots [C]. The perforated stem of the lotus [D], whole or sliced, is also used by the Chinese, as are fruits of Chinese waterchestnuts [F] and various bean sprouts [E].

3 Crushed leaves of the salad burnet (*Poterium sanguisorba*) [A] make a more unusual salad than fennel, chicory or watercress [B]. Flowers and leaves of nasturtium [C], sea beet (a variety of *Beta*) [D] and nettles (*Urtica dioica*) [E] can also be used. Valerian [F] and sorrel [G] are found on grasslands, while basil [I] grows better on dry chalk. Collard greens [H] also add interest to a salad.

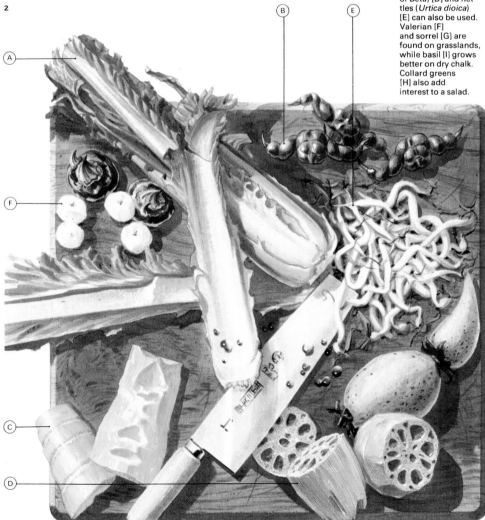

Temperate fruits

Fruit is the botanical name given to the part of a flower that persists after the blossom dies and forms a vessel enclosing the seeds until they are ripe. The seed-vessels of some plants have fleshy exteriors that form an attractive food for animals, which eat the flesh and the seeds. Most seeds pass unchanged through an animal's digestive tract and pass out in the faeces. In this way, animals become the means by which the seeds are dispersed.

Despite the botanical definition, a fruit-grower's use of the word "fruit" is limited to seed-vessels that are juicy and edible by human beings [Key]. He also recognizes nuts as fruit. His chief concerns, therefore, are the quality and size of the edible part, whether he can grow it successfully in his particular area and whether he can sell it at a profit. The fact that seeds are also produced is not relevant, although he knows that their presence is essential for the stimulation and development of the fleshy part he sells.

Fruits have always been an important part of man's diet. Initially, he gathered them directly from the wild, later he planted trees and bushes so that they grew conveniently near his dwelling-places, and finally he experimented by cross-fertilizing selected varieties, thereby improving the quality and appearance of the fruit.

Citrus fruits

The orange, lemon, lime, grapefruit and ugli are citrus fruits – an important group in which the "flesh" is formed by hairs radiating from the seeds. These hairs or pulp-vesicles are swollen and juicy, giving the fruit its characteristic texture. Citrus fruits have their origin in the East. Sweet oranges and lemons were first grown in China and limes in India, and sweet oranges did not appear in the West until the seventeenth century. The grapefruit is thought to be of hybrid origin.

In most citrus fruits the fibrous flesh is eaten and the rind discarded, although in the citron and some others the fruit is gathered mainly for the rind. The rind of citrus fruits contains tiny reservoirs of oil. Orange and lemon rinds give body and flavour to marmalade, and they can be candied and used in confectionery and for giving distinctive flavours to cakes and puddings.

Lime juice was a valuable commodity in the eighteenth and nineteenth centuries when the voyages undertaken by sailing ships were so long that vegetables could not be kept fresh. Lime juice was drunk by the sailors to ward off scurvy, a disease caused by a lack of vitamin C which fruit and vegetables, particularly citrus fruits, contain. The name "limey", given to British sailors of that time, derived from the practice.

Fruits of the rose family

A large number of fruits come from plants of the rose family (Rosaceae); these include the apple [3, 5], pear [5], cherry [4], peach [8], apricot [7] and plum [6]. Trees of this family are hardy and thrive in temperate regions. Apples and pears originated in Afghanistan, all varieties of apples being developed from the wild crab apple there. Today apples represent the second most important fruit after the grape and the major producers are the United States, France, Germany, Switzerland, Italy and the Balkan countries.

Larger trees, generally associated with apple orchards [1] planted after World War

1 Commercial apple orchards in England are found throughout the region south of a line joining the Mersey and the Wash. Most cider apples are grown in the West Midlands and in the West Country.

2 Fruit trees in gardens are grown on a smaller scale and can be trained against a wall espalier style, like this pear tree, so as to take up less space and not shade other plants.

3 The apple is one of the most important, widely cultivated fruits grown in temperate climates. Shown here is the Cox's Orange Pippin, an eating apple that is widely popular.

4 The cherry is a type of fruit known as a drupe, which takes the form of a single seed surrounded by fleshy fruit. Cherries date from Roman times and the one shown is the black Early Rivers variety.

1 Quince
2 Grapefruit
3 Ugli fruit
4 Crab apples
5 Newton apple
6 Red Richard apple
7 Conference pear
8 Cyprus lemon
9 Corinam pear
10 Sweet orange
11 Seville orange
12 Nectarine
13 Yellow plum
14 Clementine
15 Red plum
16 Rainbow-stripe cherry
17 Cyprus lime
18 Tangerine
19 Morello cherry
20 Starking cherries

5 Modern fruit is a result of thousands of years of selection of chance seedlings and mutations (sports) and, during the last 200 years, of controlled breeding. Large apples were developed from species such as the crab apple (*Malus sylvestris*). Pears (*Pyrus* sp), plums and cherries (*Prunus* sp) were derived from wild European species. The quince (*Cydonia* sp) originated in the Middle East, citrus fruits (*Citrus* sp) in the Far East. Modern techniques make fruit available at any time.

II, are becoming less economic because of the high cost of pruning and picking. For this reason most modern orchards have smaller trees planted close together so that work can be done from ground level. This has been made possible by the development of dwarfing and semi-dwarfing rootstocks.

Growers are also experimenting with mechanical pickers but the techniques so far developed damage the fruit too much for it to be usable for high-quality sales or long-term storage. Some apples for cider, however, are gathered by machines.

Pears thrive in warmer temperate regions. In Canada, Australia and South Africa a large proportion of the crop is canned but in Europe canning is less important. In France, Germany, Switzerland and England some of the crop is made into a fermented drink called perry.

Fruits of the vine
Grapes are the world's most important perennial fruit and are grown for wine, dessert use or dried as currants and raisins. Nearly all European and Mediterranean

grapes go into wine; in California, more than half of the crop is dried, one-third is used for wine and the remainder is eaten fresh or canned. Each variety of grape is grown for a specific purpose.

Most varieties of melon are derived from the musk melon *Cucumis melo* [10] and include the cantaloup melon, a dimpled and rough-skinned fruit grown in Italy in the fifteenth century from seed brought from Armenia, and the honeydew melon, which probably originated in South-East Asia and was eaten by the ancient Egyptians and Persians. The watermelon probably had its origin in Africa.

Melons and grapes are susceptible to damage by frost so their open air cultivation is restricted mainly to warm-temperate and subtropical climates.

Temperate fruit trees are nearly all deciduous. Most can resist temperatures of below −7°C (20°F) during winter dormancy and many cannot grow in tropical or subtropical climates because prolonged exposure to low temperatures in the winter is necessary for normal development.

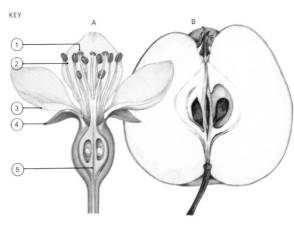

Apple blossom [A], when examined, shows features typical of an angiosperm flower – stigmas [1], stamens [2], petals [3], sepals [4] and an inferior ovary [5] (one that lies below the other flower parts). After the ovaries have been fertilized (usually by bees), the flower parts fall away leaving the receptacle (the "box" containing the ovaries) to become fleshy and swell and grow into the fruit. The flesh of the fruit [B] is formed from the receptacle wall while the ovary wall remains as the leathery core. In other fruits different parts of the ovary structure may become the fleshy portions of the plant that are edible.

6

6 The plum (*Prunus domestica*) is a hybrid of two other fruits, the cherry plum and the sloe. The various varieties that have been derived from it are the mainstay of he plum and prune industries of the world.

7 The apricot also belongs to the genus *Prunus* but is less hardy than the plum, requiring protection from frost and fungal diseases. The yellow and orange-coloured fruits are rich in sugar and contain iron and vitamin A.

7

8 The peach, like the apricot, originally came from China. Both fruits are marketed fresh, dried or canned according to the variety. There are two types of peach – the freestone and the clingstone – which differ according to the ease with which the stone is removed from the flesh.

8

9

9 The persimmon is a tomato-shaped fruit grown extensively in China and Japan and introduced to southern Europe and the United States in the 19th century where it is grown only on a small scale. The yellow-red fruit is rich in vitamins A and C and is astringent until it has ripened.

8

7

5

3

1

10

10 Melons and grapes both need a great deal of sun and water. The fruits of the melon may need to be supported in nets while growing. Most grapes are grown for wine but many fine varieties are grown as desserts or for drying.

1 Musk melon
2 Ogen melon
3 Winter melon
4 Cantaloup melon
5 Balsam pear
6 Watermelon
7 White dessert grapes
8 Black dessert grapes

4

2

6

Tropical fruits

Palm-fringed tropical islands laden with exotic fruit must have seemed like paradise to early Western explorers. Their first impressions were justified, for in the tropical regions of high annual rainfall and high temperatures the variety and luxuriance of plant life is unparalleled.

Long inhabited by man, the tropics have provided fruits of great antiquity. The great oblong leaves of the banana plant [7], sometimes 4m (13ft) in length, were said to have hidden the serpent in the Garden of Eden. The banana was known to the Arabs, and the Roman scholar Pliny mentions that Alexander the Great saw it in India in 326 BC. The Islamic conquerors carried the banana to northern Egypt in AD 650 and later on to West Africa. Fifteenth-century slave-traders took it to the Canary Islands and from there it was introduced to the Americas. The banana finally reached Mexico with the Spanish *conquistadores* in 1531. There are now more than 100 varieties of the cultivated banana. Besides the sweet dessert fruits, there are the plantains for cooking and a particular form grown for making beer.

Staff of life

The coconut not only contains an edible kernel and a natural, refreshing milk, but its fermented sap makes a heady toddy which, when distilled, becomes a potent spirit.

One of the most important oils is obtained from copra – the dried kernel, or meat, of the coconut. Coconut oil is used in the manufacture of soaps and detergents, margarines and edible oils, and a host of other products. The husk itself yields coir, a tough fibre used in making ropes, mats and baskets. Mature palm leaves are also utilized in basketry and thatching and the trunk of the coconut tree is used as rough timber for constructing native huts and fences.

The breadfruit [5A] is another invaluable

No tropical fruit, however, is as well-travelled as the coconut [2E], famed throughout the world as one of man's most bountiful providers. Growing along sandy coasts, the coconut palm's ripe nuts are swept by oceans to many tropical shores. The soft eye of the husk produces a plant which quickly takes root in new ground.

tree and is a staple food throughout the islands of Polynesia. The pulp of its fruit, roasted in its skin before it is fully ripe, tastes like freshly baked wheat bread. The first explorer to mention breadfruit was Captain William Dampier on his return from Guam in 1688. Captain William Bligh's mission in 1786 was to bring breadfruit plants from the Society Islands to the West Indies. The story of the mutiny on the *Bounty* is well known, but Captain Bligh's later voyage in 1791 was successful and the breadfruit is now cultivated from Florida to Brazil.

Symbol of fertility

One of the oldest Semitic representations of life and abundance was the multi-seeded fruit of the pomegranate tree [4C]. King Solomon had an orchard of the fruit and the pillars of his temple were decorated with carved pomegranates. In Babylon it was served at wedding feasts and it played a similar role as a symbol of love and fertility in the Far East. The fat, ripe fruits were thrown on the floor of the bridal chamber so that the seeds should burst from their smooth skins. Centuries later

CONNECTIONS

See also
192 Temperate fruits

In other volumes
70 Man and Society
70 The Natural World

1 **The importance of fruit as a food** in the tropics is evident from a glance at the stalls in any of the street markets. Oranges and bananas are important crops and they are brought to market in carts by the growers at regular intervals and are displayed in an inviting manner to the passers-by. The growers judge the timing of their visits to fit in with growing schedules and consumer demand.

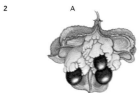  

2 **Astonishing variety** characterizes the tropical fruits. Akee [A], although a fruit, is cooked as a vegetable and tastes like a fine omelette. When the fruit ripens it bursts open to reveal

fleshy cream-coloured arils, the edible parts of the akee. Jack fruit [B], a giant which might weigh up to 31kg (70lb), may be eaten raw or cooked. It has an unpleasant odour when cut, but

the flavour of its creamy yellow pulp is more appealing. The sapodilla [C] is a Central American fruit with rough brown skin and a slightly grainy pulp. The sweet, khaki-coloured flesh

tastes like faintly astringent brown sugar. It is eaten when fully ripe. Papaya, or pawpaw [D], sprinkled with lemon juice and sugar, is eaten for breakfast in the tropics. Its smooth

skin ripens from green to yellow or orange and its succulent pulp encloses small black-brown seeds in its centre. Papain, an enzyme contained in the leaves and unripe fruit, is used to

tenderize meat. The coconut [E] has a hard, fibrous outer shell and a hollow centre. Its flesh is most familiar in the West in its desiccated form. The soft translucent flesh of an immature coconut

is enjoyed only in the tropics. One of the world's most important crops, the coconut derives its name from *coco* – a Portuguese word meaning grimace – because it resembles a grinning human face.

  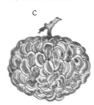

3 **Native to Malaysia**, the red fruit of the rambutan [A] is covered with soft spines. Its name comes from the Malay word for hair. It is a close relative of the lychee and resembles it in size. The fruit has a pleasant, almost acid flavour.

The citron [B] was one of the earliest fruits to arrive in the Mediterranean from the Orient. The pulp of the citron is extremely sour, but the thick, furrowed rind has an agreeable flavour. Cured in brine or sea water, this coarse outer

covering is used to make candied peel, a popular confection. The custard apple [C] or bullock's heart of the West Indies, belongs to the genus *Annona*. Its soft, sweet custard-like pulp has given the fruit its name. The seeds are poisonous

and the skin resembles a quilted fabric. The sweetsop, indigenous to tropical America, is closely related to the custard apple. The snow-white flesh of the mangosteen [D] is delicate in texture and has a rather sweet-sour taste. The

thick, dark purple rinds surround a cavity in which the pulp lies in segments like that of a mandarin. The mangosteen has been highly valued in Indonesia and the Philippines since early times but only since the nineteenth century has

it been cultivated in the West Indies. The flower of the passion fruit [E] is said to have given this fruit its name because it symbolizes the passion of Christ. The corona resembles the crown of thorns and the five sepals

and five petals have been interpreted as the ten Apostles: Judas, who betrayed Christ and Peter, who denied Him, being left out. The passion fruit is a delicately flavoured and highly aromatic fruit with a tough purplish, often wrinkled skin.

the prophet Mohammed claimed another virtue for the pomegranate. He said that eating the fruit would banish envy and hatred. The pomegranate also became the emblem of Granada and Ibn-al-Awam, a thirteenth-century Moor, recorded some ten varieties of the fruit that flourished in southern Spain at the time. Spanish colonists probably took the plant to the New World and today it is grown in gardens from the warmer areas of North America to Chile.

Fruits of hallowed legend

Throughout history the fig [4A] has played an important role in mythological tales. Adam and Eve covered their nakedness with fig leaves and it is thought that the Tree of Enlightenment which grew in Buddha's garden may have been the bo, or sacred fig. In Latin myth the fig was held sacred to Bacchus, god of wine, and was used in many religious ceremonies. It was also regarded by the Romans as a badge of prosperity because it grew over the wolf's cave where Romulus and Remus, the legendary founders of Rome, were discovered.

The fig is undoubtedly one of the earliest trees cultivated by man. It spread all over the Aegean and Levant centuries ago. It formed part of the staple diet of the Greeks and both fresh and dried figs are still so widely used in the Mediterranean that the fruit is commonly called "the poor man's food". In southern Asia the leaves, bark and fruit of the sacred fig are used in folk medicine but in India it is planted as a religious object and revered by Brahmans and Buddhists alike.

The mango [5C] is yet another fruit held in great esteem in India. Akbar, the Mogul emperor who ruled at Delhi in the sixteenth century, planted an orchard of 100,000 mango trees and one of the ancient names of the fruit stems from a Sanskrit word meaning food. Buddha also had a mango grove in which he found solitude for philosophical thought. The English name comes from *man-kay* or *man-gay*, a Tamil word adopted by the Portuguese in India. The tree did not appear in the Western Hemisphere until about 1700, when it was planted in Brazil. The juicy mango varies in colour from yellow to orange and has a succulent spicy flavour.

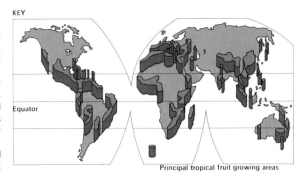

KEY

Equator

Principal tropical fruit growing areas

Tropical plantations producing exotic fruits for the world market are found in the inter-tropical zone stretching between the Tropic of Cancer and the Tropic of Capricorn. About one-quarter of the earth's land surface lies within this region. Many of the crops characteristic of tropical agriculture are also cultivated in warmer parts of the temperate zones. The plantation system is very old in tropical areas. Coffee, tea, coconut palms, pineapples, bananas and sugar cane are cultivated in this way. Tropical temperatures vary slightly from one month to the next and crops are grown the year round. The mango, citrus and breadfruit trees are intensively cultivated in South-East Asia, the fruits of the guava and papaya trees in South America. Improved transportation has sent these luxuries to countries all over the world.

4

A
B
C
D
E

4 Heavily seeded figs [A] belong to a family of more than 1,800 species. Pliny, the Roman scholar, mentions that slaves were not given the figs eaten by Roman epicures, but an inferior home-grown variety. The sweet, ambrosial figs came from ancient Ebusus but the fruit is indigenous to an area stretching from eastern Turkey to northern India. The durian [B], a favourite fruit in South-East Asia, has a repellant aroma. The delicate, creamy-white pulp of this large oval fruit has a soft, glutinous quality. But the fruit, despite its foetid smell, is sweet to the palate.

The flesh of the pomegranate [C] is densely packed with seeds that scatter when the fruit is burst. The French named the grenade after the pomegranate because of its similar properties.

Its refreshing flavour makes it popular in the arid areas where it grows. The lychee [D] is of Chinese origin and is the favourite fruit of the Cantonese. The skins are hard and scaly, turning from reddish-pink to brown. A white juicy pulp surrounds a single shiny brown seed. The lychee combines the subtle perfume of good-quality grapes with a delectable flavour. The small, oval, citrus-like kumquat [E] was originally cultivated in China and Japan and is used there mainly as a candied fruit. The mild acid flavour is excellent for making marmalade or bottled preserves.

5

A
B
C

7

5 In prehistoric times the breadfruit [A], from a tree native to Malaysia, spread to the Pacific where it became a staple food. The skin of the ripe fruit is brownish-green and the flesh white and rather fibrous. The fruit is baked or boiled, or sliced and fried like a potato. The inner bark of the tree is used to make cloth, and canoes are built from its wood. The soursop [B] is closely related to the custard apple. Its sour-sweet pulp is used to make soft drinks and sherbets. The heart-shaped, dark green fruit is covered with fleshy spines. The king of the tropical fruits, the mango [C], is an oval or kidney-shaped fruit with a delicate fragrance. Its juicy flesh surrounds a single flat seed. The mango is grown as a garden plant throughout the tropical regions.

6

A
B
C

6 "Deliciousness itself" is how American writer Mark Twain described the cherimoya [A]. Its texture and taste resemble a delicate ice-cream flavoured with pineapple and banana. Indigenous to the tropical highlands of Central and South America, it is related to the custard apple. The loquat [B] is a native to China and has been cultivated in Japan for centuries. It is often planted in parks and gardens. The fruit is borne in large, loose clusters and has an agreeably tart flavour. The flesh, whitish to orange in colour, surrounds three or four large seeds from which the plant is frequently propagated. Borassus palm or palmyra palm grows wild in South India, Burma and Sri Lanka. The immature nut [C] provides a refreshing drink and the soft kernel is edible. In India the fruit is used mainly for making sugar. A toddy is also made from the fermented sugar and liquid provided by the palm.

7 The banana was one of the first fruits to be cultivated by man and is believed to have originated in the Asian tropics. The banana "tree" is really an elongated bundle of leaf bases with a single flowering stem emerging from the top. The fruits, growing in bunches called hands in layers on the fruit stalk, are produced without pollination and so have no seeds. They require 75 to 150 days to mature and must be removed from the plant to ripen properly. After fruiting the stem is cut down or collapses and a new tree develops from younger buds on the underground stem. There are several varieties of banana grown, each suitable for a different purpose. The best flavoured is the Gros Michel variety but because of its disease susceptibility others such as the Cavendish are becoming important in the American tropics. Most bananas have yellow skins but some have skins that are red. These do not travel well and so they are eaten locally. Plantains are very large bananas that are rich in starch. They cannot be eaten raw and are used only for cooking. The Canary banana was introduced into the islands in the fifteenth century by slave traders and has now become a staple food in many other areas.

Sugar and honey

Sugars are part of a group of chemicals known as carbohydrates and are found in a wide variety of foods. The most common forms are sucrose, glucose and fructose. Less sweet varieties are lactose (found only in milk) and maltose, which is a by-product of grain germination. A well-balanced diet already contains enough sugar to provide energy and supplements are unnecessary.

Honey and its formation

The first concentrated source of sweetness available to man was probably honey, which was prized both as a food and a medicine. The association of honey with health has lingered to the present day. As it is an almost immediate source of energy its restorative reputation possibly has some slight basis in fact, but any other source of carbohydrate would do as well. Honey does have mildly antiseptic properties and can be used locally in the treatment of burns and cuts.

Honey is a sweet, viscous liquid that varies in colour and taste according to the source of the nectar collected by the bee. Single-blossom honeys have quite distinct

flavours: herbs produce a light, aromatic honey while honey from pine flowers is darker with distinctly resinous overtones. The most common single-flower honey produced by commercial beekeepers is clover, which is a pale amber and mildly flavoured, but heather honey is also common in some country districts, especially in Scotland. Most honeys sold today are derived from a mixture of blossoms since this involves less disturbance of the hive during the production season (from spring to autumn [1, 2]).

The honey bee (*Apis mellifera*) is a social insect with a rigidly organized life. There are three kinds of bee in each hive – the queen bee which lays the eggs, male bees (drones) which fertilize them, and female worker bees. The small worker bees build the wax cells in which eggs are laid or honey is stored. They collect nectar from the flowers in their honey sacs. The nectar is regurgitated and stored in the cells of the combs, where the honey forms by the conversion (inversion) of sucrose to glucose and fructose and the evaporation of moisture.

Millions of flowers are needed to produce

1kg (2.2lb) of honey but thousands of tonnes of honey are produced annually. Honey extracted from the comb may crystallize or granulate especially if it is held below 10°C (50°F). The honey can be cleared by gently warming the jar. Some honey is also sold in the comb and regarded as a delicacy.

Sugar fads and facts

The sugar extracted from sugar cane and beet [Key, 3, 4, 5] is sucrose. The juice can be refined to a point of more than 99 per cent purity to make white sugar, chemically one of the purest foods available. The brown sugars so highly prized as a health food contain a percentage of residue from the cane and beet crushing processes but are no more "natural" than white household sugar [8]. Any form of sugar, including honey, can be harmful if taken in excess, the surplus energy produced being stored in the form of fat which may contribute to heart disease. Sugar in sweets may cause tooth decay.

The average daily consumption of sugar by the developed nations in 1975 would have been unthinkable even 200 years ago. The

1 Commercial bee hives are divided into three or four sections called supers [2]. In each is a frame [3] with a wax base used by worker bees [1] as a foundation for hexagonal cells in which eggs are laid and honey is stored. The queen bee is confined to a brood or lower chamber by means of a grid [4] through which only worker bees can pass. They enter and leave the hive at the base [5].

2 Honey in the hive is capped with wax by worker bees to keep it fresh for winter food [A]. Towards the end of the season the frames are removed from the top two or three supers and the wax cap is cut away with a heated knife [B]. Honey is extracted from the comb in a centrifugal machine [C] with a fine mesh basket which is rotated by a handle at high speed. The honey expelled drips into a reservoir and is run off through a straining mesh into a ripening tank [D] where it is left to stand, thus allowing any air bubbles to escape before the wax-free product is run into jars [E].

3 The sugar cane grown commercially is *Saccharum officinarum*, a giant tropical grass growing to a height of 3–4.5m (10–15ft). The cane can be grown in most soil conditions but must have a constant temperature of about 30°C (81°F) during the main growing period. Although it is tolerant of drier conditions, 75–125mm (3–5in) of rainfall a month are needed to enable it to grow to 2m (6.5ft). In most areas the cane is propagated from sections of the main plant about 1m (3ft) long which are planted in furrows [A] where dormant buds send down roots and produce shoots [B]. Maturity is reached after one to two years when moisture and sugar content are highest. Harvesting is by hand or machine.

4 Cane sugar is extracted in factories near the cane fields. Cut cane is unloaded, washed, then passed through a shredder. Twin crushing rollers break up the cane and pass it to several mills with rollers under hydraulic pressure of up to 600 tonnes per sq cm. These press out the juice, leaving a crushed cane mixture known as bagasse which is sprayed with water before passing to the last mill to aid final juice extraction. The juice, containing up to 90% water, then passes through a clarifier where heat and milk of lime precipitate impurities. A series of evaporators and a vacuum pan reduce the juice to a syrup known as molasses which contains sugar crystals. Separation of the crystals in a centrifuge produces molasses and the raw sugar that is sent to refineries.

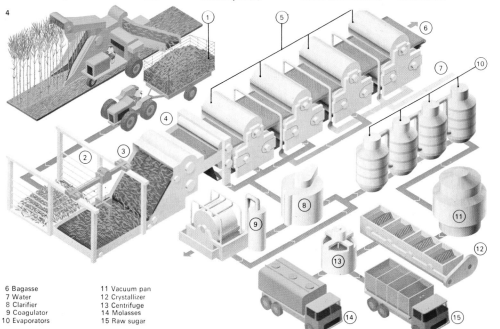

1 Cut cane
2 Washing bay
3 Shredding knives
4 Crushing roller
5 Mills
6 Bagasse
7 Water
8 Clarifier
9 Coagulator
10 Evaporators
11 Vacuum pan
12 Crystallizer
13 Centrifuge
14 Molasses
15 Raw sugar

high-yielding sugar cane was introduced into Europe during the ninth century; cultivation was not successful. Supplies came instead from the Middle and Far East, through refining factories established there by Venetian merchants at this time and the tradition continued for several centuries. The expansion of world trade during the fourteenth and fifteenth centuries encouraged the development of plantations in the New World, particularly the West Indies. Some South American countries now produce a significant proportion of the world's sugar and so does India. In the mid-nineteenth century, production from sugar beet became important in the Northern Hemisphere and about 40 per cent of the world's sugar now comes from this source [5, 6, 7]. Recent enzyme research has proved that sugar can be manufactured very economically from maize and this may prove to be an important source in the near future.

Maple syrup is the reduced sap of the maple tree and is expensive to produce [9]. The extraction and preparation of other natural syrups is equally costly and these have largely been replaced in the kitchen by syrups manufactured from cane or beet sugar. These include treacle and golden syrup, which are used to give both cohesion and taste to cakes and batter and as glazes for meats and root vegetables. Such syrups have a viscous consistency similar to honey and are artificially flavoured.

Honey-based beverages

Honey and sugar syrups have long been used in medicine to help disguise the bitterness of herbal remedies. This is still a common practice and mixtures containing honey and lemon are often taken to ease the symptoms of colds and influenza. Less common is the production of an alcoholic drink from honey. A mixture of fermented honey and water forms mead, once a popular drink in northern countries where grape vines did not flourish. During the late Middle Ages beer was introduced and this eventually relegated mead to the home brewers' shelves. The name mead is derived from the Sanskrit word for honey, *mehdu*, which may indicate that mead originated in the Middle East or Asia.

Sugar can be extracted from the stems of sugar cane or from the tap roots of the sugar beet. Sugar beet (*Beta vulgaris*) is a biennial planted as seed in spring to mature the following autumn. Mechanical drilling and thinning gives 30,000 beets per acre. Well-drained loam, free of acid, is the ideal soil. Salt and chemical fertilizers are used to increase sugar content. Selective growth has produced a single tap root from which soil is easily removed after harvesting. Sugar beet now forms part of crop rotation in most European countries, Canada, Japan and Russia. Sugar is extracted in much the same way as from sugar cane but the yield is lower.

Sugar beet   Sugar cane

5 Harvesting of sugar beet extends over a three to four month period during which time the factory works 24 hours a day. Crops are timed and harvesting regulated to achieve an even flow of sugar beet to the factory. The harvester scoops up the roots and removes the tops, which are left in the field either to be ploughed back or to provide fodder for cattle. After slicing and crushing, the juice goes through similar processes to those used in cane sugar production, although it takes about 7 tonnes of beet to produce a tonne of raw sugar. Beet pulp and molasses are used as stock food.

6

Main sugar beet areas

Main sugar cane areas

Equator

6 Sugar cane and sugar beet produce the same sugar but require completely different climatic conditions. The cane is grown as a single crop in areas between the tropics of Cancer and Capricorn. Sugar beet forms part of regular crop rotation in Europe, Russia and North America, where it is also a source of food for cattle. More recently beet has been grown in both Japan and South America.

5

7 World production of sugar is led by Russia with its vast sugar-beet growing areas. But the most concentrated sugar output in the world is achieved in Cuba with fertile plantations of cane.

7 | Cane in tonnes | Beet in tonnes

USSR 8,804,000
Cuba 7,559,000
Brazil 5,019,000
India 4,634,000
USA 4,385,000

Others 42,500,000

8 Light brown sugar

Coffee sugar crystals

Preserving sugar

Dark brown Barbados sugar

Granulated sugar
Icing or confectioner's sugar
Raw demerara sugar
Rainbow sugar
Soft brown sugar
Fine granulated castor sugar

8 Different grades of refined sugar result from selection by crystal size. They include granulated and castor sugar. Icing sugar is milled to a fine powder and a chalky powder is added to prevent lumps forming. Soft brown sugar, Demarara and Barbados sugars are taken out of the refining process earlier and thus still contain some impurities.

9

9 Syrups can be derived naturally from plants or produced artificially. A distinctively dark and resinous syrup is collected from the maple in Canada and North America, particularly the New England states. North American Indians used the sap of these trees to boil down a syrup long before the arrival of Europeans. In much the same way as rubber is obtained, the sap is tapped as it rises in early spring. Maple syrup is considered to be a luxury because it requires about 40 litres of sap to make a litre of syrup. Other plants that contain small quantities of sweet, thin syrup are the date palm, palmyra and coconut. Artificial syrups are cheaper to produce and consist mainly of sugar boiled with water to make a viscous liquid. Those produced during the refining of cane or beet sugar include "black" treacle and golden syrup.

Origins of wine

The history of wine has marched hand in hand with the history of Western civilization. In the surviving artefacts and records of the ancient peoples of the Mediterranean – Egyptian [1], Phoenician, Minoan, Greek – the vine and its juice are always present: in the evidence of daily life and worship, myth and poetry. The Mediterranean itself, a maritime market place crossed by early traders, was the "wine-dark sea" of Homer.

Greek and Roman vineyards
The arrival of the vine in those European countries with which it is now principally associated probably coincided with the spread of Greek influence several hundred years before Christ. It was then that the vine was first planted on sites where it was to settle so happily – Italy and France. It is also likely that at that time vineyards were first established in Spain and in North Africa.

The wines of Greece, themselves praised and generously documented by its poets, might not have been admirable by modern standards since they were mixed, diluted with water, and often spiced and seasoned. Such wine might have tasted like *vin rosé* and honey, perhaps with a touch of muscat, perhaps with a taste of resin and possibly with concentrated flavourings.

Italy, to the Greeks, was the "Land of Vines" and to the Romans posterity owes its greatest debt. They took the vine to Gaul.

The greatest Roman writers, including Virgil, wrote instructions to wine growers. One sentence of Virgil's – "Vines love an open hill" – might be called the best single piece of advice that can be given to a wine grower. There has been much speculation about the quality of Roman wine. It apparently had extraordinary powers of keeping, which in itself suggests that it was very good. The great vintages were drunk and discussed for longer than seems possible. The famous Opimian – from the year of the consulship of Opimius, 121 BC – was being drunk even when it was 125 years old.

Wine follows the legions
The Romans had all that was necessary for ageing wine. They were not limited to earthenware amphorae like the Greeks – although they did use these. They had barrels like modern barrels and bottles not unlike modern bottles. It is possible that the Italians of 2,000 years ago drank wines slightly similar to those of their descendants today: young, casually made, sharp or strong, according to the summer weather. The Roman practice of training of the vine up trees, in the festoons depicted in the friezes on classical buildings, is still practised, particularly in some parts of southern Italy and in northern Portugal.

In the first centuries after Christ, when the power of Rome was at its greatest in northwest Europe, the vine followed in the footsteps of the legions [7]. By the time they withdrew in the fifth century they had laid the foundations for almost all the greatest vineyards of the modern world.

Starting in Provence where vineyards had thrived for centuries (those of Marseilles – a port established by the Greeks – and the Languedoc had been planted by the Greeks), the vine spread up the valley of the Rhône and either by sea or overland to Bordeaux in the first century. It arrived in Burgundy and the

CONNECTIONS

See also
200 Vines and grapes
202 Wines of the world
206 Spirits and distilling
204 Beer and brewing

1 The vine was cultivated in Egypt in about 6000 BC. On the wall of the tomb of Nakht, an official of Thebes who lived about 1,500 years before the time of Christ, there is a mural showing a group of men treading the grapes in a vine arbour. Wine was safer to drink than water.

2 Medieval workers are seen tying up the vines and harvesting grapes in these woodcuts which were used to illustrate the 1493 Speyer edition of Piero Crescentio's *Opus Ruralium Commodorum*. These crisp and expressive woodcuts have been reprinted repeatedly ever since.

3 Wine played an important role in ancient life. The Romans used wine-making methods that have changed only little with barrels and bottles similar to those used today. A 4th-century Roman mosaic [A] shows grapes being trampled in a trough. Earlier still, a Greek feast [B] is shown on a wine vase of about 480 BC which is now in the British Museum. Such scenes were a favourite motif of Greek vase painting. The guest on the left is playing the then fashionable after-dinner game of *kottabos*. This consisted of throwing the dregs of the wine in the cup at a target – a dish delicately balanced on a pole. The game was popular enough for partygoers to be coached in the finer points of "wine wasting" sessions. Only the most determined drunkards took their wine neat, without diluting it or mixing it with other beverages.

4 In the Bayeux tapestry, Bishop Odo of Bayeux and half-brother to William the Conqueror, blesses the wine before the departure from Normandy and the historic invasion of England.

Loire [6] next and finally reached the Rhine, Mosel and Champagne.

All the early developments were in the river valleys, the natural lines of communication, which the Romans cleared of forest and cultivated. Wine is a heavy commodity to carry and boats would have provided an efficient means of transport. Of the great French wine regions, only Alsace and Champagne do not have Roman origins.

Spreading far and wide

The vineyards of France and of the Roman homeland survived the savageries of the Dark Ages following the fall of the Roman Empire. The forays of the Viking raiders, using the great rivers of western Europe – the Rhine, Seine, Loire, Gironde and Rhône – took them not only to settlements to sack, but also to the vineyards. They left their mark. Some historians believe that the Vinland of Norse sagas was a region of northeastern America in which Viking mariners found wild grapes at the end of the first millenium.

Throughout the Middle Ages vine growing flourished, most of it under the control of the powerful Church. Expanding religious orders cleared hillsides and walled round fields of cuttings; dying wine growers and proprietors bequeathed vineyards to the Church, which used them to support many "hospices" – hospitals and homes for the aged – which survive today. Wine was essential not only in ritual but also as a medicine, restorative and disinfectant. For centuries the Church owned many of the greatest vineyards of Europe. By the sixteenth century wines were being made in Mexico in the New World and in the seventeenth century they were produced in South Africa.

Until the seventeenth century wine spent its life, as it had mostly done in Roman times, in casks. If bottles were used they were simply carafes for use at the table. The rediscovery of the cork some time in the seventeenth century, and its sequel – the discovery that wine in a tightly corked bottle lasted much longer than wine kept in a barrel – brought about a wine revolution. The rise of the great estates and the evolution of modern wines dates, therefore, from the eighteenth century – the Age of Enlightenment.

The Romans interpreted the graceful Greek god of the vine, Dionysus, as a more fleshy creature, if not a bloated mannikin. In a mosaic from Pompeii, now in the Museo Nazionale in Naples, he rides his traditional mount, a tigron, and drinks from a brimming pot. The Romans called him Bacchus.

5 A late 15th-century tapestry in the Musée de Cluny in Paris shows the nobility taking an active interest in the work of the vintage as the crop of grapes is gathered and pressed, on the banks of the Loire. The region is noted not only for the wine – red, white and pink, still and sparkling – but also for some of the most magnificent of France's great châteaux. The valley of the River Loire is the longest in France – more than 960km (600 miles) long.

6 The long march west of the vine from its likely origin in the Caucasus or Mesopotamia [1] – the Fertile Crescent – began in about 3000 BC when it reached Phoenicia and Egypt [2] where it was planted. By 2000 BC it was to be found in Greece and Crete [3] and 1,000 years later it was growing in Italy, Sicily and North Africa [4]. In 500 years it had reached Spain, Portugal and the south of France [5], and eventually farther north [6].

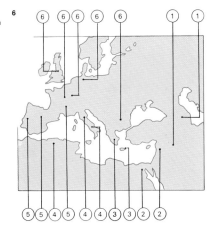

7 The most sumptuous of the famous prayer books of the Middle Ages was the *Très Riches Heures*, with 39 illustrations painted by Pol Limburg and his brothers for Jean, Duc de Berri, about 1416, at the height of the 100 Years War. In this prayer book the month of September is represented by the popular annual work at vintage time, with grapes being gathered beneath the battlements of the fairytale castle of Saumur.

Vines and grapes

Wine is the fermented juice of the grape. Every drop of wine represents rain recovered from the ground through the mechanism of the grape-bearing plant, the vine.

The classical European wine-vine, *Vitis vinifera*, one among many vine species (which include the common Virginia creeper), had its original home in Asia. Since earliest times, varieties of the species have been selected and propagated for various qualities, both for wine-making and for eating. Today there are 5,000 or more named varieties. Of these perhaps 200 and also a handful of varieties of the native American vines are commonly used for wine, and fewer than 50 are important to a wine-lover. But at least a dozen of these are all-important. They determine – along with the climate, the soil, the topography and the maker's technique – the style and character of wine.

Cultivation of the vine

A vine is a climbing plant that uses tendrils to grip and reach the tops of tall trees. Ancient wine-growers let the vine grow naturally; the Romans planted elm-trees for its conveni-ence. To this day such methods survive in parts of Italy and Portugal.

A modern vineyard uses very different techniques. The vine has become an indust-rial plant, pruned to put out a precise number of canes, which bear fruit and are then cut off each year. The grower thus limits its produc-tion to the number of bunches of grapes it can ripen fully, given satisfactory weather. As a result the exact desirable yield of each vine and each vineyard can be assessed, and in many areas legally enforced – excess wine may have to be disposed of for distilling. The general quality of wine is certainly much higher as a result of these controls.

Methods of propagation

Vines, like most other plants, will reproduce from seed, but this method of propagation is impractical since the seedlings produced can differ markedly from their parents and from each other. Seeds (pips) are used for the production of new varieties by crossing existing varieties, but new vineyards are planted with cuttings, either on their own roots or, more usually, grafted to a rooted cutting of another species. Great care is taken to ensure that the parent plant is healthy before cuttings are taken. The little slips are put into sand in the nursery for a season, until they form roots. They are then planted in rows in traditional vineyards, one metre (39 inches) apart – and increasingly two or even three metres apart; experiments have shown that the total yield of the vineyard in wine remains the same with half as many vines exploiting the same volume of soil.

As the vine grows old, its roots penetrate deeper into the earth. While it is young and near the surface, they are quickly affected by drought or floods – or the spreading of manure (which, put on the land top liberally, can affect the taste of the wine); the vine has little stability. But if the soil near the surface does not provide enough food, the vine will send its roots down deep. This perseverance is often rewarded by its discovery of valuable resources far from the surface [1].

Unfortunately the vine is subject to all manner of diseases. Some varieties are so susceptible to one particular disease, such as mildew [3], that they are gradually being

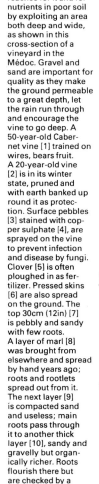

1 **Wine has its origins as water** in the soil. The vine reaches for it and for nutrients in poor soil by exploiting an area both deep and wide, as shown in this cross-section of a vineyard in the Médoc. Gravel and sand are important for quality as they make the ground permeable to a great depth, let the rain run through and encourage the vine to go deep. A 50-year-old Caber-net vine [1] trained on wires, bears fruit. A 20-year-old vine [2] is in its winter state, pruned and with earth banked up round it as protec-tion. Surface pebbles [3] stained with cop-per sulphate [4], are sprayed on the vine to prevent infection and disease by fungi. Clover [5] is often ploughed in as fer-tilizer. Pressed skins [6] are also spread on the ground. The top 30cm (12in) [7] is pebbly and sandy with few roots. A layer of marl [8] was brought from elsewhere and spread by hand years ago; roots and rootlets spread out from it. The next layer [9] is compacted sand and useless; main roots pass through it to another thick layer [10], sandy and gravelly but organ-ically richer. Roots flourish there but are checked by a compacted layer of sand at 1.2m (4ft) [11]. Below this, different colours of sand [12, 13, 14] lie in clearly defined layers. The grey areas [13] are pockets of damp sand where the rootlets can spread out to exploit the moisture.

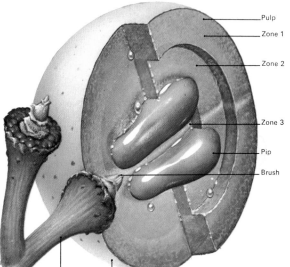

Pulp
Zone 1
Zone 2
Zone 3
Pip
Brush

Stalk   Skin

2 **A Riesling grape** is still green a month before the vintage, and will have reached about half its final size. As it ripens, it becomes translucent gold, with distinctive dark speckles upon its skin. The stalks are usually torn off before the grapes are pressed in present-day wine-making. Formerly they were left on, but they made the wine watery and sometimes bitter; in red wine they also absorbed val-uable colouring matter. The pulp divides nat-urally into three zones. Zone 2 gives up its juice in the press first, before the two zones in contact with pips and skin. This juice makes the best wine. The pips must come through the press un-scathed or they will make the wine bitter.

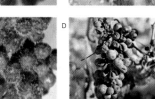

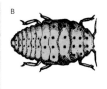

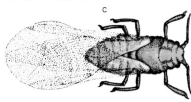

3 **The vine is prey** to a number of pests and diseases, among them the grub of the coch-ylis moth [A] which eats flower buds. Another pest is a tiny red spider [B] that sucks sap from the undersides of the leaves. Mildew [C] attacks anything green.

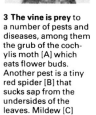

Mildewed grapes never ripen properly and have a peculiar taste. Oidium, or powdery mildew [D], is often more serious. It rots the stalks, shrivels the leaves and splits the grapes, ruining the wine and finally killing the vine.

4 **Deadliest enemy of the vine** is *Phylloxera vitifolii*. Eggs and larvae [A] develop in-to a root-eating form [B] which in the 1870s spread so much that it almost des-troyed the vineyards of Europe. It was found that the roots of the native Ameri-can vine (*Phylloxera* came from America) were immune. Virtually every vine in Europe had to be pulled up and replaced with a European cut-ting grafted on to a rooted cutting from an American vine. The pest's winged form is also shown [C].

abandoned altogether. The best combine hardiness with fine fruit, although rarely with a very generous yield.

One insect pest is disastrous: *Phylloxera* [4]. This little creature lives on the root of the vine and kills it. In the 1870s it almost destroyed the vineyards of Europe. The pest came from America, and so did the cure. By replacing the vines with European cuttings grafted on to a rooted cutting of American stock, *Phylloxera* was defeated because native American vines were found to be immune to the plague.

Natural selection

In the wine-growing districts of the Old World, the natural selection of the variety that does best, and gives the best quality combined with reasonable quantity and reasonable resistance to disease, has taken place gradually over the centuries. In many places (the Port country, Chianti, Bordeaux, Châteauneuf-du-Pape, among others) no one grape provides exactly what is needed. The tradition is either to grow a number of different varieties together, or grow them separately and then blend the resulting wines.

In the main wine districts of Europe, wine is known by the vineyard in which it is grown rather than by the kind of grape that goes into it. The choice of grape is so customary that it is taken for granted and laws usually make the traditional variety a condition of using the traditional wine name. White burgundy, for example, to be called white burgundy, must be made entirely of Chardonnay grapes [8]. Ampelography – the study of grapes – is one of the most delicate and difficult studies connected with wine.

Today, many new hybrids are made, a process that seeks to combine the best qualities of different types of wine, but none has yet achieved the quality of the old *vinifera* varieties. In the New World the choice of grapes is not a question of tradition but of judgment; a realistic balance of quality, quantity and hardiness. Hence the best wines of the New World use the grape name to specify the character of the wine, so that to an ordinary Californian, the words "Cabernet Sauvignon" [9] are more familiar than they are to a Parisian connoisseur.

Nurturing and protecting France's 1,200,000 hectares of vines is increasingly difficult. Near Saumur, on the Loire, farmers spray their vineyards in midsummer with a solution of copper sulphate (Bordeaux mixture) and sulphur powder to prevent infection by fungus pests such as mildew and oidium.

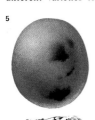

5 The Gamay grape makes first-class wine only on the granite hills of Beaujolais; in the rest of Burgundy it is an inferior variety and elsewhere it is dull, except in California where it is used for rosé. At its best it produces an incomparable wine that is light and fruity. At the present time it is inclined to be over-sugared, with the result that the wine tends to be too strong and dry on the palate.

6 The Semillon grape under certain conditions is subject to "noble rot". Given the right conditions of warmth and humidity, a normally undesirable fungus softens the skin and lets the juice evaporate, concentrating the sugar and flavouring elements and producing a creamy wine. The finest wines of Sauternes are made in this way, with some of Sauvignon. Semillon is extensively used in Australia.

7 The white grape of Anjou and Touraine on the River Loire is Chenin Blanc. It gives a "nervy" intense wine, with honey-like qualities when it is very ripe. It always has a high acidity and therefore ages well. Among its finest wines are Vouvray, Coteaux du Layon, Savennières. At Vouvray, it is also used to make a sparkling wine. It is often called Pinot de la Loire, and in California (where it is successful) White Pinot, mistakenly.

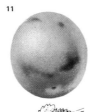
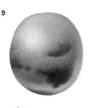

8 Chardonnay is the grape of white burgundy (Chablis, Montrachet, Meursault and Pouilly Fuissé) and also of champagne. Chardonnay gives a firm, full, strong wine with scent and character; on chalky soils it can become almost luscious without being sweet. Its wine ages well and is produced in a number of countries, notably in northern California where it rivals classic white burgundy.

9 Cabernet Sauvignon is the small, tough-skinned grape that gives distinction to the wines of Bordeaux, although it is always blended with Merlot or Malbec. The best Médocs have up to 80% Cabernet, but in St Emilion and Pomerol, Cabernet's inferior cousin, Cabernet Franc, is used. Cabernet Sauvignon is widely planted in Australia, Chile, South Africa and California. All Cabernet ages well in the bottle.

10 Pinot Noir is the single red grape of the Côte d'Or in Burgundy (Chambertin, Romanée, Corton, Beaune); it is the world's best wine grape in the right place. In Champagne it is pressed before fermentation to make white wine, which becomes the greater part of the best champagnes. When it is good, its wines are a profound pleasure. It transplants less well from France, but it does grow elsewhere in Europe and quite satisfactorily in California.

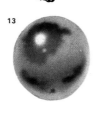

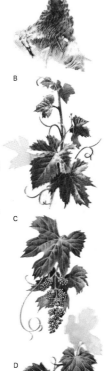

11 The Muscat grape can be black or white. Some of the first vineyards planted by the Greeks in France were Muscat. It spread from the Aegean to the Crimea, Sicily, Italy and to southern Spain. All Muscat wine, except from Alsace and Bulgaria, is sweet. The best in France comes from Beaumes de Venise near Avignon. Muscat wines are made all over the world. Today Australia makes some of the best.

12 Sauvignon Blanc is the chief white Bordeaux grape, used with Semillon and a little Muscadelle to make dry Graves and sweet Sauternes. It makes a clean, lighter wine on its own elsewhere. Its singular green and smokey character is most clearly seen in the wine made on chalk and alkaline clay at Pouilly and Sancerre on the Upper Loire; in Chile; and in the Livermore and the Santa Clara valleys of California.

13 Grenache is a sweet grape with a character of its own but not a great deal of colour. It is used in a blend to make Châteauneuf-du-Pape and, on its own, to make Tavel, the best rosé from the Rhône valley. Grenache is known as Garnacha to the Spanish of Rioja, where it is the most important red variety. It is used for dessert wines at Banyuls on the Franco-Spanish border and for light rosés in Australia and California that usually bear its name.

14 First buds [A] appear on the vine as early as April in northern Europe; the stalk, leaves and tendrils [B] come next. Flower buds [C] are formed in late May or early June and are closely followed by the flowers [D] from which the grapes develop.

Wines of the world

Crushed ripe grapes, left to themselves, ferment and make a wine of sorts. A brewer is a cook; he must heat and mix – but a wine-maker is barely more than a spectator, at the most an umpire, controlling an entirely natural process.

Thus a modern winery (a factory where wine is made) is essentially the same as the traditional small property, which in Bordeaux goes by the (usually misleadingly) grand name of *château*, and in Burgundy and Germany *domaine*. The difference is one of scale and also specialization. A private estate normally makes one, at the most two, types of wine. A winery often makes a dozen.

As to scale, a Burgundian grower (in Burgundy properties are on the whole smaller than in Bordeaux) may make as little as twenty 230-litre (50-gallon) barrels a year, while the vast E. & J. Gallo winery in Modesto, California, the world's largest, makes, bottles and distributes something in the order of 300 million bottles of wine a year.

What all growers have in common is the yearly round of duties divided between vineyard and cellar. These are described here as

they occur in the life of a small and fairly traditional French farmer, who is conscientious about the quality of his product and sells most of his own wine to private clients – a very common practice in France.

A year's hard work
The year begins with pruning. Traditionally, this was started on St Vincent's Day, 22 January, but nowadays it begins in December. Even if there is no snow, the ground is often frozen, but the sapless vines will survive temperatures down to about −18°C (−1°F). Indoors, the barrels of new wine from the September vintage must be kept full to the top and their bungs wiped every other day with a solution of sulphur dioxide as a disinfectant. In fine dry weather, bottling of the older wine can be done. The bottles are then labelled and packed for shipment to various parts of the world.

Pruning is finished in February, when cuttings are taken for grafting. They are grafted on to root-stock, then placed in sand indoors. Copper sulphate – Bordeaux mixture – is ordered for spraying later on in the year. In

fine weather, when there is a new moon and a north wind (that is, when there is high atmospheric pressure), the new wine is "racked" into clean barrels to clear it. The new wine is "assembled" in a vat to equalize the casks.

In mid-March, the vine begins to emerge from dormancy, the sap begins to rise and the brown sheaths fall from the buds. Any unfinished pruning is now completed and the tall tractor begins to move down the rows, turning over the soil to aerate it and to uncover the bases of the vines. The first racking of last year's wine is completed before the end of the month. Some mysterious sympathy between the vine and the wine is supposed to start the second fermentation when the sap begins to rise. The casks are kept topped up and bottling of last year's wine is finished.

The developing vine
Ploughing comes to an end in April, the vineyard is cleaned up and one-year-old cuttings are planted out from the nursery. Late vegetation is desirable, to escape the effect of late frosts. Indoors, topping up of casks con-

CONNECTIONS

See also
198 Origins of wine
200 Vines and grapes
206 Spirits and distilling
204 Beer and brewing

In other volumes
104 Man and Society

1 A typical small wine château in the Médoc, on which this drawing is broadly based, is a modest, specialized farm where the methods are up to date without being unusually modern.

The château was probably built about 1830 and like many in the Médoc region of Bordeaux it was given an air of importance above its station as an ordinary family house. Heavy silt

land [1] beside the stream is useless for growing vines but the stream itself is useful for draining the vineyard. The bottle cellar [2] has examples of the chateau's and its

neighbour's wines going back more than 50 years. Fodder and farming equipment, including the tall tractor used for straddling the vines, are kept in the barn [3]. In the second-year

chai [4], *barriques* (which hold 24 dozen bottles) are stored after a year to make way for a new vintage. Some of the bottling is carried out here at the château and some is done by *négociants*

in nearby Bordeaux. The bottle cellar has an important role to fill, for no good red Bordeaux is ready to drink until it has been in the bottle for at least two years. The vineyard's full-time

workman lives in a cottage on the premises [5]. Close by is the proprietor's office [6] where, on the wall, there are large-scale plans of the château and the vineyard, showing every barrel and vine.

The wine château and its vineyards are as close to a complete industry as it is possible to be. It is as if a knitwear factory bred its own sheep, sheared them, spun the yarn, wove the cloth and made garments. The ripe grapes are gathered in by pickers – often students – using

secateurs [7]. The grapes are collected from them by a tip-up cart. The *courtier* or broker [8] arrives to hear news of the vintage and to form his first impressions of its likely quality. The *maître de chai* (*chai* is the cellar) measures the sugar content of his must [9] – and hence the

alcohol content of his wine – with a hydrometer. Grapes go straight into a *fouloir égrappoir* [10] to be crushed and pumped into a fermenting vat. Stalks are ripped off and emerge separately. Sulphur dioxide in powder-or liquid form [11] is sprinkled on the grapes or into

cuve at about 10g (0.35 oz) for each 100 litres as disinfectant. The fermenting vat [12] is being filled 80% full to allow room for seething movement. Every morning and evening, the fermenting wine is pumped up and sprayed over the floating "cap" of skins [13].

This is *remontage*; in fine years it is pumped via an open tub to help aeration. The hydraulic press [14] will press the skins to extract the remaining one-fifth of the wine after the rest has been run off. The deeply coloured *vin de presse* is now mixed with the rest of

the wine. From the oak *cuves*, or vats in which it has been made, the wine now goes to *cuves* of cement [15] in which it winters, going through the gentle secondary malolactic fermentation process which rids it of malic acid, making it less harsh. In February the wine is pumped

into oak *barriques* (hogsheads) in the first-year *chai* [16] although at some châteaux the wine goes straight into barrels. The *barriques* are stoppered with loose glass bungs and here the wine stays for a year, topped up and occasionally "racked" into a fresh barrel.

tinues – five per cent of the wine evaporates through the sides of the cask in a year and there must be no empty space in the cask.

Frost danger is at its height in May. On clear nights, when it is likely to strike, small stoves are taken out among the vines and workmen sit with them to keep them fuelled. The soil is worked again and the vines are sprayed against oidium and mildew. Suckers that drain the vine's energy are removed. As the vines begin to flower, the second racking of last year's wine takes place indoors.

The vines flower at the start of June. The weather is critical at this time – the warmer and calmer, the better. After flowering, the shoots are thinned and the best ones tied to the supporting wires. The second racking of the new wine finishes.

The vines are sprayed with Bordeaux mixture in July. The ground is cultivated again to keep down weeds. Long shoots are trimmed and vine growth begins to slow down.

In August, the vintage is approaching. Black grapes turn colour. General upkeep, and the cleaning and preparation of the vats

and casks to be used, keeps everyone busy.

September is the vintage month. In about the third week the grapes are ripe and picking begins. The *cuvier* where the wine will be made is scoured out and the fermenting vats filled with water to swell the wood.

The vintage ends

The vintage continues for about two weeks into October. Once it is over, manure is spread on the ground and the land for new plantations is deep-ploughed. The new wine is now fermenting. the year-old wine is given a final racking. The barrels are now bunged tightly and rolled a quarter-turn so that the bung is at the side.

In November, manuring is completed and the vineyeard is ploughed to throw earth up over the bases of the vines to protect them from frost. Wine for bottling is racked and "fined". December arrives again; if soil has been washed down from the tops of slopes, it must be carried back and redistributed. Casks must be topped up frequently and older wine can be bottled. Out in the vineyard, pruning begins again.

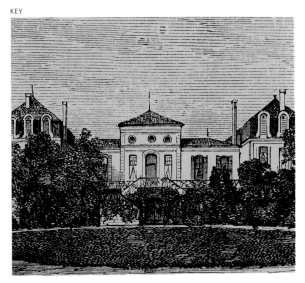

KEY

Chateau Langoa-Barton, in the Bordeaux region of western France, was built in 1758. Its wine is a typical St-Julien which is regarded as the epitome of classical claret. The region has good drainage and climate for vines.

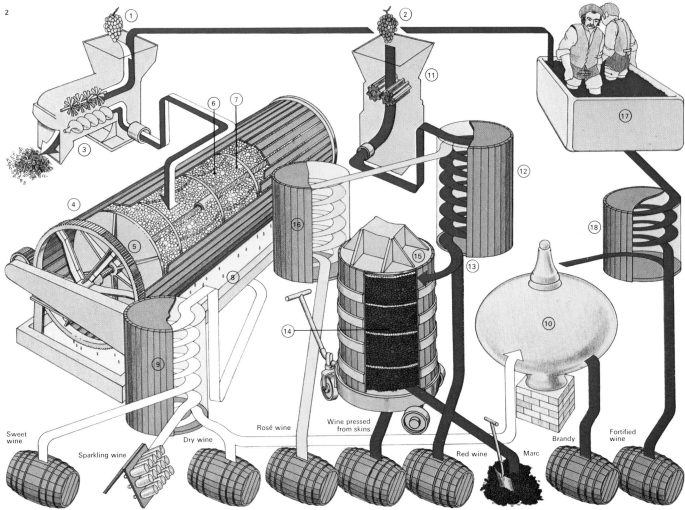

2

Sweet wine

Sparkling wine

Dry wine

Rosé wine

Wine pressed from skins

Red wine

Marc

Brandy

Fortified wine

2 To make white wine white [1] or red grapes are fed into a crusher-stemmer [3] which removes the stalks and pumps the broken grapes into a horizontal press [4]. The screw brings the end plates [5] together and the chains [6] and hoops [7] break up the caked grapes. The juice falls into a trough [8] and goes into a fermenting vat [9]. It may be fermented completely to make dry wine, the process can be halted before all the sugar is used up to make sweet wine, or the wine can be bottled before fermentation is complete to make sparkling wine. Dry wine may be distilled [10] to make brandy. For red wine, red grapes [2] are crushed [11] and fermented with their skins in a vat [12]. The colour and the tannin are absorbed from the skins and when the fermentation is complete, after about 14 days, the wine is drawn off [13]. For quicker-maturing wine the modern procedure is to take the wine off the skins after a few days to finish fermenting separately. The skins are pressed [14] by a hydraulic ram [15] and the resulting wine is usually mixed with that first drawn off. The cake of skins left in the press is called *marc* and can be used as fertilizer or distilled to make a cheap brandy. Rosé wine is put through the same process as red, but is drawn from the vat [12] when only a little pigment has been absorbed from the skins. It completes its fermentation in a second vat [16]. Port and other fortified wines are made from red grapes, which are trodden in a trough [17] to colour the juice with the skins. The juice is fermented in a vat [18] until half its sugar is converted into alcohol and brandy from the still [10] added to stop fermentation and raise the alcohol level to above 15%.

203

Beer and brewing

Beer, rather than wine, is the traditional drink of those countries in which grapes cannot flourish. The making of beer and ales – brewing – has been known for many thousands of years. Some form of brewing was carried out in Babylon about 4000 BC [Key]. The ancient Egyptians were also familiar with brewing and passed on the knowledge to the Greeks from whom the Romans learnt the skill. Brewing, using a mixture of corn and honey, developed independently in northern Europe.

The composition of beers
The basic materials used in brewing are grain (the source of sugars), which is usually barley (but wheat, rice and maize are also used) and yeast, which breaks down the sugars by fermentation to form alcohol and carbon dioxide. The use of hops as flavouring became widespread in England after the sixteenth century. It was then that the term "beer" was used to distinguish the product containing hops from ale, the product without hops.

Today there are two main types of beer.

Lagers are fermented with bottom-fermenting yeast, ales with top-fermenting yeast. Bottom-fermenting yeast moves to the bottom of the fermenting vessel, top-fermenting yeast to the top.

Ale-type beers are the most common in the British Isles, Australia and New Zealand. They include bitter or pale ale; strong ale; draught mild ale and bottled brown ale. Stout and porter (which is a weak stout) are also top-fermented.

Lager-type beers are the most widespread. They have a longer maturation time than ales (up to six months as opposed to four weeks at the most for ales). The main types are Pilsener and Dortmund.

The brewing process
Brewing begins with the production of barley malt. Grains of barley are soaked and allowed to germinate. During germination chemicals (enzymes) are produced that can convert "starch" within the grains to sugars. The germinated grains – the malt – are then dried and crushed in a grist mill [1]. Just enough rupturing of the husk occurs to

enable the grain, now called grist, to be infused with hot water in a process known as mashing. The object is to allow the conversion of the "starch" to fermentable sugars by enzymes and their extraction from the grist.

The mash is held in a vessel called the mash tun. The temperature and method of mashing define the type of beer to be produced. British beers are produced by infusion mashing, a single process in which water is added at 65.5°C (150°F) and the temperature is maintained somewhere between 65.5 and 75.5°C (150–168°F), depending on the type of beer to be produced. Other beers are usually produced by a three-mash process. Water at 38°C (100°F) is introduced and after two hours the temperature of the mash is raised to 65.5°C and then finally to 75.5°C.

The liquid from the mash tun is known as "wort" and contains a mixture of sugars and other chemicals from the malt. The wort is run off from the mash tun and is boiled, usually for about two hours, to destroy the enzymes and to sterilize and concentrate the liquid. At this stage hops are added, to give the beer its bitter tang.

CONNECTIONS

See also
180 Staple foods: grains
206 Spirits and distilling

In other volumes
40 The Natural World

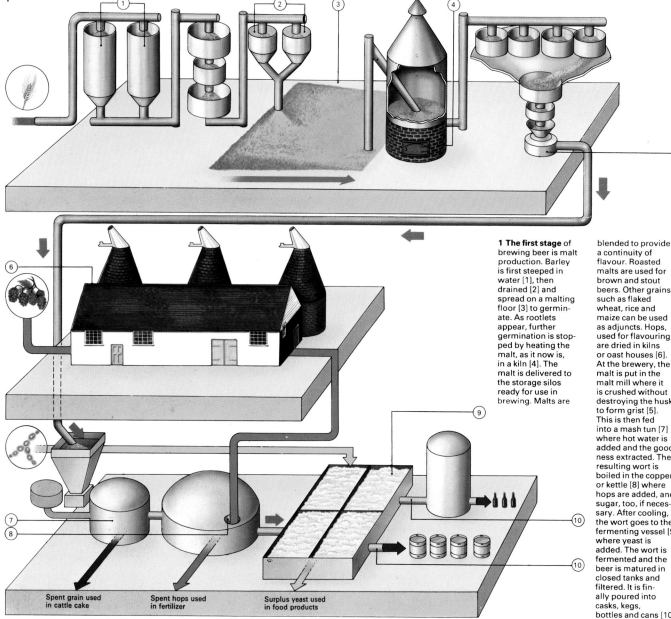

1 The first stage of brewing beer is malt production. Barley is first steeped in water [1], then drained [2] and spread on a malting floor [3] to germinate. As rootlets appear, further germination is stopped by heating the malt, as it now is, in a kiln [4]. The malt is delivered to the storage silos ready for use in brewing. Malts are blended to provide a continuity of flavour. Roasted malts are used for brown and stout beers. Other grains such as flaked wheat, rice and maize can be used as adjuncts. Hops, used for flavouring, are dried in kilns or oast houses [6]. At the brewery, the malt is put in the malt mill where it is crushed without destroying the husks to form grist [5]. This is then fed into a mash tun [7] where hot water is added and the goodness extracted. The resulting wort is boiled in the copper or kettle [8] where hops are added, and sugar, too, if necessary. After cooling, the wort goes to the fermenting vessel [9] where yeast is added. The wort is fermented and the beer is matured in closed tanks and filtered. It is finally poured into casks, kegs, bottles and cans [10].

Spent grain used in cattle cake

Spent hops used in fertilizer

Surplus yeast used in food products

Any solid objects, including spent hops in the wort, are now filtered off. The liquid is cooled and transferred to the fermenting vessel. Living yeast (*Saccharomyces* sp), which is a type of fungus, is added at this stage. The character of the finished product depends in large measures upon the absolute purity of the yeast culture.

During fermentation the yeast reproduces and increases in bulk by about five times and the excess is skimmed off. Stringent precautions are taken to ensure that no impurities invade the brew. The temperature of the fermenting wort is also carefully controlled because too low a temperature may stop the fermentation and too high a temperature may cause it to race, with disastrous effects upon the character of the beer.

After about eight days, when most of the sugar has been converted to alcohol, the beer is removed from the fermentation vessel and separated from the bulk of the yeast. The new or "green" beer is then stored in maturation tanks, again at strictly controlled temperatures, so that it may mature in flavour and character. At this stage a secondary fermentation occurs with lager beers and with some ale-type beers.

Basically, the brewing process is the same for both lager and ales, but in brewing lager the yeast is different and fermentation and maturation take longer to complete.

Packaging the brew

After brewing comes packaging: filling casks or kegs for the draught beer trade or transferring the brew to a bottling hall for packing into bottles or cans. Cask draught beer is filled through nozzles on simple gravity-feed machines. For filling kegs a more complicated system is required. Kegs are sterilized and mounted on a filling head, which displaces air from the keg by means of CO_2. The CO_2 is in turn displaced by the in-flowing beer until the correct quantity has been filled. This leaves the keg full of beer but with a small head space filled with CO_2, again to prevent air entering. Bottles and cans are filled in much the same way.

In some countries private persons are permitted to brew beer at home [5], not for resale, without paying excise or similar duty.

Brewing is an extremely ancient art. It probably originated some 6,000 years ago in Babylonia. This clay seal was made in the kingdom of Mitanni in northern Mesopotamia around 1500 BC. It depicts a hunting scene, mythological animals and a man drinking beer through a straw. The art of brewing spread westwards to Europe through North Africa and eastwards to China and Japan.

2

3 A

B

2 Barley may have come originally from central Asia. Eventually traders carried it as far north as Scandinavia and Britain. The art of brewing accompanied the geographical spread of the grain.

3 Hops are climbing plants of the family Caunabinaceae. They are grown in hop gardens [A], on inclined strings and wires. The female flowers [B] are used to impart a bitter flavour to beer.

4 The basic process of brewing can, with variations in materials and detail of production, give rise to many different types of beer. The flavour, sweetness, colour and gaseous content can all vary.

5 Home brewing is an art that can be practised with little more than ordinary domestic equipment for mashing, boiling, fermenting and bottling. The best containers are of glass or stainless steel.

4

5

Sugar

Hot water (8 litres)

Hop and malt extracts

Yeast (sprinkled on when all has settled)

Loosely covered, left 10 days to ferment

Tested with hydrometer

Siphoned into bottles

One spoonful of sugar added

Tightly sealed

Stored 2-3 weeks

205

Spirits and distilling

Spirits, in common with all other alcoholic beverages, are produced basically by alcoholic fermentation (the decomposition of sugars into ethyl alcohol by yeasts) of sugar-rich liquids. The diversity of spirits and related alcoholic beverages is a reflection of the widely differing sources of fermentable sugar. This sugar is directly available in some raw materials such as grapes and molasses, but in other raw materials (cereals, potatoes) starch has to be converted into fermentable sugars by adding enzymes (biological catalysts). The traditional source of such enzymes is barley malt and the process of starch conversion is known as "mashing".

Distillation and maturation

Yeast fermentation can lead to a maximum alcoholic content of about only 15 per cent in beverages such as wine. But in the production of spirits, fermentation is followed by distillation, which concentrates alcohol in the distilled spirit. Distillation [Key, 1, 4, 6, 7] is the heating of alcohol-containing liquids, so that the alcohol boils off and can be condensed, collected and added as desired to increase

strength. The degree of concentration and distillation of volatile flavouring substances or "congenerics" depends on the type of still and the way it is operated. Simple "pot" stills [6] distil alcohol at lower concentration to give a spirit with a characteristic flavour, whereas "continuous column" stills [7] are more efficient, yielding spirits of higher alcohol concentration but with little or no congeneric flavour.

Newly distilled spirit is seldom suitable for immediate consumption. It is often stored in wooden casks [5] for a number of years to achieve a mellow and developed flavour such as that of whisky or brandy. The quality of spirits is frequently related to the period of storage in wooden casks. For bottling and commercial sale, distilled spirits are normally reduced with pure water to an alcohol content of about 40 per cent by volume.

Malt and grain whiskies

Scotch malt whisky is made entirely from a mash of malted barley. The fermented mash produced in a mash tun [2] is first distilled into what are known as "low wines" in a mash

still [4] and the low wines are re-distilled in a similar pot still, the spirit still. In the second distillation "heads and tails" fractions (the first and last parts obtained in the distillation) are returned to the low wines. The middle fraction is the new whisky. Malt whisky is aged in reusable oak casks for eight, twelve or more years before being sold as single malt whisky – an individual and distinct product from each distillery.

Malt whisky is blended with Scotch grain whisky to produce blended Scotch whisky. Grain whisky is made from a mash of pre-cooked maize and a proportion of barley malt. The wash is distilled in a semi-continuous, two-column still. The grain and malt whiskies are aged separately and blended shortly before bottling. Irish whiskey is made from unmalted cereals, including barley, wheat and rye, which are mashed together with malted barley. Bourbon whiskey is made in the American states of Kentucky and Illinois from a grain mash of at least 51 per cent maize. The distillate is matured for four years in new oak casks which are charred internally before fil-

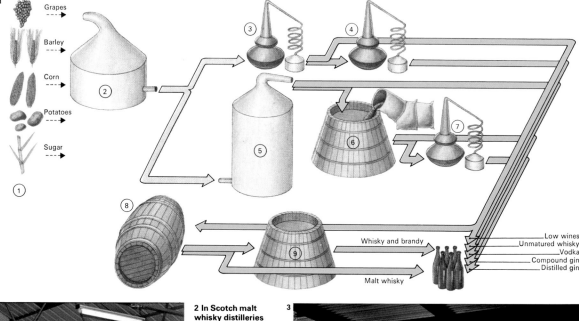

1 The distilling process varies according to the initial ingredients and the desired product. The raw ingredients [1], crops that will supply the sugar for fermentation, are fermented in the mashing tun [2]. For whisky and brandy the mash is distilled in a mash still [3] and the spirit produced re-distilled in a spirit still [4]. It is then aged in oak casks [8] and may be blended [9] before bottling. Mash for other spirits may be distilled continuously in a distillation column [5] and the product usually has flavouring added to it [6]. The flavoured spirit may now be distilled once more in a pot still [7]. Gin and vodka are manufactured in this manner.

Grapes
Barley
Corn
Potatoes
Sugar

Whisky and brandy
Low wines
Unmatured whisky
Vodka
Compound gin
Distilled gin
Malt whisky

2 In Scotch malt whisky distilleries mashing is carried out in a vessel called a "mash tun". This is filled with a mixture of milled barley malt and hot water. The liquid "worts" (infusions of malt before they are fermented), charged with soluble substances extracted from the malt, are drawn off through perforated plates in the base of the mash tun.

3 Worts from the mash tun are cooled and piped to the tun room for fermentation. The stainless-steel covered fermenters, called "wash backs", are filled with worts and yeast is added to start a vigorous fermentation. To prevent foam overflowing during fermentation, propellers are rotated inside the wash backs.

ling. Canadian whiskey is a blend of a strongly flavoured whiskey distilled from rye and a lighter maize-based whiskey.

Spirits and liqueurs

Brandy is the distillate of wine. Cognac, the best known brandy, is made from the wine of the Charente (a region in southwest France) by two successive distillations in characteristic pot stills. The first distillate or "brouillis" is returned to the still for a re-distillation, or "bonne chauffe". Cognac is matured in Limousin oak casks and the finer qualities attain ten, 20 or even more years. The quality of cognac is related to the situation of the vineyards from which the wine is produced around the town of Cognac. Armagnac is made near Agen in southwest France. It is singly distilled in a more efficient pot still and has a distinctive flavour.

Rum is the distillate of fermented molasses, the mother liquor of crystallized cane sugar. Flavoured rums, dark in colour, are traditionally made by pot still distillation and aged in wood, but most white rums are distilled to a light flavour, which is modified by an additional treatment with charcoal. Other spirits are made from local sources of fermentable materials: arrack from palm juice, rice or molasses; calvados from apples; kirshwasser from cherries; and tequila from cactus. Vodka is a spirit that has no flavour apart from that of ethyl alcohol. It is made by purifying a neutral spirit by means of contact with absorbent charcoal.

Spirits such as gin, ouzo and liqueurs derive their flavour from added ingredients and not, in general, from the congenerics of the original spirit. They are made from a base spirit of relatively neutral and flavourless character, produced from a variety of materials, including grain, molasses and potatoes. Gin is flavoured by juniper berries [8] and spices, Dutch gin is also flavoured with juniper and spices, although its base spirit has a characteristic flavour of its own. Ouzo is made by flavouring a base spirit with aniseed. Aquavit is a base spirit flavoured with caraway. Liqueurs are usually sweetened by the addition of sugar. Gin, vodka and liqueurs do not generally require to be aged, although it is usual practice to age Dutch gin.

Some form of alcoholic beverage was developed by many early civilizations. By 800 BC the Chinese were distilling a beverage from rice beer and the Romans also produced distilled beverages. Production in Western Europe gained impetus after contact with the Arabs – the word "alcohol" is of Arabic derivation. This engraving of about 1480 from Salerno, Italy, is one of the earliest illustrations of a still.

4 Fermented mash is heated in a mash still until all the alcohol is distilled and collected as low wines. A similar but smaller spirit still is charged with a mixture of low wines and "feints" (impure spirits produced in the first and last stages of distillation), and the new whisky is selected as the middle portion of this distillation. The rest of the feints are collected separately and then returned for re-distillation in the spirit still.

5 New whisky is put into oak butts, hogsheads and barrels and stored on steel racks in warehouses until mature. After emptying, the casks are examined and repaired by coopers and refilled with whisky.

6 Copper pot stills charged with a batch of wash or wine are heated directly by a furnace or indirectly by steam. On boiling, the more volatile constituents vaporize and rise to the top of the still. The vapour that reaches the top of the column is then led over into the condenser, liquefied and collected. Alcohol is progressively removed by distillation and the watery residue that remains in the pot still is discharged.

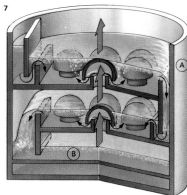

8 Juniper berries are the principal flavouring of gin. In addition to other ingredients such as coriander seed, angelica root, cassia and cinnamon bark, orange peel, cardamon, liquorice, nutmeg and others, the juniper berries are distilled in a pot still with a charge of spirit. Dutch gin or "genever" is also re-distilled with juniper and other botanicals but the base spirit or "moutwyn" has a characteristic flavour. Liqueurs are brandies containing large proportions of sugar and alcohol, and can be made from raspberries, cherries and plums, or even tea, cocoa and coffee.

7 In a continuous still the columns are vertical copper cylinders [A] with horizontal, circular plates [B]. "Bubble caps" cause the rising vapour and the falling liquid to mix in the column and this action promotes the separation of alcohol at the top and water at the bottom of the column. Through regulation of the steam supply and product feed, a dynamic balance is set up and high-strength, high-purity spirit is drawn continuously from the top and waste water from the bottom of the column.

Coffee, tea and cocoa

Coffee is the most widely drunk caffeine beverage in the world. It is enjoyed not only for its rich flavour and aroma, produced by its caffeol content, but also for the caffeine it contains, which has a stimulating effect on the nervous and blood systems. Leading coffee-drinking countries include the United States, West Germany, France and Italy. Large plantations are in Brazil, Costa Rica, Mexico, Java, India, Arabia and Kenya.

Growing and processing coffee

The common coffee plants are *Coffea arabica* [2, 3] and *C. robusta*. Coffee plants reach maturity after four years and remain productive for about another 30 years, but there is usually a drop in quality in about the fifteenth year, after which the trees are generally replaced.

After picking, the coffee seeds or beans are extracted from the ripe berries by machine and the residual pulpy mass is used as fertilizer. The seeds are cleaned and then sun- or machine-dried, after which they are hulled in revolving drums where their tough outer coatings and silvery skins are removed.

Finally they are sorted by size, graded and shipped to the world's markets.

Most coffee beans are roasted before being sold; this gives them a dry, brittle texture and a deep brown colour. They also acquire their typically rich coffee aroma during roasting by undergoing a complex chemical change. The freshly roasted coffee is cooled and ground to appropriate size, although some customers prefer to grind and even roast the beans themselves. So-called "instant coffee" may be powdered or freeze-dried. In each process the coffee is first brewed and the moisture removed from the liquor. More than 100 different kinds of coffee are marketed, each with its own flavour. The differences arise mainly from climate and soil types in the various regions in which the plants are grown.

The story of tea

Tea, like coffee, is a stimulant, because of its caffeine content. Tea remained unknown to the Western world until the great explorers of the sixteenth century who travelled to China and other Far Eastern countries returned with a host of new foods, spices and beverages. Very soon a thriving commerce in China teas was established between Europe and the Far East. In 1826 the Dutch established plantations on Java, followed some ten years later by the British who set up tea estates in India. The production of tea has since spread rapidly to regions in Sri Lanka, Bangladesh, China, Iran, Japan, the Soviet Union and parts of Africa and South America. Among Western nations Britain and the USA are the two greatest consumers.

Tea is made from the leaves of an evergreen tropical and subtropical plant, *Thea sinensis* [5]. There are three major kinds of finished tea. Black tea, which makes up more than 90 per cent of the tea trade, is produced by first allowing tea leaves to wither. This reduces their water content and prepares them for rolling. The leaves are fed through rolling machines to release the juices. The rolled leaves are then placed in cool, extremely humid rooms and left to ferment. Fermentation is stopped by drying the leaves over fires in pans, trays or baskets, a process that also seals in their final flavour.

1 Coffee houses flourished in most large European towns in the 17th and 18th centuries and were the centre of much of the political, commercial and literary activity of the day. The first one in London opened in 1652. The custom of drinking coffee spread westwards from Arabia into Europe. Coffee itself is believed to have come originally from Ethiopia. According to legend a 9th-century goatherd noticed that his animals became livelier soon after eating some crimson coffee berries. He tasted some himself and experienced a similar feeling of exhilaration. The first reliable reference to coffee occurs in Arabian literature of the 10th century AD. But it was not until several hundred years later that the Arabs learned how to make a delicious drink from the roasted ground beans of the coffee berry.

2 Coffee plantations flourish only where temperatures average a modest 21°C (70°F). Altitude influences the quality of coffee, the best crops coming from plantations that are between 600 and 2,000m (2,000 to 6,000ft) above sea-level. Coffee trees grow best in partial shade and for this reason are often planted on the east- or west-facing slopes of volcanic mountains, where they thrive in the well-drained, potash-rich soil. Once the berries have ripened they are harvested by hand. Each tree yields an average of about 700 grammes (1.5lb) of ground beans each year.

3 Arabian coffee *Coffea arabica*

3 The coffee plant is a small evergreen tree with leaves some 7.5 to 15cm (3 to 6in) long. The Arabian species is the most common kind of coffee plant. Coffee trees are able to grow to 7.5m (25ft) high but are pruned to 3m (10ft) on plantations. The fragrant white blossoms are followed by tiny green berries, each holding two tough-skinned, greenish beans. The berries ripen to a deep red after six or seven months and are then ready for picking.

A second variety of tea is oolong or semi-fermented tea which is prepared from a special kind of China tea plant. The leaves are heated before fermentation progresses very far, then they are rolled and, finally, dried.

Lastly there are the green or non-fermented teas made by first steaming the leaves or else heating them to sterilize them and kill the enzymes responsible for fermentation. The leaves are then rolled and roasted until they acquire a blue-green tint. Teas are graded for size, age and quality, and classified according to leaf size such as orange pekoe, pekoe and souchong.

Paraguay tea or maté, which is widely drunk in southern South America, is made from infusions of the leaves of a species of holly (*Ilex paraguariensis*).

From cacao to chocolate

Cacao trees [6] had been cultivated in Central America for centuries before Columbus arrived there. The Spanish *conquistadores* were sufficiently impressed with *cacauatl*, the bitter cocoa drink the Indians made from the seeds of the cacao tree [7], to take it back to the court of Spain. The secret of cocoa was jealously guarded by the Spaniards for more than a century until the seventeenth century when the rest of Europe was introduced to it. It became a highly fashionable drink virtually overnight. So great was the rage for cocoa that "chocolate houses" sprang up in every sizeable town.

The cacao tree is an evergreen plant, *Theobroma cacao*, that grows in tropical regions. The pods are cut down, split open with a heavy knife or mallet and the pulp and seeds scooped out. The entire pulpy mass is left to ferment for a week, during which time the beans change from purple to reddish brown and acquire a pungent, chocolate aroma. When fermentation is complete the beans are either sun- or kiln-dried then cleaned and shipped to processing plants. The manufacture of cocoa continues with the roasting of the beans. Their shells are then cracked and removed and the nibs or kernels are pressed to release the oil (cocoa butter). The remaining paste is known as chocolate liquor and this is the raw substance from which cocoa powder and chocolate are produced.

The world's tea crop is traded at great auctions such as the one in London through which the harvest of some 25 nations passes. Before the tea is bought for blending and packaging, tea-tasters examine the dry leaves and analyse the aroma, taste and appearance of the actual brew.

4 Tea is mostly cultivated in monsoon Asia – India, Sri Lanka and China are the principal producers. Three original varieties of tea – China, Assam and Cambodia – are the ancestors of the many hundreds of varieties grown today. Tea grows best in warm, well-watered climates where the soil is slightly acid, well-drained and rich in organic compost. Like coffee, it seems to thrive at higher altitudes, and can be grown at a maximum height of 2,300m (7,000ft). Planters using hybrid plants and modern methods of agriculture can expect an annual yield of about 1.7 tonnes (3,700lb) per hectare.

5 Tea
Thea sinensis

5 Tea is a small evergreen shrub that is kept pruned to a maximum height of 1.5m (5ft) on plantations. Its leathery, oblong leaves grow to a maximum of 25cm (10in) in length. It takes roughly 40 days for a tea plant to produce a full "flush" of leaves ready for picking. From the top downwards the leaves are named pekoe tip, orange pekoe, pekoe, first and second souchong and first and second congou.

6 The cacao tree, a tropical plant, sprouts pendulous pods 15–35cm (6–14 in) long from its branches and trunk. Each pod contains 30 to 40 beans from which cocoa and chocolate are made.

7 Aztec and Mixtec paintings show early American peoples drinking a beverage made from cocoa beans, which they cultivated, mixed with maize flour, spices, aromatic herbs and water.

8 Coffee, tea and cocoa are grown in tropical and subtropical areas. The coffee trade is valued at more than $2,000 million a year and represents nearly 4.25 million tonnes of beans. The tea crop averages some 1.25 million tonnes, the largest producers being India and Sri Lanka. Nearly 70% of the world's cocoa, 1.36 million tonnes, comes from Ghana, Nigeria, Ivory Coast and the Cameroons.

Cacao
Theobroma cacao

8

Coffee
Cocoa
Tea

Seeds and spices

Throughout Europe and the Middle East, seeds from plants in the fields, forests and hedgerows yield a wide variety of both subtle and startling flavouring. Their value as a means of varying the tastes of the common staple foods such as beans, peas, rice, wheat and root vegetables was widely appreciated in biblical times: since then a great number have been cultivated in kitchen gardens large and small all over the world.

Indigenous species still provide the dominant flavours in northern Europe, the Mediterranean and the Middle East, but they are by no means confined by national or natural boundaries. Fenugreek seeds from the herb *Trigonella foeunum-graecum*, a native of the Middle East and Asia, have a musky aromatic flavour and are used in vegetable dishes and sweetmeats in Egypt, Turkey and Iran. They also form part of the curry mix of herbs and spices in Asia. The cumin seed (*Cuminum cyminum*) has similar origins and gives curry its characteristic pungency. Fresh eucalyptus-scented cardamom seed (*Amomum cardamomum*) is popular in Scandinavia. Little used in Eng-

land, the cardamom seed, which is encased in a parchment-coloured pod, gives a lift to coffee, cakes and fruit, particularly when simmered with pears and quinces. Coriander (*Coriandrum sativum*) is a native of southern Europe and the Middle and Near East and is one of the oldest known members of the family Umbelliferae, which provides many other species of commonly used seeds.

The prize of princes

Spices were products of tropical climates. Until new sea routes broadened trade between Europe, Asia and the Far East, spices such as peppers [Key], cloves [3G] and cinnamon [3C] were as rare as gold dust and just as valuable in Europe. During the sixteenth and seventeenth centuries wars were fought over footholds in Asia and a share in the wealth brought by the spices of the Orient. Perhaps the most prized among the spices was pepper. Now so common and cheap as to be taken for granted, a pound of pepper could cost several weeks' wages for the average European so that a "peppercorn rent" was far from a derisory sum.

Spices such as nutmeg [3A], cinnamon and turmeric [3H], luxuries in the kitchens of Europe, were used sparingly and on special occasions – to add sweet aromas to mulled wines in the festive seasons and as ingredients for special cakes celebrating the end of Lent – and only a little more freely at the tables of princes. In the Near and Middle East and Asia they were and still are used lavishly to make the staple diet of pulses and rice less monotonous and to disguise the fact that fresh meat, fish and vegetables deteriorate rapidly in relentless heat.

Fashions in food

The use made of aromatics and spices has varied with the passage of time and geographical location. Aniseed, once commonly used in cakes, pastries and bread throughout Europe, is now confined to sweets and the anise-flavoured drinks for which the Mediterranean is famous. The French pastis, Greek ouzo and Turkish raki are now internationally popular aperitifs. In the Middle East aniseed is used as a flavouring for root vegetables.

CONNECTIONS

See also
212 Culinary and medicinal herbs
214 Berries, nuts and olives

In other volumes
70 The Natural World

1

1 Many seeds can be used for flavouring. Both the leaf and seed of the dill plant (*Anethum graveolens*) [A] have a flavour similar to caraway but are a milder complement to fish and vegetables.

Anise (*Pimpinella anisum)* [B] produces oval light brown seeds tasting of liquorice. Caraway (*Carum carvi*) [C] has strongly flavoured seeds which are used in bread and cakes. The root is also edible. Celery

seeds, the dried fruit of the plant *Apium graveolens* [D] are more strongly flavoured than the stems and are used to add interest to root vegetables, meat and fish dishes. Fennel seeds have

a stronger flavour than the bulbous stems of the fennel plant (*Foeniculum vulgare*) [E]. Sweeter than anise, fennel goes well with fish or pork. Mustard seed [F] is used mainly as a condiment, the red being hotter than the yellow variety. Edible poppy seeds come from the annual *Papaver rhoeas* [G]. They are often used in cooking and are not narcotic (poisonous). The plump, white seeds from *Sesamum*

indicum [H] are a source of nutty-flavoured oil and protein and form an important food crop in the Middle East. Sesame seeds ground to a paste are called tahina and used as a spread and a sauce popular from Greece to Syria and Israel. Star anise [I] is the seed from the star-shaped fruit of *Illicium verum*, an evergreen tree found in China and Japan, and is used widely in those countries.

In the East, ginger [3D] is used as a spice for both fish and meat dishes where the sweet, rich root provides a striking contrast of texture and taste in an otherwise savoury dish. Powdered Jamaican ginger is more often used in the West, usually in sweet foods. Almost every European country has a variety of gingerbread. Yet a dusting of ginger, nutmeg and coriander turn buttered rice into a subtly different vegetable to serve with chicken or veal. Ginger was very widely used in Roman times, but is rarely found in modern Italian cuisine.

One fashion that has hardly changed in centuries is the Middle European use of poppy seeds [1G] sprinkled on bread to flavour and decorate the more fancy varieties. A stronger flavour is provided by the tang of caraway seeds in rye bread and pumpernickel. In the Middle East bread is more usually topped with sesame seeds before cooking, so that they brown and impart a rich, nutty flavour to the bread. Small cakes can also be prepared with a meal made from sesame which is mixed with lemon juice and honey. Middle Eastern meals are often accompanied by a paste made from chick peas, sesame seed pulp (tahina), lemon juice and garlic. It is traditionally eaten with unleavened bread called pitta.

Innovations and revivals

Sesame seeds [1H] are becoming more popular in Western cooking. They can be added to most cream sauces for poultry and meat dishes with impunity, creating a subtle difference without overpowering other flavours. Sesame seed oil makes a perfect addition to a French dressing to be served with the strong varieties of lettuce such as cos, or with endive or chicory. Nutmeg and mace were once much used as condiments for members of the brassica family and can turn plainly cooked everyday cabbage into a much more worthy vegetable. Saffron [3E], the dried stamens from the mauve crocus of Asia Minor, has always been the most expensive of spices, since the stamens are collected by hand and about 200,000 are needed to make up a pound of saffron. The colour and aroma saffron gives to rice dishes such as the Spanish paella cannot be imitated.

Pepper is the dark round fruit of the Asian vine *Piper nigrum*. The fruit, when fresh and un-ripe, is green and pungent. It forms the basis of a popular French condiment. Black pepper is the whole dried fruit with its un-mistakable aromatic taste. Fresh-ground, it enhances the flavour of all vegetables, meats and fish, even fresh fruits. One famous meat dish using coarsely crushed peppercorns is steak *au poivre*. Long cooking can spoil the aroma so that coarse ground black pepper is best added at the end. White pepper is the seed used with its outer skin (pericarp) re-moved. It has a hotter taste than black pepper but adds less flavour.

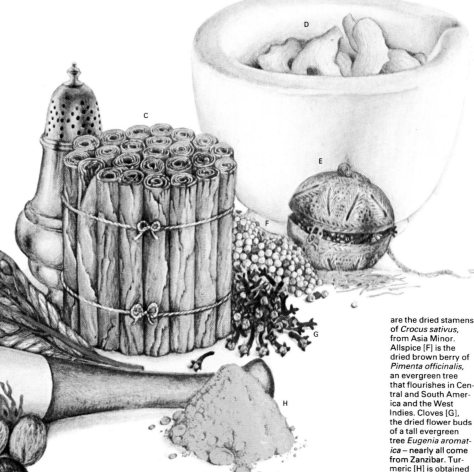

2 Indian merchants in a market sell spices from shallow brass and copper dishes, carrying on a thousand-year-old tradition. The chilli powders in the foreground are relative newcomers.

3 The search for spices, a saga of triumph and bloodshed, covered every corner of the globe and their discovery brought great wealth. Nutmeg and mace [A] both come from the fruit of *Myristica fragrans*. The former is the kernel and the latter the net-like membrane, or aril, which surrounds it. Sassafras [B] is obtained from root bark of the tree *Sassafras albidium*. Cinnamon [C], the dried rolled bark of a small evergreen tree, *Cinnamomum zeylanicum*, is native to Sri Lanka, the Malabar coast of India, and Burma. Ginger [D] is the rhizome of the tropical plant *Zingiber officinale*. Saffron strands [E] are the dried stamens of *Crocus sativus*, from Asia Minor. Allspice [F] is the dried brown berry of *Pimenta officinalis*, an evergreen tree that flourishes in Central and South America and the West Indies. Cloves [G], the dried flower buds of a tall evergreen tree *Eugenia aromatica* – nearly all come from Zanzibar. Turmeric [H] is obtained from the dried rhizome of a plant of the ginger family, *Curcuma longa*.

Culinary and medicinal herbs

Herbs have been used in the culinary arts, in medicine and in the cosmetic industry since ancient times – so the excavations of sites in Greece, Italy and the Mediterranean islands tell us. Hippocrates (of the 5th century BC), known as the "father of medicine", cultivated some 400 herbs and made up herbal remedies that continued in common use for centuries for treating diseases.

Medicinal herbs
The medicinal use of herbs probably originated after man had discovered them for cooking purposes. Perhaps he first noticed the aromas when throwing plant waste onto the cooking fires. Thereafter he may deliberately have sought them out so that the scent and flavours would impart new zest to roast meat and fish. Later still, herbs were used to disguise meat and game that were no longer in their prime.

Herbs were introduced into most of Europe by the Romans who in turn had adopted them from the Greeks. Many were endowed with mystical powers and they were used for ceremonial purposes. Triumphant Roman emperors are depicted wearing crowns of bay leaves [1C]; during the Middle Ages current superstition held that hyssop and garlic would ward off the witches who added concoctions of wild herbs and other plants to their brews. Peasants in some remote eastern European countries still protect their homes against evil spirits and vampires by hanging garlic wreaths over their front doors. Even in the late twentieth century some civilized people insist on sowing parsley [1A] only by candlelight on Good Friday. Many others believe that transplanting parsley seedlings will bring bad luck.

The cultivation of herbs and the use of them became a speciality of the Christian monks and every monastery has its own extensive herb garden. The plants were grown for their medicinal qualities and as late as the eighteenth century physicians still relied heavily on herbal medicines in the treatment of disease. The famous London herbalist John Gerard (1545–1612) published a huge volume in 1597 listing thousands of plants with healing properties. Herbs have dwindled in importance for the medical profession but many are still used in the preparations of soap, cosmetic creams and lotions and skin tonics.

The heyday of the herb garden came in the sixteenth century when wealthy landowners laid out intricate herb gardens whose patterns were picked out with low hedges of lavender or box. An average herb garden of the famous Elizabethan knot garden type might include more than 50 different herbs, some grown for cooking, others for medicine and soothing balms and tonics.

The nineteenth century saw a decline in the use and growing of herbs although nowhere as drastically as in Britain where the Victorian cook used few herbs but parsley, sage [3B] thyme [3D] and mint [2E]. With the expansion of the tourist trade in the mid-twentieth century many people have a new appreciation of herbs as a result of their eating foreign dishes. Today herbs are used increasingly in everyday meals.

The best flavour is obtained from fresh herbs although dried herbs often have a stronger taste than fresh ones and should therefore be used in much smaller quantities

1A   B   C   D   E

1 Parsley (*Petroselinum crispum*) [A] is the classic herb of European cooking. Plain and curly leaved varieties are available, both rich in vitamin C. The stalks have as much flavour as the leaves. Parsley is always used fresh in *bouquet garnis* and chopped to flavour sauces, salads, *fines herbes* and *maître d'hôtel* butter; small fresh sprigs are used for garnishing a variety of dishes.

Sweet basil (*Ocimum basilicum*) [B] is native to India and Iran. Young, sweetly clove-scented leaves have the best flavour and are used, chopped, in dishes containing tomatoes. Bay (*Laurus nobilis*) [C] is an evergreen tree from the Mediterranean. The aromatic leaves are used, fresh or dried, in *bouquet garnis* and as flavouring for meat and fish stews, pâtés and terrines. Chervil (*Anthriscus*

cerefolium) [D] comes from eastern Europe. Its leaves have a slight resemblance to parsley but with a delicate aniseed flavour. It is used chopped in soups, sauces and egg dishes. Coriander (*Coriandrum sativum*) [E] is sometimes known as Chinese or Japanese parsley because the feathery leaves are as popular in Eastern cooking as parsley is in Europe. It is one of the oldest herbs known and the seeds are an essential ingredient of *garam masala*, the spicy flavouring for Indian curries. In Europe and North America dried seeds are used in fish and meat dishes, bread and cakes.

2A   B   C   D   E

2 Dill (*Anethum graveolens*) [A] is a European herb particularly popular in Scandinavian, German, Russian and Balkan cookery. Both leaves and seeds are used to impart an aniseed flavour to sauces and vinegars, salads and pickles.

Fennel (*Foeniculum vulgare*) [B] comes from the Mediterranean and is much used in flavouring Provençal dishes. Stalks, leaves and seeds have a liquorice flavour that associates well with pork and veal. It is traditional with fish and as an addition to forcemeats and sauces. Grilled sea bass and red mullet are flamed on a bed of dried fennel. Finely chopped leaves may also be used in small quantities to season soups, salads, mayonnaise and *sauce vinaigrette*.

Wild marjoram (*Origanum vulgare*) [C] is a pungent-flavoured herb much used in Mediterranean cooking to season meat, poultry, soups and omelettes. It is known in the dried form as oregano. It may be used instead of thyme. Camo-

mile (*Anthemis nobilis*) [D], both wild and sweet, grows in Europe and Asia. The aromatic flower heads are used dried as infusions and in herbal teas and sometimes in the manufacture of vermouth and other aperitifs, as well as lotions.

Mint (*Menthá* sp) [E] has many species and varieties. In many parts of the world the strongly flavoured leaves are cooked whole with young summer vegetables and used, finely chopped, in jellies and chutneys. It is also used to flavour cool summer drinks, cups and juleps. It is relatively unknown in French cuisine but is a common seasoning in the Middle East where it is often added to chutneys and yogurts.

– the very essence of a herb, and in particular parsley, chervil and mint, is lost when it is deprived of its water content. The cook can always have fresh herbs at hand for they can be grown in a window box or even in a few pots on a sunny kitchen sill. Some herbs, such as thyme, rosemary and bay, tolerate prolonged cooking but others, including chervil [1D], dill [2A] and fennel [2B], are best added to the dish at the last minute so that their aromatic qualities are not lost.

Fines herbes and their uses

The French term *fines herbes* is applied to a mixture of very finely chopped herbs. It usually consists of parsley and chives, and sometimes of parsley only, but it should correctly also include chervil and tarragon [3D]. In former times burnet, chopped mushrooms and shallots were part of the *fines herbes* mixture. These herbs are usually added to quickly cooked dishes, particularly omelettes, or sprinkled as a garnish over meat and fish dishes and young vegetables. They may also be incorporated in butters and sauces to serve with meat and fish.

Gremolata, an Italian version of *fines herbes*, is made up of finely chopped parsley and anchovy, crushed garlic and grated lemon rind. *Gremolata* is used both as a seasoning and a garnish with many Milanese dishes, notably the veal and tomato dish *osso buco* (stewed veal knuckle).

Potpourri and pomanders

All their other uses apart, aromatic herb mixtures can be enriched with dried, scented flowers to make up a potpourri. Sweet marjoram [2C] and sprigs of rosemary [3A] can be mixed with clover and petals and small buds of roses, verbena and pelargoniums. They are sometimes enclosed in linen satchets but are most often seen in decorative china containers that may be placed in cupboards and on table tops to release their scent throughout the room.

Pomander balls are made from oranges, lemons and limes impregnated with a mixture of herbs and spices, including rosemary, nutmeg and cinnamon, and stuck with whole cloves. They were once believed to be effective protection against infections.

Bouquet garni is the classic flavouring for stocks, soups and casseroles; it usually consists of parsley sprigs, thyme and bay leaves tied in muslin or between celery sticks. Rosemary, basil, marjoram or oregano may be added to give a distinctive flavour. Garlic, peppercorns and or orange peel can be used as an addition to traditional Provençal casseroles.

3A B C D E F

3 Rosemary (*Rosmarius officinalis*) [A] is a strongly flavoured herb that can be reminiscent of camphor. Small sprigs are used to season roast lamb, pork, veal, rabbit, kid and grilled fish, particularly in Italy.

Sage (*Salvia officinalis*)[B] comes from the Mediterranean. The aromatic leaves are used with onions in stuffings for poultry and meat and to flavour sausage meat and, in Germany and Belgium, eels also.

Tarragon (*Artemisia dracunculus*) [C] is popular in French cooking. The aromatic leaves are used, finely chopped, in sauces, butters, soups, salads and vinegars. Thyme (*Thymus vulgaris*) [D] has a pungent aroma and when dried retains much of its flavour. It is a favourite of Mediterranean cooks for casseroles, vegetable stews and soups, with fish and in stuffings for meat and game. It is an

essential ingredient of a *bouquet garni*. Celery (*Apium graveolens*) [E] is a favourite salad vegetable, and the leaves are equally useful. The feathery, pale-green foliage can be used, finely chopped, to flavour soups, salads, veal and chicken stews; small, whole leaflets make attractive garnishes when watercress is scarce. The leaves are sometimes dried and used instead of celery salt. Summer

savory (*Satureia hortensis*) [F] is a member of the mint family. The pungent leaves should be used young, before the herb flowers, as flavouring for salads, soups, fish and vegetable dishes.

4 A B C D E

4 Sorrel (*Rumex acetosa*) [A] leaves may be cooked whole like spinach or made into a purée with butter to be served cold with fish, fatty meat and poultry; the young bitter leaves can be used in small quantities to season soups and salads.

Myrtle (*Myrtus communis*) [B] is an evergreen aromatic shrub from the mountains around the Mediterranean and is used to season lamb. The scented, purple-black berries which follow the white flowers were formerly dried and used like pepper.

Borage (*Borago officinalis*) [C] is native to southern Europe and was introduced to Britain by the Romans. The hairy leaves have a cucumber scent and flavour and are most often used in iced drinks; young leaves, finely chopped, may be added to salads, yogurts and cream cheese. In Italy borage is used as a stuffing for ravioli, boiled like spinach or fried in batter. The pale blue flowers are sometimes candied and used to decorate cakes, desserts and confectionery.

Angelica (*Angelica archangelica*) [D] has many uses in cookery. Its stems, leaves and seeds have a characteristic musky flavour. Young stems are candied or crystallized and used for decoration of cakes and trifles. The leaves are cooked as vegetables in many northern countries and they can also be used to flavour orange marmalade. Young shoots can be blanched and added to salads. Roots and seeds are used in the manufacture of liqueurs, vermouth and gin.

Salad burnet (*Poterium sanguisorba*) [E] is another herb with a cucumber flavour. It is used fresh to flavour salads and salad dressings. The outer leaves are bitter and thus only young, tender, centre leaves should be used.

Berries, nuts and olives

Of all the fruits eaten by man the strawberry is one of the most prized. The natives of Chile were cultivating plump, aromatic strawberries [1] long before Europeans reached South America. In the early eighteenth century these plants were brought to Europe and crossed with the already established North American sweet variety. A new strawberry (the one eaten today) was produced, combining the size of the one with the excellent flavour and colour of the other, and is known botanically as *Fragaria ananassa*.

The red raspberry (*Rubus idaeus*) is native to most European countries, although the best crops come from temperate areas where summer days are long. Different varieties of the fruit have been developed for other conditions. A yellow raspberry is also grown in England as is the hybrid loganberry – a cross between the raspberry and blackberry (*Rubus fruticosus*) – which originated in California. Other types of hybrid blackberry are the boysenberry and the phenomenal berry, similar to the loganberry but with slightly larger fruit.

The hardy blackberry with its trailing brambles and clusters of fruit is a familiar wild plant in many places. The black mulberry (*Morus nigra*) has been cultivated for centuries both for its dark red fruits and for its ornamental attraction. The white mulberry (*M. alba*) originated in China and its leaves are used in silkworm culture.

Fruits for the epicure

Bilberries, blueberries and whortleberries are all species of *Vaccinium*. Cranberries, native to Europe and America, also belong to this genus. Their crimson, acid fruit make an agreeable sauce or jelly to accompany venison or turkey. The scarlet berries of the rowan or mountain ash are inedible when raw, but, cooked and jellied, they lend a tart contrast to game and venison. Red, black and white currants (*Ribes* spp) are familiar as garden plants and have been grown in Europe since the fifteenth century. Rich in vitamin C, blackcurrants were once regarded as a cure for sore throats but are now gathered to make fruit syrups, jams and jellies. Redcurrant jelly is used in France with game stews and in Britain with roast lamb.

Closely related to the currants is the gooseberry (*Ribes grossularia*) whose greenish fruits have been a common sight in English gardens since 1600. This deciduous plant flourishes in northern climates even up to the Arctic Circle. The red, green or yellow berries can be sweet or acid, smooth or hairy, some making reasonable dessert when fully ripe, but are excellent for bottling, jam, and pies and of course for making gooseberry fool in which the fruit is stewed, made into a purée and mixed with cream or custard.

Nuts for the confection box

The sweet almond (*Prunus amygdalus*) [5B] is one of the world's most popular nuts and is indigenous to the eastern Mediterranean. Those grown in Jordan are particularly fine for making sugared almond. Sweet almonds are an essential ingredient of Florentines, marzipan and dainty *petits fours*, and bitter almond provides the essence to flavour confectionery. One of the most delicious sweetmeats is the candied *marrons glacés* of southern France, made from the large European chestnut (*Castanea sativa*) [5C]. Chest-

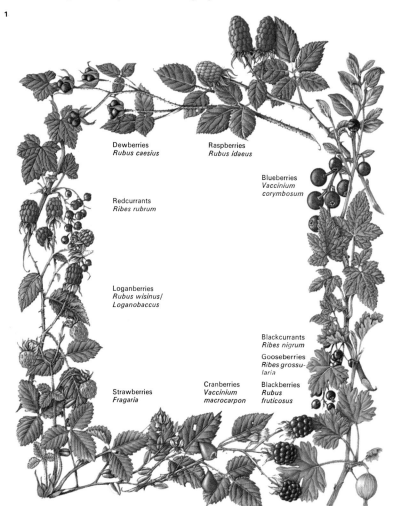

1 Succulent fruits such as the raspberry have been popular and widespread since ancient times. The soft red flesh of this berry makes it an ideal fresh dessert. Blueberries and blackcurrants are both rather acid and these fruits are better cooked or made into jellies. Gooseberries make a delicious filling for tarts and pies. The soft purplish-black fruit of blackberries does not keep for long and these should be used as a pie filling soon after picking. The cranberry is most often used to make preserves and sauces or to add flavour to meat. Juicy red strawberries, eaten fresh with cream and sugar, are one of the most popular berries. Sweet loganberries are served stewed with sugar. Redcurrants are also eaten fresh as a dessert, or as an excellent jelly. The dewberry is similar to the blackberry but has a trailing stem and the fruit is generally thought to possess a finer flavour.

Dewberries
Rubus caesius

Raspberries
Rubus idaeus

Blueberries
Vaccinium corymbosum

Redcurrants
Ribes rubrum

Loganberries
Rubus wisinus/ Loganobaccus

Blackcurrants
Ribes nigrum

Gooseberries
Ribes grossularia

Strawberries
Fragaria

Cranberries
Vaccinium macrocarpon

Blackberries
Rubus fruticosus

2 The olive is one of the oldest crops in the world. Both immature green and ripe black olives are bitter and inedible before processing. But treated with an alkaline solution and soaked in brine – an early process still in use today – the olive becomes a noble delicacy. The olive tree grows quite slowly, but may live for more than 1,000 years. There are hundreds of named varieties of olive.

3 Oil from fresh olives was extracted by a primitive press. Today, mills clean the fruit and crush it coarsely between grinders. The pulp is collected in cloths and pressed between heavy racks. Called virgin oil, the product of this first pressing is of fine quality. But as the pulp is subjected to further pressings the liquid obtained becomes inferior. The oil is washed to remove bitterness and filtered for bottling. Some varieties of olive tree are grown primarily for their oil, a commodity prized not least by the ancient Greeks who used it to anoint their bodies. Sometimes the oil-bearing fruit is beaten from the trees, although in many Mediterranean groves it is picked by hand. Today the processing of olive oil is an extremely important industry. The fruity green olive oils of Tuscany, Greece and Cyprus are strong in flavour. More delicate virgin oil from the Provençal olive is used for salads and mayonnaise. Tunisia is another major olive oil producing region. There are other vegetable oils in use for cooking and salads. Walnut oil, even more expensive than olive oil, is kept by cooks to dress salads for special occasions. Groundnut or peanut oil, largely produced in West Africa, is tasteless and is therefore used as a general-purpose oil for frying and for salads.

nuts are roasted or used to make preserves.

The pecan (*Carya illinoensis*) [4C], one of the finest of oily nuts, has a mild but distinctive flavour much in demand for baking and confectionery. In the United States, where it is extensively cultivated, it is also popular as a dessert nut. The elongated, brown shell of the pecan holds a fruit similar to a walnut, to which it is related. The Brazil nut [Key B] (*Bertholletia excelsa*), indigenous to the Amazon valley, is among the choicest dessert nuts seen on European Christmas tables. This slightly oily nut, tasting faintly of coconut, also features in confectionery.

Salt and savour delicacies

Walnuts are also popular nuts. Although the black walnut (*Juglans nigra*) of America is used in sweet dishes, the pronounced flavour of most walnuts is an ideal contrast to the richness of caramelized sweets, layer-cakes and sugary pastries. In France, fresh walnuts, accompanied by a glass of new wine, are eaten with hot bread, coarse sea salt and sweet butter. Roasted hazel nuts [4A] are served with the local wine at Avellino, a small

town near Vesuvius, in Italy, hence their specific name *Corylus avellana*. These slow-roasted, pale golden-brown nuts have been famous since Roman times. The shiny, round hard-shelled Macadamia or Queensland nut is yet another variety of walnut that is served with drinks.

The pistachio (*Pistacia vera*) a pale green, finely textured nut, is prized as an ingredient in pâtés and sausages and as a colouring for confections. The waxy, ivory-coloured seeds of a considerable number of species of pine are invaluable as a seasoning for savoury dishes. The fruits of Roman or stone pine (*Pinus pinea*) [5B] are used by the Italians as a sweetly sour stuffing for sardines.

Peanuts or groundnuts (*Arachis hypogaea*), native to Brazil and Peru, were cultivated by the ancient Incas and Mayas. Peanuts, which are not true nuts but pulses, like peas, are a major source of edible oil. They are used for their rich dominating flavour in Oriental cooking, and in the West salted peanuts are served as snacks. The pleasant taste of cashew nuts (*Anacardium occidentale*) [4B] is enhanced by roasting.

KEY

Black mulberries *(Morus nigra)* [A] grow on short, squat trees cultivated in Eurasia and North America for their fruits. Brazil nuts (*Bertholletia excelsa*) [B] grow on a large Amazonian tree. Olives (*Olea europaea*) are picked green [C] before they ripen or as black olives [D] after they ripen.

4 Ripe, fresh hazel nuts [A] are partially covered with leafy husks. These nuts grow wild but cobnuts and filberts are two common varieties cultivated in England. Cashew nuts [B] are grown in the tropics. Each bean-shaped nut is formed beneath an apple-like fruit and has an inner and an outer shell which are removed before the nut is roasted. Mottled brown shells of the pecan [C] burst apart to release the ripe nut. The pecan, a relative of the walnut, grows on large trees which are found in the temperate parts of North America. Pecans make excellent dessert nuts and are also salted.

4 A
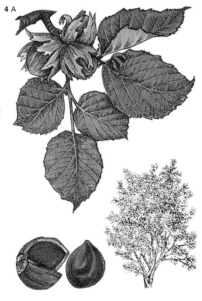

B
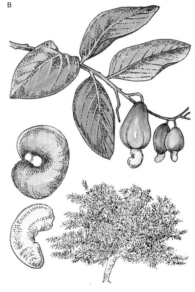

C
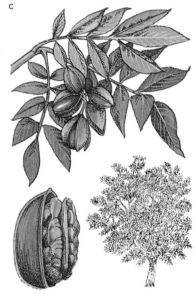

5 A

5 Raw or roasted, seeds of the stone pine [A] may be eaten or used for cooking. The pine cones open in the heat of the sun to reveal a waxy kernel. Many species grow in the Mediterranean area. Bitter and sweet almonds [B] related to the stone fruits like the peach are cultivated in temperate climates. Sweet almonds are edible but the kernel of the bitter variety is inedible and used only for the extraction of its oil. Sweet chestnuts [C] have had a variety of uses since Roman times. They may be roasted, boiled, ground into flour or fed to livestock. The best quality chestnuts grow in southern Europe, and are used in France's candied *marrons glacés*.

B

C
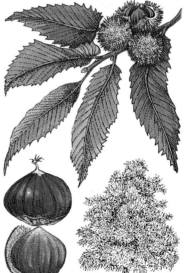

Fibre and oil crops

Men cultivate plants not only for food, but also for their oil and fibre content. Vegetable fibres are obtained from the fruits, leaves and stems of plants. These fibres consist largely of cellulose, a long organic molecule of the sugar or carbohydrate family which forms the walls or membrane of plant cells and makes up as much as 90 per cent of the cell. Fibres are characteristically strong and pliable, allowing plants to bend without snapping.

Various types of plant fibres

Archaeological evidence from caves in Mexico and the American southwest reveals that fibres were already being used more than 10,000 years ago. Apart from using flax hemp and other fibres to make ropes, nets or sacks, early man made crude fabrics by pounding the fibrous tissues of certain trees and plants into flat sheets of "bark-cloth", such as the Polynesian *tapa* still made today. By 3000 BC cotton was being spun into yarn and woven into cloth in India while the manufacture of linen was well developed in Egypt even earlier.

Fibres from different parts of a plant have varying characteristics. Bast fibres come from the inner tissues beneath the bark of dicotyledons (plants having two seed leaves). Made up of long overlapping cells, they are bedded in a cementing material that must be removed before the fibres can be peeled apart. Bast fibres are known as "soft fibres" and are particularly flexible. They can be made into rope or twine or woven into coarse, heavy-duty sacking. Jute [4], hemp [3] and flax [5] are the most commercially important and flax can also be woven into a soft, fine fabric known as linen [6].

By soaking, or retting, the long stems in water, the material binding bast fibres together is partially decomposed by micro-organisms in the water. The stems are then beaten and passed through rollers that separate the softened fibres from the mucilaginous matter. Jute fibres are stripped away by hand.

Leaf fibres, mostly obtained from perennial plants, are part of the vascular structure of leaves. They have a stiff texture and are much shorter than bast fibres. In a process called decortication, rolling machines crush the leaves, scrape off non-fibrous matter and wash it away with jets of water. These "hard fibres" are generally too stiff to be made into fabrics. Their major use is as cord, twine, brush bristles and coarse sacking. The most important leaf fibres are sisal, hennequen and abaca.

Seed fibres grow as fine hairs on the seeds of certain plants [1]. Each individual fibre is a single, elongated cell. Cotton [Key], coir and kapok are the only seed fibres of any major commercial value. Of the three, only cotton is suitable for spinning into fine yarn [2]. Kapok is mainly used as stuffing or insulation, and coir is made into ropes, sacks and brushes.

Vegetable oils for the human diet

Vegetable oils are essential to the human diet and contain more energy per unit of weight than any other food. Unlike mineral oils, they consist primarily (over 95 per cent) of triglycerides – combinations of glycerol and stearic, lauric, oleic, linolenic and similar fatty acids.

Extraction of oil from the seeds and fruits of plants that have commercially significant

Cotton *Gossypium* sp

1 The cotton plant is a shrub-like annual native to subtropical regions the world over. After rapid flowering, small green seedpods (bolls) develop. The cotton seeds within the bolls sprout a mass of fine fibre hairs. When mature, the bolls rupture and a soft cloud of cotton erupts. The crop is harvested either by hand or machine and then taken to be ginned (separating the seeds from the fibres), cleaned, carded and spun into yarn. Of four different species of cotton now cultivated, American Upland and Egyptian are the most important. The former accounts for some 85 per cent of the world's cotton production.

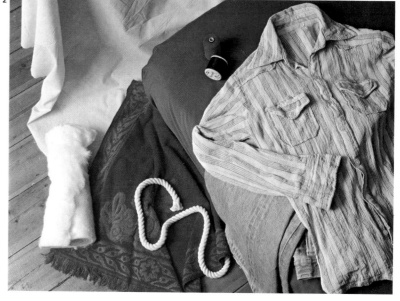

2 Cotton products date back at least to 3000 BC in India and Central America. The highly durable and versatile fibre has been used for countless purposes ever since. The longest, high-quality fibres are made into laces and fine fabrics while coarser fibres are used to make sheets, carpets, blankets, sailcloth and industrial products such as thread, film, plastics, paper and cotton wool.

3 A

Hemp *Cannabis* sp

B

3 Hemp is a well-known source of bast fibre. First used in Central America nearly 5,000 years ago, it has since found its way to most temperate regions. It grows up to 3m (10ft) in height [A]. After harvesting, the hemp is dried and then ripped apart for the fibres found in its long, woody stem. These yellowish to grey coloured fibres are stronger than flax but far coarser. They are easily twisted into rope or twine or else woven into sacking and other coarse fabrics. Hemp was in demand in the days of sail when it was used for rope [B]. Today, sisal and synthetic fibres are cheaper and more durable. Oil is extracted from hemp seeds. Hemp sap or resin contains a narcotic drug, cannabis.

4 Jute is native to India. This tall annual plant yields a fine bast fibre that is cheap, easy to bleach and dye and can be readily woven into coarse fabrics, sacking, ropes and twine.

4 Jute *Corchorus* sp

5 Flax *Linum* sp

5 Flax thrives in moist, temperate climates. Only one of several species is cultivated for its fibre and rich oil seeds. After harvesting, flax stems are retted (soaked in water) to soften the fibres which are then spun into yarn.

6

6 Flax and jute products exploit the different qualities of their fibres. Bed and table linens are fabrics made from flax which combine strength with fineness and pliability. Linen is also highly absorbent and is used for tea towels. Flax can be made into rag pulp for cigarette papers. Jute is a coarse, strong fibre which can be cheaply produced and is therefore made into rope or materials such as hessian, scrim and burlap for sacking and furnishing uses.

amounts is possible only if the oil-bearing cells are first ruptured by heat and pressure. Typically, seeds are hulled then coarsely ground and "cooked" to reduce their moisture content as much as possible. They are then fed into either hydraulic or screw presses that crush the seeds into a pulp and release their oil content. Alternatively, cooked, flaked seeds are passed through a solvent bath consisting of hexane, carbon disulphide or similar volatile solvents. Oil and solvent are separated by distillation.

Vegetable oils, which can be broadly distinguished from fats by their liquid physical state at room temperatures, have both edible and nonedible uses. Early Mediterranean civilizations, for example, not only consumed olive oil but also used it as a grease and lubricant. Other oils were made into soaps, burned in lamps or used to make paints and varnishes.

Demand for vegetable oils has risen spectacularly. Today soya beans are the single most important source of vegetable oil. The oil (16–18 per cent by weight) is extracted from the beans by the solvent process. If it is to be made edible, the oil is further refined to be sold as an odourless, almost colourless cooking oil with free fatty acids removed. If it is to be turned into margarine (an emulsion of oil, skim milk, flavouring and preservatives) or shortening, it is also passed through a hydrogenation process. This turns unsaturated fats into saturated ones with a higher melting-point and greater stability. A variety of industrial uses for soya bean oil includes paints, varnishes and the manufacture of synthetic alkyd resins.

Industrial oils and their uses

Apart from soya beans, the most important edible oils [7] are peanut, corn, cotton seed, rape seed, sunflower, coconut [9] and olive.

The main industrial oils [8] are castor, tung and linseed, all of which contain a greater amount of free fatty acids and other contaminants such as resins and sterols than edible oils. Industrial oils are used in the manufacture of soaps (for which purpose they are treated with caustic soda or potash), detergents, plasticizers, cosmetics, paints and a variety of chemicals.

Cotton is a valuable plant, providing both fibre and oil. The fibre grows on the seeds and is separated from them by ginning. The oil is extracted from the seeds. Cotton cultivation is the chief activity of such regions as the southern United States, where ripe bolls were once picked by hand.

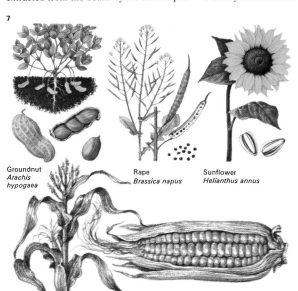

7 Seeds and fruits of many plants provide edible oils. Sunflower seeds yield a light, high-quality oil used in margarine, shortening and confectionery. Rape seed and corn germ both have an oil content of some 40-50%. Groundnuts, or peanuts, are the next most important source of vegetable oil after soya beans.

8 Industrial uses are found for castor and linseed oil in lubricants, chemicals, plasticizers, paints, varnishes and printer's inks. Volatile oils such as almond and pine which bear the essence of a plant's fragrance are used as scents in soaps and perfumes and as flavourings in food.

Groundnut *Arachis hypogaea*

Rape *Brassica napus*

Sunflower *Helianthus annus*

Maize (corn) *Zea mays*

8 Linseed *Linus* sp

Almond *Prunus amygdalus*

Castor *Ricinus communis*

Pine *Pinus* sp

9 To produce coconut oil the coconuts are first harvested and dried [1]. The dried flesh, or copra, is loaded into a silo [2] and then fed through a cleaning and crushing mill [3]. The pulp is then heated [4] to rupture the cell structure and free the oil, and then passed through an expeller [5] in which the oil is squeezed out by a screw. This oil is then filtered [6] and is ready for refining. The pulp is dried [7] and broken [8] again and passed to a solvent extractor [9] where the remaining oil is removed by a petroleum solvent and the residue taken off as cattle feed [10]. The solvent is distilled out [11] and the oil joins that from the expeller to be bleached [12], filtered [13] and deodorized [14].

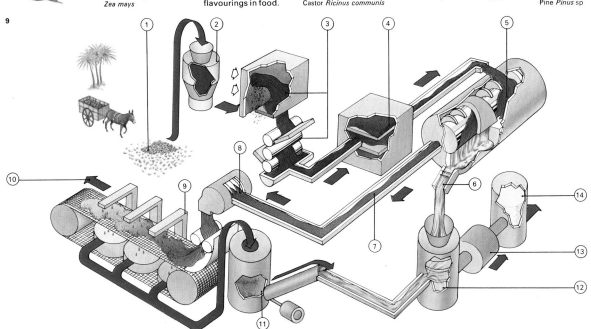

Forestry

Forestry is the art and science of managing stands of growing trees and their associated soils – and, by extension, plant and animal life – so that they yield the greatest possible benefits to mankind. Timber, firewood and pulpwood for papermaking are the main economic forest produce. Woodlands are also valued for soil conservation, the safeguarding of water supplies, their scenic attractions and many forms of recreation.

Every good forest manager normally aims to bring these objectives into balance by applying the concept of multiple-use forestry. Profitable timber production may become his sole objective.

In most countries much, or even all, of the forest area is state owned. Where woodlands are privately owned it is usual to find some state control, because the long-term objectives are too far-reaching to be left entirely to individual owners.

Natural woodland
Natural forests once covered nearly two-thirds of the world's land surface, but clearances for farming have reduced this figure to barely one-third today. Forests grow mainly on mountainous or remote regions on slopes or soils unsuited to agriculture. The world's 4,035 million hectares (10,000 million acres) of forest yield 2,000 million tonnes of timber annually, equivalent to half a tonne of wood for each of the earth's inhabitants. This yield is only one per cent of the volume of standing timber. Foresters always aim at sustained yield, so that future needs can be met, in perpetuity, by maintaining adequate areas of forests and volumes of standing timber, that is living, growing trees. Although these stocks are already seriously depleted.

Natural forests still dominate vast areas of the northlands, particularly those of Canada, northern Europe, Scandinavia and the USSR, and many other mountainous regions of the world. In the north temperate zone most forest trees are conifers which yield the softwoods in great demand for building, packaging and paper pulp. Farther south, the temperate-zone broad-leaved trees, such as oak, ash, beech, birch and sycamore, dominate the native woodlands and provide a source of temperate-zone hardwoods, each valued for specialized uses, such as shipbuilding or furniture making.

In many tropical regions of the Americas, Africa, Asia and northern Australia, broad-leaved trees may grow in extensive, close-ranked rain forests or, where the rainfall is lower, on open savannas. Most of the many and varied hardwoods they produce are mainly used locally, but the best of them, such as teak, mahogany, greenheart and rosewood, are exported and used worldwide.

Selection forestry
Selection forestry [2] provides the simplest means of producing useful timber with the least disturbance to the environment and at the lowest cost. After an exacting study of the trees growing in the area, their rate of growth and replacement by self-sown seedlings, a skilled, experienced forester draws up a working plan. Under this, each portion of the forest is tackled in turn, possibly one-tenth being dealt with each year for ten years, after which the working cycle is repeated. The workers fell selected mature trees, so creating gaps that will be filled gradually by

1 Natural forests, untended by man, can attain equilibrium with their surroundings and endure for thousands of years. As individual trees age and fall, through windthrow, fungal decay or lightning strikes, they are replaced by younger trees springing from self-sown seed. In this view of a North American forest the principal trees are Scots pine [1], Norway spruce [2] and birch [3].

3 Selection forests are managed to secure regular yields with minimum disturbance to the environment. Because trees of different kinds and ages stand side by side, they are also called multi-species and uneven-aged. In Oregon, USA, there are lodgepole pine (*Pinus contorta latifolia*) [2], shore pine [3], tamarack

(*Larix laricina*) [4], Douglas fir (*Pseudotsuga menziesii*) [5], ponderosa pine (*Pinus ponderosa*) [6] and red pine (*Pinus resinosa*) [8]. Trees are selected every few years for felling [1] and the logs are hauled out by a skidding tractor [7]. A tree must be felled and removed while it is still healthy.

As the trees compete for sunlight, water and nutrients, the weaker trees are suppressed by the stronger ones. Seed from the overcrop produce young trees which then grow up in the gaps caused by felling. Large logs are sold to sawmills, and smaller ones are used for paper pulp.

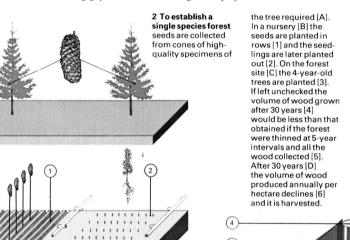

2 To establish a single species forest seeds are collected from cones of high-quality specimens of the tree required [A]. In a nursery [B] the seeds are planted in rows [1] and the seedlings are later planted out [2]. On the forest site [C] the 4-year-old trees are planted [3]. If left unchecked the volume of wood grown after 30 years [4] would be less than that obtained if the forest were thinned at 5-year intervals and all the wood collected [5]. After 30 years [D] the volume of wood produced annually per hectare declines [6] and it is harvested.

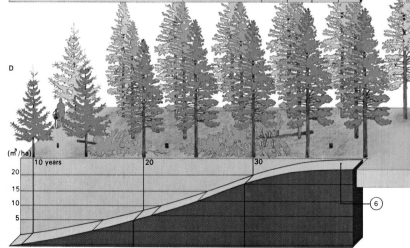

natural seedlings, and thin out groups of smaller, still immature specimens. This method causes minimum disturbance to the forest canopy, soil, wildlife and scenery. It is favoured in national parks and on mountainsides where soil erosion might follow clear cutting or avalanches occur should the slopes be exposed.

Even-aged, single-species forests [3] are the obvious choice where people have to plant bare land to increase a country's timber supplies, replace uneconomic scrub with worthwhile trees, or make good devastation due to past clear cutting or forest fires. By this method, trees are grown like a farm crop, though over a much longer period of time, to meet the needs of expected future markets at fair profits. A careful choice is made of the most profitable tree for the land available and this often proves to be a foreign one. For example, Monterey pine (*Pinus radiata*), native only to California, is widely grown in South America, South and East Africa, Australia and New Zealand, and Australian eucalyptus trees, such as *Eucalyptus globulus*, are planted in India.

Selected seed, which can be cold-stored if required for several years, is sown in well-cultivated and fertile soil in nurseries, usually in spring. The resulting seedlings are transplanted, when one or two years old, to transplant beds, partly to give them more growing space but also to promote vigorous growth of fibrous roots and check too rapid shoot growth. One or two years later the saplings are planted, either by hand or on easy ground by machine, at their final positions in the forest. As they grow taller, the forester protects them against damage by disease, insect pests, weeds, browsing animals and fire.

Thinning out and harvesting

After 15–25 years the first harvests begin. A proportion, often about one quarter, of the trees are harvested as thinnings, to give the others more growing space. This thinning-out process is repeated every few years until the crop is considered mature. Then, at an age of, from 40 years for spruce and pine, to perhaps 200 years for oak, the whole forest is harvested in the final felling [8]. The land is then replanted with its next tree crop.

KEY

Planting trees by hand is the usual method because most forests are established on sloping rocky ground or land carrying the stumps of a previous crop where machines cannot operate. This planter carries four-year-old Douglas fir transplants in a bag that keeps their roots moist. Using his spade, he cuts a deep V-shaped notch in the soil and opens the gap. He then inserts the tree's roots into this to their correct depth and firms the soil around. Planting is done in late autumn or spring, before active growth starts; thereafter the roots are able to make vital contact with soil moisture.

4 European oak (*Quercus robur*) is a traditional source of exceptionally strong, durable and beautifully grained hardwood.

5 Teak (*Tectona grandis*) occurs in the jungles of southern India, Burma and Indonesia. Outstanding strength, resistance to chemicals, workability and attractive golden-brown colour ensure its use internationally in shipbuilding and sturdy furniture.

6 American mahogany (*Swietenia macrophylla*), native to rain forests of the Caribbean region, attracted early Spanish explorers with its firm, red-brown, lustrous and readily worked timber. It is used to build strong high-grade furniture and in the shipbuilding industry.

7 Douglas fir (*Pseudotsuga menziesii*), which forms vast forests in western North America, is now extensively planted in Europe. Its fast-growing timber holds strong, dark summer-wood bands, making suitable for construction work, joinery and heavy-duty plywood.

8 Felling with power-driven chain-saws has generally superseded the use of hand axes and saws. To bring down this tall Douglas fir, the lumberjack first makes an under-cut, slicing out a wedge of timber from the side towards which he wants the tree to fall. Removal of support causes the tree to lean a little, putting remaining fibres under tension. Next, he makes a clean cut from the opposite side, breaking fibres that spring away from his saw, so that it does not jam. Now, as the last slender segment of wood breaks, the tree keels over safely to land in the desired direction.

10 Floating logs down rivers is the traditional, cheapest way of transport from upland forests to sawmills sited at river mouth ports. These logs are felled, during winter, in the Canadian forest. They are then drawn on sledges pulled by tractors over ice-bound roads to lie on the ice of a frozen lake. When the spring thaw comes, boatmen guide them downstream and they float freely. At the sawmill, they are halted by a boom, then stored, still floating, in a pool until each in turn is drawn to the endless chain that lifts them ashore.

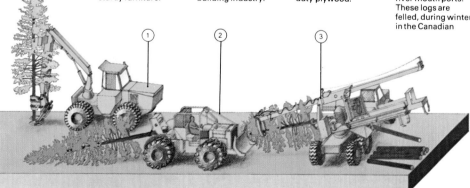

9 In Sweden, Canada and the USSR, even-aged crops are now sawn or sliced down by a felling machine [1] which cuts the tree near the base, using a movable grab to lay the trunks in conveniently placed bundles. These are then dragged from the felling site by a skidder [2] and delivered to the processor [3]. The telescopic grab of the processor feeds the tree, butt first, into a ring of rotating blades which remove the branches. It also saws the resulting logs into standard lengths.

The modern tree nursery

Botanist-explorers have brought to garden, park and arboretum thousands of species that once lived in distant and perhaps largely unapproachable habitats. In a more prosaic undertaking, nurserymen have also contributed to the huge range of trees available.

An explorer brings home a new species or a new variety of a known species. A nurseryman frequently crosses one species with another to produce a hybrid with a better commercial "performance" – that is to say, one that is easier for an amateur to establish, perhaps with a faster growing period or more attractive blossoms than either of its parents.

Special cultivation
The nurseryman also keeps a careful watch on his seed beds for any interesting deviant that nature may send his way. Many of the most-planted trees today have been selected from mutants that normally have to be maintained by man. A few "cultivars" (the varieties selected for their desirable qualities) reproduce themselves faithfully from seed, but a great number can be kept in being only through the arts of propagation that bypass

the sexual systems: rooting, cuttings, layering or grafting. Cultivars produced by interspecific hybridization are usually sterile and this means that the various vegetative means of propagation are the only ones available to the nurseryman.

The names of many of the famous nurseries of the past are preserved in the names of plants they have bred or selected: Lucombe of Exeter, Hillier of Winchester, Späth of Berlin, Vilmorin of Paris (*Sorbus vilmorinii*), Veitch of Chelsea (*Ampelopsis veitchii*). These names are as familiar to the committed gardener as the great names of today – Hillier, Treseder of Truro, Hesse of Bremen, Germany, or Gulf Stream of Virginia.

The greater part of the work of a commercial nursery, however, is simply the production of a wide range of plants up to the packaged stage for the customer. The modern trend inevitably is to confine production to species and cultivars that have the greatest public demand. The nursery illustrated below [3] is modelled on one – Hillier of Winchester – that sets out to do the opposite. Their *Manual of Trees and Shrubs* is also their

catalogue, listing almost 8,000 species of trees, shrubs, climbers and bamboos hardy in the Northern Hemisphere.

Moving established trees
Trees grown in nursery conditions are the easiest to replant; much of the cultivation that has gone into them has been to this end. It is possible, with care, to move established trees and replant them in a more suitable situation. Large trees are clearly the most difficult and require either a large labour force or a huge machine. Even then the tree will not survive unless laborious preparations, beginning perhaps two years or more before, are made. Trenches are dug, first around one side of the tree, then the other, after a year's interval, and filled with leafmould. The tree then fills the leafmould with fibrous feeding roots to make a compact root system that can be moved without half of it being lost. In nurseries on light sandy soils trees may be moved each year so that deep roots do not develop and transplanting is easy.

The easiest trees to move, and ones that amateurs can shift around within their own

CONNECTIONS

See also
218 Forestry
222 Trees: problems of climate and disease

1 The tree nursery, with machines and scientific methods to supplement the more natural methods of plant propagation, has played a major part in introducing new species of plants to many areas.

2 Many new cultivars of trees cannot be grown from seed and it is necessary to propagate them vegetatively (avoiding the normal sexual processes) by means of cuttings or grafts, for example.

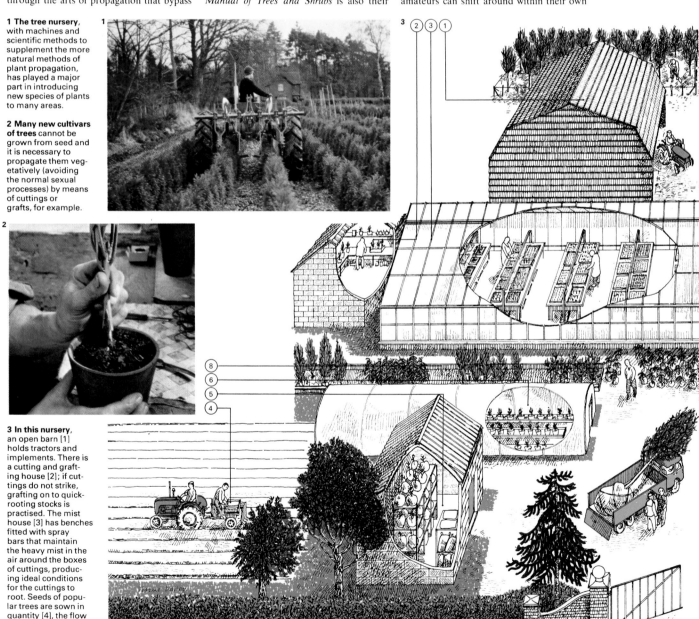

3 In this nursery, an open barn [1] holds tractors and implements. There is a cutting and grafting house [2]; if cuttings do not strike, grafting on to quick-rooting stocks is practised. The mist house [3] has benches fitted with spray bars that maintain the heavy mist in the air around the boxes of cuttings, producing ideal conditions for the cuttings to root. Seeds of popular trees are sown in quantity [4], the flow being regulated by an operator sitting on the gardens, are typical garden conifers, the cypresses and *Thuja* spp which have small zones of dense root close to the soil surface. It is easy for two men in a day to prepare a new hole and then unearth and transfer a tree of up to 6m (20ft), losing hardly any root. With a strong stake or guy ropes, and regular watering for at least a year, success is likely.

Planting a tree can be simple – and is, for example, in forestry plantations where tens of thousands of trees are planted. Holes are made, the trees inserted and the soil made firm. But care and attention improve the chances of success with garden trees. "Basin planting", with the tree placed in a little hollow, will encourage moisture flow in dry areas as, in contrast, "mound planting", with the tree raised slightly above ground level, helps where the ground is very wet.

Best planting times

In regions where winter conditions are severe, all planting is best left to early or mid-spring, when the ground is thawed and not too sodden. In more temperate areas, deciduous trees are best planted about the time their leaves fall. Their roots can grow surprisingly well in mild winter weather, giving them a better chance to become established than if they were planted in the spring. And cold spring winds are also less likely to dehydrate them if they have had a winter in the ground in which to make roots [Key].

In the same regions, evergreens can be planted either in the early autumn or spring, but if in the spring, not until the ground has a temperature of not less than 5°C (41°F) and the danger of cold, drying winds is over. Broadleaved evergreens can have some of their leaves cut off to reduce water loss through transpiration.

Trees in pots or containers are, in theory, safe to plant at any time. Obviously the best planting season for them is the same as for any other trees. If they are planted in mid-summer, they must be watered well and often. A tree that appears to be flourishing well may have been in its container too long so that the soil is filled with a dense tangle of roots. To ensure a good start, the soil should be shaken off and the roots untangled and spread carefully in the hole when planting.

The most important thing for a newly planted tree is security from having its roots disturbed. Semi-standard and larger trees must be tied firmly to a stake for the first two years after planting. A practical form of tie is a plastic belt with a buckle passing through a band that keeps the tree and stake apart [inset]. Damage by animals can be prevented by means of a special plastic sleeve or wire guard which covers at least the bottom 60cm (2ft) of the trunk. In areas with hot summers, the bark of young trees may be scorched after they have been planted out. Paper-wrapping is one method of shielding the trunk from the sun.

trailer. Seeds in bulk and cuttings awaiting handling are kept in a cool 3°C (37°F) store [5] or a cold frame. Cuttings can be kept in good condition for months. Some have survived after being carried across the world, their stems stuck in a potato. Tunnel houses [6] of polythene sheets are economical greenhouses for the protection and nurturing of young plants just struck from cuttings for a "weaning" period. Modern potting is carried out by machine in the potting shed [7]. Rooted plants from the mist house or frames are potted up in their hundreds, once a laborious task for human hands. Container-grown trees are stored in container beds full of damp sand and peat [8]. These trees, and the bigger ones in pots or cans in front of the mist house, are ready for sale at any time. A tractor with spray bar [9] sprays lines of seedlings with insecticides or selective weed-killer. Another tractor [10] undercuts the young trees with a blade to prevent formation of tap roots. Some beds have permanent irrigation pipes and in these [11] seed is hand sown. The cloches at the back of the bed have mist pipes for striking cuttings in summer. An area is permanently planted with stock trees [12] from which cuttings and seeds are collected. Administrative buildings adjoin a locked store where chemicals are kept [13].

Trees: problems of climate and disease

Trees include the largest and oldest living things in the world and in this sense represent the culmination of plant life on earth. Most of them are so huge, dignified and apparently everlasting that they are often taken for granted as more or less permanent features of the landscape. Yet despite their beneficent role as of helping to provide oxygen in the air that he breathes and making life tolerable for him on this planet, trees are menaced today by man as never before. They face their other problems – pests, fire, diseases and the vagaries of climates – with greater equanimity.

The effects of climate

Climate is one of the most crucial factors in deciding what trees grow where and its broadest movements in geological time have governed the evolution of tree species. Relatively recent climate changes have settled the present natural distribution of most of these species round the globe. The trees, in their turn, have enabled modern science to establish prehistoric patterns of climate – even to plotting the prevailing winds, which helped to distribute their seeds.

The ancestors of all our trees were tropical plants. In the tropics the seasonal temperature changes are usually small; what alters most from one time of the year to another is rainfall. Most tropical plants are evergreen and can grow either continuously or intermittently whenever there is enough moisture. Exceptional regions within the tropics are the montane zones – the equatorial Andes, for example, or Mounts Kenya and Kilimanjaro in Africa, where temperatures decline with altitude and the high slopes and "meadows" are characterized by weirdly shaped, low-growing alpine plants.

Temperate-zone trees are precisely adapted to the changing seasons. They are called "hardy" because they are able to withstand long spells of freezing temperatures and fluctuating conditions.

Ecological niches

Trees evolve in such a way as to be best adapted to the local environment. This effect can be seen among trees growing on a mountain range that lies in a north-south direction. In the western ranges of the United States

and Canada [2], the tapering pattern of each tree's range from south to north tells the story. In the Sierras a species may find its ecological niche at high altitudes. In the Cascades (and farther north still, in the Coast Range) similar basic climatic conditions, the length of time that the snow lies, the number of days during which the temperature rises above 4.4°C (40°F), in which it can grow and ripen new wood, force the tree right down to sea-level.

Once a tree is moved (or its seed is planted) out of its accustomed zone, it is in potential danger. A larch from Siberia, if moved to a milder climate, might be expected to luxuriate in the longer growing season while still being totally hardy; whereas what happens in practice is that it is lured out of its safe dormancy too early in the spring by higher temperatures than it expects. As a result, it starts growth, only to be cut back eventually by late spring frosts. This happens repeatedly, and it may die.

The converse happens when a southern tree moves north. It may be relatively safe in the spring, for bud-break will be delayed, but

CONNECTIONS

See also
218 Forestry
220 The modern tree nursery

1 Cause and effect of frost: where clouds or a tree canopy insulate the layer of warm air at ground level from the open sky heat cannot escape, so frost is unlikely [A]. On a night without clouds [B] heat is free to escape and the soil temperature falls below that of the air. [C] The soil takes heat from the air near the ground and as a result produces radiation frost. [D] Cold air, which gathers in a thin layer at ground level, flows downhill, collecting in hollows and valleys, which thus become frost "pockets". The height to which the hollows fill with frost is marked by the dead lower branches of trees. The house and garden [E] are protected from frost because it is stopped from flowing downhill by the trees.

2 The mountains of the west of North America [A] reach from British Columbia's coast range in the north, through the Cascades of Washington and Oregon to the Sierras of California. The foreshortened cross-section [B] plots the altitudes at which the same tree species occur. In the south, western white pine grows at 2,743m (9,000ft); in the north it has come down to 762m (2,500ft) to find the same growing conditions. These conditions are, in the main, a question of the length of the growing season, at temperatures of above some 4.4°C (40°F). There are exceptions to the rule. These are brought about by purely local rather than general conditions – chiefly where competition from other species is increased. The species described in the diagram are mainly conifers – the firs and pines of temperate mountain environments.

Alpine fir	Engelmann spruce
Bigleaf maple	Grand fir
Black cottonwood	Mountain hemlock
Douglas fir	Ponderosa pine

Sugar pine
Western juniper
Western red cedar
Western white pine

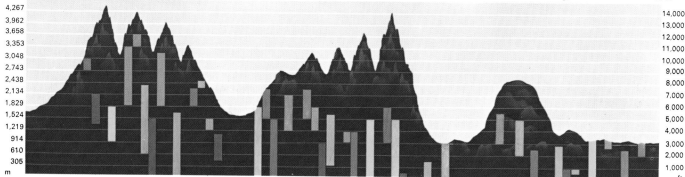

if growth continues late in the northern summer, its new wood may be still soft and immature when the autumn frosts strike.

More surprising is the difficulty trees experience in moving from the west coast to the east of North America – or from the Orient to Europe. Western conifers are as unhappy in New England as oaks from Ohio are in Britain or France. On the other hand, spruce trees native to subarctic conditions in Canada and Norway are widely used for reafforestation in western Europe.

Chances of survival

The most extreme example of upset is that of a cold climate tree that is moved to the sub-tropics. What happens here is that its buds may fail to open at all. Built into its schedule is the need for a cold spell (winter) to break its dormancy. If there is perpetual warmth, it is stuck: in all probability it will die.

In forestry the question of provenance (ie exactly where the seed comes from) is clearly of the greatest importance. The forester's object is to extend the growth period of his trees as far as he can without putting them in

danger of frost damage. He has little room for manoeuvre, but if he can add even a week to the growing season by getting his seed from 160km (100 miles) farther south without the trees suffering, he may add a whole year's growth in 20 years.

Many temperate zone trees can stand being frozen solid while they are dormant. What is more harmful, and can sometimes kill, is winter drought. When the ground is frozen, and no water is available to the roots, high winds and, often, low humidity continue to evaporate water from the branches. As a result, the tree begins to dry out. On ever-greens it shows in the browning of the leaves by the end of even a normal winter.

If the climate – the weather and its vagaries – provides the principal influence on a tree, a more immediate effect upon its existence may be posed by some microscopic vector of death – an insect or bacterium forming part of a population to which a tree is home and food. Most of these tiny creatures do no harm at all; others, such as the elm bark beetle, distribute a fungus that kills off centuries-old trees over wide areas.

Fir trees completely shrouded by snow resemble an army of spectres high in the mountains of Hokkaido, northern-most of Japan's large islands. Snow, which protects tiny plants from the more severe effects of cold, dry winds, provides vital pro-tection to evergreens in the intense cold and the dry sunshine

of the mountains. A prevailing wind deforms trees in an exposed situation by stunting or by preventing the growth on the wind-ward side. New shoots can grow only in the shelter of old ones and the tree takes on a characteristic shape, with its branches and foliage trailing to leeward.

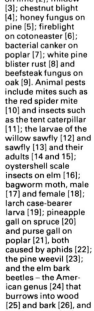

3

3 Trees can be attacked by numerous organisms. Fungal and bacterial pests in-clude anthracnose on plane [1]; rust fungus on lime [2]; mildew [3]; chestnut blight [4]; honey fungus on pine [5]; fireblight on cotoneaster [6]; bacterial canker on poplar [7]; white pine blister rust [8] and beefsteak fungus on oak [9]. Animal pests include mites such as the red spider mite [10] and insects such as the tent caterpillar [11]; the larvae of the willow sawfly [12] and sawfly [13] and their adults [14 and 15]; oystershell scale insects on elm [16]; bagworm moth, male [17] and female [18]; larch case-bearer larva [19]; pineapple gall on spruce [20] and purse gall on poplar [21], both caused by aphids [22]; the pine weevil [23]; and the elm bark beetles – the Amer-ican genus [24] that burrows into wood [25] and bark [26], and the European [27] that also damages wood [28] and bark [29], infecting them with a killer fungus.

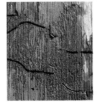

Farm stock breeding and management

Until mechanical power replaced the ox and the horse, animal husbandry was a part of every farm's working life. Farms entirely without livestock are still a minority. In the United Kingdom, for instance, farm revenue from livestock is normally more than twice that from all crops. About three-fifths of this comes from cattle and sheep and the remaining two-fifths from pigs and poultry.

Improving the stock

Breeding and management have a combined aim – the most economic production of meat, milk, wool or eggs, in environments ranging from entirely natural to wholly man-made. Thus, domesticated species have been spread, partly by selection of types best suited to conditions (for example, hill cattle and sheep breeds) and partly by changing the environment – for example, providing shelter and an improved diet. The more such changes are made, the more breeding aims have to be changed to suit the new conditions that are brought into being.

Cattle (synonymous with wealth in some languages) are managed in both ways. They may live entirely by grazing or be completely confined and even have fresh grass brought to them from the field. The beef cow has to produce and rear a calf a year, while the dairy animal has to provide a profit from her milk. The whole routine of the farm revolves round the milking [5], which takes place twice a day. For winter production the cow must be mated to calve in autumn instead of, as normally, in spring. Her feed is rationed to her yield of milk and the calf is reared away from her.

Sheep, in all but the most extreme conditions (for example, in Icelandic winters), are seldom housed and live by grazing – sometimes on crops specially sown for them. Flocks grazing on hill and mountain in summer may be moved to lower ground for the winter and lambing in the spring. Lambing and shearing remain the shepherd's biggest tasks in the year.

Except in the tropics, the pig – once a forager in woods and wastelands and housed only for fattening – has moved under cover. The new-born pig thrives only in warmth. The main cost of rearing thereafter is food; a pig that is cold burns food to keep warm instead of using it to put on weight. Modern pig systems are designed to speed up growth by regulating temperature and restricting movement. Much the same considerations apply to modern poultry farming.

Genetic resources of livestock

Breeding for improved production is easiest with species that mature early and produce many young in a year [1]. However, an outstanding male animal, whose attributes are inherited by his descendants, can have a vast effect within a few generations. This process has been greatly accelerated by the wide adoption of artificial insemination [3], especially in cattle. The semen of an outstanding sire can be stored in deep freeze and used for matings after he is dead or in countries on the other side of the world, to which the transfer of live animals would be too expensive or be a disease risk.

Ancestors of more than one type have usually contributed to the make-up of existing breeds. The new type has then been fixed by a period of close inbreeding. This cannot be carried too far without a consider-

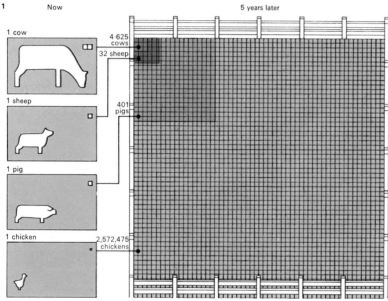

1 Rapid change in a livestock population is easiest where generations are short and numbers in each large. This accounts for the big improvement in pig and poultry performance over recent years, by the development of in-bred lines for crossing and recrossing to provide commercial stock. Upgrading of cattle is facilitated by using superior sires in successive generations and selecting progeny on performance.

2 New sheep breeds may be produced by introducing desired attributes from foreign types. The basis of the Colbred sheep was the Friesland, a Dutch milk breed with a high rate of reproduction but unsuitable for meat. Crossing involved three different British breeds. The Colbred, used for fat lamb production, retained some of the extra prolificacy. Many breeds are similarly created.

Border Leicester · Friesland · Dorset Horn · Clun Forest · Colbred

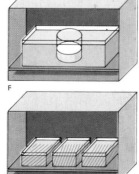

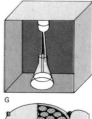

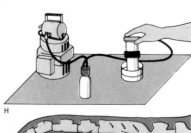

3 Artificial insemination, dating from the late 1930s, is now in world-wide use. Semen, collected from the bull on a dummy or "teaser" cow [A], is kept warm in water [B] and a second ejaculate added. It is diluted [C] and put into glass "straws" [D], which are corked and prepared for storage [E]. The straws are then cooled for seven hours at 4°C (39°F) [F] and stored in deep-freeze units [G]. When needed, the semen is thawed and introduced into a cow's uterus through a pipette [H].

able risk of diminished vigour and inherited defects. Crossbreeding to overcome this is a regular commercial practice, particularly with sheep and beef cattle, where the hardy hill breed is mated to a less hardy but quicker-growing breed.

Improvement of performance within breeds has come to depend increasingly on detailed records: yields of milk and eggs, and speed of growth and feed economy in meat animals. This entails elaborate testing procedures and statistical analysis of records. The computer has become a necessary tool of the larger breeding organizations which sell stock or semen to commercial farms.

The greater the emphasis on economic performance, the more the stock farmer has to rely on specialist advice and the results of scientific research. Animal nutrition has become a large study in itself, as has the design of housing and equipment.

Veterinary care

Above all, increasingly dense populations of animals have to be kept healthy. Consequently, the role of the veterinary profession is now very important and the veterinarian must be an expert in preventive medicine [7] as well as the diagnosis and treatment of clinical disease when it appears. The history of stock farming in Europe is punctuated by recurrent attacks of plagues and "murrains" (infectious diseases in animals) that are now controlled by closing frontiers, strict quarantine and in some diseases slaughtering of the affected animals and their contacts [8].

Vaccination is still considered the most effective method of disease prevention in countries where infection is rare and frontiers easily closed. Elsewhere, vaccination has reduced the risk, although for success the strain of the infection must be effectively typed. Vaccination has, for instance, greatly reduced the impact of foot-and-mouth disease in Europe. Other diseases, including bovine tuberculosis, brucellosis and swine fever, which once caused serious stock losses and even human illness, are now wholly or partly wiped out in most advanced countries. But there are always new risks and a close watch has to be kept on those areas where husbandry is primitive and disease endemic.

A championship at one of the big shows is the culmination of many breeders' dreams. Prize-winning animals, such as this Friesian cow (854), are greatly valued for breeding.

4 "Transhumance", the mass movement of sheep and cattle from summer mountain pasture to winter quarters and back again in spring, is widely practised. Flocks and herds move up the slopes as the season advances, always feeding on fresh grass. This is the reason for the high reputation of hill lamb from southern France, Norway and Wales. Rich mountain sheep's milk is also the basis of many famous, special cheeses.

5 Cows move on a roundabout in a labour-saving milking parlour. The milker need not move to fix or remove the unit on the teats of each cow. The rotary parlour's speed is set at the average milking time of the herd and slow milkers may be sent round again. Yields are automatically measured and the milk is piped to the dairy. The milker can adjust the quantity of feed given to each cow while the milking takes place.

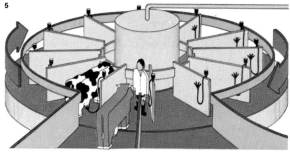

6 A 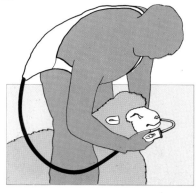 **B**

6 Internal and external parasites are a cause of unthriftiness and discomfort in farm animals and may carry more serious consequences. To control them sheep are given routine doses of a drench [A] against those that infest the digestive tracts and the lungs. A number of skin parasites, which thrive in the shelter of the fleece, have to be dealt with by dipping completely in a disinfectant bath [B]. This must cover all the animal including its head. The bath is now often replaced by a race, through which the sheep are driven between a series of jets that thoroughly soak the fleece. This is much quicker.

7 Vaccination against infections, which can spread fast among a herd kept in buildings under intensive systems, is now a matter of routine and here young pigs are being treated by injection. For pigs and poultry oral vaccines, which can be mixed in rations or drinking water, have also been developed. The period of protection is sometimes limited and they may need to be highly specific against particular strains of the organism that is liable to cause the disease.

8 Foot-and-mouth disease in Britain and North America is still dealt with by the slaughter of all possible contacts. The carcasses have to be burned or buried to avoid further infection. Where the disease is more frequent, susceptible stock are vaccinated. The disease is seldom fatal to adult animals but its effects on production can be catastrophic. Endemic in some parts of the world, there are several virus strains against which vaccines must be prepared.

Cattle for beef and milk

Cattle, first domesticated by man in Neolithic times, can be selected to serve a wide range of needs and to adapt to many environments. In early settled agriculture the same beasts served as power units for cultivation and haulage, as milkers and, finally, as a source of meat. The first use has almost disappeared in developed countries and the aim of modern breeders is to improve the yield and composition of milk, and to increase the speed and economy of weight gain for beef.

Ancient lineage and modern stock
Improvement of cattle on a regional basis has a long history, especially in areas where feed was abundant. From the sixteenth century onwards the big cattle of the Low Countries had an important influence on breeding in many parts of Europe. But detailed records of pedigrees were not available until towards the end of the eighteenth century. Then a growing demand from the urban population for meat stimulated such pioneer British improvers as Robert Bakewell (1726–96) who bred Longhorns in Leicestershire and Robert Colling (1749–1820) who, with his

brother Charles (1750–1836), bred what were the first modern Shorthorns at their farm in Yorkshire. They opened the way for the establishment, over the next 100 years, of breed societies and herdbooks covering all the major breeds. The result was that outstanding animals could be more easily identified and the influence of these beasts extended by a wider use of their descendants, particularly in the male line.

This process has been greatly accelerated by the general adoption, in the post-war years, of artificial insemination. As a result of this process an outstanding sire may be the parent of several thousand offspring. Using frozen semen "matings" are possible in any part of the world, even after a bull is dead. His genetic influence on such factors as milk and butterfat yield in his daughters can be precisely measured by comparing their performances with those of cows in the same herds sired by other bulls. Expert inspection of these cattle will also indicate the bull's influence on body conformation. Measuring the genetic contribution of a bull used for siring beef cattle is rather more difficult but

progeny records of growth rate and feed consumption give useful information.

A good beef animal must not only grow quickly and economically but must also carry the main weight of lean muscle in those parts of the carcass preferred by consumers – the hindquarters and back. Heavy bone is wasteful to the butcher and fat, although some is needed for flavour and tenderness, is now little in demand. The characteristics of the three most prominent beef cattle types are quite different: the ubiquitous Hereford is docile and able to thrive under fairly rough conditions; the Aberdeen Angus is a smaller animal but regarded by many as the producer of the best beef, while the French Charolais has great size, speed of growth and broad, heavily muscled hindquarters.

Dairy and dual-purpose animals
The pure dairy cow presents a strong contrast to cattle bred purely for beef. As much of her food as possible must be converted into milk, not muscle. She needs a large digestive system and a well-shaped and capacious udder. Two specialist breeds of this kind are

1 Cattle vary widely in size as well as in conformation and colour. A mature Charolais bull may weigh 1.5 tonnes, a Jersey no more than 375kg (825lb). Mature Friesian cows weigh up to 600kg (1,320lb) and may give ten times their bodyweight in milk annually. The Ayrshire, also a high yielder, weighs about 500kg (1,100lb) when mature and the Dairy Shorthorn about 550kg (1,210lb). The Aberdeen Angus is naturally polled (hornless). It is now general practice in commercial herds of other breeds to remove the horns by destroying the horn buds of young calves. The Hereford is a common beef breed in temperate regions. The hump of the Brahman, a characteristic of tropical breeds, enables it to store fat against drought and famine. The West Highland is bred to live and thrive in an upland habitat.

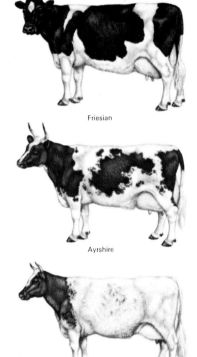

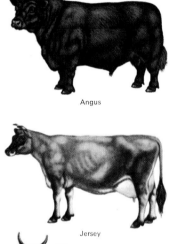
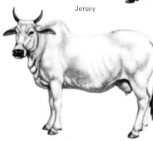
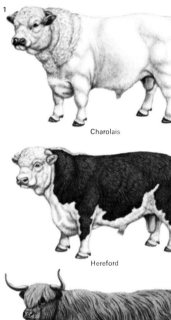

Charolais

Friesian

Angus

Hereford

Ayrshire

Jersey

West Highland

Shorthorn

Brahman

2 African cattle such as the long-horned Ankole are the mainstay of many pastoral tribes outside tsetse fly infested areas. Progressive eradication of this pest has widened their range.

3 Mountain meadows provide rich summer pasturage in the Swiss Alps up to the snowline. The cattle are brought up from the valleys where they spend the winter under cover. The main breeds are the Simmental and the Brown Swiss (illustrated).

the Ayrshire from Scotland and the Jersey from the Channel Islands; the latter is the smallest of all widespread modern breeds, giving milk containing one and a half times the average content of butterfat.

Much of the world's meat and milk supply comes from cattle that can be regarded as dual purpose. Of these the most numerous are the black and white Friesian or Holstein cattle, that originated in The Netherlands but which now show local variations of type. The Friesian is a large animal that produces heavy yields of milk whose composition has been greatly improved since the advent of artificial insemination. It has displaced the once dominant Shorthorn.

In a beef-breeding herd the sole revenue is from the calves. In a pure dairy herd most of the male calves have little value for rearing. To overcome this, in many commercial herds only the very best cows are mated with a dairy bull – to provide replacement heifers – while the remainder are mated to a beef bull to provide crossbred calves suitable for beef production. In order to make these crossbreeds easily identifiable, breeders use a

bull that will pass to the progeny some distinguishing features of the breed, such as the white face of the Hereford or the creamy coat of the Charolais.

Specialized breeds
There is still a great demand for breeds of cattle that are adapted to particular environmental niches. Examples include the somewhat slow-growing but extremely hardy Highland cattle of Scotland, which can survive the northern winter in the open, and a wide range of European hill breeds. In Africa cattle must be capable of surviving long periods of drought and also be resistant to local conditions that European breeds cannot tolerate. In tropical and subtropical areas much use is being made of the humped Brahman type of India and South-East Asia. This has been widely used in breeding programmes for the hotter areas of Australia and the Americas, most notably in the development of the Santa Gertrudis of Texas. Some work has also started on the improvement of the wild American bison, or buffalo, and has resulted in the "beefalo" crossbreed.

The English engraver Thomas Bewick (1753–1828) depicted this Holstein in about 1800. This breed was already noted for its yield and has been bred for milk for about 2,000 years. It originated in The Netherlands.

4 Autumn calving is necessary in areas where fresh milk is in demand all year. An average European cow, giving about 2,700 litres (815 gallons) over 10 months, will have a lactation curve that rises to a peak about six weeks after calving, falls gradually during the winter and rises again when she goes out to grass in the spring. Although milk is usually about 87 per cent water, its fat and sugar content are an important source of energy in the diet and its protein is of high quality and easily digested. Calcium and other minerals are left as ash. Selective breeding of milk cattle aims at increasing the total yield of milk solids.

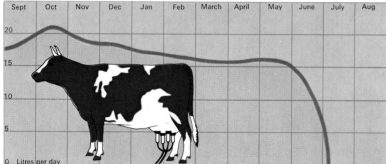

50 litres milk | Water 87·1% | Sugar 19% | Ash 0·7% | Protein 3·4% | Fat 5·9%

5 In machine milking the cow is conditioned to let down milk as if she were feeding her own calf. A partial vacuum holds the rubber teat cup in place and a varying pressure of air through the valve presses [A] and then releases [B] each teat in turn. The milk flow can be directed into a separate container or, as in large modern installations, piped direct to a cooling tank in the dairy.

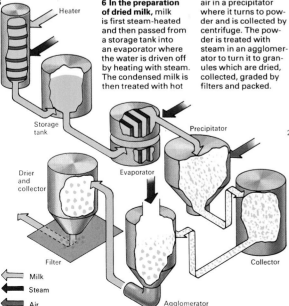

6 In the preparation of dried milk, milk is first steam-heated and then passed from a storage tank into an evaporator where the water is driven off by heating with steam. The condensed milk is then treated with hot air in a precipitator where it turns to powder and is collected by centrifuge. The powder is treated with steam in an agglomerator to turn it to granules which are dried, collected, graded by filters and packed.

Milk
Steam
Air

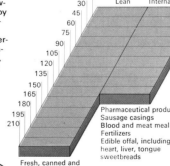

Beef cattle
Average weight 500kg

7 Slaughter products include many items besides meat itself. Young beef steers of about 500kg (1,100lb) live weight produce about 210kg (460lb) of lean beef. The internal organs, weighing in all about 130kg (285lb), include edible offal such as heart, liver, tongue, sweetbreads and tripe. The gut is cleaned for sausage casings and the contents used to make fertilizer. Internal glands are collected and used as raw material by the pharmaceutical industry. Fat, accounting for perhaps 70kg (155lb) in all, may be partly sold with lean meat joints, but much is processed – often for margarine. Bone – about 50kg (110lb) – is boiled to extract raw material for glue and the residue ground for mixing in animal feed, or as fertilizer. Skin goes to make leather and blood is collected for animal feed and fertilizer. The economic and efficient collection of these by-products explains the trend towards slaughtering in large factory abattoirs.

227

Dairy produce and cheese

Man began supplementing his diet with the milk of animals before he was even capable of recording the fact. References to milk and dairy products in both the Bible and the Hindu Vedas indicate that they were traditional foods long before the birth of Christ. Today, although milk is consumed as a nutritious drink, more than two-thirds of world milk yields are converted into other products, principally butter, cheese, yogurt, ice-cream, dried and condensed milks. Man is, however, not the only consumer of dairy products; milk can also be used as animal feed. It can even be converted into plastics. Before the recent intensive development of petrochemical plastics, casein (the principal protein in milk) was used as the raw material in the manufacture of a wide range of thermoplastic products, from buttons to billiard balls.

Milk and milk treatment methods

Dairy animals vary from culture to culture. The United States, Europe, New Zealand and Australia, the major dairy producers, use cows' milk for their dairy products. In Asia, both cow and buffalo milk are used and around the Mediterranean sheep and goats' milk is the basis of local dairy products. Reindeer, mares, yaks and camels are further sources of milk for human consumption.

Apart from its by-products, milk itself now comes in a variety of forms. Consumption of raw milk, which goes sour quickly and is easily contaminated, is less and less common in urban communities and pasteurization, in which milk is heated to kill all disease-carrying organisms and most of those that cause souring, is now a widespread practice. Milk may also be homogenized, a process by which the fat is distributed evenly and does not rise to form a layer of cream.

A large industry concerned with preserving milk has developed. Drying is the commonest method of preservation; for this full cream or skimmed milk, with a minimal fat content [1], may be used. (The fats removed in skimming are used to make cream or butter.) Condensed and evaporated milks are those in which the water content has been reduced. Condensed milk is often sweetened to improve its taste.

Cream is the fatty part of milk, which can be separated with a centrifuge. Some cream is retained for sale or used in the preparation of cream cheese, butter and proprietary desserts. The skimmed milk is either dried or sold as animal feed.

Butter, buttermilk and yogurt

Butter is the major dairy product, using about a third of all milk yields. It was probably the earliest milk product to be made and is manufactured from full cream which is agitated until the fat globules and solids clump together. Although it has little protein, its high fat content [1] makes it an excellent energy food.

Buttermilk is the liquid residue from butter-making. Its slightly sour taste makes it a popular and refreshing drink. Yogurt, originally a Middle Eastern food, is a semi-solid fermented milk food, characterized by a smooth texture and slightly sour taste. Traditionally made from goats' milk by a process of fermentation (a small amount of yogurt from a previous batch is used to "start" a new ferment), yogurt is now usually made from cows' milk and dried milk solids

CONNECTIONS

See also
226 Cattle for beef and milk

In other volumes
36 The Natural World

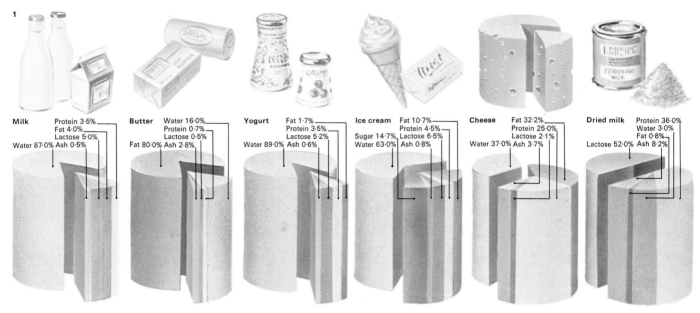

Milk Protein 3·5% / Fat 4·0% / Lactose 5·0% / Water 87·0% Ash 0·5%

Butter Water 16·0% / Protein 0·7% / Lactose 0·5% / Fat 80·0% Ash 2·8%

Yogurt Fat 1·7% / Protein 3·5% / Lactose 5·2% / Water 89·0% Ash 0·6%

Ice cream Fat 10·7% / Protein 4·5% / Lactose 6·5% / Sugar 14·7% Ash 0·8% / Water 63·0%

Cheese Fat 32·2% / Protein 25·0% / Lactose 2·1% / Water 37·0% Ash 3·7%

Dried milk Protein 36·0% / Water 3·0% / Fat 0·8% / Lactose 52·0% Ash 8·2%

1 Milk and milk products have the same basic constituents – water, protein, fat, lactose (milk sugar) and minerals (ash) – but their percentages vary considerably according to the finished product.

2 Butter moulds were very popular in the nineteenth century and are still used in exclusive restaurants. Their function was to turn a foodstuff into an attractive table decoration.

3 Ice-cream, once a luxury food, was popularized by Italian vendors who took their barrows and recipes all over Europe and America.

4 Dried milk is one major contribution that the world's richer nations, who are often large dairy producers, can make to the Third World. Despite intensive use of milk, in America, Europe, Australia and New Zealand, surpluses often arise. Dried milk is easily stored and transported and, when reconstituted, can provide a significant proportion of protein and nutritional requirements.

and fermented with cultured bacteria. Yogurt's lactic acid content can make it a valuable aid to digestion.

The cheese-making process
Cheese is made from ripened milk curds and if made from full cream milk will contain most of the food properties of the milk. Although there are really only three categories of cheese (soft, hard pressed and blue), variations in the process of making it produce over 2,000 different kinds. The main variables are the type of milk used and the conditions under which the source animal was fed. In addition, the methods of maturing the cheese greatly affect it [5].

Despite these variations, the basic method of cheese-making is the same everywhere. A curdling agent is added to the milk to precipitate the solids. (The solidified casein, fat and other water insoluble constituents are called the curds, and the remaining liquid is known as whey.) The curds are broken or cut to release most of the whey and left to drain. They are then broken up, salted and put into moulds. Finally the

cheese is ripened. This is one of the most important stages in cheese-making, for the conditions and length of time under which a cheese is matured are critical. The action of bacteria at this stage will create the cheese's unique characteristics of both taste and appearance.

Cream cheese differs slightly in that it is rarely ripened. A "home-made" cream cheese is made by allowing milk to sour naturally, adding rennet (a solution containing rennin) and hanging the curds up in muslin to drain. The cheese can be eaten 24 hours later. Processed cheese is a factory-made product in which the cheese is sterilized rather than matured. When tinned or vacuum packed it can be stored indefinitely.

Until the advent of refrigeration, ice-cream, a favourite dish in the courts of seventeenth-century Europe, graced only the tables of the rich. Nowadays it is a popular food [3] for all classes. Although the trend in recent years has been to replace dairy products with vegetable oils in the manufacturing process, traditionally made dairy ice-cream is still preferred in most countries.

KEY

Unhygienic dairies of the early nineteenth century bear little resemblance to the sanitary units now achieved in developed nations as a result of improved technology and health laws.

5 Health laws have altered the character of many cheeses; the best ones are still made by traditional methods. Here, Roquefort is matured in caves to induce the "blueing" caused by bacteria.

6 Factory-produced cheeses are easily stored and marketed. They are made in large moulds, then cut up and vacuum packed. Hygienic conditions tend to produce rather tasteless cheeses.

7 Many cheeses are instantly recognizable by their distinctive appearance.

Some of the more famous cheeses are illustrated here.

1 Provolone
2 Parmesan
3 Samsoe
4 Edam
5 Gouda
6 Mimolette
7 Blue Cheshire
8 Fontina (Danish)
9 Stilton
10 Cheddar (Canadian)
11 Gloucester
12 Cheddar (English)
13 White Wensleydale
14 Ricotta
15 Bleu de Bresse
16 Dunlop
17 Mozzarella
18 Jaalsberg
19 Danish Blue
20 Caciocavallo
21 Leicester
22 Feta
23 Fontina (Italian)
24 Gruyère
25 Monterey Jack
26 Tome au Raisin
27 Lancashire
28 Caerphilly
29 Edelpilzkäse
30 Limburger
31 St Nectaire
32 New England Sage
33 Red Windsor
34 Brick
35 Port Salut
36 Gorgonzola
37 Vacherin
38 Epoisses
39 Emmenthal
40 Dolcelatte
41 Tilsiter
42 Pont l'Evêque
43 Livarot
44 Quargel
45 Roquefort
46 Banon
47 St Marcellin
48 Camembert
49 Münster
50 Brie
51 Bel Paese
52 Maroilles

Pigs and sheep

Throughout the world pigs and sheep exist as both wild and domesticated animals. Wild pigs are largely found in damp, open woodlands in the Old World from northern Europe to South-East Asia. Wild sheep live in mountainous regions of Asia, North America and the Mediterranean but the many different breeds of domesticated sheep thrive at all latitudes and in nearly all habitats from hot lowland desert to the high altitudes beyond the mountain snowlines.

Pig breeding

The pig is one of nature's most efficient and omnivorous scavengers. The first records of the domestication of the pig – one of man's most efficient providers of meat – date back to about 3000 BC in China. The wild boar was probably first domesticated in Europe in about 2900 BC.

More recently pig breeders have concentrated on producing animals with the ability to convert feed into lean meat rather than unwanted fat and on rearing females capable of giving birth to larger litters. Today's "improved" females are ready to breed when they are seven or eight months old and will produce litters of more than ten piglets within 16 weeks of conception. If properly fed, managed and housed a healthy sow will produce more than 20 piglets a year.

Before being slaughtered piglets are fed to a variety of weights. These depend on the different market requirements – fresh pork, cured bacon and hams and a wide variety of processed foods such as sausages, pies and tinned meat products [2]. For pork and bacon production, the most important uses of the pig, most farmers and factory buyers now demand the white breeds of pig. Throughout Europe most pigs are now crossbreeds or planned hybrids between two or three pure white breeds. The most popular crossing breeds are Large Whites of British origin and the Landrace found in Europe.

Pig rearing and pork production

It is a popular misconception that pigs are dirty animals. Given well-planned living conditions pigs will keep themselves cleaner than most other domestic animals. But if they are crowded together in pens that do not provide easily identified sleeping, eating and dunging areas their natural preference for cleanliness is upset. In hot conditions pigs always need water in which to cool themselves because they have few and inefficient sweat glands. The pig uses its snout to forage for food and a large herd of pigs kept on free range will quickly turn a wet field of grass into a sea of mud. But good grass and forage root crops can provide adult breeding pigs with a large proportion of their diet and open air pig-keeping methods are still popular and profitable in areas where the climate is mild and the soil free-draining.

Because pigs dislike extremes of heat and cold most farmers now keep fattening pigs in intensive housing conditions. Well-designed pig houses are equipped with mechanical feeders and dung disposers designed to provide maximum comfort and encourage rapid growth rates on scientifically balanced rations [4]. In such conditions well-bred pigs will reach light pork weight in 120 to 150 days, bacon weight in 160 to 180 days and heavy manufacturing weights at an average age of six months. Pigs are fast becoming big

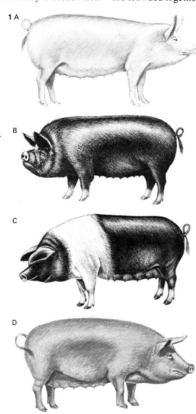
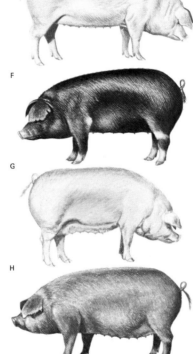

1 Pure-bred pigs vary widely in the quality of meat that they produce. The Large White [A], known as the Yorkshire in countries where it has been exported, is the predominant pure breed in Britain. The squat Berkshire [B] is too fat for modern tastes and is no longer used commercially. Black and White Saddlebacks [C] are popular free-range stocks in England but are mainly used for crossbreeding, as the pure breeds are too fat for modern markets. The sandy-coloured Tamworth [D], originally a forest pig, is now a minor breed in Britain. Different types of Landrace pigs [E] are found throughout Europe. The most famous is the Danish type, bred specifically for lean bacon. Poland, China [F] and Chester White [G] pigs are popular breeds in the USA along with the long red Duroc [H] but are uncommon in other countries.

2 Manufacturers of pig products claim that "nothing is wasted but the squeal". Pig meat is used in numerous ways including fresh pork and cured bacon. The poorer cuts are incorporated in pies and sausages and the congealed blood in black puddings. Pig bristle is used for a variety of brushes and the tough hide, with its characteristic grain, is converted to a variety of leather goods.

3 Pigs have played an important part in human culture, from the "unclean" animal of Islamic and Jewish culture to being the most common meat in China and a member of the family in parts of Melanesia. Many Oceanic tribes measure wealth in terms of the number of pigs owned, like cattle-herding tribes elsewhere. They are most reluctant to eat them and if food is scarce a mother will even suckle a pig instead of her own child. Slaughtering is usually a ceremonial act.

4 A modern pig farm ensures that its animals are fed carefully balanced rations. The trend towards scientific pig feeding has been accelerated by rearing to meet the human demand for low-fat pork and bacon. In the wild, pigs are efficient scavengers and their natural diet consists of roots, fallen fruits and nuts. And although pigs are valuable consumers of domestic "swill" and other waste food, today's pigs require diets that are more exactly controlled.

business in the form of large factory-scale units. Some of the largest such units are to be found in Eastern European countries including Romania, Hungary, Bulgaria and Yugoslavia, where units with up to 10,000 sows are currently in operation.

Sheep for wool and meat

Sheep are multi-purpose animals [8] that were first domesticated about 7,000 years ago but were not bred for their fleeces until a thousand years later. The wealth of medieval England was built on sheep kept for their wool but today's largest sheep producers are Australia and New Zealand. Spanish Merinos were taken to Australia from Europe and fine Australian Merino wool is still considered to be the best in the world. Pure-bred Merino flocks are still important in Australia but breeders have introduced other stock to improve meat content at the expense of wool quality.

In New Zealand, where there were nearly 56 million sheep in 1974, wool takes second place to meat production. Crosses between the English Romney and Border Leicester breeds have helped lay the foundation for a thriving fat lamb export trade [7].

The British breeds of sheep show more variation than in most other countries and there are sound geographic reasons for this. Hardy upland breeds, including the Scottish Blackface, Cheviot, Swaledale and Welsh Mountain, thrive on high rough grazings unsuited to cattle and crop production; at intermediate and lowland level crossbred sheep are preferred. A popular crossbred ewe is obtained by putting the small hill ewes to larger longwool breeds such as the Border Leicester. This ewe is in turn often put to a lowland ram like the Suffolk or Dorset Down to produce early maturing fat lambs.

Sheep have evolved in different ways in different countries. The Finnish Landrace and Russian Romanov breeds are noted for their ability to produce twins, triplets and even quins. Breeders in Europe are now using these sheep to increase production in their native stock. Texel sheep from The Netherlands, which produce compact carcasses, are now being used for crossbreeding in Britain to boost yield and quality.

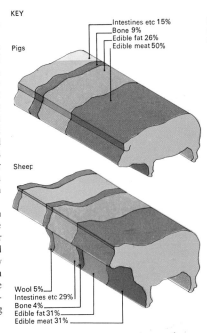

KEY

Pigs

Intestines etc 15%
Bone 9%
Edible fat 26%
Edible meat 50%

Sheep

Wool 5%
Intestines etc 29%
Bone 4%
Edible fat 31%
Edible meat 31%

Domestic breeds of pigs and sheep are constantly being improved by selective breeding to increase the proportion of usable parts of the carcass. Efficiency at converting food into meat is a prime consideration in pig-breeding but sow productivity and the viability of a litter is also important. There are hundreds of varieties of sheep, mostly bred for a particular characteristic such as wool or meat, and so it is difficult to give a breakdown of the weight of an average sheep. Many sheep, however, are bred for a number of different purposes and an all-round sheep can show a compromise between high meat and wool production.

5 Many sheep breeds have been based on European breeds such as those illustrated here. Merinos [A], originally bred in Spain, are famed worldwide for the fine quality of their wool. Karakul sheep [B] are the source of Persian lamb. Dutch Texels [C] produce heavy white fleece and are much used in crossing programmes for their compact carcasses. Blackface sheep [D] are native to Scotland and noted for their hardiness and quality mutton; their fleeces are used to make tweed and carpets. Welsh Mountain sheep [E] are lighter and smaller but thrive on their native hills. Borders [G] are also hardy. English Romneys [F], crossed with Leicesters, helped found the New Zealand export trade. The Suffolk [H], which originated from crosses of Norfolk Horn ewes and Southdown rams, is a popular crossing ram for fat lamb production in Britain.

6 High-speed mechanical sheep-shearing methods were pioneered in Australia and New Zealand. On large stations with small staffs they are a necessity. Good facilities for holding large numbers of sheep and preparing them for shearing are essential if the skilled worker is to achieve maximum output. To meet seasonal needs, itinerant gangs of contract shearers serve many large farms.

7 Fertile soils and an equable climate make New Zealand excellent country for intensive sheep husbandry on hills and lowlands. Crossbred stock produce fat lambs – mainly for export – and also wool.

8 Sheep provide meat and wool plus milk and a wide variety of manufactured products. Hides are used for coats, hats and gloves, for book bindings and bags. Parchment, an early type of writing paper, was made of untanned sheep hide.

Cuts of meat

Most meat eaten in the Western world is bought cut and prepared from butchers' shops, although with the increasing availability of domestic freezers many people now buy half or quarter animals and prepare the meat themselves. This costs less, saves time spent in shopping and provides a store of meat for unexpected demands.

Commercial butchering
Butchers' meat is, broadly, beef, veal, mutton, lamb, and pork. After an animal has been slaughtered, the viscera are removed to be sold separately as offal or variety meats (liver, kidneys, heart, brains and so on). Some organs are sold to commercial food processors to be used in pies and puddings or packaged as pet food.

Offal meats are highly nutritious but are neglected by many people. Liver and kidneys, in particular, are easily cooked and supply many of the minerals and trace elements essential to a proper diet.

Brains and sweetbreads – the pancreas – are easily digested and are excellent as the basis for restorative meals for invalids. Tripe needs special attention – blanching and scrubbing – but that is normally done by the butcher. The classic dish *tripes à la mode de Caen* is simmered in a very low oven for 24 hours. Ox and lamb tongue is usually boiled, but ox tongue can be roasted.

All meat, with the exception of pork, benefits from conditioning, or hanging, a matter of literally hanging the carcass for a few days while enzymes in it break down connective tissue and make the meat tender. The process (a week for lamb or mutton, 10–14 days for beef) also improves flavour.

Meat cutting and hanging
Butchers prepare meat to produce cuts that will command the best prices. Most people equate quality with tenderness (although in fact some relatively tough cuts carry more flavour). Hindquarters of beef, legs of lamb and pork and mutton chops are preferred.

Beef hindquarters (prepared as steaks or roasts) should be strongly red in colour, firm, fine-grained and marbled with streaks of fat. The outside fat should be a rich cream in colour and slightly crumbly.

Veal [2], the meat of young calves between two weeks and a year old, is rarely hung, but most of the blood is removed to give the meat its characteristic pale colour. The flesh should be plump, fine-grained and marbled with white, almost transparent fat.

Mutton is at its best when the sheep is at least four years old before it is killed. It is a highly flavoured meat, and should be red with white fat and firm, fine-grained flesh. As mutton can be tough, it should be marinated before it is cooked to avoid dryness. Lamb refers to sheep less than 12 months old. It is very tender and, owing to its relatively high fat content, easily cooked. Lamb is popular throughout the world, especially around the Mediterranean basin. Very young milk-fed lamb is regarded as a great delicacy.

English lamb, increasingly popular because of its tenderness and the ease with which it is cooked, is largely exported to Europe where its delicate flavour is appreciated. A good deal of lamb is also exported by Australia and New Zealand.

Lamb is at its best when the animal is between three and five months old [5]. It should

CONNECTIONS

See also
234 Pork, ham, sausages and cold meats
226 Cattle for beef and milk
230 Pigs and sheep

1 Beef is preferred by the Western world to almost all other meat. The classic English dish is roast beef with Yorkshire pudding, a batter cooked around and under the roast. The tradition of eating it roasted on Sunday began in the last century; for the poor, it was the only day on which they could afford to eat meat and for the middle class it was the day on which the whole family came together. Beef is best when rare.

1 Brisket : used for pot roasts or salt beef
2 Shoulder and neck : the basis for stews, casseroles and curries
3 Ribs : roasted or cut into steaks for frying
4 Sirloin : makes the best steaks and roasts, including fillet, porterhouse and T-bone
5 Topside and silverside : cut from the round end and either roasted, or spiced and salted
6 Rump steak : highly flavoured ; best grilled or roasted whole
7 Cuts from the flanks : usually stewed or pot roasted though whole joints are often boiled
8 Shin of beef : stewed, casseroled or minced
9 Cow heel : boiled to provide aspic, a clear jelly

2 Veal is far more popular in Europe, especially Italy, than in Britain. It is a tender meat, easily cooked.

1 Breast of veal : can be pot roasted, braised or boiled
2 Shoulder : usually boned, stuffed and roasted
3 Scrag end of neck : used for pies, stews and goulashes
4 Middle neck cuts : best pot roasted or braised
5 Cutlets : from the best end of the neck
6 Loin : best roasted or cut into chops
7 Fillet : provides the tenderest and tastiest portions
8 Leg : can provide an excellent roast although it is far better when thinly sliced into escallopes and fried or grilled
9 Knuckle : makes excellent fricassés and casseroles

3 The oldest recognized breed of cattle, and the heaviest, is the Italian Chianina [A]. Bred in Tuscany since ancient times, it is grown for its enormous size. The Chianina grows fast and is slaughtered at an earlier age than other cattle. The finest French beef cattle are the white Charolais [B], named after the area in Burgundy where they originated. They are always fed on grass and as a result their meat is extremely lean and yet tender. Red and white Herefords [C] are the result of generations of breeding by cattle farmers in Hereford, England. The breed has been exported all over the world, principally to the USA, Canada, South America and Australia.

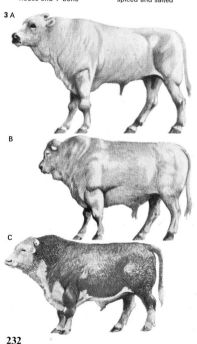

4 The cow, sheep and pig are the world's major food animals. This chart is a proportional representation of the beef, lamb and pork eaten in pounds per person per year in various countries. The main factor influencing consumption is the standard of living in the countries.

Beef and veal consumption
Pork consumption
Mutton and lamb consumption

Africa | Japan
Argentina | Mexico
Australia | The Netherlands | New Zealand
Austria
Belgium–Luxembourg | North America
Canada | Norway
Caribbean Islands | Portugal
Central America | South Africa
Denmark | South America
Finland | Rhodesia
France | Spain
Germany | Sweden
Greece | Switzerland
Haiti | United Kingdom
India | United States of America
Ireland | USSR | Uruguay
Israel | Venezuela
Italy | Yugoslavia
Jamaica

be plump and pale brownish-pink in colour with fine, hard opaline fat.

Pork, a good source of protein and vitamin B, must be carefully chosen, not kept too long, and carefully cooked because of the danger of disease. The meat should be pale pink, lean, with white, firm fat, and have a thin skin. It should yield to pressure, springing back readily. Pork can be eaten at any time of the year provided it is kept refrigerated, clean and is thoroughly cooked. If the meat still looks raw after it has been cooked, it should be thrown away.

Cooking meat

Meat should always be wiped with a damp cloth before it is cooked, and surplus fat removed. The most common methods of cooking meat are roasting or baking, stewing, braising, casseroling, frying and grilling.

The less tender portions of meat can often be improved by marinating before cooking. A marinade – a seasoned liquid (a cheap one can be made of onions, water and vinegar) – helps to tenderize the meat and gives it more flavour. The best marinades are made from brandy, wine and herbs and if meat is left in them overnight the improvement in flavour and tenderness is dramatic.

Roasting is done in an oven (wrapping the meat in aluminium foil helps seal in the flavour and prevents the oven getting too dirty) and the meat is basted with its own juices. Small joints are apt to get dry and a glass of wine or meat stock is often added to them halfway through the cooking process.

Pot roasting is a method of cooking a roast in a heavy closed pan on an open element. The meat is first sealed by heating in a frying pan and then placed in the saucepan with the fat yielded by the frying. It is then left to cook normally in its own juices.

Stewing and braising are similar – the meat is cut into pieces and cooked in water – but when it is to be braised it is first sealed in a frying pan. Meat may be stewed or braised with vegetables and if it is cooked with them in a closed oven-dish a casserole is produced.

Only the most tender cuts of meat should be grilled or fried. Steaks, whether they are fried or grilled, should be sealed quickly first to preserve the natural juices.

A

B

C

D

Crown roast [C] and Guard of Honour [D] are elegant ways of preparing a traditional roast. The starting point for both are two best end necks of lamb with the bones cleaned [A]. To make the crown roast, the two necks are stitched together [B] and the centre stuffed; to make the Guard of Honour, the necks are interlocked and the hollow stuffed.

5 Lamb is immensely popular in the Middle East where it is often cooked on a spit over an open fire. It benefits from the use of herbs; the tradition of serving it with mint sauce is an Arab innovation. Three-month-old lamb is the tastiest, but all lamb, even deep-frozen, is tender.

1 Breast of lamb: though fatty, it is succulent; it is usually roasted after being boned and stuffed
2 Shoulder: often roasted whole but can be boned and stuffed
3 Head: used as a basis for stock, broths and soup
4 Scrag end of neck: best stewed or casseroled
5 Middle neck cuts: excellent for braising but can be grilled
6 Best end of neck: cut into chops or roasted whole
7 Saddle: prime roasting meat which can be spit-roasted whole

8 Loin: usually cut into chops but can also be roasted
9 Leg: roasted whole or cubed for shish kebab

10 Shank: usually sold with the leg; even though mainly bone, the meat portions have a pronounced flavour

11 Lamb's trotters: can be boiled to provide thickening for gravies and sauces or slowly baked for eating

6 Milk-fed pork gives a meat so tender that nearly all the carcass can be roasted or fried. A great delicacy is roast suckling pig. Unweaned piglets are usually roasted with chestnut or herb stuffing.

1 Leg of pork: usually roasted and the skin scorched so that crackling forms
2 Loin: can be roasted or cut into chops for frying
3 Foreloin: best roasted whole or can be cut into chops
4 Spare ribs: grilled, fried or braised

5 Hand or shoulder: casseroled whole or boiled
6 Belly of pork: it is fatty and best used for flavouring other dishes
7 Trotters: still considered a great delicacy, they are boiled and can be served either hot or cold or used to flavour casseroles and pâtés

7 A

B

7 Kebabs are a favourite method of serving lamb in the Middle East. The word is Arabic for "skewered". The tradition began when soldiers cooked freshly killed lamb on their swords over open fires. In modern shish kebab [B], the meat is marinated in wine then threaded on skewers with thin layers of fat and grilled, usually over charcoal. Doner kebab [A], originally a whole leg, is now often minced lamb mixed with herbs into the shape of a leg. It is roasted on a vertical spit and slices are carved off. Kebabs are usually served with rice and raw vegetable salads. Lamb, veal and pigs' kidneys are also excellent when cooked on skewers with peppers and onions. In classical French cuisine they are soaked in brandy and served flaming on a skewer at the table.

8

8 Goats are bred mainly in countries where the pasture is too poor for sheep: the desert areas of the Middle East, for example, and the mountainous regions of Greece and the Balkans. Old goat needs long cooking; indeed, many recipes suggest slow simmering overnight. Very young kid can be sweet and tender and is often roasted whole, like suckling pig. In the Middle East, kid is a great delicacy and is often stuffed with rice, raisins and nuts. The goat has been a general purpose animal since time immemorial, giving man both meat and milk.

233

Pork, ham, sausages and cold meats

Pork, the fresh meat of the domestic pig, is one of the most versatile of meats. High in calories and a good source of vitamin B_1(thiamin), pork products are highly nutritious. The pig was first domesticated in the Near East in about 7000 BC and its omnivorous habits and undemanding nature make it a relatively easy animal to rear. It produces more edible meat per unit of carcass weight than most other food animals and, in spite of its proscription under Judaic and Islamic law (according to which the meat of the pig is deemed unclean), it has long been a staple of both Western and Eastern cuisines. Indeed, when the Chinese say "meat", they mean pork. They refer to beef, chicken and lamb by name.

The curing of pork and ham
Pork may be eaten fresh or cured. Great care must be taken when cooking fresh pork; it should never be eaten very rare (underdone) because the larvae of the parasite *Trichinella spiralis* can infest pork meat in temperate climates and can be killed for certain only by long, slow cooking.

The best cured pork and ham [1] comes from the prime parts of the pig – the hind legs and the loin. The curing and salting of pork is an ancient process that began in Roman times or even earlier and its purpose was to preserve the meat for the winter. The comparatively recent introduction of winter feed for the animals, and of refrigeration as a means of storage, have lessened the need for curing as a method of preservation. Flavour is now a main reason for curing. The ham is removed from the body of the pig and is cured wet or dry, usually in salt or in brine, but sometimes in molasses or maple syrup. The cured meat may then be smoked (Irish hams are smoked over peat which gives them a very distinctive taste) and left for varying periods of time to mature; in the Huelva region of Spain, the jamón de serrano is traditionally left in the snow to mature.

Nutrition plays an all-important part in the ultimate flavour of the specialist hams – Parma pigs are fed on parsnips, Virginia pigs on peanuts and peaches, and Kentucky pigs on wild acorns, beans and clover. Some of the world's most famous hams, including Bayonne, Parma and Westphalia, may be eaten uncooked. Lesser-known German, French and Polish country hams such as Lachsschinken and Losoiowa can also be eaten raw. Other fine hams such as Czechoslovakian Prague ham, Pražská šunka, are cured, smoked and left to mature in preparation for final cooking.

Bacon comes from the body of the pig and is cured in brine. It may be bought smoked or unsmoked ("green"), boned and rolled in joints, or in rashers and steaks. Scandinavian countries, particularly Denmark, have long specialized in the intensive production of high-grade lean bacon, primarily for export. The pigs are specially bred for this purpose and both age and breeding play an important role in the taste of the final product.

"Little bags of mystery"
Pork is also the basic ingredient of many types of sausage [3]. The origins of this food (the word derives from the Latin *salsus*, meaning salted) are obscure and the earliest reference to its consumption appear in Homer's *Odyssey*. In the Middle Ages the

CONNECTIONS

See also
232 Cuts of meat
230 Pigs and sheep
226 Cattle for beef and milk

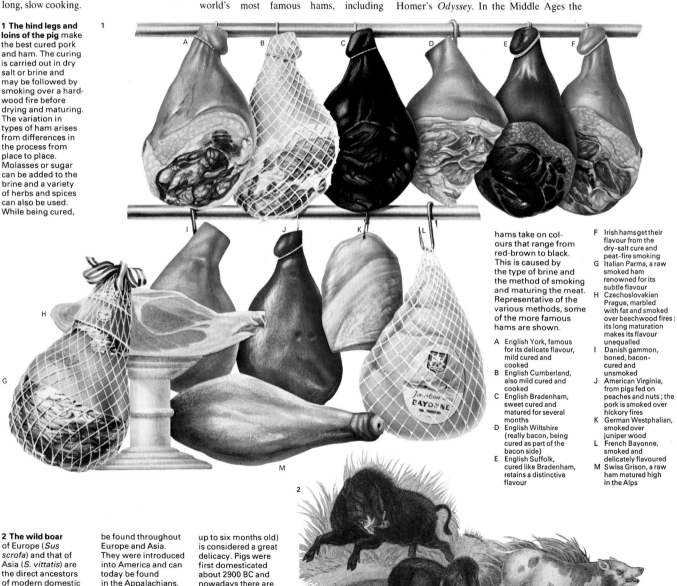

1 The hind legs and loins of the pig make the best cured pork and ham. The curing is carried out in dry salt or brine and may be followed by smoking over a hardwood fire before drying and maturing. The variation in types of ham arises from differences in the process from place to place. Molasses or sugar can be added to the brine and a variety of herbs and spices can also be used. While being cured, hams take on colours that range from red-brown to black. This is caused by the type of brine and the method of smoking and maturing the meat. Representative of the various methods, some of the more famous hams are shown.

A English York, famous for its delicate flavour, mild cured and cooked
B English Cumberland, also mild cured and cooked
C English Bradenham, sweet cured and matured for several months
D English Wiltshire (really bacon, being cured as part of the bacon side)
E English Suffolk, cured like Bradenham, retains a distinctive flavour
F Irish hams get their flavour from the dry-salt cure and peat-fire smoking
G Italian Parma, a raw smoked ham renowned for its subtle flavour
H Czechoslovakian Prague, marbled with fat and smoked over beechwood fires: its long maturation makes its flavour unequalled
I Danish gammon, boned, bacon-cured and unsmoked
J American Virginia, from pigs fed on peaches and nuts; the pork is smoked over hickory fires
K German Westphalian, smoked over juniper wood
L French Bayonne, smoked and delicately flavoured
M Swiss Grison, a raw ham matured high in the Alps

2 The wild boar of Europe (*Sus scrofa*) and that of Asia (*S. vittatis*) are the direct ancestors of modern domestic pig breeds. Wild boars are extinct in Britain because of over-hunting and the loss of the forests, but they can still be found throughout Europe and Asia. They were introduced into America and can today be found in the Appalachians. The boar is still hunted for game, usually with dogs. In France the flesh of the *marcassin* (a term for a wild boar up to six months old) is considered a great delicacy. Pigs were first domesticated about 2900 BC and nowadays there are two main varieties – the long-backed Chinese pig and its heavier cousin, the European or the Danish variety.

plain, rather humble sausages of classical times were transformed, by the use of mixed meats and exotic flavourings and spices, into the forerunner of today's products.

Known rather derisively in Victorian times as "little bags of mystery", sausages consist of a filling of chopped or minced meat or meats, plus seasonings, preservatives and sometimes a cereal, all encased in an edible skin made either from the animal's intestine or from man-made cellulose. Sausages may be divided into two main categories. The "dry" sausages or charcuterie (from the French chair cuite, meaning cooked meat of the pig) are ready to eat. They include the ubiquitous salami, traditionally made from pork meat and lard, although Hungarian salami should contain donkey meat; Italian mortadella, which is often studded with pistachio nuts; and various types of liver and garlic sausages. The second type of sausage known as "wet" must first be cooked. It is sometimes allowed to cool, but more often forms an integral part of a stew or hot meat dish. "Wet" sausages include the German Frankfurter and Bratwurst, Italian zampone

(encased in the skin from a pig's trotter), Spanish chorizo, the famous saucisson de Toulouse from France, and regional specialities such as the haggis from Scotland, blood or black puddings from France, Germany and Britain and the andouilles and andouillettes of provincial France.

The sausage, which began as one of the world's first convenience foods and a means of economically using up odd scraps of meat, now, in a variety of forms, occupies an important and integral part of international cuisine.

The pig stripped naked
The fact that there is very little wastage from the carcass of a pig has contributed to its worldwide popularity and exploitation as a food animal. Liver, heart, tongue and brains are all sold as delicacies. Trotters, snouts, heads, tails, ears and chitterlings (small intestines), although in declining demand, are still widely bought for human consumption. Pig skin is used for leatherwork such as handbags and purses; back and caul fat are used to moisten lean cuts of meat; and the bones are ground for animal feed.

A **medieval print** shows a swineherd knocking down acorns for fodder. Pigs have been reared in Europe since 2900 BC.

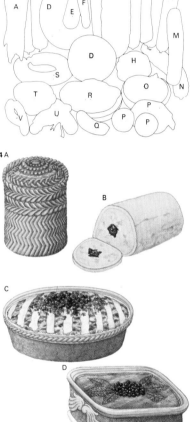

3 Sausages made from chopped meat mixed with other foods and stuffed into a gut, are part of the culinary tradition which is centuries old and stretches back to ancient Greece, being mentioned in Homer's *Odyssey* in the 9th century BC.

Pork is particularly suited to sausage-making and all parts of the pig can be used, including the liver – in Leberwurst – and the blood – in boudin or black pudding. Beef is often mixed with pork or is used alone. Sausages are also made from

other animals and even fish. Finally sausage is a very convenient presentation for the new "non-meat" proteins such as soya bean protein, either alone with a filler or mixed with traditional sausage meats. Some European sausages are shown.

A Italian *zampone*
B Polish *krakowska*
C Polish *debowiecka*
D Italian *mortadella*
E German *Leberwurst*
F Polish *kabanosy*
G Italian *salami*
H French *saucisson sec*
I Hungarian *salami* with donkey meat
J German *cervelat*
K Italian *crespone*
L Italian *cotechino*

M German *Bratwurst*
N German *Mettwurst*
O French *boudin*
P Scottish *haggis*
Q French *boudin blanc*
R English *Cumberland sausage*
S Polish *wiankowa*
T German *Blut Zungenwurst*
U Spanish *chorizos*
V German *Frankfurter*

4 Pâtés are made of minced pork, veal, rabbit, game or liver flavoured with herbs, spices and wine or brandy. A terrine is a pâté cooked and kept in an earthenware pot such as the hare pâté [A], the pâté de campagne [C] and the pheasant pâté

[D]. The French pâté de foie gras [B] is made from the extra-rich and enlarged livers of force-fed geese and always includes large pieces of truffles. An even more exotic pâté is that made in New Caledonia from fruit bats.

Poultry and egg production

The modern trend throughout livestock farming is towards more intensive methods. In this, specialists in poultry have set the pace; since the 1920s poultry production has changed out of all recognition. Barnyard fowls have given way to egg and broiler factories controlled by a decreasing number of large, integrated organizations. The largest broiler chicken companies control their own feed-production, house-building, slaughtering and freezing plants.

As a result of the revolution in poultry raising, smallholders who once made a comfortable living out of 1,000 laying hens have been driven out of business. Poultry as a sideline, too, has been discontinued on many farms, except for direct sale of fresh free-range eggs for which a premium must be charged to cover high production costs. The old-fashioned five- to six- month-old cock chicken reared outside on corn stubbles has been replaced by the broiler, a bird that is mass produced in dim, controlled-environment houses. These birds are ready for market in eight to ten weeks. Turkey and duck farming [6] is becoming equally inten-

sive and today only small flocks of geese and some water fowl and game birds are still produced in traditional conditions.

Breeding poultry for profit

Among livestock, poultry were the first to receive serious attention from geneticists, who had discovered that planned crossbreeding of specially selected pure and inbred lines of plants could give exciting increases in production. Tailor-made hybrids for egg and broiler chicken production quickly ousted the old pure breeds [1] such as the Rhode Island Red, White, Black and Brown Leghorns, Light Sussex and crosses between them.

Hybrid poultry breeding [5] started seriously in Britain in the mid-1950s and today only a few small independent breeders are left. Poultry breeding for egg and broiler production is dominated by large companies based in the USA and Canada. These organizations sell grandparent stock from their closely controlled breeding programmes to licensed multiplier breeders and hatcheries farther down the production

pyramid in many different countries. Royalties are charged on day-old chicks produced and these are then sold to commercial producers of eggs, broilers and turkeys.

The genetic make-up of modern hybrids is changing all the time and most are now described purely by code name. They are advertised and sold on their performance specifications proved in trials [3]. These include eggs per bird; size and weight over a given period; and the ability to convert expensive balanced rations into eggs or poultry meat. Resistance or immunity to disease will also be in the sales specification.

The technology of intensive rearing

The other reasons for the rapid changes in poultry farming methods lie in twentieth-century technology. The development of the incubator to replace the mother hen sitting her seasonal clutch of eggs was the first major step towards factory farming of poultry [2]. Artificial incubation led to an immediate increase in egg production.

Compared to other farm livestock the life cycle of poultry is short and the number of

1
A Australorp   E New Hampshire
B Cornish   F Plymouth Rock
C Dorking   G Rhode Island Red
D Leghorn   H Light Sussex

1 Commercial hybrid poultry, bred specifically for eggs or meat, are descended from a variety of pure breeding birds. The most productive egg-laying strains have been derived from Leghorns and Rhode Island Reds. Pure White, Brown or Black Leghorns are lightweight birds. The object in the breeding of table chickens is efficient food conversion and rapid growth, plus the maximum amount of leg and breast meat. Today's broilers are based on heavier, quieter breeds like the New Hampshire, Plymouth Rock and Light Sussex. The Dorking and Cornish are old English meat breeds while the Australorp is a new Australian strain.

2 **Mechanical incubation** of fertile hens' eggs in heated and ventilated cabinets has turned poultry production into an industry. Each egg is kept in the incubator for 21 days during which time the temperature of 37.65°C (99.75°F) is maintained. Humidity is kept at 60% then raised to 70% for the last three days before hatching.

Temp °C
Days
7   14   21
Humidity %

3

20 weeks growing period

Production year, each bird producing 240-250 eggs

Exceptional birds are given 16 wks rest for moulting before going on to second season

Second production season, 6 months of greatly reduced output

Months of life   1 2 3 4 5 6 7 8 9 10 11 12 13 14 15 16 17 18 19 20 21 22 23 24 25 26 27

No. of eggs produced

Birds usually sold off for soup at this point

3 Modern egg production is a highly intensive and carefully costed business. The efficiency of different laying strains is compared since producers need to know how long it takes different breeds and hybrids to mature, the number and sizes of eggs laid over a given period and the amount of scientifically balanced food needed to maintain the bird and produce the eggs. Most modern hybrid strains grow to maturity in about 20 weeks and are then kept in laying cages for approximately 12 months before they go into moult. During this period they produce 240-250 eggs. Occasionally they are allowed a second season – six months of greatly reduced output – but usually they are slaughtered.

eggs and potential offspring produced is much higher. A modern hen reaching maturity in five months can lay well over 200 eggs a year. Fertile hen, duck and turkey eggs can be moved long distances to a central incubator plant and carefully boxed day-old chicks are then sent all over the world.

Except for water fowl, most poultry lend themselves to highly intensive housing and production methods. Egg production was the first aspect to become really intensive on a large scale and most laying birds are now housed in cages of different types. The latest laying cages have automatic feeding, watering and egg collection. Waste droppings are removed mechanically from three or four tier cages, or allowed to build up in pits under single-tier flatdecks or staggered tiers of cages. Some farmers still prefer their replacement egg layers reared on open free range, but once again the trend is towards more easily managed, intensive housing methods.

It has been proved that poultry, like plants, respond to variations in light length and intensity. Therefore, the light pattern for laying birds in windowless houses is gradually increased from the onset of lay, to simulate a perpetual spring when birds in their natural state lay most eggs. Well insulated and fan-ventilated houses provide an even temperature and in addition artificial heat is used for replacement chick rearing and broiler chicken production. Most broilers are still produced on comfortably littered solid floors, but more controlled cage-rearing methods have now appeared. The main drawback to wire-floored cages for table birds has been that they cause breast blisters, which look unsightly in the carcass.

The case for intensive methods
Intensive poultry-rearing methods, which have cut the cost and increased the efficiency of egg and poultry meat production and changed chicken meat from an expensive luxury to a competitively priced source of meat, are condemned as unnatural and cruel by advocates of more natural open range farming methods. However, poultry producers claim that their methods must meet all the birds' comfort and nutritional requirements or production would suffer.

The days of the traditional farm where poultry survived on scraps from the kitchen and pickings from the farmyard have largely disappeared with the advent of intensive rearing methods. Nowadays the poultry industry is based on large numbers of birds housed under one roof.

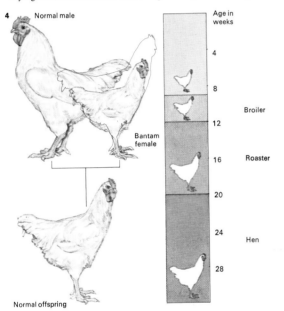

4 Various techniques are used for breeding poultry for the table. The female of a dwarf or bantam strain of White Plymouth Rock is economical on feed but when crossed with the male of a normal-sized strain it produces normal-sized offspring. Desirable traits in table birds include early feathering, which aids immediate sexing, and white plumage, which gives a cleaner carcass when the bird is dressed. A dressed bird can be marketed under different names according to its age when killed. A bird 9-12 weeks old is a broiler or fryer, 12-20 weeks is a roaster and older than that is a stewing bird or hen.

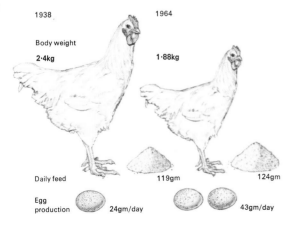

5 Increase in egg production has been dramatic since selective poultry breeding started in earnest in the post-war years. The increase has been achieved by selecting smaller, more fertile hybrid layers. Although their eggs are smaller the hens eat less food per egg produced than birds laying large eggs, and are twice as efficient at converting food into eggs compared with their pre-war ancestors. The feeding of scientifically balanced rations has also raised efficiency.

Geese
A Chinese
B Roman
C Egyptian
D Greylag
E Toulouse
F Embden

Ducks
G Buff Orpington
H Muscovy
I Khaki Campbell
J Aylesbury
K Pekin

L Indian Runner White
M Rouen

Turkeys
N Beltsville
O Broadbreasted Bronze
P Black Norfolk

6 Ducks and geese were domesticated over 3,000 years ago. Turkeys reached Europe in the 1500s.

Poultry and game

Today's most universally eaten meat is probably the domestic chicken, which is raised almost everywhere in the world, in places ranging from back yards to huge factories. The Romans bred chickens for the table and are reputed to have drowned the birds in red wine to impart a finer flavour. In the Middle Ages, chicken was fare fit for kings and lords only; the poor kept the birds for eggs and new chicks and killed a hen only when it became too old to lay eggs. Chicken [2A, 2B] is now one of the cheapest of all meats. Most chickens are bred in batteries to be marketed as roasting birds, varying from tiny poussins to family-sized broilers. Factory-reared chickens lack taste, because they are killed young and also because they are not hung to let the meat mature, even for a short period.

Goose, turkey and duck
Goose [2C, 2D] was considered the finest poultry of all for festive occasions until ousted by the turkey in the sixteenth and seventeenth centuries. Roast goose was a favourite dish of Queen Elizabeth I of England and was eaten on Michaelmas Day, 29 September. In the nineteenth century, goose was the principal Christmas bird, sometimes in the form of a goose pie, which consisted of a boned goose stuffed with smaller boned birds and encased in a rich pastry crust. In Scandinavia, Germany and parts of France, goose is still eaten on festive occasions. The Strasbourg goose is specially fattened and the large liver used for *fois gras*; the meat is salted, cooked and potted to make *confit d'oie*, which keeps for months.

Turkey [2E] was introduced to Europe from America in the sixteenth century via the Gulf of Mexico. It quickly replaced goose in popularity because of its larger size and its closer, less fatty meat. In Britain and North America it is now traditional at Christmas and weddings and in America also on Thanksgiving day.

Duck [2H] is the most expensive type of poultry; it is difficult to breed under factory conditions and the proportion of meat to bone is small. The French Rouen duckling is prepared and cooked with its own blood, which gives a special flavour and red colour to the meat. Nantes duckling is plumper and the meat paler; it is often cooked *à l'orange*. Gourmets maintain that the Chinese methods of preparing and cooking duck, such as Peking and Szechwan duck, are superior to any other recipes. These involve spicing the exterior of the bird with honey and wind-drying it before finally roasting.

Game animals and birds
Quails and squabs are domesticated varieties of game birds and are classified as such in many countries, notably the United States, France and Italy. In Britain, quails [2F] for the table are imported frozen from Japan although a few farms are now producing them commercially. Squabs [2G] are a type of pigeon, bred to be eaten young while tender.

Man's first food consisted of birds' eggs, roots, berries and wild vegetables, a scant diet that he had to supplement by killing game animals. These included elephants, bison, moose [1A], reindeer and wild pigs. As man's nomadic existence in search of food was replaced by community life, with tilling of fields and breeding of cattle, game hunting became less important as a source of food.

CONNECTIONS

See also
236 Poultry and egg production

1 Game animals and birds of the world include the North American moose [A], one of the largest and rarest of game animals; the black bear [B]; the black-buck [C]; a typical antelope from India, is prized for its meat and for trophies; the red deer [D]; the wild boar [E]; such ground game as the rabbit [F] and the hare [G] which are prolific throughout Europe; the grey squirrel [H], hunted in parts of America; mallard duck [I], the largest of the wild duck; the wild Canada goose [J]; the wild turkey [K], indigenous to North and Central America; the mourning dove [L], hunted in North America; partridge [M], woodcock [N], wild guinea fowl [O] and pheasant [P], regarded as the finest game birds; the golden plover [Q], now protected in many countries; the great bustard [R], one of the largest flying birds, found chiefly in Europe; and the lark [S], regarded as a great delicacy in France.

although furs and skins were still needed for clothing. But big game became scarcer and many of the original species are now extinct in many areas.

Red deer [1D] are still distributed throughout Europe, Asia and North America and they provide the favourite game meat. Deer's meat is known as venison.

The wild boar [1E] no longer roams the dense forests of Europe as it did when it was the target of medieval royal hunts. Then the elaborately decorated boar's head, complete with polished tusks, formed the centrepiece at banquets.

Venison and rare game birds remained the prerogatives of the rich, while the poor satisfied their craving for meat with wild rabbit [1F], hare [1G] and squirrel [1H], most often made into stews and pies.

In Europe and China, pheasant [1P] is the favourite and most expensive game bird; the meat of the hen is more tender than that of the cock, which is larger and distinguished by its brilliant tail feathers. Partridge [1M], woodcock [1N], wild duck [1I] and guinea fowl [1O] are also popular. In most countries,

game birds are protected by law during certain months of the year; they may not be killed in the so-called "close seasons". These vary from country to country and from state to state, but usually include nesting periods. In Britain, "the Glorious Twelfth" – 12 August – opens the grouse-shooting season, which ends again in December.

Hanging of game birds

All game – birds and animals – should be hung for a time to tenderize the meat and develop the flavours. The hanging period depends on individual taste (some people prefer their game "high"), on the weather and on the age of the game. Young game birds, best roasted and grilled, are hung for a shorter time (usually seven days) than older game, which is more suitable for casseroles.

Peacocks and swans are now prized for their ornamental features and not for the gastronomic qualities with which the Middle Ages endowed them. In most countries, small song-birds are protected by law, but in France they are not and the blackbird, lark [1S] and thrush are considered delicacies.

This 19th-century engraving shows a busy English market day. The poulterer supplied whatever game was in season as well as chicken, duck and giblets. The upper classes of the 18th and 19th centuries were prodigious eaters. A typical menu of the late 18th century included "A couple of rabbits smothered with onions, a neck of boiled mutton, and a goose roasted with currant pudding and a plain one"

2 Most popular chicken breeds like the White Leghorn [A] and Rhode Island Red [B] are for table use and egg production; few chickens are free-range nowadays, most being battery-reared for expanding markets. Domestic geese, now in dwindling supply, include Chinese goose [C], bred from the Siberian swan goose, and the greylag goose [D]. The turkey [E] has replaced the goose in popularity and availability and is the traditional festive bird in Britain and North America. Quail [F] is a migratory bird coming to Europe from Africa for the summer, but the birds seen in poulterers' shops have been bred on poultry farms. The squab or dovecot pigeon [G] is a domesticated version of the wild species; it is becoming rare and consequently an expensive delicacy. Species of farmyard ducks vary from country to country: in Britain the Ayles- bury duck [H] is the favourite breed, and was originally bred from the wild mallard.

Commercial fishing

The sea is a vast hunting ground for food for man. But fish are caught in commercial quantities in relatively small areas, where the conditions are favourable to the growth of phytoplankton (minute plants) on which many marine organisms feed. Regions where cold water wells up to the surface from deeper levels are particularly rich in phytoplankton and although they make up only about one per cent of the area of the sea, half the annual catch of fish is taken in them. Fin fish make up about 90 per cent of the catch. Among the remaining ten per cent are seafoods of high commercial value, such as molluscs and crustaceans. Of the total catch, about half goes to make animal feed-stuffs.

The danger of overfishing

Throughout his history as a seafarer, man has until recently assumed that the seas would provide fish in unlimited quantities; the size of the catch appeared to be directly related to improvements in the fishing industry's technology. But in the modern world developed nations have become so efficient at catching fish that they have disturbed the delicate balance between the hunter and his prey. The rate at which major known fish stocks are being replaced has now fallen behind the rate at which fish are being caught.

Fish catches rose rapidly from the mid-1950s but statistics show that on a world basis a period of positive decline has now begun. The peak year was 1970 when a world catch of 69.6 million tonnes was recorded. Since then the annual catch has been cut back to about 65 million tonnes [7].

Diminishing stocks and fishing control

In recent years, fish stocks have been depleted by various natural events as well as by overfishing. The Peruvian anchoveta fishery – which provided the world's largest annual catch of a single species – was destroyed by the appearance of El Niño, a warm sea current that invaded the area, disturbing the environment for the fish with the result that the shoals disappeared. The loss of the anchoveta supply has been partly offset by the sudden development of a market for Alaskan pollack, regarded as a cheap alternative to cod. Caught in the northeast and northwest Pacific Ocean, Alaskan pollack has now replaced anchoveta as the world's biggest single species catch.

Because supplies of fish have decreased, estimates of future production have been revised. At one time it was thought that 200 million tonnes of fish could be caught each year without drastically depleting stocks. Now, even assuming development of unexploited resources and effective international control of fishing, the maximum potential yield of marine species of fish, crustaceans and molluscs is estimated at about 118 million tonnes.

Fish stocks are a food resource capable of self-replenishment but they can continue to provide valuable protein for the world's growing population only with careful management. The near-extinction of whale stocks should provide a grim example of the folly of uncontrolled hunting of fish. Until there is international agreement about the perils of such overfishing, too many boats will continue to chase too few fish.

Progress towards international agreements on the control of fishing has been

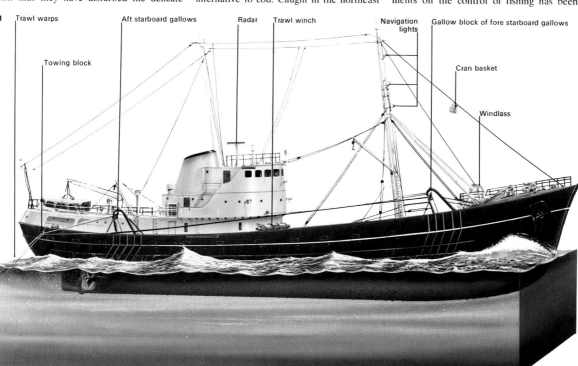

1 Trawl warps · Aft starboard gallows · Radar · Trawl winch · Navigation lights · Gallow block of fore starboard gallows · Towing block · Cran basket · Windlass

1 Otter trawling is the most widely used technique of mass catching. The method requires powerful engines to drag the net over the sea-bed where flatfish such as plaice and sole live. While the most modern trawlers are designed to trawl from the stern, the side-fishing method is still extensively used. Side trawlers are all built to the same pattern, but their sizes are dictated by the durations of their voyages. The warps (cables attached to the trawl) are led from steam-operated winches on the deck through rollers suspended in frames (known as gallows) sited on the side of the boat. The freeboard is low to make it easier to haul the catch inboard for cleaning and sorting.

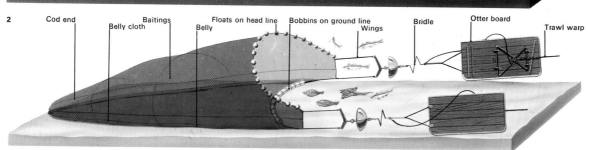

2 Cod end · Baitings · Floats on head line · Bobbins on ground line · Belly cloth · Belly · Wings · Bridle · Otter board · Trawl warp

2 The warps of an otter trawl are connected to otter boards – heavy steel or wooden panels that are linked to each wing of the net by bridles. When the net is under tow, the otter boards are forced apart, keeping the mouth of the net open horizontally. Floats on the head line and bobbins (heavy rollers) on the ground line keep the net open vertically. The underside or belly of the net is lined with hide "chafers" to protect it from obstructions on the sea-bed. The tapered cod end of the net is secured by a rope that permits quick release of the catch after it is hauled aboard.

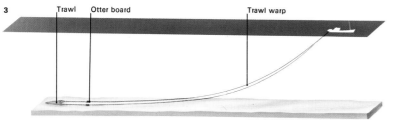

3 Trawl · Otter board · Trawl warp

3 Fishing gear is under great strain when in use. On a deep-sea trawler, a single tow under full power can last as long as three hours, with the trawling gear streaming out as far as 1,600m (1 mile) from the ship. With the trawl extended that far, there is a danger of other trawlers in the area cutting across the gear, which can result in a lost net. To minimize this risk, two cones (or, more usually, a cran basket) are hoisted on the forestay to warn other vessels.

intermittent and no effective, concerted scheme for managing the oceans' food stocks has been worked out.

Apart from the effort to achieve direct, internationally agreed controls on fishing through catch quotas and net-size regulations, the best hope of conserving fish stocks lies in the establishment of wider territorial limits around national coastlines. Present moves to widen these limits will restrict the activities of highly developed fishing nations. As the great ocean-roaming fleets (mainly Soviet, Japanese, West European and American) are forced to accept catch limitations within many of their traditional fishing grounds the pattern of fishing will change. Nations will restructure their fleets to concentrate their efforts within new limits, but the pressure they put on stocks within their own waters will still need strict supervision.

Finding new commercial species

Important as conservation measures are, an increase in food production from the sea also depends on the marketing of species that have previously been ignored and on fish

farming techniques. Already the potential of fish farming is becoming recognized, but it will be many years before higher production from such farms significantly increases the total world catch.

Some nations are beginning to exploit fish species that are not widely eaten now. In the waters of the Antarctic, for instance, krill, a shrimp-like creature that is the food of the baleen whale, can be found in massive swarms and is attracting trawlers to this remote region. Although krill is plentiful, the question of how it is to be marketed has to be resolved. As well as finding ways to make krill attractive as a food, experts are investigating the possibilities of using it as a feed for farmed fish and animals.

Alternative species found in very deep water [9] also present a marketing problem. The unattractive appearance of some of these species tends to make them hard to sell. The profitability of catching them will probably depend largely on their marketability in other forms and on a shortage of more familiar fish, but new commercial species could in turn be quickly overfished.

Stern trawling, like most modern fishing methods, is so efficient that the survival of some fish stocks is at risk. There are also destructive side-effects: trawl nets wreck the sea-bed and catch species that, although not eaten, are an integral part of the ocean's food chain.

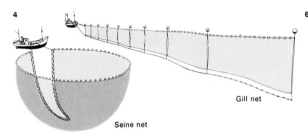

Seine net

Gill net

4 Purse seining and gill netting are two common fishing methods. The purse seine is used to surround shoals of pelagic (open sea) species. A rope running around the bottom of the net is drawn tight, trapping the fish. The gill net is held in a vertical position by floats and weights and is "shot" across the expected path of the fish, which are caught by the gills as they try to swim through.

5 Beam trawls are bottom nets kept open by a beam and weighed down by a mesh of heavy chains that drags the sea-bed, disturbing its ecology. Use of these nets is now restricted.

8 Fishing grounds in many parts of the world are affected by recent moves by some nations with coastlines to extend out to 320km (200 miles) the zones in which they may control the size of catches that are taken. If restrictions are imposed on the size of fleets of nations seeking access to these zones major changes could be seen in present patterns of fishing. The dangers of over-fishing will remain and the International Fisheries Commission has done little yet to protect endangered species; international supervision is of limited value unless it is accepted by all fishing nations.

Anchoveta *Cetengraulis mysticetus*

Cod *Gadus callarias*

Herring *Clupea harengus*

Haddock *Melanogrammus aeglefinus*

European pilchard (sardine) *Sardinops* sp

6 Large-scale commercial fishing is concentrated on relatively few species. Mainstays of the world catch have been anchoveta, herring, haddock, cod and sardine. Until 1972, anchoveta provided the largest catch. In 1970, the peak year, more than 13 million tonnes of anchoveta were landed. But in 1973 the catch declined to just less than two million tonnes, although in 1975 stocks showed signs of recovery. Catches of Alaskan pollack have recently overtaken other major species.

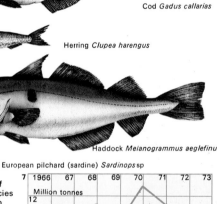

7 World catches of several major species of fish have shown a decline since the 1960s. Catches of anchoveta (used mainly as a feed) showed a particularly spectacular fall between 1971 and 1973. During this period, Peru's lead in fish production was overtaken by nations such as Japan, active in the northwest Pacific.

9 Some deep-water species have proved to be very palatable although their gruesome appearance may deter some people from buying them. Fished at depths of 800-1,100m (2,600-3,600ft), species such as the grenadier or rat-tail [A] and red director [B] could provide an alternative to familiar fish with depleted stocks.

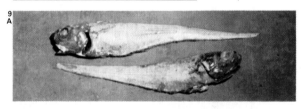

Industrial fishing grounds   Subsistence fishing

Fish farming

In one form or another, man has been farming fish for more than 4,000 years but the development of scientific fish farming (or aquaculture) began only recently. It has come at a time when most of the major commercial fish species are either at or near the limit of their potential yield.

Importance of aquaculture
The total annual production from aquaculture is still more than five million tonnes, but is increasing as research and modern technology provide new techniques for the fish farmer. In some developing countries aquaculture is an important part of the economy, providing both food and work.

In terms of output China leads the world in fish farming. Estimates of production vary, but it is well over two million tonnes annually – about 40 per cent of China's total fish and shellfish supply. Aquaculture is also important in India and Indonesia, respectively providing about 38 and 22 per cent of the fish eaten. The Soviet Union, the Philippines, Thailand, Japan and Taiwan are also fish-farming nations. In Europe and the United States, much of the development in fish farming – apart from shellfish culture – has so far been concentrated on technological research rather than production.

Mangrove swamps, estuaries, lakes, fresh and salt water lagoons, shallow coastal waters and artificial ponds are all being utilized for fish farming. According to recent estimates about three million hectares (seven million acres) of water surface are under cultivation. Some experts believe this could be increased to 30 million hectares (70 million acres) in the foreseeable future and that world farmed fish production could reach 40 to 50 million tonnes by the year 2000.

The choice of animals available to the fish farmer includes molluscs, crustacea and fin fish. Mollusc culture (mussels [1], oysters [2], clams and so on) is well established, especially in Europe, North America and Japan; and many countries have some method of oyster farming. The development of intensive hatchery systems during the last 40 years has made available mass rearing techniques for many mollusc species.

Farming of crustacea (which include shrimps and prawns) [3] has developed more slowly and is mainly found as a semi-culture that makes use of the water in artificial ponds formed when rice fields are flooded. Mass rearing of shrimps or prawns for general sale has not yet been achieved, but almost certainly will be in the near future.

Many species of freshwater fish can be farmed, with varying degrees of success. These include carp, tilapia, catfish, trout [4], salmon [5] and eels. The only salt-water fish farmed in any numbers are milkfish and sea bream, although experiments are currently being carried out to increase the numbers of farmable marine fish to include the flatfish sole, turbot, halibut and plaice.

Techniques employed in aquaculture
Aquaculture techniques are many and varied, but fall into four categories. First the young animals can be reared through the most vulnerable stages of their lives for release into lakes or the sea, to supplement natural stocks (for example trout and sturgeon). A second method is to capture immature wild fish and marine animals. Confined

1 The edible mussel (*Mytilus edulis*) is cultivated on ropes hanging from stakes or similar structures driven into sea-beds (as here in northern France), or on ropes suspended from floating rafts (as in Spain and Britain). Both methods first involve the collection by settlement of mussel "seed" or "spat". The "seed" may then be transferred to farming areas free from predators or pollution.

2 Cultivation of the European "flat" oyster (*Ostrea edulis*) is one of the earliest examples of aquaculture, known since Roman times. It involves the collection of larval animals, "spat", and their subsequent growth in protected tidal bays and estuaries. The sexually mature oyster [1] begins to spawn in the spring as the seawater temperature rises. When a spawning or "spat fall" occurs, the growers set down artificial spat collectors in the water to which the tiny larval oysters [2] can attach themselves. Around the Brittany coast of France spat collectors are usually curved, lime-covered roof tiles [3] stacked just clear of the sea-bed at right-angles, to facilitate water circulation [4]. The young (about 50 to a tile) are left until winter when, as orange-coloured "seed oysters" [5], they are prised [6] from the tiles and planted out in growing beds [7] or "parcs" where they feed on microscopic particles brought in by the tides. Parcs may be protected by nets [8] to help prevent attack by predators. After about 18 months the beds are dredged and the oysters sent by barge to fattening grounds, also in the estuary. The oysters are allowed to grow for about five years, each year adding a new layer to their shell [9]. Then they are gathered from the estuary and, before marketing, left for a few days in sterile seawater to cleanse them [10].

Sea level

Sea-bed

3 Several species of prawns and shrimps can be reared under controlled environmental conditions. Some fast-growing tropical prawn species are considered particularly suitable since they provide three or four crops a year. Eggs are obtained either from captive stock or from egg-bearing females caught wild. Mass-rearing techniques are used through the juvenile stages. Then the young prawns are fattened in an enclosed circulation system, shown here, where water at a controlled temperature is recycled by pump through the rearing tank. Dangerous toxic waste products excreted by the prawns must continuously be removed by biological filters like those used for purifying sewage.

Insulated room to control temperature — Biological filter, column of gravel colonized by micro-organisms — Pump — Prepared food made from mussels, crabs and shrimps — Juveniles bred in another tank or reared from eggs collected in the sea — Mature prawn — Prawns at a density of 200 per m² of tank floor — Aeration unit to remove excess CO_2 — Temperature control in enclosures, they may be left to forage for themselves or be provided with supplementary feed. Those best adapted to this kind of cultivation include milkfish, mullet, shrimp, oysters and mussels.

Third, eggs from wild parents can be gathered, incubated and farm-reared to a marketable size. In the most sophisticated form of fish farming, eggs are hatched and the young reared so that a growing brood stock is maintained. The operator therefore achieves full control over the life cycle of the animals and may be able to breed them selectively. Trout, salmon, catfish, carp and oysters are successfully farmed in this way.

Technological developments
Although most fish farming is still based on techniques of partial culture, it is in the carefully controlled intensive-rearing system that most technological developments are taking place. For example, closed-circuit water recycling enables the farmers to remove waste products – mainly ammonia, urea and feed or waste solids – and return clean water to the rearing facility [3].

Fish grow more rapidly in warm water. For this reason, some fish farms are sited near power stations and farmers can make use of the warm waste water from cooling towers, previously regarded as a pollutant, in the recycling process to accelerate growth rates of fish and shellfish, thus saving on heating costs. In the United States the effluent from a thermal power station has been put to good use in the rearing of molluscs and lobsters, and in Britain flatfish have been grown in heated water from a nuclear power station.

Successes in catfish culture systems [Key] and in farming the Atlantic salmon demonstrate the exciting possibilities for fish farming. However, many factors must be considered. These include feed costs in relation to product value, disease control in intensive systems, increasing costs where labour-intensive methods are used and, for some species, consumer resistance.

Looking ahead, however, fish farming will grow, spurred on by the over-fishing of useful wild stocks in the sea and by the need for new sources of protein to feed the ever-increasing populations of the world.

Probably one of the most impressive examples of industrial-scale fish farming is the pond culture of catfish, seen here in Arkansas. It has achieved small industry status in the US. Production in 1973 was about 40,000 tonnes.

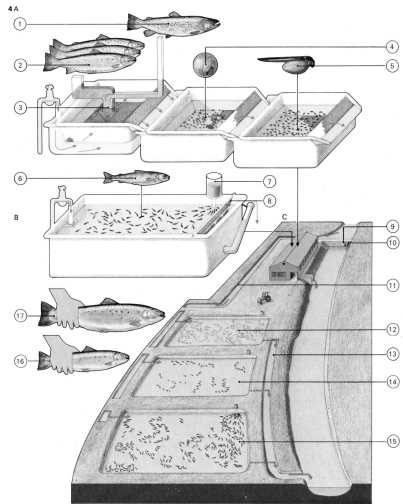

4A
B
C

4 In trout farming the operator can achieve complete control over his stock throughout its life cycle. The hatchery process [A] involves the removal of milt from male fish [1] and eggs from female fish [2]. The eggs are fertilized [3, 4]
and placed in an incubator and when the alevins (larval fish) [5] have absorbed their yolk sacs they are transferred to the fry tank [B]. Here the fry [6] are fed automatically [7] and are prevented from escaping by a perforated screen [8]. When the
fish are sufficiently developed they are removed to the fish farm [C]. Water from a river is diverted by a dam [9] incorporating an eel pass and fish ladder [10], and flows to the tanks by a channel [11]. The first pond [12] accommodates
fish up to one year old. Water is returned to the river by an outlet channel [13]. Fish are held in the second pond [14] up to two years old [16] and in the third [15] until they are three years old or sexually mature [17]. These are kept for breeding.

5 The salmon is migratory; it must move from fresh to salt water. Salmon eggs are incubated and the young fish reared throughout their early life stages in fresh water. But when the important "smolt" stage is reached they must be transported to seawater where they are normally kept in floating pens. Here Atlantic salmon (*Salmo salar*) are harvested from cages in a Scottish sea loch.

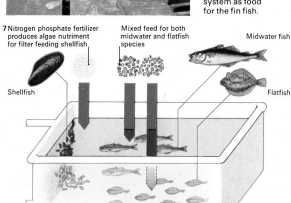

6 Flatfish, particularly turbot and Dover sole, have been shown to be suitable for farming. After being reared in a nursery they are fattened in cages in sheltered tidal waters.

7 Two or more species of fish or shellfish can be reared together in a "polyculture" system. The shellfish may either be marketed or returned to the system as food for the fin fish.

7 Nitrogen phosphate fertilizer produces algae nutriment for filter feeding shellfish

Mixed feed for both midwater and flatfish species

Midwater fish

Shellfish

Flatfish

Fish of ponds and rivers

Rivers and other inland waters throughout the world support a rich and varied fish population, but it is chiefly in the Northern Hemisphere that freshwater fish are favoured as a popular source of food. This is probably due to the much larger areas of continental land mass in the Northern Hemisphere, which result in a greater and more varied distribution of fish in the inland waters of the northern temperate regions and even across the Arctic.

Fish of northern and southern rivers

All members of the salmon family – including trout, whitefish and char – are native to northern rivers; sturgeon inhabit rivers of North America although the species of this highly prized fish from which caviare is obtained is found in the rivers of the USSR and the Balkans and in coastal waters of the Black and Caspian Seas.

Apart from introduced species, these popular freshwater fish are absent from most waters south of the Equator but many Southern Hemisphere rivers are inhabited by native species of good food value. The New Zealand river whitebait, for example, which are the young of several *Galaxias* species, are much appreciated delicacies; and the golden perch, also known as callop, is indigenous to Australian rivers where it is fished for its delicate flesh.

The perch family of northern waters is represented in the Southern Hemisphere by a similar freshwater family, the cichlids, widely distributed throughout Africa and most of South America. The cichlids are, confusingly, known as bream in southern Africa. Many cichlids are popular aquarium fish.

The carp family numbers more than 1,500 species. Many of these are minute and of no culinary interest, but the mirror carp, which may weigh up to 6kg (15lb), is highly valued in Continental European countries. The carp [3] inhabits rivers and ponds of Europe, Asia and North America.

The migratory eel

Freshwater eels (zoologically related to the carps) have been valued for their rich flesh since ancient times. Eels are distributed throughout the temperate zones of the world, apart from South America and the west of North America. The European eel, like the salmon, is a migrant. It begins its life in the Sargasso Sea and drifts on the ocean currents towards European and American rivers.

Eels spend most of their young and adult lives in fresh waters. American eels are thought to return to the Sargasso Sea to spawn, after which they die. The Japanese eel, widely eaten in Japan, is similar to the European species.

While eels migrate from their spawning grounds in the sea to spend their adult lives in fresh waters, the Japanese ayu does the opposite. It spawns in rivers of Japan and Korea where the currents are strong and the water temperature low. After mating the adult fish dies. The fry float on the currents down to the sea where the young fish remain until sexually mature. At the age of about 18 months, the ayu return to the rivers in order to spawn.

Freshwater food is not as abundant in the Northern Hemisphere as it once was. Many species have been almost destroyed by over-fishing or pollution; once prolific, the stur-

CONNECTIONS

See also
246 Fish of the ocean deeps
248 Shellfish and other seafoods
242 Fish farming
240 Commercial fishing

In other volumes
230 The Natural World
130 The Natural World

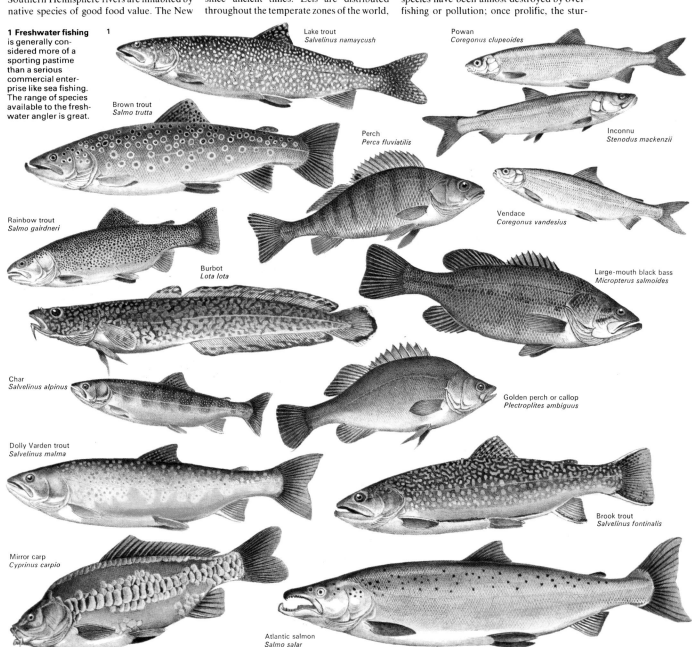

1 Freshwater fishing is generally considered more of a sporting pastime than a serious commercial enterprise like sea fishing. The range of species available to the freshwater angler is great.

Lake trout
Salvelinus namaycush

Powan
Coregonus clupeoides

Brown trout
Salmo trutta

Perch
Perca fluviatilis

Inconnu
Stenodus mackenzii

Rainbow trout
Salmo gairdneri

Vendace
Coregonus vandesius

Burbot
Lota lota

Large-mouth black bass
Micropterus salmoides

Char
Salvelinus alpinus

Golden perch or callop
Plectroplites ambiguus

Dolly Varden trout
Salvelinus malma

Brook trout
Salvelinus fontinalis

Mirror carp
Cyprinus carpio

Atlantic salmon
Salmo salar

geon is now so rare that its price puts it beyond the reach of most people.

The Atlantic salmon [Key], once very common in London's River Thames, was eliminated from these waters by 1870. It has, however, returned in recent years. A new hazard for this species has been the discovery, by Danish and Greenland fishermen of the salmon's main feeding grounds in the Atlantic, not far from the coast of Greenland. Massive and indiscriminate netting of immature salmon in these waters led to an outcry from land-based net fishermen and rod-and-line anglers in the 1960s when catches declined. At the same time a fungal disease reduced the numbers of salmon still further.

Game and coarse fishing

Game fishing (fishing for trout and salmon) and coarse fishing (angling for other freshwater fish) are becoming increasingly popular. Game fish tend to be scarcer and therefore game fishing is more expensive than coarse fishing which is usually obtainable in most local rivers, lakes and canals. Often coarse fish (which are not always easy

to cook or good to eat) are returned to the water after being caught and (in competitions) weighed, while the trout and salmon caught by the game fishermen are delicious to eat. Freshwater salmon can grow as heavy as 16kg (40lb) or more in European rivers.

Brown trout inhabit all kinds of freshwater environments; the rainbow trout tends to prefer a still water home such as a lake or pond. Rainbow trout can thrive in certain river conditions and some brown trout occasionally adopt the habits of migrating salmon and swim down the rivers to estuaries where they manage to survive in briny conditions and can grow to very great weights: a 6.5kg (14.5lb) brown trout of this sort was caught in Scotland in 1956. Neither the salmon nor the trout (brown or rainbow) should be confused with the sea trout, which is sometimes wrongly called salmon. The sea trout is a cousin of the Atlantic salmon, sharing with the salmon the same life cycle of being born in rivers and going out to sea to grow to maturity. But, unlike the salmon, the sea trout never swims far from the mouth of the river where it was born.

A
B

The Atlantic salmon is born in fast-flowing rivers. At this stage it is known as a fingerling and after two years it becomes a parr. It journeys down-river as a smolt, maturing in the estuary as a grilse before setting out to live in the ocean in an area from the Arctic southwards to latitude 40°N. The return migration from the Atlantic begins after four years when the salmon has reached its full maturity [A]. It may cross more than 4,800km (3,000 miles) of sea to reach its spawning grounds in the fresh water of the very river [B] where it was born. After the laying and the fertilizing of the eggs the salmon usually dies.

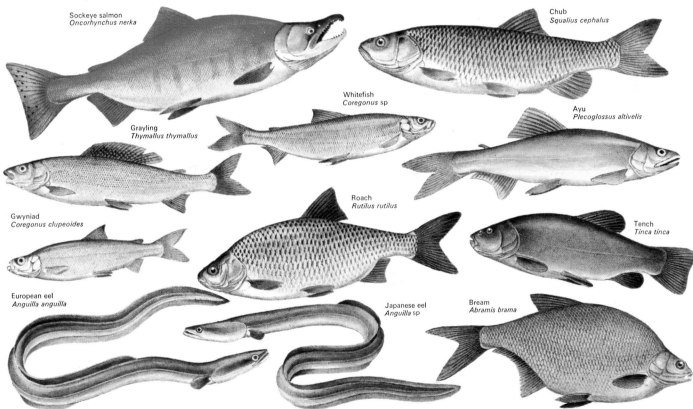

Sockeye salmon
Oncorhynchus nerka

Chub
Squalius cephalus

Whitefish
Coregonus sp

Ayu
Plecoglossus altivelis

Grayling
Thymallus thymallus

Roach
Rutilus rutilus

Tench
Tinca tinca

Gwyniad
Coregonus clupeoides

European eel
Anguilla anguilla

Japanese eel
Anguilla sp

Bream
Abramis brama

2

2 The carnivorous pike of the family Esocidae is one of the longest surviving species of freshwater fish. It inhabits rivers and lakes where it lives among dense colonies of water plants and weeds. Young pike, also known as jacks, live in shallow water, feeding on larvae and the fry of other fish. When the pike matures it moves into deeper water. The elongated, flecked body is camouflaged by the plants from which the pike darts out to attack and devour fish, waterfowl and even small mammals. One of the most vicious predators of fresh waters, the pike is equipped with a large and voracious mouth bearing backward-curving, sharp teeth from which no prey can escape once it is caught. The pike can grow to a length of 1.4m (4.5ft) and attain a weight of 21kg (46lb). It is widely fished for sport and also for food as the firm, white flesh is greatly valued for its flavour.

3

3 Carp originated in Asia and spread throughout the Northern Hemisphere. Carp bones excavated from Greek Stone Age sites establish the importance of freshwater fish in prehistoric times. The golden carp of Japan symbolizes bravery inherent in the carp's struggle to spawn upstream. It is popularly believed that by eating carp this strength can be passed on to man. It is reared for food and ornament.

Fish of the ocean deeps

Rich in protein, and in some cases more easily obtained than meat animals, fish have for centuries formed part of the diet of many peoples. More recently fish have been processed as feedstuffs for animals. The fish of the seas are not, however, merely a supply of nourishment for man; on an even larger scale they provide food for each other. Many of the species of fish that are caught by man for food or for processing are part of natural food cycles in the oceans. They may be predators that feed on smaller fish, and may in turn be devoured by even larger fish.

Man has fished the shallow margins of the seas from the earliest times, but not until he found methods of preserving catches did deep-sea fishing, using ocean going ships, get under way. Various nineteenth-century developments included the large-scale building of trawlers, the introduction of railway transport from fishing ports to inland cities, and the establishment of canning factories. Today the fleets of trawlers that fish the grounds off Greenland, Newfoundland and Japan for months at a time are floating factories with deep-freeze equipment.

The scientifically developed fishing methods and the easier (and different) means of preserving catches and transporting them today – leading to overfishing in many areas – have caused the fishing industry to become an international rather than a national concern. In spite of the prolific breeding habits of marine fish, indiscriminate fishing of some established grounds has led to a falling population in some species, and action is now being taken on an international scale to safeguard the inhabitants of the oceans.

Cold northern waters

Fewer species of fish inhabit the colder waters of the North Atlantic than the warmer Mediterranean and Pacific areas. But these fish, particularly the herring [8] and the members of the cod family, are caught in vast quantities. A female herring is less fertile than a female cod but still lays about 50,000 eggs each year. The herring population is held in check by other fish which feed on the numerous eggs. In spite of this, a large proportion of the eggs survive and develop through various stages to the mature herrings

that are fished for throughout the year by fleets following their migrations.

The herring is among the most nutritious of all fish. In addition to protein, it contains vitamins A and D as well as minerals such as iron and calcium. It is marketed fresh or frozen, salted, cured, smoked, canned and pickled, and some people regard fresh or canned herring roes as a delicacy.

Cod [3] is another important species for the fishing industry of northern Europe. A related species lives in the North Pacific but most commercial cod fishing is confined to deep cold waters, such as those of the northeast and northwest Atlantic. For some nations, the cost involved in reaching such far-off fishing grounds has in recent years made cod an expensive food fish, although it is still cheaper than its close but less abundant relative the haddock [20]. Dogfish [18], halibut [13] and saithe (also known as the pollack or pollock) are other species of fish found in cold seas.

Skate [2] and whiting [9] are also fish of cold waters, although they do not inhabit the extreme North Atlantic. Farther south – as

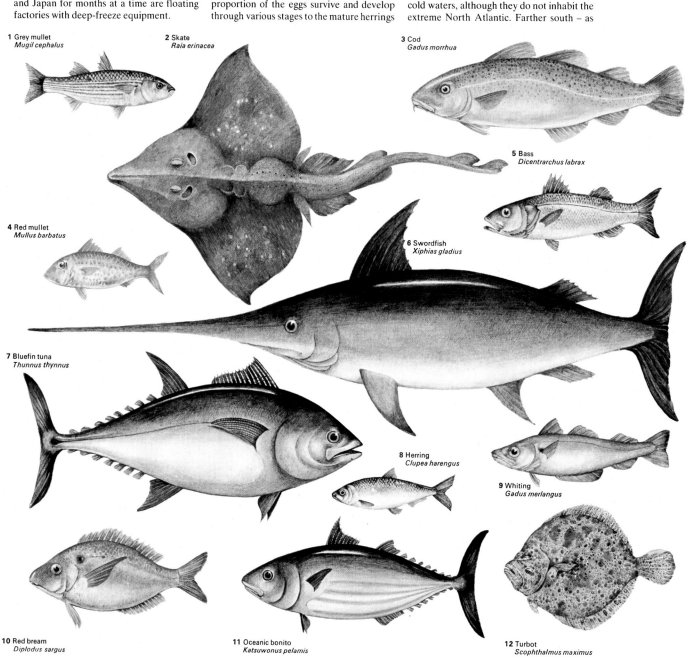

1 Grey mullet
Mugil cephalus

2 Skate
Raia erinacea

3 Cod
Gadus morrhua

4 Red mullet
Mullus barbatus

5 Bass
Dicentrarchus labrax

6 Swordfish
Xiphias gladius

7 Bluefin tuna
Thunnus thynnus

8 Herring
Clupea harengus

9 Whiting
Gadus merlangus

10 Red bream
Diplodus sargus

11 Oceanic bonito
Katsuwonus pelamis

12 Turbot
Scophthalmus maximus

far as the African coast – John Dory [14], mackerel [15] (related to the tuna), sole [17] and turbot [12] are the most important.

Mediterranean waters

The temperate Mediterranean Sea supports numerous species of fish. Several types of sea bream [10] are caught for their delicate flesh, along with sardines and anchovies.

The red mullet [4], known in France as *rouget*, is considered by many to be a Mediterranean speciality and was in classical times as expensive as Scottish salmon is today. The name "red mullet" is often given to another Mediterranean fish, the gurnet or gurnard, which while similar in colouring has coarser flesh.

The true red mullet can also be found in parts of the Atlantic off southern England and the southern United States. The goatfish, abundant in the Indian and Pacific Oceans, is related to the red mullet.

The ugly angler-fish is regarded as a delicacy in Venice and used as the main ingredient of *coda di rospa*. It is also one of the components of a Marseilles bouillabaisse.

Yet another plentiful fish of the Mediterranean Sea is the silvery bass, a similar species of which is found off the shores of the United States. Swordfish [6] and tuna (tunny) are frequently caught during the summer months in Mediterranean waters, although their breeding grounds are located in warmer seas.

Fish of tropical waters

Both the bluefin [7] and bonito [11] tuna belong to warm oceans, but many related species inhabit the Mediterranean Sea and South Atlantic Ocean (although they are rarely sighted north of the Bay of Biscay). The swordfish has a similar distribution pattern.

The red snapper [16] is a native of the Gulf of Mexico whereas the croaker [19] lives off the coasts of the southern United States. Like many other sea fish that live in similar conditions, the croaker is known throughout the world but by different names. In South Africa it is the *kabeljou* (from the Dutch name for cod) and in Australia it is the mulloway or jewfish.

KEY

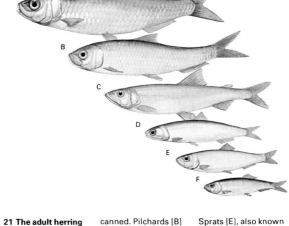

Trawlers range far and wide in their search for fish, using technological aids such as echo-location. Consumer demand can lead to international incidents: as catches fall off, trawler go farther afield, "invading" seas that are claimed as the exclusive territory of other nations. Recently, many coastal countries have extended their national limits in order to conserve remaining fish stocks, and "fish wars", with damage to nets and ships, have been common. The so-called "cod war" between Britain and Iceland in the mid-1970s was typical.

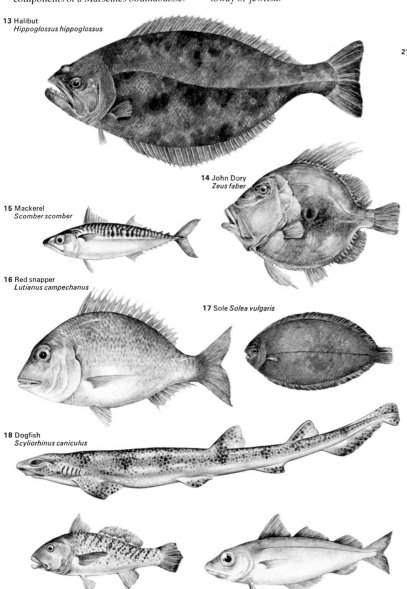

13 Halibut
Hippoglossus hippoglossus

14 John Dory
Zeus faber

15 Mackerel
Scomber scomber

16 Red snapper
Lutianus campechanus

17 Sole *Solea vulgaris*

18 Dogfish
Scyliorhinus caniculus

19 Croaker
Micropogon undulatos

20 Haddock
Gadus aeglefinus

21 A, B, C, D, E, F

21 The adult herring at three years of age has matured through a series of stages, all of which are used by man as food. Adult herrings [A] are sold fresh or salted, cured whole (as bloaters, bucklings and kippers) and pickled as fillets (matjes and rollmop herrings). Soft and hard roes are sometimes sold fresh but are usually canned. Pilchards [B] may be found fresh locally but are mainly used for canning; sardines [C] now appear fresh but many people think their flavour becomes more pronounced after canning in oil or tomato, either whole, skinned or boned. Young herring [D] may be smoked whole but are more usually sold as whitebait. Sprats [E], also known as brislings, may be lightly smoked before being canned in oil or a piquant sauce. The tiny anchovies [F], fried whole when freshly caught by Italian fishermen, are also exported as pastes, concentrated sauces, or as fillets salted and dried or canned in oil, sometimes stuffed with olives or pimentos.

22 Tora fugu
Fugu rubripes rubripes

22 The Japanese puffer fish is poisonous like other species of puffers and blowfish. Although poison from a badly prepared puffer fish can be fatal, the flesh is delicate, appetizing and perfectly safe once the harmful liver and ovaries are removed competently. Indeed the puffer fish is regarded as a gourmet's choice in a country which abounds with a variety of tropical seafood, including tuna, ayu (sweet fish), carp and squid. They are often served fresh in slivers and eaten raw. Puffer fish may be one of the dishes in a Western style hors d'oeuvres.

247

Shellfish and other seafoods

Coastal and inland waters the world over are inhabited not only by fish but also by numerous other creatures that have become an increasingly important source of food. Among these sea creatures are the highly prized shellfish, easily obtained when uncovered at low tide or when washed up on to beaches and the banks of estuaries.

Shellfish fall into two separate categories: the crustaceans with jointed shells, which include lobsters [1], shrimps [5], crawfish and crabs [1]; and the molluscs such as mussels, oysters, clams and cockles [2] and squid [Key D] and octopus.

Today a steadily rising demand coupled with limited or falling supply has led to increasing prices for all types of shellfish in Europe and North America, where lobsters, shrimps and oysters have become luxury foods. The traditional American clambake, has now become an expensive feast.

King of the shellfish

Many gastronomes consider that as salmon is the king of fish, so lobster is supreme among the crustaceans. There are two main types of lobster – the larger *Homarus americanus*, which lives off the eastern coast of the United States (chiefly in the area of Maine in New England) and the lobster of European waters, *Homarus vulgaris*. The American species is dark green when alive, whereas the European lobster is dark blue.

Lobsters are caught alive in "pots". These are actually wooden or metal cages baited with food. The lobster can enter easily, but once inside cannot escape. The pots are weighted down on the sea-bed with rocks or stones and attached to a float on the surface to mark their location. The spectacular appearance, the delicious white, firm flesh and the scarcity of lobster combine to make this shellfish one of the most prized of all food delicacies. The female lobster is generally preferred to the male, partly because it has a larger body, with more tender flesh, and partly because of the eggs (known as coral) that are carried in the tail and are considered an added delicacy.

The crawfish or crayfish, sometimes also called rock or spiny lobster, resembles the true lobster, but its colour is greeny-brown and it lacks the prominent claws. The meat is contained in the large, fan-shaped tail. Crawfish are larger and heavier than lobsters and their flesh is sometimes more tender. Freshwater species predominate in the USA. The main sea catches come from the Atlantic and the Mediterranean, but a related species, the Martinique lobster, abounds in the western Indian Ocean and the Caribbean Sea. The Florida crawfish, which is also found in the Pacific, although smaller than its other relatives, is said to be of superior quality.

Oysters and crabs

Oysters have been eaten since the earliest times. They were certainly known to the ancient Greeks, who used the shells to cast votes at elections. The Romans supplemented local supplies by importing oysters from Britain. Until the mid-nineteenth century oysters were commonplace fare in some areas, where they were used as ingredients in peasant cooking until the seemingly inexhaustible oyster beds were badly depleted. Since then, the price of oysters has risen astronomically. Most of the oysters

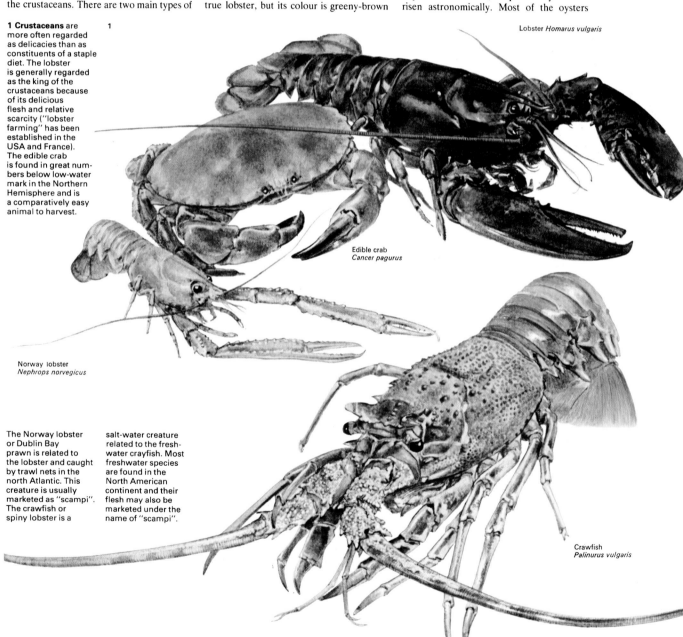

1 Crustaceans are more often regarded as delicacies than as constituents of a staple diet. The lobster is generally regarded as the king of the crustaceans because of its delicious flesh and relative scarcity ("lobster farming" has been established in the USA and France). The edible crab is found in great numbers below low-water mark in the Northern Hemisphere and is a comparatively easy animal to harvest.

Lobster *Homarus vulgaris*

Edible crab
Cancer pagurus

Norway lobster
Nephrops norvegicus

The Norway lobster or Dublin Bay prawn is related to the lobster and caught by trawl nets in the north Atlantic. This creature is usually marketed as "scampi". The crawfish or spiny lobster is a salt-water creature related to the freshwater crayfish. Most freshwater species are found in the North American continent and their flesh may also be marketed under the name of "scampi".

Crawfish
Palinurus vulgaris

marketed today come from protected beds and undersea farms established round parts of Europe, the USA, Japan and Australasia. Oysters are valued more for their flavour or rarity than for any particular food content.

The clam is similar to the oyster and is extremely popular in North America, where it may be served and eaten raw. It is also cooked in traditional New England chowders. Crabs, on the other hand, are so abundant in all seas that they have remained among the cheapest of shellfish. The Kamchatka [4], also known as the Japanese crab, is larger than the European species and, like the giant Tasmanian crab, commands higher prices because of its comparative rarity.

Mussels are even more common than crabs and are much used in Mediterranean cuisines. Other less expensive shellfish include cockles, whelks and winkles [2], popular on stalls at the seaside and in some street markets in Britain.

Squid, octopus and sea-urchin
In northern Europe and North America there is little demand for molluscs of the cephalopod order, which include cuttlefish, squid and octopus. These are all tentacled, rather ugly-looking creatures, distinguished by the ink sac placed near the heart. From this sac they squirt a dark fluid into the water in order to confuse pursuers. But around the Mediterranean, squid and cuttlefish are highly valued specialities; cuttlefish is also among the national dishes of China, Japan and some of the Pacific islands.

Octopus is popular in Greece and Cyprus where dried and grilled pieces of the tentacles are served as appetizers with pre-dinner drinks (the tentacles of squid are used for the same purpose in Italy). The flesh of octopus is extremely dry and tough and must be beaten to break down the tough fibres before it is cooked for eating.

Sea-urchins are spine-covered marine animals that cling to rocks in warm waters such as those of the South Pacific. The long, sharp, movable spines stick out from a hard lime shell that protects the soft body inside. Popularly known as sea eggs, the urchins are a speciality of Marseille. They are cut open, cleaned and served raw with fingers of bread to dip into the yellow "yolk" of the shells.

A wide range of seafood is found in most of the world's oceans. Shellfish varieties include the crustaceans and the molluscs. A typical crustacean is the edible crab [A]. When the animal is cooked the shell is broken open and the flesh inside is eaten. The most important

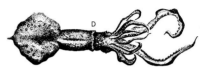

edible molluscs are the bivalves, such as the scallop [B], which have a pair of hinged shells containing the edible flesh. Gastropods like the whelk [C] are molluscs with a single shell. The flesh is extracted through the aperture. Cephalopods, the tentacled molluscs such as the squid [D], usually have no shell and there is little wastage.

2

Scallop
Pecten maximus

Oyster
Ostrea edulis

Common mussel
Mytilus edulis

Frilly clam
Tridacna squamosa

Cockle *Cardium edule*

Whelk *Buccinum undatum*

Winkle
Littorina littorea

3 A

2 Bivalves and gastropods are the most important molluscs to be harvested. They are mostly found in shallow coastal waters and are quite easily collected. Scallop shells have been used as decorative embellishments since Roman times and were taken as emblems by the pilgrims to the church of St James of Compostela; in France the scallop is known as the pilgrim shell or *Coquilles Saint Jacques*. Oysters, once a cheap seafood, are today one of the most expensive luxury foods, often eaten raw and on ice with lemon and brown bread, usually during winter. The common mussel is an inexpensive shellfish, particularly appreciated in French and Spanish cooking as the basis for soups and stews. Cockles are popularly known as the poor man's oyster and like them may be eaten raw, although they are more often marketed already cooked and shelled. Whelks are also sold cooked and shelled and are traditionally eaten with vinegar and brown buttered bread; they are also used as fish bait. Many species of clam may be eaten raw, or alternatively can be grilled or used in any of the well-known American clam recipes. Winkles or periwinkles are smaller relatives of whelks; they may be eaten raw or cooked. A long pin is needed to extract the flesh from the spiral shell.

3 The sea turtle [A] is a marine reptile that is rated as one of the world's delicacies. It lives offshore in warm waters such as in the Pacific, but comes ashore to lay eggs. Turtles are exported from Australia, South America and Africa, but the finest come from the West Indian islands and are used in the famous turtle soup. Sea fig [B] or *figue de mer* is a sea squirt, a primitive creature which, after a free-swimming larval stage, settles to a sedentary existence on rocks near the shore. It is gathered and eaten at its adult stage and despite its leathery covering it is considered a delicacy in France. Sea cucumber [C] is an elongated relative of the starfish. About two dozen species are harvested in the Indian and Pacific Oceans where a food product, *bêche-de-mer*, is prepared from the body wall of the animal.

4

4 Kamchatka crab is caught in the muddy waters and sand banks of the North Pacific. The meat is chiefly used in the canning industry on a large scale and is usually marketed by American firms under the name "king crab".

5 A

5 Many less well-known crustaceans are edible and the two Mediterranean examples shown here can provide the base for excellent soups. The mantis shrimp [A] lives in shallow water and has an average length of 20cm (8in). Large specimens of the creature known in France as *petite cigale* [B] may be prepared and served like lobsters.

249

Food preservation

Almost all food, whether plant or animal in origin, is prone to deterioration and eventual decay. If it is not properly preserved, it soon begins to dry out, bacteria and moulds start to grow, fats begin to oxidize and turn rancid, and enzymes within the food cause tissues to break down.

The rate at which food decays depends on its composition and on where it is stored. Soft berries, shellfish and milk all begin to spoil within a few hours. On the other hand chocolate, root crops and nuts will usually keep for weeks without any special precautions for storage. Food that is kept in cool, dry conditions keeps much longer than food stored in warm, humid surroundings.

Methods of preservation

The practice of preserving food has been known since ancient times. One of the earliest methods was to dry food in the sun [2] or beside a fire, thereby evaporating much of its water content and reducing the rate at which chemical and biological processes of decomposition (which are water-dependent) make it unfit to eat. An adaptation of drying

is smoking [3]; the chemicals in the smoke greatly increase the storage life of foods. Drying and smoking are widely employed, especially among nomadic peoples for whom dried meat and cured cheese are staple foods.

Salt [1] has been used as a preservative for thousands of years. It extends the storage life of foods by inhibiting the growth of bacteria. Salt and spices are both used to make food more palatable.

Almost all agricultural societies are familiar with fermentation and its preservative effects. The chemical products, alcohol, acetic acid and sometimes lactic acid, that are formed during fermentation limit the growth of decay-causing organisms. A similar process is that of pickling in vinegar, alcohol or brine [4]. Foods that are placed in baths of brine or some fermented liquid absorb the liquid and, as a result, become virtually immune to decomposition.

One of the major modern methods of food preservation is canning [6]. Although the principle of heat-treating food to destroy the pathogens (disease-carrying organisms) was explained by the French scientist Louis

Pasteur (1822–95) in the mid-1800s, the art of canning had already been discovered. In 1809, after 14 years of experiment, the Frenchman Nicolas Appert (1749–1841) found that certain foods would keep for months at room temperature if they were first sealed in glass bottles and then heated in boiling water. In the following year the same process was carried out using steel cans. Canning was a tedious process at first because it required up to five hours' heating.

Today, canning is generally an automatic process during which raw food is washed, sorted, cooked, canned and packaged by machine. Heat treatment of the can takes little more than half an hour and even less if foods are acid, salty or contain preservatives. Canning is a good means of preserving the food's nutritional value.

Refrigeration and freezing

Other common methods of food preservation are cooling and freezing, based on the principle that biochemical processes gradually come to a halt at low temperatures. Cooling is an ideal way of preserving the freshness of

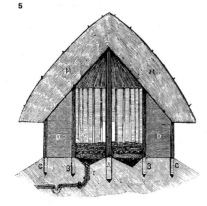

1 Fresh meat and vegetables were formerly not available during long sailing voyages. Before the invention of canning, travellers had to eat dried, smoked or pickled foods. Salted beef was a staple in the seafaring diet. Slabs of beef were packed in dry salt or a saline brine inside large wooden casks. In general a salt solution of 10-15% was effective in preserving the meat throughout a long voyage.

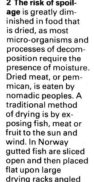

2 The risk of spoilage is greatly diminished in food that is dried, as most micro-organisms and processes of decomposition require the presence of moisture. Dried meat, or pemmican, is eaten by nomadic peoples. A traditional method of drying is by exposing fish, meat or fruit to the sun and wind. In Norway gutted fish are sliced open and then placed flat upon large drying racks angled to face the sun.

3 Preservation by smoking is an age-old technique used to cure meat, fish and cheese. Smoking serves to dehydrate food and to coat its surface with chemicals that retard decomposition and inhibit the growth of moulds and bacteria. In a typical smokeroom the food is hung by hooks from the ceiling and the floor covered with a smouldering layer of sawdust. A smoking lasts for 24 hours behind sealed doors.

4 In pickling, a salt solution (brine) controls the changes that the food undergoes. Instead of decay taking place, fermentation occurs within the food tissues and produces acids that act as preservatives. Pickled foods may remain edible for many months. Today pickling is more important as a flavouring process. Fruit may be preserved by bottling in sugar syrup or by cooking with sugar to make jam.

5 Ice houses were traditionally used to store perishable meats and dairy products for short periods. Food that is merely chilled, at 4°C (40°F), instead of being frozen, will deteriorate, but at a slower rate than normal. It is possible to keep food for up to two weeks in this fashion, especially if insulating material such as sawdust is packed around it. Vegetables and fruits with high water contents keep less well than meat.

6 Canning is the process in which food is placed in containers (traditionally tin-lined steel, but today often plastic-lined aluminium cans), heated to 115°C (240°F) and then sealed under a vacuum. A brief heating of the can kills off any bacteria that cause decay or that might otherwise contaminate the food. Modern canning machinery will sort, wash, cook and can foods fully automatically, producing many cans a minute.

food [5]. As early as 250 BC in India and Egypt, food containers were wrapped in wet rags. The heat loss due to the water's evaporation was an efficient cooling technique. Long before this, however, cellars and caves were being used as storage rooms for food, and inhabitants of cold regions have always been familiar with freezing techniques. Even the process of freeze-drying [7], involving dehydrating and freezing, was known to some ancient peoples.

Food will keep fresh if it is stored in a cool, dry place. Below 4°C (40°F) most fruits and vegetables will keep for up to a week. Freezing denies water to micro-organisms, bringing their growth to a standstill although not necessarily killing them in the process. They are revived when the food is again thawed out.

If food is frozen gradually as in a domestic freezer, large slivers of ice form within it. These crystals rupture the cell tissue so that the food loses its moisture and texture when it is thawed out. The flavour and texture of delicate fruits and vegetables can be altered by slow freezing.

A range of sulphates and benzoates are common preservatives. In small amounts, under 0.1 per cent, they check the growth of bacteria, moulds and yeasts. In addition some foods are treated with anti-oxidants such as ascorbic acid, and with bleaching chemicals, neutralizers and stabilizers to preserve their "garden fresh" appearance.

Recent progress

A fairly recent process of food treatment uses irradiation techniques [7]. Although complete sterilization by this method requires doses of radiation so high that food tissues are destroyed, lower pasteurizing doses are extremely effective in killing many organisms and halting sprouting.

In the past 200 years, great strides have been made in the techniques of food preservation. In recent years it has become possible for consumers in most parts of the world to obtain meat, vegetables and fruit in "fresh" condition although they may have been grown on the other side of the world and stored for weeks or even months before being eaten [Key].

The larder and refrigerator in the average household contains foods from almost every corner of the globe. Modern methods of preserving and packaging food have made it possible to eat fruit and vegetables all year round.

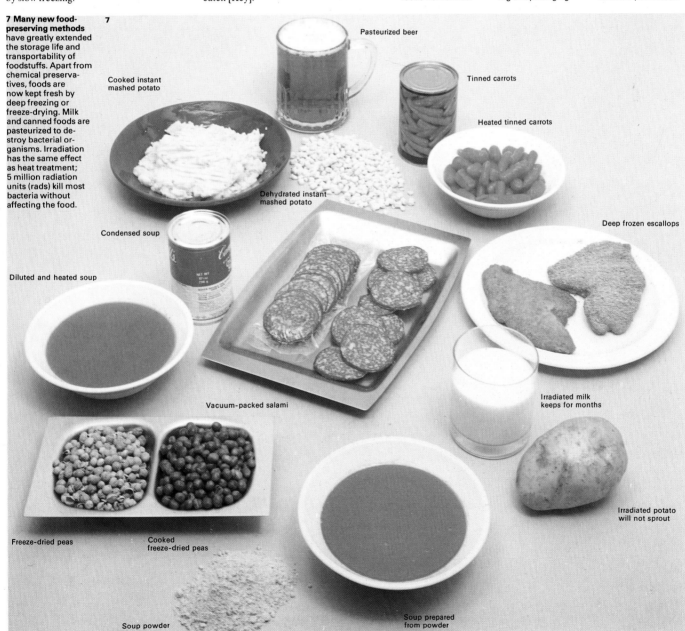

7 Many new food-preserving methods have greatly extended the storage life and transportability of foodstuffs. Apart from chemical preservatives, foods are now kept fresh by deep freezing or freeze-drying. Milk and canned foods are pasteurized to destroy bacterial organisms. Irradiation has the same effect as heat treatment; 5 million radiation units (rads) kill most bacteria without affecting the food.

Cooked instant mashed potato

Pasteurized beer

Tinned carrots

Heated tinned carrots

Dehydrated instant mashed potato

Condensed soup

Deep frozen escallops

Diluted and heated soup

Vacuum-packed salami

Irradiated milk keeps for months

Freeze-dried peas

Cooked freeze-dried peas

Irradiated potato will not sprout

Soup powder

Soup prepared from powder

The future of food

In his *Essay on the Principle of Population* published in 1798, Thomas Malthus (1766–1834) put forward the idea that the human population was growing faster than the supplies of food needed to keep it alive. He concluded that if population were permitted to increase unchecked a food crisis would eventually result and war and disease would reduce the population to a level compatible with existing food supplies [9].

To date, the crisis that Malthus predicted has not materialized. In spite of several serious famines in the last 200 years and the persistence of widespread malnutrition, people of affluent countries are better nourished today than ever before.

Why Malthus was wrong

Malthus was wrong because he could not possibly foresee the dramatic changes that would greatly increase man's ability to produce more food. The first of these changes began with the rapid expansion in agricultural land that occurred when vast areas of fertile land were opened up in North America, Australia and, to a lesser extent, South America during the nineteenth century. The invention of the steam engine, and later the discovery of the principles of refrigeration, enabled the produce grown on those continents to be shipped to the newly industrialized nations in Europe, thus greatly increasing their food supplies.

Perhaps more important were the scientific advances that eventually led to man's being able to grow more food from each hectare of land. In 1840, Baron Justus von Liebig (1803–73) discovered that man could replace the nutrients extracted by crops from the land. This important discovery led to the development of the fertilizer industry and to the eventual widespread use of inorganic fertilizers by farmers. Another major advance was the discovery by Gregor Mendel (1822–84) of the laws of inheritance (how characteristics are passed from parents to offspring). The growth in the knowledge of genetics that followed eventually led to its practical application. New, higher-yielding crops were bred [5], as were increasingly productive livestock [4].

These and other developments such as the discovery of chemical pesticides enabled man to grow more food on the land. At the same time he began to look more and more to the sea as a source of food. The sea-going trawlers, together with the development of mechanical refrigerating machines, meant that people living away from the coast could purchase and eat fresh fish [3].

Food for the future

Population projections suggest that a great effort will have to be made over the next 50 years if widespread starvation is to be avoided. The United Nations predicts that by AD 2000 there will be a total of about 6,500 million mouths to feed. How can we make sure they will all be satisfied?

Obviously, food production is going to have to increase above its current level. Following the conventional approach, this might possibly be done either by bringing more land under cultivation or by further intensifying production on land already being cultivated, or by doing both these things. Unfortunately, this approach presents several problems. Most of the best agricultural land in the world is already under cultivation. To bring the

CONNECTIONS

See also
156 World food resources

In other volumes
242 The Natural World
250 The Natural World

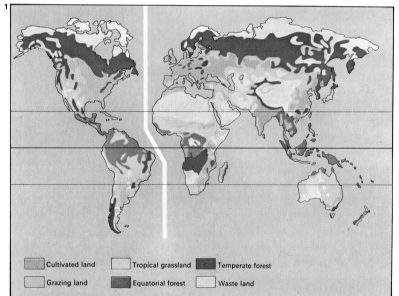

1 Potential arable land of 1,500 million hectares (4,000 million acres) is as yet unexploited. The most fertile land has long been under cultivation and what remains is marginal land where the cost of bringing it into production is high. Once in production, this new land will only with difficulty produce high crop yields because of the severe limitations imposed by soil fertility, topography and adverse climatic conditions.

2 Intensifying production by increasing the use of artificial fertilizers may initially increase total food production, but there are limits to the ability of the crop to respond to higher levels of inputs.

Cultivated land
Grazing land
Tropical grassland
Equatorial forest
Temperate forest
Waste land

No fertilizer applied

A little fertilizer applied

Too much fertilizer applied

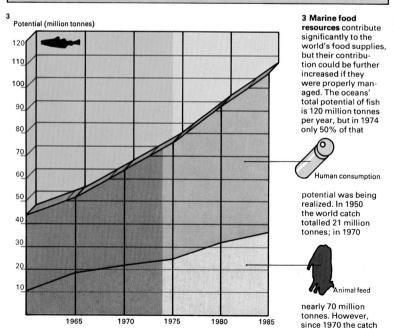

Potential (million tonnes)

120
110
100
90
80
70
60
50
40
30
20
10

1965   1970   1975   1980   1985

Human consumption

Animal feed

3 Marine food resources contribute significantly to the world's food supplies, but their contribution could be further increased if they were properly managed. The oceans' total potential of fish is 120 million tonnes per year, but in 1974 only 50% of that

potential was being realized. In 1950 the world catch totalled 21 million tonnes; in 1970 nearly 70 million tonnes. However, since 1970 the catch has fallen, largely as a result of overfishing the commercial table species and also because of the disappearance of the anchovies (used in large quantities for animal feed) off the coast of Peru in 1972. Conflict over fishing rights is increasing and several nations are proposing an extension of the existing 19km (12 mile) limit to 320km (200 miles). Despite overfishing there still remain regions such as the Indian Ocean that have not yet been exploited for their food stocks.

4 The beefalo is an experimental hybrid of American buffalo and a domestic beef animal. It can convert grass and other roughage to meat better than conventional beef breeds, and may help to increase world food production.

remaining 1.5 thousand million hectares (4 thousand million acres) of potential agricultural land into production may prove costly [1]. Much of this land is found in tropical areas such as the Amazon and Congo basins, which are currently covered with thick jungle. If the protective canopy of trees were indiscriminately removed and the land ploughed up to grow crops, heavy tropical rains would soon transform the soils into barren laterite (a soil rich in iron and aluminium oxides but devoid of mineral nutrients such as potassium and nitrates).

The use of higher levels of inputs such as fertilizers would be limited by their availability and cost [2]. Because of the law of diminishing returns, higher applications of fertilizer do not result in correspondingly higher yields. Consequently, unless increased fertilizer production were earmarked for land that currently receives very little fertilizer (and most of this land is found in the developing world), its use could result in more expensive food. In addition, the application of more fertilizer could lead to severe pollution of water supplies.

Given the population projections and the limitations imposed on increasing food supplies by conventional methods it appears that continued efforts including novel approaches must be adopted.

Simultaneous attacks on the problem

It is important to work simultaneously on three fronts: to increase food production; to improve the global distribution of food supplies; and, finally, to limit the growth in demand for food by curbing the expansion in the world's population.

An increase in food production requires both conventional and non-conventional methods. Non-conventional methods include the development of new crops, animals and foods. If the increased food production is to benefit the world as a whole, then its distribution must be improved from the present situation in which about 33 per cent of the world's population eats much more than half the food. Initially this can best be done by ensuring that the increase in food production occurs primarily in the developing nations of the world.

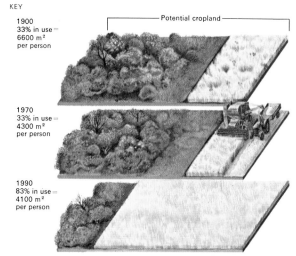

World food production has increased four-fold since 1900 although the area of potential cropland under cultivation has remained the same. This was made possible through mechanisation and the increased use of fertilizers. By 1990, however, 83% of potential cropland will have to be cultivated to yield sufficient food to support the increased population.

5 Opaque 2 — Ordinary maize

Tyrosine 453 — 365
Valine 569 — 461
Tritophan 151 — 67
— 254
Lycine 489 —

Leucine 977 — 1252

Other amino acids 1627 — 1485

Milligrams per 100g maize

5 High protein grains could improve nutrition, particularly in the poorer nations of the world where grain is a major source of protein. The development of Opaque 2 varieties of maize has already increased the effective protein yield of these varieties 25 per cent above that of ordinary maize. This has become possible because of the higher lycine and tritophan content of the Opaque 2 varieties.

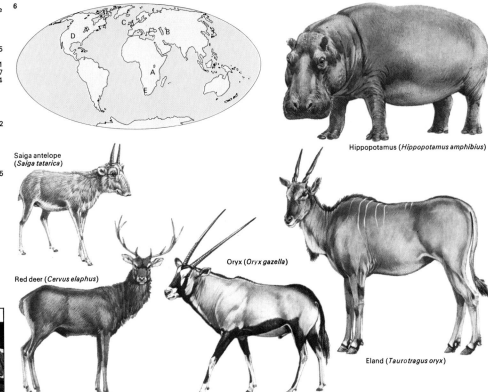

Hippopotamus (*Hippopotamus amphibius*)

Saiga antelope (*Saiga tatarica*)

Oryx (*Oryx gazella*)

Eland (*Taurotragus oryx*)

Red deer (*Cervus elaphus*)

6 Protection and sensible culling of wildlife stocks enable species to thrive and at the same time provide meat. In Ugandan game parks the hippopotamus has been culled for meat.

The saiga antelope is found in the USSR and is culled for its meat, leather and horns. The red deer is more efficient than any domestic animal at converting the poor moss and grasses of the Scottish Highlands into meat. The eland is a native of East Africa but is now bred in Africa and the USA for meat. The oryx ia one of the few meat producing species that can live in deserts.

7 New foods that make use of crops not generally eaten by people in the Western world are being developed. One is textured vegetable protein (TVP) made from soya bean. Protein may be extracted from leaves to feed animals and, possibly, humans.

8 New animal foods are being developed by the growing of bacteria, algae, fungi and yeasts on hydrocarbons. If successful, this could release large quantities of high-protein products (from fish and pulse crops) for human consumption.

9 Malthus suggested that population tends to expand more rapidly than food supplies and that a food crisis is inevitable. This crisis has been avoided so far, but it may materialize unless ways are found to curb world population growth.

253

INDEX

Picture Credits

Every endeavour has been made to trace copyright holders of photographs appearing in *The Joy of Knowledge*. The publishers apologize to any photographers or agencies whose work has been used but has not been listed below.

Credits are listed in this manner: [1] page numbers appear first, in **bold** type; [2] illustration numbers appear next, in parentheses; [3] photographers' names appear next, followed where applicable by the names of the agencies representing them.

16–17 Jon Gardey/Robert Harding Associates. **18** Leonard McCombe/T.L.P.A. © Time Inc. 1976/Colorific. **19** Mats Wibe Lund. **24–5** [Key] Scripps Institute of Oceanography. **26–7** [3] Trans Antarctic Expedition; [6] C. E. Abranson. **28–9** [1] Bill Ray: Life © Time Inc. 1976; [9] Popperfoto. **30–1** [5] Picturepoint; [8] Mats Wibe Lund; [9] C. E. Abranson; [10] Heather Angel. **32–3** [6] David Strickland. **34-5** [Key] Photri. **36–7** [1] Jon Levy; [2] NASA; [3] NASA. **44–5** [1, 2, 3, 4, 5, 6A, 6B] NASA/Sachem. **46–7** [1, 2, 3, 4, 5, 6] NASA. **52–3** [1, 2, 3, 4, 5, 6] NASA. **56–7** [1] NASA; [2] NASA; [3] NASA; [4] NASA; [5] NASA; [6] NASA; [7] Jon Levy; [8] NASA. **60–1** [1, 2, 3, 4, 5, 6, 7, 8] NASA. **64–5** [1] NASA; [2] Nasa; [3] NASA; [4] NASA; [5] NASA; [6] NASA; [7] Jon Levy; [8] NASA. **68–9** [Key] Barnabys Picture Library. **70–1** [Key] Jon Levy; [2A] Ken Pilsbury; [2B] Martyn Bramwell; [2C] Ken Pilsbury; [7] Bettman Archive. **72–3** [Key] BBC Copyright by kind permission of Michael Fish [1] C. E. Abranson; [6] Jon Levy/NASA; [7] The Controller HMSO: The Director General of the Meteorological Office, photographs taken at East Hill Dunstable by W. G. Harper. **74–5** [2A] Janine Wiedel/Robert Harding Associates; [5B] Spectrum Colour Library; [6A] F. Jackson/Robert

Harding Associates. **76–7** [2] ZEFA; [8] Martyn Bramwell. **78–9** [16] Robert Cundy/Robert Harding Associates; [4] Dr J. Wilson; [6A] Alan Durand/Robert Harding Associates; [6B] ZEFA. **80–1** [2] Picturepoint; [5] Bill Ray: Life © Time Inc. 1976/Colorific; [8A] Tony Stone Associates; [8B] Tony Stone Associates. **82–3** [Key] Institute of Oceanographic Sciences [4] Dr Kempe/British Museum [Natural History] [7A] C. E. Abranson. **90–1** [Key] Courtesy National Oceanic and Atmospheric Administration National Marine Fisheries Service [1] C. E. Abranson; [6A] O.S.F./Bruce Coleman Ltd.; [6B] O.S.F./Bruce Coleman Ltd.; [7] O.S.F./Bruce Coleman Ltd. **92–3** [7] Vickers Oceanics Ltd. **94–5** [Key] C. E. Abranson; [2] C. E. Abranson; [3] C. E. Abranson; [4] C. E. Abranson; [5] C. E. Abranson; [6] C. E. Abranson; [7] C. E. Abranson; [8] C. E. Abranson; [12] Institute of Geological Sciences. **96–7** [Key] Spectrum Colour Library; [1A] C. E. Abranson; [1B] C. E. Abranson; [1C] C. E. Abranson; [7] Basil Booth; [8] Basil Booth; [9] Basil Booth; [10] Basil Booth; [11] C. E. Abranson; [12] C. E. Abranson; [13] C. E. Abranson; [14A] Basil Booth; [14B] Basil Booth; [14C] Basil Booth. **98–9** [Key] Ron Boardman; [2] Courtesy of De Beers Consolidated Mines Ltd.; [6A] Institute of Geological Sciences; [6B] C. E. Abranson; [6C] Basil Booth; [6D] Institute of Geological Sciences; [7A] Institute of Geological Sciences; [7B] Basil Booth; [8A] Institute of Geological Sciences; [8B] C. E. Abranson; [9A] Institute of Geological Sciences; [9B] Institute of Geological Sciences; [10A] Institute of Geological Sciences; [10B] Institute of Geological Sciences; [11A] Institute of Geological Sciences; [11B] Institute of Geological Sciences; [12A] Institute of Geological Sciences; [12B] C. E. Abranson; [13A] Institute of Geological Sciences; [13B] Institute of Geological Sciences; [14A] Institute of Geological Sciences; [14B] Institute of Geological Sciences; [15A] Institute of Geological Sciences; [15B] Institute of Geological Sciences; [16A] Institute of Geological Sciences; [16B]

Institute of Geological Sciences; [17A] Institute of Geological Sciences; [17B] Institute of Geological Sciences; [18A] Institute of Geological Sciences; [18B] Institute of Geological Sciences. **100–1** [6, 7A, 7B, 7C, 7D] Basil Booth. **102–3** [3] Basil Booth; [4] Basil Booth; [5A] Basil Booth; [5B] C. E. Abranson; [5C] C. E. Abranson; [5D] C. E. Abranson; [5E] Basil Booth; [5F] Basil Booth. **106–7** [7] Picturepoint; [8] Australian Tourist Commission. **108–9** [6] Spectrum Colour Library. **110–11** [3] Dr A. C. Waltham; [4] Ardea Photographics; [5] C. E. Abranson; [6] C. J. Ott/Bruce Coleman Ltd.; [7] C. E. Abranson. **114–15** [Key] David Strickland; [4] P. Morris; [5] P. Morris; [6] Picturepoint. **116–7** [Key] Spectrum Colour Library; [3] Picturepoint; [4] Barnabys Picture Library; [5] Picturepoint; [6] Barnabys Picture Library; [7] Barnabys Picture Library. **118–9** [3] Barnabys Picture Library. **120–1** [Key] Chris Bryan/Robert Harding Associates; [6] C. Walker/Natural Science Photos; [7A] Paul Brierley; [7B] Paul Brierley; [12] Picturepoint. **122–3** [2] Picturepoint; [6] G. R. Roberts. **124–5** [Key] Picturepoint; [2] Picturepoint; [5] Isobel Bennett/Natural Science Photos. **128–9** [6] D. Dixon. **130–1** [4] F. Jackson/Robert Harding Associates; [6] C. E. Abranson; [7] A. J. Deane/Bruce Coleman Ltd. **132–3** [Key] C. E. Abranson; [4] C. E. Abranson; [5] C. E. Abranson; [6] C. E. Abranson; [7] Basil Booth; [8] Basil Booth. **134–5** [Key] M. F. Woods & Associates [3] W. Bockhaus/ZEFA; [4] Spectrum Colour Library; [6] C. E. Abranson; [7] Weir Group Ltd. **136–7** [Key] Institute of Geological Sciences; [5] Photri; [6] Picturepoint. **138–9** [Key] C. E. Abranson. **140–1** [5] Barnabys Picture Library; [7] C. E. Abranson; [8] Basil Booth. **142–3** [3] J. Nuyton/Robert Harding Associates. **144–5** [Key] George Hall/Susan Griggs Picture Agency; [3] Picturepoint; [5] John G. Ross/Susan Griggs Picture Agency. **148–9** [Key] Spectrum Colour Library; [3] Source unknown. **150–1** [6A] Picturepoint; [6B] Picturepoint. **154–5** [Key] Daily

Telegraph Colour Library; [2] Fairey Surveys; [3] KLM Aerocarts. **156–7** [8] Picturepoint. **158–9** [1] Picturepoint; [5] Picturepoint; [7] Museum of English Rural Life, University of Reading; [8] Aerofilms [9] Picturepoint. **160–1** [2] Picturepoint; [4] Adam Woolfitt/Susan Griggs Picture Agency; [5] Picturepoint. **164–5** [1] Photri; [6] David Strickland. **168–9** [Key] Spectrum Colour Library; [1] J. Edwards/Robert Harding Associates; [2] Leonard Freed/Magnum; [3] Ronald Sheridan; [4A] C. E. Abranson; [4C] Photri. **170–1** [Key] Shell Photographic Library; [5] New Zealand High Commission. **172–3** [Key] Fisons Agricultural Division; [6] Tropical Products Institute [Crown Copyright]; [8A] Glasshouse Crops Research Institute; [8B] Ministry of Agriculture & Fisheries; [8E] National Vegetable Research Station. **174–5** [Key] Farmers Weekly/Philip Felkin; [4] Basil Booth. **176–7** [Key] Plant Breeding Institute Trumpington, Cambs. **180–1** [Key] ZEFA. **182–3** [Key] Ron Boardman; [4] Tim Megarry/Robert Harding Associates; [5] Picturepoint. **184–5** [Key] Michael Francis Wood & Associates. **186–7** [Key] Potato Marketing Board. **192–3** [1] Bruce Coleman Ltd./John Markham; [2] E. W. Tattersall. **194–5** [1A] William MacQuitty; [1B] Picturepoint. **196–7** [5] British Sugar Corporation; [8] Source unknown; [9] Photri. **198–9** [Key] Scala/Napoli Museo Nazionale; [1] Mansell Collection; [2] Source unknown; [3A] Michael Holford; [3B] Michael Holford/British Museum; [4] Michael Holford; [5] Photographie Giraudon/Musée de Cluny; [7] Photographie Giraudon/Musée Conde Chantilly. **200–1** [Key] John Bulmer; [3A] Pierre Mackiewicz; [3B] Pierre Mackiewicz; [3C] Pierre Mackiewicz; [3D] Pierre Mackiewicz; [4A] Pierre Mackiewicz. **202–3** [Key] Source unknown. **204–5** [Key] Michael Holford; [2] Fisons Photo Studio; [3A] Spectrum Colour Library; [3B] Brewers Association. **206–7** [Key] Source unknown; [2] International Distillers and Vintners; [3] International Distillers and

Vintners; [4] International Distillers and Vintners; [5] Streets Financial Ltd./Highland Distilleries; [8] C. E. Abranson. **208–9** [Key] ZEFA; [1] Mary Evans Picture Library; [2] C. E. Abranson; [4] Spectrum Colour Library; [7] C. E. Abranson/Museum of Mankind. **210–11** [2] Jon Gardey/Robert Harding Associates. **214–15** [3] Bill Holden; **216–7** [Key] C. E. Abranson; [2] David Strickland; [3B] C. E. Abranson; [6] David Strickland. **218–19** [Key] Forestry Commission; [8] Barnabys Picture Library; [10] Picturepoint. **220–1**[1]

Hilliers Nurseries Winchester; [2] Hilliers Nurseries, Winchester. **222–3** [Key] Jeffrey Craig/Robert Harding Associates [Key] A–Z Botanical Collection. **224–5** [Key] Ron Boardman; [4] Spectrum Colour Library; [7] Picturepoint; [8] Camera Press. **226–7** [Key] C. E. Abranson; [2] C. E. Abranson; [3] Spectrum Colour Library. **228–9** [Key] Museum of English Rural Life; [3] Mr Pampa; [4] Oxfam/Nick Fogden; [5] Daily Telegraph Colour Library; [6] Express Dairies. **230–1** [2] David Strickland; [3] Mike Holmes; [4] Daily

Telegraph Colour Library; [6] Ian Sumner/Robert Harding Associates; [7] G. Riethmeier/ZEFA; [8] David Strickland. **232–3** [7A, 7B] David Strickland. **236–7** [Key] Photri. **238–9** [Key] Mansell Collection. **240–1** [Key] Harry Barrett/Fishing News; [9A] Michael Francis Wood & Associates; [9B] Michael Francis Wood & Associates. **242–3** [Key] Photri; [1] Photo Fratelli Fabbri Editori; [5] Marine Harvest Ltd.; [6] Michael Francis Wood Associates. **244–5** [Key A] Photri; [Key B] R. Thompson/Frank W. Lane; [3] Frank

W. Lane/F. W. Lane. **246–7** [Key] Keystone. **250–1** [Key] David Strickland; [1] C. E. Abranson/National Maritime Museum; [2] John Massey-Stewart; [3] Spectrum Colour Library; [4] David Strickland; [5] Mansell Collection; [6] H. J. Heinz & Co. Ltd.; [7] David Strickland. **252–3** [4] Texas Meat Brokerage Inc., Burlingame, California; [7] Courtaulds Ltd.; [8] B.P. Proteins/British Petroleum Ltd.; [9] Mansell Collection.

Artwork Credits

Art Editors
Angela Downing; George Glaze; James Marks; Mel Peterson; Ruth Prentice; Bob Scott

Visualizers
David Aston; Javed Badar; Allison Blythe; Angela Braithwaite; Alan Brown; Michael Burke; Alistair Campbell; Terry Collins; Mary Ellis; Judith Escreet; Albert Jackson; Barry Jackson; Ted Kindsey; Kevin Maddison; Erika Mathlow; Paul Mundon; Peter Nielson; Patrick O'Callaghan; John Ridgeway; Peter Saag; Malcolme Smythe; John Stanyon; John Stewart; Justin Todd; Linda Wheeler

Artists
Stephen Adams; Geoffrey Alger; Terry Allen; Jeremy Alsford; Frederick Andenson; John Arnold; Peter Arnold; David Ashby; Michael Badrock; William Baker; John Barber; Norman Barber; Arthur Barvoso; John Batchelor; John Bavosi; David Baxter; Stephen Bernette; John Blagovitch; Michael Blore; Christopher Blow; Roger Bourne; Alistair Bowtell; Robert Brett; Gordon Briggs; Linda Broad; Lee Brooks; Rupert Brown; Marilyn Bruce; Anthony Bryant; Paul Buckle; Sergio Burelli; Dino Bussetti; Patricia Casey; Giovanni Casselli; Nigel Chapman; Chensie Chen; David Chisholm; David

Cockcroft; Michael Codd; Michael Cole; Gerry Collins; Peter Connelly; Roy Coombs; David Cox; Patrick Cox; Brian Cracker; Gordon Cramp; Gino D'Achille; Terrence Daley; John Davies; Gordon C. Davis; David Day; Graham Dean; Brian Delf; Kevin Diaper; Madeleine Dinkel; Hugh Dixon; Paul Draper; David Dupe; Howard Dyke; Jennifer Eachus; Bill Easter; Peter Edwards; Michael Ellis; Jennifer Embleton; Ronald Embleton; Ian Evans; Ann Evens; Lyn Evens; Peter Fitzjohn; Eugene Flurey; Alexander Forbes; David Carl Forbes; Chris Fosey; John Francis; Linda Francis; Sally Frend; Brian Froud; Gay Galfworthy; Ian Garrard; Jean George; Victoria Goaman; David Godfrey; Miriam Golochoy; Anthea Gray; Harold Green; Penelope Greensmith; Vanna Haggerty; Nicholas Hall; Horgrave Hans; David Hardy; Douglas Harker; Richard Hartwell; Jill Havergale; Peter Hayman; Ron Haywood; Peter Henville; Trevor Hill; Garry Hinks; Peter Hutton; Faith Jacques; Robin Jacques; Lancelot Jones; Anthony Joyce; Pierre Junod; Patrick Kaley; Sarah Kensington; Don Kidman; Harold King; Martin Lambourne; Ivan Lapper; Gordon Lawson; Malcolm Lee-Andrews; Peter Levaffeur; Richard Lewington; Brian Lewis; Ken Lewis; Richard Lewis; Kenneth Lilly; Michael Little; David Lock; Garry Long; John Vernon Lord;

Vanessa Luff; John Mac; Lesley MacIntyre; Thomas McArthur; Michael McGuinness; Ed McKenzie; Alan Male; Ben Manchipp; Neville Mardell; Olive Marony; Bob Martin; Gordon Miles; Sean Milne; Peter Mortar; Robert Morton; Trevor Muse; Anthony Nelthorpe; Michael Neugebauer; William Nickless; Eric Norman; Peter North; Michael O'Rourke; Richard Orr; Nigel Osborne; Patrick Oxenham; John Painter; David Palmer; Geoffrey Parr; Allan Penny; David Penny; Charles Pickard; John Pinder; Maurice Pledger; Judith Legh Pope; Michael Pope; Andrew Popkiewicz; Brian Price-Thomas; Josephine Rankin; Collin Rattray; Charles Raymond; Alan Rees; Ellsie Rigley; John Ringnall; Christine Robbins; Ellie Robertson; James Robins; John Ronayne; Collin Rose; Peter Sarson; Michael Saunders; Ann Savage; Dennis Scott; Edward Scott-Jones; Rodney Shackell; Chris Simmonds; Gwendolyn Simson; Cathleen Smith; Lesley Smith; Stanley Smith; Michael Soundels; Wolf Spoel; Ronald Steiner; Ralph Stobart; Celia Stothard; Peter Sumpter; Rod Sutterby; Allan Suttie; Tony Swift; Michael Terry; John Thirsk; Eric Thomas; George Thompson; Kenneth Thompson; David Thorpe; Harry Titcombe; Peter Town; Michael Trangenza; Joyce Tuhill; Glenn Tuttsel; Carol Vaucher; Edward Wade; Geoffrey Wadsley; Mary Waldron;

Michael Walker; Dick Ward; Brian Watson; David Watson; Peter Weavers; David Wilkinson; Ted Williams; John Wilson; Roy Wiltshire; Terrence Wingworth; Anne Winterbotham; Albany Wiseman; Vanessa Wiseman; John Wood; Michael Woods; Owen Woods; Sidney Woods; Raymond Woodward; Harold Wright; Julia Wright

Studios
Add Make-up; Alard Design; Anyart; Arka Graphics; Artec; Art Liaison; Art Workshop; Bateson Graphics; Broadway Artists; Dateline Graphics; David Cox Associates; David Levin Photographic; Eric Jewel Associates; George Miller Associates; Gilcrist Studios; Hatton Studio; Jackson Day; Lock Pettersen Ltd; Mitchell Beazley Studio; Negs Photographic; Paul Hemus Associates; Product Support Graphics; Q.E.D. [Campbell Kindsley]; Stobart and Sutterby; Studio Briggs; Technical Graphics; The Diagram Group; Tri Art; Typographics; Venner Artists

Agents
Artist Partners; Freelance Presentations; Garden Studio; Linden Artists; N.E. Middletons; Portman Artists; Saxon Artists; Thompson Artists